SOIL
PHYSICAL
CHEMISTRY

Second Edition

Edited by
Donald L. Sparks, Ph.D.
Department of Plant and Soil Sciences
University of Delaware
Newark, Delaware

CRC Press
Boca Raton Boston London New York Washington, D.C.

Acquiring Editor:	John Sulzycki
Project Editor:	Carol Whitehead
Marketing Manager:	Becky McEldowney
Cover design:	Dawn Boyd
PrePress:	Carlos Esser
Manufacturing:	Lisa Spreckelsen

Library of Congress Cataloging-in-Publication Data

Soil physical chemistry / edited by Donald L. Sparks. --2nd ed.
     p. cm
     Includes bibliographical references and index.
     ISBN 0-87371-883-6 (alk. paper)
     1. Soil physical chemistry. I. Sparks, Donald L., Ph. D.
S592.53.S65   1998
631.4'1--dc21   1001569497   98-10571
                                                    CIP

© 1999 by CRC Press LLC
Lewis Publishers is an imprint of CRC Press LLC

No claim to original U.S. Government works
International Standard Book Number 0-87371-883-6
Library of Congress Card Number 98-10571
Printed in the United States of America 1 2 3 4 5 6 7 8 9 0
Printed on acid-free paper

Preface

Since the publication of the first edition of *Soil Physical Chemistry* in 1986, there have been many new and exciting developments in the field of soil chemistry. The emphasis on understanding the rates and mechanisms of soil chemical reactions and processes has greatly intensified. This is in large part due to the importance of accurately predicting the fate, mobility, speciation, and bioavailability of plant nutrients, metals, metalloids, organic chemicals, and radionuclides in soil and water environments. Significant advances in surface complexation modeling, kinetic measurements, and the use of *in situ* spectroscopic and microscopic surface techniques have occurred in the last decade. These innovations have enabled soil chemists to precisely determine the mechanisms of important soil chemical reactions at the molecular scale.

The second edition of *Soil Physical Chemistry* includes discussions on advances in modeling surface complexation reactions at the mineral/water interface (Chapter 2), kinetics of soil chemical reactions (Chapters 4 and 5), and the use of *in situ* spectroscopic and microscopic techniques to elucidate reactions mechanisms and to ascertain the structure and chemistry of soil organic matter (Chapters 4 and 6). Updated chapters on the electrochemistry of the double-layer (Chapter 1), thermodynamics of the soil solution (Chapter 3), kinetics of soil chemical processes (Chapter 4) and soil redox behavior (Chapter 7) are contained in the second edition. New chapters on chemical modeling of ion adsorption in soils (Chapter 2), precipitation/dissolution reactions in soils (Chapter 5), and the chemistry of soil organic matter (Chapter 6) are also found. The book should be of interest to students and professionals in soil chemistry, environmental chemistry, geochemistry, environmental and chemical engineering, and material and marine sciences.

I greatly appreciate the first-rate chapters by the authors. I am also indebted to the referees of each chapter for their careful and thoughtful assessments, and to my secretary, Muriel Toomey, and research associate, Jerry Hendricks, for their fine assistance in preparing the manuscript. Special thanks are extended to my graduate students and postdoctoral associates for their ideas and comments on the book's content. I am also most grateful to my wife Joy for her wonderful support and encouragement.

Donald L. Sparks
Newark, Delaware

About the Editor

Dr. Donald L. Sparks is Distinguished Professor of Soil Science, Francis Alison Professor, and Chairperson, Department of Plant and Soil Sciences at the University of Delaware at Newark. He also holds joint faculty appointments in the Departments of Civil and Environmental Engineering and Chemistry and Biochemistry. He received his B.S. and M.S. degrees at the University of Kentucky, Lexington, and his Ph.D. degree in 1979 from the Virginia Polytechnic Institute and State University, Blacksburg, Virginia. He served as Assistant Professor at the University of Delaware from 1979 to 1983, Associate Professor from 1983 to 1987, Professor in 1987, and was named Distinguished Professor and Francis Alison Professor in 1994 and 1996, respectively. He served as Assistant Department Chairperson from 1983 to 1985 and as Chairperson beginning in January, 1989.

Dr. Sparks is internationally recognized for his research in the areas of kinetics of soil chemical processes, surface chemistry of soils and soil components using *in situ* spectroscopic and microscopic techniques, and the physical chemistry of soil potassium. He is the author or coauthor of 145 publications; these include 22 edited or authored books, 25 book chapters, and 98 refereed papers. He is the author of 2 widely adopted textbooks, *Kinetics of Soil Chemical Processes* and *Environmental Soil Chemistry* published by Academic Press. He is the editor of a textbook, *Soil Physical Chemistry;* 2 monographs, *Rates of Soil Chemical Phenomena* and *Methods of Soil Analysis: Chemical Methods;* and the most prestigious serial review in the fields of crop and soil science, *Advances in Agronomy.* He has presented his research findings at national and international symposia and has served as an invitational speaker at 35 universities and institutes in the U.S., Canada, Asia, and Europe.

Dr. Sparks has been the recipient of over $3,000,000 in grants and contracts from a number of governmental, academic, and industrial sources including the Agency for International Development, Potash and Phosphate Institute, Departments of Agriculture, Interior, and Energy, University of Delaware Research Foundation, U.S. Borax Corporation, U.S. Israel Binational Research and Development Fund, Bikini Atoll Rehabilitation Committee, the DuPont Corporation, USDA, the UNIDEL Foundation, and NASA.

Dr. Sparks has served on the editorial boards of the *Soil Science Society of America Journal* (as Associate Editor, 1985 to 1987, Div. S-2, and as Technical Editor, Div. S-2, 1988 to 1993). He currently serves as Co-Editor-in-Chief of *Geoderma* and serves on the editorial boards of *Soil Science, Advances in Agronomy, Trends in Soil Science,* and *Advances in Environmental Toxicology and Chemistry.*

At the University of Delaware, Dr. Sparks has been very active in graduate education. He has served as major professor and mentor to 42 graduate students and postdoctoral fellows who hold positions in academe, government, and industry in the U.S. and throughout the world. He has also served as host to visiting scholars and professors from around the world. His graduate students have been the recipients of 29 awards and honors including Potash and Phosphate Institute International Fellowships (4 students), Theodore Wolf Dissertation Prize in the Physical and Life Sciences, University of Delaware Fellowships, Outstanding Northeastern Regional Agronomy Society Graduate Student Awards, and a NASA Fellowship. Two recent Ph.D. students, Drs. Cristian P. Schulthess and Scott E. Fendorf, were the recipients of the Emil Truog Award from the Soil Science Society of America. This award is given for the most outstanding dissertation in the soil sciences.

Dr. Sparks has been the recipient of numerous awards and honors. In 1982, he was the recipient of the University of Delaware Sigma Xi Distinguished Scientist Award. In 1983, he was cited by the International Potash Institute for his outstanding research on soil potassium. In 1985, he was recipient of the American Society of Agronomy's Visiting Scientist Award, and in 1986 he received the Research Award from the Northeastern Branch, American Society of Agronomy. In 1987, Dr.

Sparks was elected Chairman of the Soil Chemistry Division (Div. S-2) of the Soil Science Society of America. In 1989, he was named a Fellow of both the American Society of Agronomy and the Soil Science Society of America, the most prestigious honors given by both societies. In 1991, he received the F. D. Chester Distinguished Performance Award from the College of Agricultural Sciences and the M. L. and Chrystie M. Jackson Soil Science Award from the Soil Science Society of America. In 1994 he was named Distinguished Professor of Soil Science, the first such professorship in the College of Agricultural Sciences at the University of Delaware. In 1994, Dr. Sparks also received the Soil Science Research Award from the Soil Science Society of America and was elected Vice-Chair (Commission II, Soil Chemistry) of the International Society of Soil Science. In 1996 Dr. Sparks was the recipient of the University of Delaware's Francis Alison Award, which is the highest award that a faculty member can receive. The award is given for distinguished achievements in scholarship and one's profession, dedication, and mentoring of students. In 1998 he was named a Fellow of the American Association for the Advancement of Science (AAAS).

Dr. Sparks has served as a consultant to the DuPont Corporation, U.S. Borax Corporation, Research Triangle Institute, Bikini Atoll Rehabilitation Committee, and the Department of Energy.

Dr. Sparks has been instrumental in building the internationally recognized soil science program at the University of Delaware. He has also served on numerous university committees, as a Faculty Senator, and as a member of the University Budget Council.

He is a member of the American Society of Agronomy, Soil Science Society of America, Clay Minerals Society, American Association for the Advancement of Science, American Chemical Society, and American Geochemical Society and the honorary societies Gamma Sigma Delta, Sigma Xi, and Phi Kappa Phi. He has served on many committees of both the Soil Science Society of America and the American Society of Agronomy.

Contributors

Richmond J. Bartlett
Department of Plant and Soil Science
The University of Vermont
Burlington, Vermont

Elisabetta Loffredo
Istituto di Chimica Agraria
Universita di Bari
Bari, Italy

Wayne P. Robarge
Department of Soil Science
North Carolina State University
Raleigh, North Carolina

Nicola Senesi
Istituto di Chimica Agraria
Universita di Bari
Bari, Italy

Upendra Singh
IFDC Headquarters
Muscle Shoals, Alabama

Donald L. Sparks
Department of Plant and Soil Sciences
University of Delaware
Newark, Delaware

Donald L. Suarez
U. S. Salinity Laboratory
Riverside, California

Goro Uehara
University of Hawaii
Honolulu, Hawaii

John C. Westall
Department of Chemistry
Oregon State University
Corvallis, Oregon

John M. Zachara
Battelle, Pacific Northwest Laboratories
Richland, Washington

Contents

Dedication

For Joy, Betty, and Carolyn

CHAPTER 1

Electrochemistry of the Double Layer: Principles and Applications to Soils

Upendra Singh and Goro Uehara

CONTENTS

0-87371-883-6/99/$0.00+$.50
© 1999 by CRC Press LLC

There has been a minimal revision in this chapter compared to the first edition. A section on simulating nutrient dynamics has been included to illustrate a recent application of the electrical double-layer theory in modeling and in transport of interacting solutes. The objective of the chapter remains — providing students a simple and complete mathematical explanation of double-layer theory. A thorough discussion of the strengths and weaknesses of double-layer theory has been sacrificed for completeness of theory and its derivation. Many steps that are omitted in advanced papers on the subject are included for clarity and continuity in this chapter. The aim of this chapter is to entice a larger number of scientists to adopt and apply the double-layer theory. Double-layer theory remains underutilized in aquatic chemistry and particularly in soil science because potential users have not learned the basic relationships that comprise it.

It is our premise that there exists a sizable group that is ready to use double-layer theory in its work. Soil science is a good case in point. In soil science, the use of double-layer theory is largely confined to soil chemists, soil physicists, and soil mineralogists. However, the greatest potential for practical application of double-layer theory is in the fields of soil fertility and pedology. Many soil scientists continue to operate on the assumption that a soil's cation exchange capacity is constant. They make soil fertility recommendations and classify soil on the basis of this assumption. Papers on soil classification that equate the sum of bases and extractable aluminum with permanent surface charge appear unabated in journals.

Double-layer theory has much to offer, but those who can profit from it cannot do so because they do not have access to an unabridged version of its derivation. This chapter alone will not convert nonusers to users, but it is an attempt to fill a critical need. It does so by providing the missing steps that so often discourage the neophyte from pursuing the subject to its logical conclusion. We hope that by illustrating the application of double-layer theory in simulating nitrate leaching, potential uses of the theory in soil nutrient dynamics will be realized.

I. INTRODUCTION

A particle or surface with acquired charge which is different from the surrounding solution accumulates countercharge in order to preserve electrical neutrality. The countercharge may consist simply of a diffuse atmosphere of counterions or it may take the form of a compact layer of bound charge and a diffuse atmosphere as well. The surface charge and the sublayers of compact and diffuse counterions are called the electrical double layer.

Originally, the double layer was assumed to be built up by the "monoionic" layer of opposite charges touching each other in the boundary plane. This very simple picture, which neglects the thermal agitation of the ions, is often ascribed to von Helmholtz (1835).[1] The true concept of electrical double-layer theory is due to Quincke (1861),[2] although he did not use the term. Gouy (1910) and Chapman (1913) independently described the diffuse distribution of counterions accumulated near the surface of mercury electrodes in response to an externally applied potential.[3]

Stern and Grahame refined this, recognizing that counterions are unlikely to approach the surface more closely than the ionic radii of anions and the hydrated radii of cations.[2] They also introduced the concept of binding energy for specific adsorption. The historical background to the development of electrical double layer is reviewed by Verwey,[1] Grahame,[2] and Overbeek.[4] A schematic representation of the Gouy-Chapman model and Stern-Grahame model is presented in Figure 1, to illustrate the fundamental difference between the two models. These and further modifications of Gouy-Chapman theory will be discussed in the later sections of this chapter.

II. ORIGIN OF SURFACE CHARGE IN SOILS

Three main types of colloidal charge constitute the mineral assemblage of soils. Based on origin of charge, these are categorized in the following ways: (1) isomorphic substitution or crystal lattice defects in the internal structure of the mineral; (2) ionic dissolution on surfaces of soil colloids; and (3) ionization of active organic functional groups.

A. Isomorphic Substitution

The most commonly acknowledged source of surface charge on soil colloids is from the structural imperfection in the interior of the crystal structure. Structural imperfections, due to ion substitution or site vacancies, frequently result in permanent charge on the soil colloidal particle. In theory, this charge may assume either a positive or negative value.[5] However, due to ion size limitation, the substitution is generally restricted to a lower valence element for one with a higher valency.[6,7] For example, Al^{3+} substituting for tetrahedral Si^{4+} and Mg^{2+} or Fe^{2+} substituting for octahedral Al^{3+} leads to a deficiency of positive charge on the crystal lattice. The resulting charge is generally negative on the clay structure.[8,9] Such colloids have permanent charge; however, they may not display nonamphoteric behavior because their edges may be amphoteric.

B. Ionic Dissolution

Surface charges also develop as a result of chemisorption of H_2O, i.e., water splitting into H^+ and OH^- during adsorption to form a hydroxylated surface. Establishment of charge on such a surface is then viewed as either an adsorption of H^+ or OH^- or as dissociation of surface sites which can then assume either a positive or a negative charge.[10] The sign and magnitude of the surface charge are determined solely by the ion that is adsorbed in excess onto the hydroxylated surface; such ions are termed potential determining ions. The sorption and eventually the surface charges are dependent on the activity of the potential determining ions in the bulk solution. The mechanism by which a colloid acquires its surface charge is schematically presented in Figure 2. Many minerals in soils typify this type of surface charge. The most notable of these are the oxides and/or the hydroxides of Al, Fe, Mn, Si, and Ti. Nevertheless, this type of charge is not confined to the oxides only. Equally illustrative is kaolinite which by its broken edges reflects similar charge behavior.

(a) Gouy-Chapman Model   (b) Stern Model

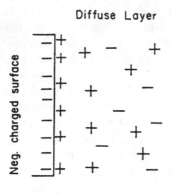
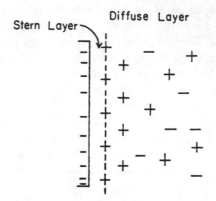

Charge distribution as related to distance.

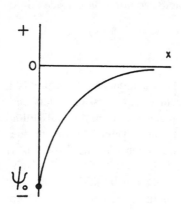
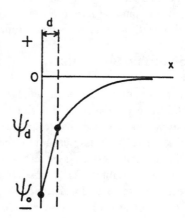

Potential distribution as related to distance.

Figure 1   Distribution of electrical charges and potential in double layer according to (a) Gouy-Chapman theory and (b) Stern theory. ψ_o and ψ_d are surface and Stern potential, respectively; d is the thickness of the Stern layer.

C. Ionization

Confined mainly to the organic fraction, ionization is viewed as a process by which a colloid acquired its charges by dissociation of H^+ either from or onto the active functional group. The charge generated could be either positive or negative:

$$R-COOH \overset{H_2O}{\rightleftharpoons} R-COO^- + H^+$$

$$R-NH_2 + H_2O \overset{H_2O}{\rightleftharpoons} R-NH_3^+ + OH^-$$

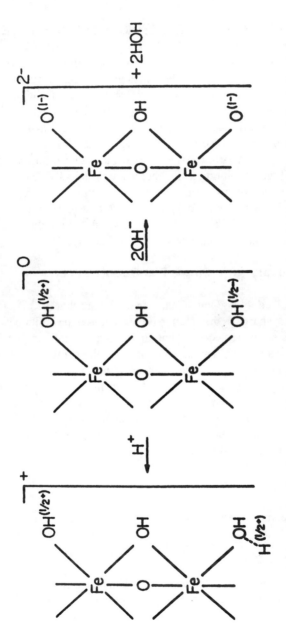

Figure 2 Schematic representation showing the charge reaction and zero point of charge of a colloid in which ion dissolution is the charge generating process.

In some ways this is comparable to ion dissolution. The charge is dependent on the dissociation constant of each functional group and pH. Examples of functional groups generating this type of charge are the carboxyl, phenolic, and amino groups.

Hereafter, the charged colloids will be considered as permanent or variable charged or permanent-variable charged mixtures.

III. THEORY OF THE ELECTRICAL DOUBLE LAYER

Irrespective of the origin of the surface charge on the colloidal surface, electrical neutrality demands that an equal amount of charge of the opposite sign must accumulate in the liquid phase near the charged surface. For a negatively charged surface, this means that positively charged cations are thus, by electrostatic forces, attracted to the charged surface.

At the same time, due to diffusion forces, the cations are also drawn back towards the equilibrating solution. An "atmospheric" distribution of cations in a "diffuse layer" is established where the concentration of cation increases towards the surface, the concentration increasing from a value equal to that of the equilibrating solution to a higher value principally determined by the magnitude of the surface charge. On the contrary, ions of equal sign (anions) are repelled by such a surface with diffusion forces acting in an opposite direction, such that there is a deficit of anions near the surface. Based on such a theory, different models are formulated relating the density of charge on the surface and the distribution of counterions in the diffuse layer.

A. Gouy-Chapman Theory for a Single Flat Double Layer

The assumptions involved in the Gouy-Chapman model are that the charge is uniformly spread over the surface. The space charge in the solution is considered to be built up by unequal distribution of point charges. The solvent is treated as a continuous medium, influencing the double layer only through its dielectric constant which is assumed invariant with position in the double layer.[4] Further, it is assumed that ions and surface are involved only in electrostatic interactions. The present derivation is for a flat surface infinite in size and at an infinite distance from the surface. The double-layer theory applies equally well to rounded or spherical surfaces.[4] The derivation that follows is based on van Olphen,[7] Overbeek,[4] Bolt,[11] and El-Swaify.[12]

At equilibrium between any point in the double layer and the bulk solution for ionic species i are electrochemical potentials $\bar{\mu}_i$ and $\bar{\mu}_{oi}$ respectively.

$$\bar{\mu}_i = \bar{\mu}_{oi} \tag{1}$$

The electrochemical potential is defined as

$$\bar{\mu}_i = \mu_i + z_i F \psi_i \tag{2}$$

where μ_i is the chemical potential, z_i is the valence, ψ_i is the electrical potential in the double layer, and F is the Faraday constant. One could relate the chemical potential, μ_i, to activity, a_i, as

$$\mu_i = \mu_i' + RT \ln a_i \tag{3}$$

where μ_i' is the standard chemical potential, R is the gas constant, and T is the absolute temperature. Thus, Equation 1 can be rewritten as

$$\mu_i' + RT \ln a_i + zF\psi_i = \mu_i' + RT \ln a_{oi} + zF\psi_{oi} \tag{4}$$

Therefore,

$$\psi_i = \psi_{oi} = \frac{RT}{zF} \ln \frac{a_{oi}}{a_i} \tag{5}$$

If ψ_{oi} is assumed to be zero and activity is assumed to be equal to molar concentration,[6] then Equation 5 becomes Boltzmann's equation:

$$c_i = c_{oi} \exp\left(\frac{-z_i F \psi}{RT}\right) \tag{6a}$$

$$c_- = c_{o_-} \exp\left(\frac{z_- F \psi}{RT}\right) \tag{6b}$$

$$c_+ = c_{o_+} \exp\left(\frac{-z_+ F \psi}{RT}\right) \tag{6c}$$

where c_+ and c_- are the local molar concentrations of the cations and anions, c_{o+} and c_{o-} are their molar concentrations far away from the surface in the equilibrium liquid, concentration is expressed as $kmol/m^3$, z_+ and z_- are the valences of cations and anions, respectively, F is the Faraday constant, R is the gas constant, and T is the absolute temperature. Thus, when equilibrium is established in the double layer, the average local concentration of ions can be expressed as a function of the average electrical potential ψ at that distance according to the Boltzmann theory.[4,13,14]

The space charge density (C/m^3) is given by

$$\rho = \sum c_i z_i F$$
$$= c_+ z_+ F - c_- z_- F \tag{7}$$

The coulombic interaction between the charges present in the system is described by the Poisson equation

$$\Delta \psi = \frac{-4\pi\rho}{e} \tag{8}$$

where ψ is the potential which changes from a certain value ψ_o, at the colloid-water interface to zero in the bulk solution, and ρ is the space charge density, e is the dielectric constant, and Δ is the Laplace operator, which in Cartesian coordinates is equal to $d^2/dx^2 + d^2/dy^2 + d^2/dz^2$.[10]

At an infinitely large interface (flat surface), Δ simplifies to d^2/dx^2. Thus, Equation 8 becomes:

$$\frac{d^2\psi}{dx^2} = -\frac{4\pi\rho}{e} \tag{9}$$

Substituting Equation 7 into 9 one obtains:

$$\frac{d^2\psi}{dx^2} = -\frac{4\pi}{e}\left(c_+ z_+ F - c_- z_- F\right) \tag{10}$$

Using the assumption for a symmetrical electrolyte ($z_+ = z_- = z_i$, then $c_{o+} = c_{o-} = c_o$) and substituting Equation 6b and 6c, respectively for c_+ and c_- in Equation 10 one obtains:

$$\frac{d^2\psi}{dx^2} = -\frac{4\pi}{e}\left[zFc_o \exp\left(\frac{-zF\psi}{RT}\right) - zFc_o \exp\left(\frac{zF\psi}{RT}\right)\right] \tag{11}$$

$$= -\frac{4\pi}{e}zFc_o\left[\exp\left(\frac{-zF\psi}{RT}\right) - \exp\left(\frac{zF\psi}{RT}\right)\right] \tag{12}$$

By rearrangement Equation 12 becomes:

$$\frac{d^2\psi}{dx^2} = \frac{4\pi}{e}zFc_o\left[\exp\left(\frac{zF\psi}{RT}\right) - \exp\left(\frac{-zF\psi}{RT}\right)\right] \tag{13}$$

By definition:

$$\sinh x = \frac{\exp(x) - \exp(-x)}{2}$$

Therefore,

$$\frac{\exp\left(\dfrac{zF\psi}{RT}\right) - \exp\left(\dfrac{-zF\psi}{RT}\right)}{2} = \sinh\frac{zF\psi}{RT} \tag{14}$$

Thus, by substituting the sinh term in Equation 13, the basic Poisson-Boltzmann differential equation, the basis of the diffuse double-layer theory is obtained:

$$\frac{d^2\psi}{dx^2} = \frac{8\psi zFc_o}{e}\sinh\left(\frac{zF\psi}{RT}\right) \tag{15}$$

It is convenient to rewrite this equation in terms of the following dimensionless quantities:[7]

$$y = \frac{zF\psi}{RT} \tag{16}$$

$$\xi = \kappa x \tag{17}$$

Here for a single symmetrical electrolyte,

$$\kappa^2 = \frac{8\pi z^2 F^2 c_o}{eRT}m^{-2} \tag{18}$$

Then Equation 15 becomes simply:

$$\frac{d^2y}{d\xi^2} = \sinh y \tag{19}$$

The steps involved in arriving to Equation 19 are based on El-Swaify,[12] and Overbeek[4] and are shown to be:

$$\frac{dy}{dx} = \frac{zFd\psi}{RTdx} \tag{20}$$

Thus:

$$\frac{d^2y}{dx^2} = \frac{zF}{RT}\frac{d^2\psi}{dx^2} \tag{21}$$

Substituting Equation 15 into 21:

$$\frac{d^2y}{dx^2} = \frac{zF}{RT}\left[\frac{8\pi zFc_o}{e}\sinh\left(\frac{zF\psi}{RT}\right)\right] \tag{22}$$

Further substitution of Equations 16 and 18 gives:

$$\frac{d^2y}{dx^2} = \kappa^2 \sinh y \tag{23}$$

Differentiating Equation 17 with respect to distance $\times$ and substituting, one obtains the following relationships:

$$\frac{d\xi}{dx} = \kappa \tag{24}$$

$$\frac{dy}{dx} = \frac{dy}{d\xi} \cdot \kappa \tag{25}$$

$$\frac{d^2y}{dx^2} = \frac{d}{dx}\left[\frac{dy}{d\xi} \cdot \kappa\right] \tag{26}$$

$$= \kappa\frac{d^2y}{d\xi} \cdot \frac{1}{dx} \tag{27}$$

From Equation 24:

$$1/dx = \kappa/d\xi \tag{28}$$

Substituting into Equation 27 we obtain:

$$\frac{d^2y}{dx^2} = \kappa^2 \frac{d^2y}{d\xi^2} \tag{29}$$

Thus:

$$\frac{d^2y}{d\xi^2} = \frac{1}{\kappa^2} \frac{d^2y}{dx^2} \tag{30}$$

Therefore, substituting Equation 23 into the above gives Equation 31:

$$\frac{d}{d\xi}\left(\frac{dy}{d\xi}\right) = \frac{d^2y}{d\xi^2} = \sinh y \tag{31}$$

Now assign:

$$p = \frac{dy}{d\xi} \tag{32}$$

Thus:

$$\frac{d}{d\xi}\left(\frac{dy}{d\xi}\right) = \frac{dp}{d\xi} \tag{33}$$

Therefore, by the chain rule

$$\frac{dp}{d\xi} = \frac{dy}{d\xi} \cdot \frac{dp}{dy} \tag{34}$$

or

$$\frac{dp}{d\xi} = p \frac{dp}{dy} \tag{35}$$

Hence, Equation 35 becomes

$$p\,dp = \sinh y\,dy \tag{36}$$

By integrating both sides:

$$1/2\,p^2 = \cosh y + b \tag{37}$$

or

$$p^2 = 2\cosh y + b_1 \tag{38}$$

Therefore,

$$p = \frac{dy}{d\xi} = -\left(2\cosh y + b_1\right)^{1/2} \tag{39}$$

in which the negative root is chosen indicating the decline in magnitude of the potential with increasing ξ.[12] With the boundary conditions for $\xi = \infty$ (i.e., infinite plate-plate distance), $dy/d\xi = 0$, and $y = 0$, then $b_1 = -2$ because $\cosh 0 = 1$.

$$dy/d\xi = -\left(2\cosh y - 2\right)^{1/2} \tag{40}$$

Using the hyperbolic equality:

$$\sinh y/2 = \left[1/2\left(\cosh y - 1\right)\right]^{1/2} \tag{41}$$

Rearranging

$$dy/d\xi = -\left\{4\left[1/2\left(\cosh y - 1\right)\right]\right\}^{1/2}$$

$$dy/d\xi = -2 \sinh y/2 \tag{42}$$

$$dy/\left(\sinh y/2\right) = -2d\xi \tag{43}$$

or

$$\left(csch\, y/2\right)dy = -2d\xi \tag{44}$$

Using the integration formula

$$\int csch\frac{u}{a}\,du = a\ln\left|\tanh\frac{u}{2a}\right|$$

and integrating Equation 44, one obtains

$$2 \ln \tanh y/4 = -2\xi + b \tag{45}$$

Dividing by 2 and rearranging:

$$\ln \tanh y/4 - b_2 = -\xi \tag{46}$$

Using the boundary condition, Equation 16

$$y_o = \frac{zF\psi_o}{RT}\, at\, x = 0\, \left(hence\, \xi = 0\right) \tag{47}$$

Here ψ_o is the potential at the particle surface, then $b_2 = \ln \tanh y_o/4$ and Equation 46 is then rewritten as:

$$\ln \tanh y/4 - \ln \tanh y_o/4 = -\xi \tag{48}$$

or

$$\frac{\tanh y/4}{\tanh y_o/4} = \exp^- \xi \tag{49}$$

To solve for y, one would proceed as follows:

$$\tanh y/4 = a \, \exp(-\xi) \tag{50}$$

Here, $a = \tanh y_o/4$; thus, $y = 4$ arctanh a exp $(-\xi)$. Using the identity for the inverse hyperbolic function:

$$arctanh \, a = 1/2 \ln \frac{1+a}{1-a} \left(a^2 < 1\right) \tag{51}$$

Then:

$$y = 2 \ln \frac{1 + a\exp^{-\xi}}{1 - a\exp^{-\xi}} \tag{52}$$

and by multiplying the denominator as well as the numerator of the right hand side by exp (ξ), one obtains

$$y = 2 \ln \frac{\exp(\xi) + a}{\exp(\xi) - a} \tag{53}$$

or substituting Equation 17 for ξ, and Equation 47 for the y_o term in a:

$$y = 2 \ln \frac{\exp(\kappa x) + \tanh \dfrac{zF\psi_o}{4RT}}{\exp(\kappa x) - \tanh \dfrac{zF\psi_o}{4RT}} \tag{54}$$

Thus, the above is an expression for potential distribution as a function of the distance from the plate surface. It describes the decay of potential with distance at a given surface potential, electrolyte concentration, and for a given counterion valence. This decay is approximately exponential.[12]

Finally, the total double-layer charge is given by

$$\sigma = -\int_0^\infty \rho \, dx \tag{55}$$

The charge of the colloid σ is equal in magnitude but opposite in sign to the excess charge in solution. Substituting:

$$\rho = \frac{-e}{4\pi}\frac{d^2\psi}{dx^2}\ (from\ \ Equation\ 9) \tag{56}$$

$$\sigma = \frac{e}{4\pi}\int_0^\infty \frac{d^2\psi}{dx^2}dx \tag{57}$$

$$= \frac{e}{4\pi}\left[\frac{d\psi}{dx}\right]_{x=0}^\infty \tag{58}$$

However:

$$\frac{d\psi}{dx} = 0\ as\ x \to \infty \tag{59}$$

Thus,

$$\sigma = \frac{e}{4\pi}\left[0 - \left(\frac{d\psi}{dx}\right)_{x=0}\right]$$

$$= \frac{-e}{4\pi}\left[\frac{d\psi}{dx}\right]_{x=0} \tag{60}$$

This indicates that σ is proportional to the initial slope of the electrical potential in the double layer.[12] Recalling (Equation 42):

$$dy/d\xi = -2\ \sinh\ y/2$$

and

$$dy = \frac{zFd\psi}{RT}\ and\ d\xi = \kappa dx \tag{61}$$

Thus, Equation 42 can be rewritten with the above substitutions as

$$\frac{zF}{RT}\frac{d\psi}{\kappa dx} = -2\ \sinh\ y/2 \tag{62}$$

$$\frac{d\psi}{dx} = -2\frac{RT}{zF}\kappa \sinh\ y/2 \tag{63}$$

For

$$x = 0, \left(\frac{d\psi}{dx}\right)_{x=0} = -\frac{2RT}{zF}\kappa \sinh\ y_o/2 \tag{64}$$

and by substitution of Equation 64 into 60 one obtains:

$$\sigma = \frac{-\varepsilon}{4\pi}\left(\frac{-2RT}{zF}\kappa\sinh y_o/2\right) \tag{65}$$

By further substitutions of definitions of κ (Equation 18) and y_o (Equation 47) one obtains:

$$\sigma = \frac{\varepsilon RT}{2\pi zF}\left(\frac{8\pi z^2 F^2 c_o}{\varepsilon RT}\right)^{1/2}\sinh\frac{zF\psi_0}{2RT} \tag{66}$$

$$\sigma = \left(\frac{2c_o\varepsilon RT}{\pi}\right)^{1/2}\sinh\frac{zF\psi_o}{2RT} \tag{67}$$

This fundamental charge-electrolyte concentration-electrical potential relationship (Equation 67) can be related to many chemical and physical properties of soils.

The Gouy-Chapman equation (Equation 67) has limited quantitative application, due to the assumption that ions in solution behave as point charges and can approach the surface without limit.[13] Thus, they may reach excessively high concentrations at the liquid interface. It also does not consider specific interactions between the surface, the counterions, and the medium. This limitation particularly applies to variable charge surfaces as will be discussed later. Even at moderate surface potential (e.g., 250 mV), absurdly high values for the amount of counterions adsorbed into the diffuse layer are predicted.

B. The Stern Theory

In his theory, Stern (1924) has divided the region near the surface into two parts: the first consisting of a layer of ions adsorbed at the surface forming a compact double layer (the Stern layer) and the second consisting of the diffuse double layer (the Gouy layer) (Figure 1). He has accounted for the finite size of ions and the possibility of their specific adsorption.

The Stern theory of the electrical double layer then assumes that the surface charge is balanced by the charge in solution, which is distributed between the Stern layer at a distance (d) from the surface and a diffuse layer which has a Boltzmann distribution.[4] The total surface charge, σ, is therefore due to the charge in the two layers

$$\sigma = -(\sigma_1 + \sigma_2) \tag{68}$$

σ_1 is the Stern layer charge and σ_2 is the diffuse layer charge. The charge in the Stern layer, as outlined by van Raij and Peech[15] is

$$\sigma_1 = \frac{N_i zF}{1 + (N_A w/Mc)\exp\left[-(zF\psi_d + \phi)/RT\right]} \tag{69}$$

here
σ_1 = Stern layer charge in C/m²
N_i = number of adsorption sites available per m² of surface to ionic species i
N_A = Avogadro's number = 6.0×10^{23}
M = molar mass of the solvent (kg)

w   = solvent density (kg/m^3)
c   = electrolyte concentration (kmol/m^3)
ϕ = specific adsorption potential (J)
ψ_d = Stern potential in V
d   = thickness of Stern layer (m)

Other terms defined previously are
z   = ionic valence
F   = Faraday constant (9.6487×10^7 C/kmol)
R   = gas constant (8.314×10^3 J kmol^{-1} K^{-1})
T   = absolute temperature (K)

The Gouy or diffuse layer charge in C/m^2 can be expressed as:

$$\sigma_2 = \left(\frac{2c\epsilon RT}{\pi}\right)^{1/2} \sinh\frac{zF\psi_d}{2RT} \tag{70}$$

here ϵ = dielectric constant (C^2/Jm). The rest of the symbols have the same meaning and value as indicated above.

Since a linear drop in potential across the Stern layer is assumed, the surface charge is also given by the Gauss equation for a molecular condensor:

$$\sigma = (\epsilon'/4\pi d)(\psi_o - \psi_d) \tag{71}$$

Here ϵ' is the average dielectric "constant" of the Stern layer.[13]

C. Theory of Interacting Flat Double Layers

The interacting plate method acknowledges the fact that the double layers of particles may influence each other. The schematic presentation of potential distribution between two interacting flat double layers of the Gouy-Chapman model is shown in Figure 3.

In the midway plane, i.e., at distance d, the field strength:

$$\left(\frac{d\psi}{dx}\right)_{x=d} = 0$$

is zero, and the potential is ψ_d
Also:

$$y = u = \frac{zF\psi_d}{RT} \tag{72}$$

The fundamental differential (Equation 19) for computing potential distribution is:

$$\frac{d^2 y}{d\xi^2} = \sinh y$$

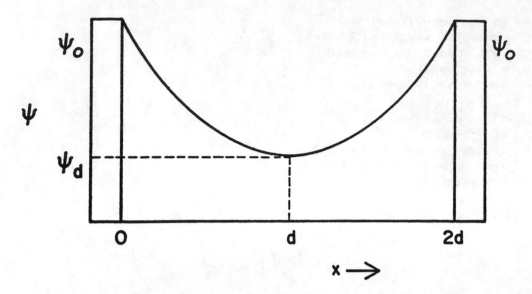

Figure 3 Distribution of surface potential between two interacting flat double layers.

As for the fully extended case

$$\frac{dy}{d\xi} = -\left(2\ \cosh\ y + b_1\right)^{1/2}$$ (73)

and for x = d,

$$\frac{dy}{d\xi} = 0 = -\left(2\ \cosh\ u + b_1\right)^{1/2}$$ (74)

thus,

$$b_1 = -2\ \cosh\ u$$ (75)

Therefore, Equation 73 becomes:

$$\frac{dy}{d\xi} = -\left(2\ \cosh\ y - 2\ \cosh\ u\right)^{1/2}$$ (76)

The total charge is obtained in a manner similar to that of the extended layer. Thus,

$$\sigma = -\int_0^\infty \rho dx$$

or

$$\sigma = \frac{-e}{4\pi}\left(\frac{d\psi}{dx}\right)_{x=o}$$

where

$$\left(\frac{d\psi}{dx}\right)_{x=o} = \left(\frac{dy}{d\xi}\right)_o \kappa \frac{RT}{zF} \tag{77}$$

Inserting dy/d as given by Equation 76:

$$\left(\frac{d\psi}{dx}\right)_{x=0} = \frac{-\kappa RT}{zF}\left(2 \cosh y_o - 2 \cosh u\right)^{1/2} \tag{78}$$

Substituting into Equation 60

$$\sigma = \frac{e\kappa RT}{4\pi zF}\left(2 \cosh y_o - 2 \cosh u\right)^{1/2} \tag{79}$$

Finally, putting in the definition of κ we obtain:

$$\sigma = \left(\frac{c_o eRT}{\pi}\right)^{1/2}\left(\cosh y_o - \cosh u\right)^{1/2} \tag{80}$$

As expected, the above equation predicts that the double-layer charge is very slightly affected if the distance between the two plates is large (i.e., interaction between the two double layers is small).

D. The Gouy-Chapman-Stern-Grahame Model

There have been a number of further modifications to double-layer theory, primarily to explain experimental observations. Grahame[2] suggested that specifically adsorbed anions can approach the inner Helmholtz plane or IHP (i.e., adsorbed into the Stern layer) when they lose their water of hydration, whereas the hydrated cations are only electrostatically attracted to the surface (outer Helmholtz plane, or OHP).

Bolt[13] introduced the following corrections to account for the limitation in Gouy-Chapman theory: the effect of ion size, dielectric saturation, polarization energies of the ion, Coulombic interaction of the ions, and the short-range repulsion between the ions. The dielectric constant of water surrounding an ion (8.9×10^{-9} C^2/Jm) is virtually constant from infinity to 5×10^{-9} m from the center and then drops to a value of 0.33 to 0.67×10^{-9} C^2/Jm. With the above modifications, he successfully described the distribution of Na^+ and Ca^{2+} around an "illite" particle without having to invoke any specific adsorption mechanism. Bolt's extension of Gouy theory further enabled distinction of ions with the same valence.

However, dielectric saturation and ionic polarization seem to offset each other almost completely if the field strength does not equal 5×10^8 V/m or the charge density on the colloids is ≤ 0.2 C/m^2 (or 2×10^{-7} me/cm²).[13] For larger values, the polarization term increases very rapidly, leading to overestimation of ionic concentrations by the Gouy-Chapman equation. Generally, the Coulombic interaction exceeds the influence of the short-range repulsion up to concentrations of 1 or 2 M. Thus, in colloids with charge density of 0.2 C/m^2 or less, the Coulombic interaction of the ions will be the main correction factor. The simple Gouy-Chapman theory gives fairly reliable results, for colloids with a charge density not exceeding 0.2 to 0.3 C/m^2 and especially if the colloid under consideration has constant charge density.

IV. APPLICATIONS OF ELECTRIC DOUBLE-LAYER THEORY
TO SOIL-WATER SYSTEMS

Soil covers a wide range of chemical and structural components; thus, it is unreasonable to expect any one single electrical double-layer model to apply. Despite its simplicity, the Gouy-Chapman model gives an invaluable though not always quantitatively applicable description of the configuration of the double layer. Although electrical double-layer theory has been applied to the soil system for some time, only recently has it been extensively used. In this section, the applicability and the validity of the theory to soil systems will be discussed.

A. Ion Exchange

The excess of ions of opposite charge (to that of the surface) over those of like charges are called exchangeable ions because any ion can be replaced by an ion with the same charge by altering the chemical composition of the equilibrium electrolyte solution. The maximum number of counterions (cations for a negatively charged surface and anions for a positively charged surface) present in the electric double layer per unit of exchanger under a given set of solution conditions is the cation exchange capacity of the system.[16] The deficit of charge is

$$\sigma_- = \left(\frac{ceRT}{\pi} \right)^{1/2} \exp\left(\frac{zF\psi_o}{2RT} - 1 \right) \tag{81}$$

while total charge is

$$\sigma_t = \left(\frac{2c_oeRT}{\pi} \right)^{1/2} \sinh \frac{zF\psi_o}{2RT}$$

Therefore, the excess of charge is

$$\sigma_+ = \left(\frac{ceRT}{2\pi} \right)^{1/2} \left[2 \sinh \frac{zF\psi_o}{2RT} - \exp \frac{zF\psi_o}{2RT} + 1 \right] \tag{82}$$

The Gouy-Chapman theory explains the exchange capacity concept in the range of surface charge density normally encountered in soil clays. However, corrections are needed if the ratios of two or more ions adsorbed at a surface from a mixed electrolyte are considered. Thus, Bolt[13] was successful in describing Na^+ and Ca^{2+} distribution, around an "illite" particle. Similarly, ion size effects can be theoretically introduced in the Stern theory approach by allowing for variation in the specific adsorption potential, ϕ; according to Equation 69

$$\sigma_i = \frac{N_i zF}{1 + \left(N_A w/Mc \right) \exp\left(\dfrac{-\left[zF\psi_d + \phi \right]}{RT} \right)}$$

Selectivity would then be related to the different values of ϕ. A more recent approach introduced by Bowden and coworkers[17] limits the use of the specific adsorption potential ϕ and the Stern Equation 69 to only those cases where true specific adsorption occurs.

When ions of different valence are involved, the simple Gouy-Chapman model predicts the different ion distribution and therefore selectivity.[6,18-20] The fraction of the surface charge neutralized by monovalent cations in an electrolyte system of two symmetrical mono- and divalent salts was derived by Erikson,[19] Babcock,[21] and given in Gast[6] as:

$$\frac{\Gamma_1}{\Gamma} = \frac{r}{\Gamma\beta^{1/2}} \, arcsinh \frac{\Gamma\beta^{1/2}}{r + 4v_d(Mo_2)^{1/2}} \qquad (83)$$

where

Γ_1 = charge neutralized by monovalent cations, (kmol/m²)
Γ = total surface charge density (kmol/m²)
r = "reduced ratio" = $Mo_1/(Mo_2)^{1/2}$ where Mo_1 and Mo_2 are the molar concentrations of mono- and divalent cations in the bulk electrolyte solution, respectively, (kmol/m³)
β = 8 F²/εRT, where F = 9.648 × 10⁷ C/kmol and ε is the dielectric constant, (C²/Jm)
v_d = Cosh u at midplane between clay particles

$$u = \frac{zF\psi_d}{RT}$$

For materials such as cation exchange resins, negative adsorption of anions could be accounted for satisfactorily, except in dilute solutions, on the basis of the Donnan equilibrium equation (thermodynamic basis). However, for nonhomogeneous potential distributions, such as clay suspensions, the Donnan equation is not as satisfactory as Schofield's equation for calculating negative adsorption.[22] Schofield's Equation 84 is based on the Gouy theory of the diffuse double layer:

$$\frac{\Gamma_-}{N} = \frac{q}{(z\beta N)^{1/2}} - \frac{4}{z\beta\Gamma} \qquad (84)$$

here

Γ_- = negative adsorption in kmol/m²
Γ = surface charge density in kmol/m²
z = valence of counterion
β = 8 F²/εRT (m/kmol)
N = concentration of repelled ions in kmol/m³
q = tabulated factor dependent on P (= z_+/z_-)

Bolt and Warkentin[22] reported that for purified clay (ultrafiltration), measured negative adsorption agreed well with that calculated from diffuse double layer over the entire concentration range.

The Gouy-Chapman theory predicts that the ratio of the concentrations of two ion species of the same valence in the counterion atmosphere is the same as that in the bulk solution. Further, the theory predicts that multivalent ions are concentrated in the double-layer to a much larger extent than monovalent ions. Thus, the ratio of Ca^{2+}/Na^+ is much higher in the proximity of negatively charged surfaces than in the medium. These predictions have been quantitatively confirmed by experiments with soils in which specific interaction effects are absent, for example, for Na-H exchange and Ba-Na exchange in AgI sols.[15] The fact that the selectivity for greater concentrations of divalent cations decreases with increasing ionic strength can also be derived from the Gouy model. Thus, experimental data are in accord with these generalized predictions.

B. Application of Electric Double Layer Theory to Clay Suspensions

The Gouy theory may be used to describe the properties of soil and clay suspensions.

1. Flocculation

One of the most important supports for the double-layer theory of stability is the quantitative agreement between the theoretically predicted ratios of flocculation values for different kinds of electrolytes and the empirical Schulze-Hardy rule.[7] The degree of stability or instability may be defined or measured in terms of coagulation rates.[23] Whenever clay particles approach each other, repulsion between the particles occurs, because the outer part of the double layer has the same type of charge (positive). This is the case especially at low electrolyte concentration, when attractive energy is low (Figure 4). In this respect, the Stern-Gouy model would be more applicable to the clay double layer than the Gouy model because in many clays there is specific adsorption of counterions. For example, specific layer to counterion interactions occur in "illite" clay in which K^+ ions provide a strong link between layer surfaces.

When two particles approach each other in suspension, their counterion atmospheres interfere. Due to compression of the double layers at increasing electrolyte concentrations, the range of repulsive force is considerably reduced (Figure 4). The repulsive energy or the repulsive potential is the amount of work required to bring the particles from infinite separation to a given distance between them. Repulsive force can also be manipulated by changing the valency (z) of the counterion.

Repulsive force is

$$P = 2c_o RT(\cosh u - 1) \tag{85}$$

and repulsive potential

$$U_R = \frac{64c_o RT}{\kappa} \exp(-2\kappa d) \tag{86}$$

where
P = repulsive force SI (N/m^2)

$U = \dfrac{zF\psi_d}{RT}$

U_R = repulsive potential SI (joules)

$$= \frac{\exp(y_o/2) - 1}{\exp(y_o/2) + 1} \; and \; y_o = \frac{zF\psi_o}{RT} \tag{87}$$

and ψ_d is the potential midway between the particles. Other terms are as defined earlier.

Theoretically for 2:1 layer charges (nonamphoteric system) the van der Waals attractive energy is the main criterion for causing flocculation,

$$U_{vdw} = \frac{-A}{48\pi}\left[1/d^2 + 1/(d+\Delta)^2 - 2\Big/\left(d+\frac{\Delta}{2}\right)^2\right] \tag{88}$$

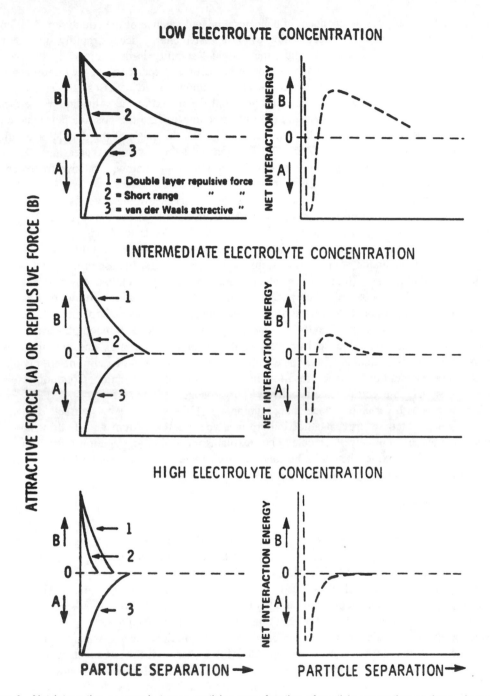

Figure 4 Net interaction energy between particles as a function of particle separation at three electrolyte concentrations. (From van Olphen, H., *Introduction to Clay, Colloid Chemistry,* Interscience, New York, 1977. With permission.)

Here

U_{vdw} = van der Waals energy of attraction

A   = constant (10^{-12})

d   = half distance between the plates

Δ = thickness of the unit layer measured between the same planes (0.66 nm)

The attractive energy is independent of concentration and valence of the counterion. Maximum flocculation is induced at high concentration, when repulsive energy decays rapidly and attractive force becomes dominant (Figure 4). Thus, the colloidal stability is determined by the balance between the repulsive and attractive forces which the particles experience as they approach each other. Measurements of zeta (ζ) potential are also commonly used to assess the stability of a colloidal soil.[24] The ζ potential is the electric potential developed at the solid-liquid interface in response to movement of colloidal particles in one direction (toward the positive pole) and counterions into another direction (toward the negative pole). The thickness of the double layer affects the magnitude of the ζ potential. Thus, the ζ potential decreases with increasing electrolyte concentration. At ζ potential equal to zero (isoelectric point) and below, repulsive forces are no longer strong enough to prevent flocculation of colloidal particles.

However, on the edge surfaces of clay plates the electrical double layer is created by the adsorption of potential determining ions. These surfaces, in contrast to the flat layer surfaces, carry positive charge, hence a positive electrical double layer. This leads to three different modes of particle association: face-to-face, edge-to-face, and edge-to-edge. The applicability of the Schulze-Hardy rule in the flocculation of clay suspensions implies that the flocculation process is dominated by a negative double layer (face-to-face interaction). Nonetheless, detailed observations of the effects of salts on the properties of clay suspensions indicate that, under certain conditions, the presence of a positive double layer on the edge surfaces must be assumed to explain the flocculation phenomenon. Thus, even predominantly constant charge clay particles exhibit amphoteric behavior.

Deflocculation can be achieved by reversing the positive edge-charge and creating a well-developed negative edge double layer. The edge-to-face attraction would be eliminated, giving rise to repulsion between all the surfaces, resulting in a breakdown of gel structure. Commonly sodium hexametaphosphate is used to create negatively charged edges.

For minerals with variable surface charge (amphoteric), the van der Waals attractive energy as well as the electrostatic energy of attraction are involved. Such a system with both positive and negative charges are mutually flocculated. This explains the stable aggregation of soils with 1:1 clays and Fe and Al oxides. The stability and other characteristics of soils with variable charge clays will be considered in greater depth in later sections.

2. Osmotic Swelling

In addition to intracrystalline swelling or swelling due to hydration energy, clay particles undergo osmotic swelling. This occurs at plate distances beyond about 1.0 nm, where surface hydration is no longer important.[7] The osmotic swelling is due to double-layer repulsion by which particles or the layers are pushed farther apart, resulting in large volume changes. The swelling pressures of clay pastes as directly measured is identical with double-layer repulsion as long as it is the only operating force or by far the dominant force between the plates.[25]

Since midway between two interacting plates potential gradient, $d\psi/dx$, is zero, it follows that the osmotic pressure must be constant and equal to $RT\epsilon c_d$, where R is the gas constant, T is absolute temperature, and c_d is the molar concentration (kmol/m³) of any given ion midway between the plates. The "swelling pressure" of the suspension, P, is the difference between the osmotic pressure of the suspension and the osmotic pressure of its equilibrium dialyzate

$$P = RT(c_{+d} + c_{-d} - 2c_o) = 2c_o RT(\cosh u - 1) \tag{89}$$

where c_{+d} and c_{-d} are cation and anion molar concentrations midway between the plates.

It is evident that swelling pressure is a direct measure of the interaction between the particles as it is related to the central potential. Peech and coworkers[26] showed that the swelling pressure measurements on a Na-montmorillonite suspension agreed with the values obtained by Gouy theory.

A close agreement has also been reported by Warkentin and others[27] for Na-montmorillonite and Na-"illite" suspension in equilibrium with dilute NaCl. However, the behavior of Ca clay is usually very different from that of Na-dominated clay, and shows that forces in addition to those normally found in diffuse double layers are operative.[27]

In general, measured swelling pressures as a function of water content (or plate distance) are of the magnitude of the calculated double-layer repulsion, which very likely dominates the swelling process.

3. Schiller Layers

The parallel orientation of the plates or rods (Schiller layers) at large distances can be explained by the counteraction of the gravity forces by double-layer repulsion forces. This explanation is supported by the observation that the particle distance decreases with increasing electrolyte concentration in a manner which is predicted from the computed compression effect of electrolytes on the diffuse double layer. Therefore, the phenomenon of Schiller layers supplies direct evidence for the long range character of the electric double-layer repulsion.[7]

C. Application of Electrical Double-Layer Theory to Constant Surface Charge Soils

Most of the examples looked at so far were for soils with permanent (constant) charge clays. Now an attempt will be made to explain the behavior observed in such soils using the Gouy model. However, for quantitative evaluation and cases where specific adsorption may be predominant, the Gouy-Chapman-Stern-Grahame model would be more appropriate.

In this system surface charge density is controlled by lattice defects in the interior of the crystal, so that double-layer potential, concentration of electrolyte, dielectric constant of the solvent, temperature, and counterion valence are not able to influence the sign or magnitude of the surface charge. Therefore, Equation 67 becomes

$$\left(\frac{2ceRT}{\pi}\right)^{1/2} \sinh\frac{zF\psi_o}{2RT} = constant \tag{90}$$

so that change in electrolyte concentration, dielectric constant, counterion valence, or temperature is counteracted by reduction or increase of surface potential. The increase in potential is accomplished by the compression of the double layer. The degree of double-layer compression is governed by the concentration and valence of the counterions, whereas the effect of co-ions is comparatively small.

The Gouy-Chapman theory shows that the "thickness" of the double-layer decreases, i.e., surface potential increases, with reduction in the dielectric constant of the medium. Such a reduction is attained by the addition of alcohols, acetone, or other water-miscible solvents. The counterion distribution as a function of distance is given from Equations 6a and 16

$$c_i = c_o \exp\left(\frac{-zF\psi}{RT}\right) = c_o \exp(-y) \tag{91A}$$

and also from Equation 53,

$$y = 2\ln\frac{\exp(\xi)+a}{\exp(\xi)-a} \tag{91B}$$

and from Equations 17 and 18, where

$$\xi = \kappa x$$

$$\kappa = \left(\frac{8\pi z^2 F^2 c_o}{\varepsilon RT} \right)^{1/2} \tag{91C}$$

Thus, a decrease in dielectric constant causes an increase in κ which in turn increases ξ and y (or ψ), hence c_i approaches c_o faster, implying the compression of the double layer. On addition of a water-miscible solvent, the range of particle repulsion is reduced and size of the energy barrier becomes smaller, resulting in the well-known enhancement of the flocculation power of an electrolyte.[7]

However, if alcohols are adsorbed on the surface in competition with water, they may influence the capacity of the Stern layer potential. In addition, the point of zero charge (variable surface charge soils) and van der Waals forces may be affected. This explains why alcohol does not always have the flocculating effect predicted by Gouy theory.

D. Application of Electrical Double-Layer Theory to Variable Surface Charge Soils

In spite of growing awareness, few studies have applied the electrical double-layer theory to explain the charge behavior of soils exhibiting variable surface charge. The work that was the cornerstone of modern soil chemistry in this regard was reported in 1972, where an effort was made to study and explain the electrochemical properties of soils developed from tropical areas using double-layer theory.[15] Recently, further studies have been undertaken[28,31] and a complete bibliography is provided by Bowden et al.[32]

In a variable charge system, the surface potential is controlled by the adsorption of potential determining ions, which in turn depends on the activity of these ions in the equilibrium solution. The magnitude of this potential is not affected by the addition of an indifferent electrolyte, so long as the concentration of the potential determining ions or their activity is not affected by the presence of electrolyte.[33] Thus, in such systems Equation 67 becomes

$$\sigma_v = \left(\frac{2c\varepsilon RT}{\pi} \right)^{1/2} \sinh \frac{zF}{2RT} (constant\ \psi_o) \tag{92}$$

If the potential determining ions are H^+ or OH^- the constant surface potential, ψ_o, is related to H^+ by a Nernst equation

$$\psi_o = RT/(zF) \ln H^+/H_o^+ \tag{93}$$

or

$$\psi_o = RT/F\ 2.303(pH_o - pH) \tag{94}$$

where
ψ_o = surface potential (volts)
R   = gas constant (J K^{-1} kmol^{-1})
T   = absolute temperature
z   = valence of potential determining ion

F   = Faraday constant (C/kmol)
H^+ = activity of the hydrogen ion
H_o^+ = the hydrogen ion activity when $\psi_o = 0$

Substituting ψ_o by Equation 94, we obtain:

$$\sigma_v = \left(\frac{2ceRT}{\pi}\right)^{1/2} \sinh 1.15z\left(pH_o - pH\right) \tag{95}$$

Thus, the surface charge density is influenced by the valency of the counterion (z) dielectric constant (ϵ), temperature (T), electrolyte concentration (c), pH of the bulk solution, and the pH_o (zero point of charge) of the soil.

1. *Factors Influencing Surface Charge*

a. pH_o *point of 0 charge*

This is the pH value of a hydroxylated surface when it exhibits zero net surface charge. At pH_o the adsorption of H^+ and OH^- as potential determining ions is considered to be equal. The pH_o is determined by the point of intersection of potentiometric titration curves at different electrolyte concentrations, provided that the electrolyte is not specifically adsorbed. Electrokinetic methods can also be employed to determine the zero point of charge of a soil mineral. However, the equilibrium methods (potentiometric titration and adsorbed ion charge balance) have been favored, because these methods are simpler to apply to heterogeneous assemblies of solids, such as soils, than are the electrokinetic methods.

As evident from the Gouy-Chapman Equation 96, pH_o is an important parameter in a variable charge system because it determines the sign of net surface charge. It should be possible to increase the cation retention capacity of a soil by lowering the pH_o value. The lowering of pH_o of goethite by silicate has been reported by Hingston et al.[34] Similar effects have been obtained by application of phosphate, which resulted in increased negative surface charge density of a Gibbsihumox, a Hydrandept,[28] and a Torrox.[30] Aluminum, on the other hand, has an opposite effect.[35]

b. *Soil pH*

Equation 96 also shows that soil pH determines the magnitude of net charge as well as the sign because of its relationship to pH_o:

$$\sigma_v = constant\left(pH_o - pH\right) \tag{96}$$

Thus, the cation exchange capacity is increased by raising the pH. But, as will be shown in a later section, increasing pH above 6.5 to 7.0 is difficult in soils with variable charge clays.

c. *The Electrolyte Concentration*

The Gouy-Chapman equation further predicts that the net surface charge is directly proportional to the square root of the electrolyte concentration (Figure 5).[31]

In a constant potential (or variable charge) system, the exchange capacity will be even more dependent upon electrolyte concentration because of the greater dependence of total double-layer charge density on concentration. By substituting Equation 94 in 82, the exchange density of a variable surface charge system is given as

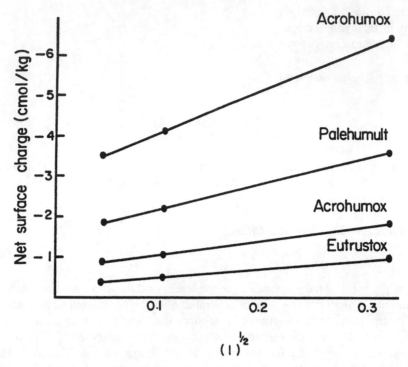

Figure 5   Variation in net surface charge with (ionic strength)[12] for selected highly weathered soils from tropical Queensland (cmol/kg = meq/100 g).

$$\sigma_+ = \left(\frac{ceRT}{2\pi}\right)^{1/2} \left[2 \sinh 1.15z(pH_o - pH) - \exp 1.15z(pH_o - pH) + 1\right] \tag{97}$$

where σ_+ is the cation exchange density when $pH > pH_o$ and the anion exchange density where $pH < pH_o$.

d. The Valence of the Counterion

The theory indicates that the surface charge and surface potential also vary with counterion valence. This effect has been demonstrated on hematite.[33]

Both temperature and dielectric constant influence the net surface charge by a square root relationship.

2. Experimental Evidence

Potentiometric results of van Raij and Peech,[15] Uehara and Keng,[35] El-Swaify and Sayegh,[28] Gillman,[36] and Tama and Ef-Swaify[37] all have shown that soils behaved in a manner predicted by the Gouy-Chapman theory.

From Figure 6 it can be seen that Gouy-Chapman theory explains qualitatively the variation of surface charge with pH, electrolyte concentration, and valency. From the potentiometric titration curves, it is evident that an increase in electrolyte concentration results in a higher value (absolute) for the net electric charge on soil colloids. However, the curves for the highest electrolyte concentration, particularly in Figures 6A and 6B, do not intersect at a common point. This is related to displacement of adsorbed aluminum ions and exposure of the surface charge formerly balanced by them.[35] Strongly adsorbed (high-affinity specific adsorption) cations shift the pH_0 to a higher value. Hence, desorption of such cations results in lower pH_0 (Figure 6). The increase in slope with increasing electrolyte concentration and the effect of pH on net electric charge can also be explained by the Gouy-Chapman theory. The slope of the potentiometric titration curve indicates the buffering capacity of soil (discussed later in this chapter). The pH_0, determined using an indifferent electrolyte, is an intrinsic property of the soil.

Table 1 shows that even quantitatively there was a general agreement between the values calculated using Gouy-Chapman theory and the experimental values, especially for an Oxisol (Halii). However, for the Hydrandept (Akaka), phosphated samples gave better agreement.[28] When van Raij and Peech[15] compared their experimental results with calculated values, they found in using the Gouy-Chapman theory and Stern theory that the latter was more accurate at high potential and high ionic strength.

3. Ion Adsorption

Any ion whose adsorption at the surface is influenced by forces other than simply by electrical potential can be regarded as being specifically adsorbed. Specifically, adsorbed ions can be recognized by their ability to reverse the sign of zeta (ζ) potential, whereas indifferent ions can only reduce ζ asymptotically to zero (Figure 7). Figure 7 also illustrates the mechanism by which reversal of the ζ potential occurs for a simple Stern model at the interface. By way of definition, ζ potential is the average potential in the surface of the shear.

Physically adsorbed counterions (cations) in the Stern layer (low affinity specific adsorption) would induce additional negative charge on oxide surfaces by proton desorption, lowering the pH_0 value. Conversely, the low-affinity specific adsorption of anions would result in increased pH_0 values. Such effects have been observed with Ca^{2+} and SO_4^{2-} for hematite[33] and soil minerals.[35]

Further evidence comes from Figure 6. With a 1:1 indifferent electrolyte (NaCl), the zero point of charge is 4.6 and with $CaCl_2$, as the electrolyte, where Ca^{2+} is specifically adsorbed, the pH_0 drops to 4.3. When anion adsorption occurs, as with Na_2SO_4 as the supporting electrolyte, the zero point of charge value increases to 5.8. A similar but less marked increase in pH_0 occurs when $MgSO_4$ or $CaSO_4$ is the supporting electrolyte.

On the other hand, low-affinity specific adsorption of cations raises the isoelectric point (iep), whereas the opposite occurs with anions. The isoelectric point is the pH at which $\zeta = 0$, i.e., the pH where the charge on the plane separating the Stern layer and diffuse layer is zero. For an indifferent electrolyte (in the absence of specific adsorption in the Stern layer) iep is the same as pH_0 in a purely variable charge system. On specific adsorption, not only is ζ reversed (Figure 7) but the specifically adsorbed anion tends to make ζ more negative, and a more positive surface potential is required to offset the effect. The opposite effect occurs with a cation. This explains the shift in pH_0 (and iep) on specific adsorption of ions. The increase in positive charge at a fixed concentration of potential determining ions with increasing concentration of specifically adsorbed ions (anions), is equivalent to a shift in the pH_0 to a high pH.[38]

Thus, the model predicts a situation where, despite an increase in pH_0 due to low-affinity anion adsorption, the cation retention capacity in the diffuse layer is increased because the isoelectric point is lowered. On the other hand, adsorption of low-affinity cations (Ca^{2+}) reduced the capacity of the diffuse layer to hold cations and even resulted in charge reversal in an Oxisol.[30] Ayres and Hagihara[39] showed that SO_4^{2-} retarded K^+ leaching because the isoelectric point was lowered.

However, silicate and phosphate anion adsorption (high-affinity specific adsorption or chemisorption) leads to reduction in pH_0 values.[28,30,34] High-affinity specific adsorption refers to chemical coordination to the surface metal ion. With high-affinity specific adsorption, the adsorbed ion becomes part of the surface and translates its charge to the solid, hence the lowering of pH_0 as well as iep on chemisorption of an anion.

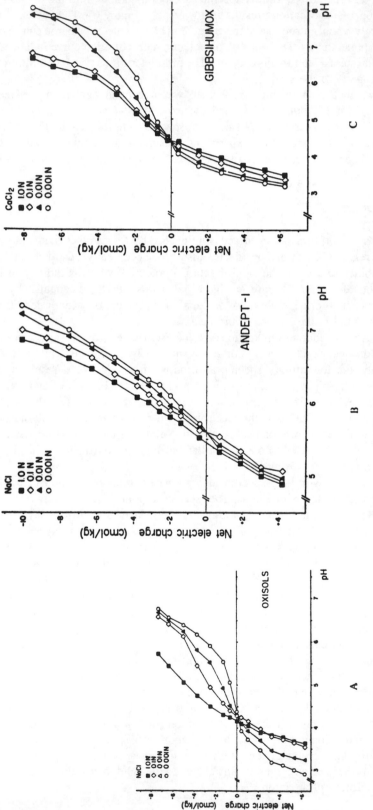

Figure 6 (A to F). The net charge of soils as determined by potentiometric titration (cmol/kg = meq/100 g, kmol/m³ = M).

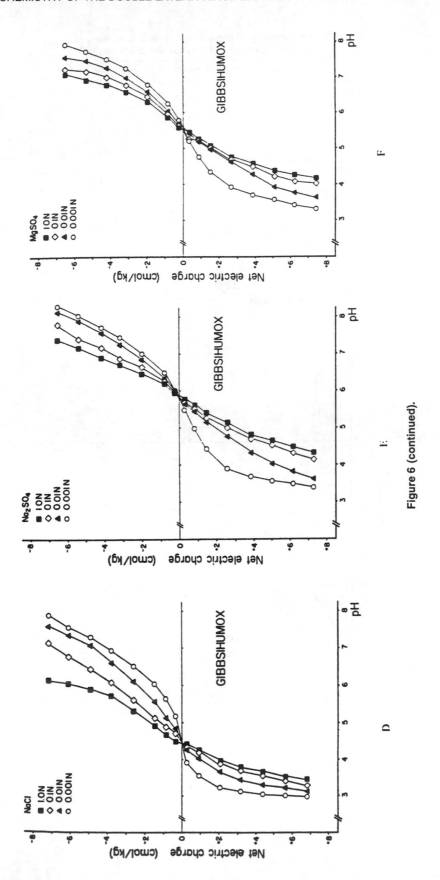

Figure 6 (continued).

Table 1   **Calculated and Experimental Values of Net Negative Charge Density for the Soils at pH 7**

Soil and treatment	Electrolyte concentration (kmol/m³)	kmol/m² – 10 measured m² × 10	kmol/m² – I0 calculated m² × 10
Halii	0.2	1.92	1.75
	0.02	0.92	0.90
Halii-P	0.2	3.68	3.30
	0.02	2.00	1.85
Akaka	0.2	1.45	3.40
	0.02	0.45	1.90
Akaka-P	0.2	4.35	4.35
	0.02	2.20	2.55

From El-Swaify, S. A. and Saygeh, A. H., *Soil Sci.*, 120, 49, 1975. With permission.

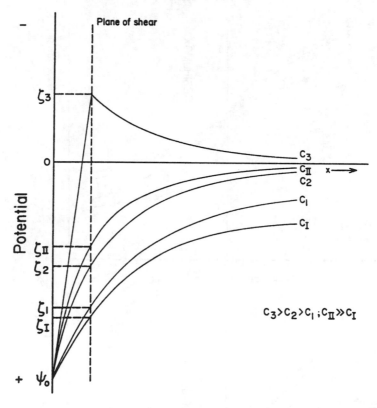

Figure 7   Effect of concentration of indifferent electrolyte (C_I, C_{II}) and specifically adsorbed anion (C_1, C_2, C_3) on zeta potential, using a Stern model.

4. *Buffering Capacity of Soils with Variable Charge Clays*

The buffering capacity (BC) of a soil with variable charge clays is given by

$$B.C. = \frac{Sd\sigma}{dpH}$$

(98)

Table 2   Effect of pH and Valence on the Buffering Capacity

	Buffering capacity (cmol/kg)						Org C	CEC
	pH 5.5		pH 6.0		pH 7.0			
Soil family	z = 1	z = 2	z = 1	z = 2	z = 1	z = 2	(%)	(cmol/kg)
Hydric Dystrandepts								
PUC-H	3.1	14.3	6.4	18	3.1	40	9.03	52.5
lole-K	3.4	13.3	6.3	20	40	50	8.12	34.5
Typic Paleudults								
NAK-A			1.9	2.6	7.7	18.2	2.50	18.5
BPI-H				3.0	5.0	14.3	1.50	20.8
Tropeptic Eutrustox								
Wai-G			2.4	4.0	10	16.7	2.30	18.2
Mol-I			2.3	2.5	25	20	3.70	17.6

From Singh, U. and Uehara, G. unpublished data, 1982. With permission.

where BC is buffering capacity, S is the specific surface, and $d\sigma/dpH$ is the derivative of Equation 92 or the slope of the potentiometric curve. Thus,

$$BC = S\left(-1.15z\left(\frac{2ceRT}{\pi}\right)^{1/2}\cosh 1.15z\left(pH_o - pH\right)\right) \tag{99}$$

From the above equation it is evident that buffering capacity, like the surface charge density, is dependent on pH_o, pH, electrolyte concentration, counterion valence, dielectric constant of the solvent, and temperature. The increase in buffering capacity with increase in valence (z) is more marked than the increase in surface charge. This is quite obvious from Equations 95 and 99. The practical implication of the above is that lime requirements determined with a monovalent electrolyte like NaOH underestimates the amount of $CaCO_3$ required to produce a certain pH change.

Also, other things being constant, an increase or decrease in pH from the pH_o value causes an increase in buffering capacity of the soil. This explains the common problem encountered in tropical soils where an exorbitant amount of lime may be required to raise the soil pH to neutrality. Experimental support for the applicability of Gouy-Chapman theory to buffering capacity of the soils is given in Table 2 and Figure 8. The figure represents a hyperbolic cosine function, as evident from Equation 99.

E. Application of Electrical Double-Layer Theory to Soils with Mixtures of Constant and Variable Charge Surfaces

True as it may be for some extreme cases, in general most soils do not have purely variable charge surfaces. Similarly, soils in the temperate regions tend to be dominated by the constant charge clay minerals; however, their edge surfaces and oxide/organic matter coatings generally carry pH-dependent charge. Generally, all soils contain a mixture of both constant and variable charge surfaces, even though one type might tend to dominate over the other. These mixtures may result from any one or all of the following:

1. Edge effects on crystalline clay minerals (already discussed)
2. Isomorphous substitution or site vacancies in Si-Al cogels[14]
3. Oxide coatings or interlayers on crystalline clay minerals

The total net surface charge density σ_t in a mixed system is

$$\sigma_t = \sigma_p + \sigma_v \tag{100}$$

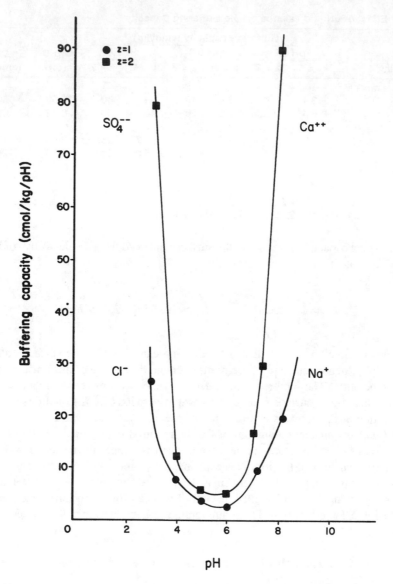

Figure 8   Effect of pH and valence on buffering capacity of variable charge soil.

where σ_p, representing permanent surface charge density, is obtained from Equation 67 and σ_v, the variable surface charge density, from Equation 95.

By substituting Equation 95 in the above we obtain:

$$\sigma_t = \sigma_p + \left(\frac{2ceRT}{\pi}\right)^{1/2} \sinh 1.15z(pHo - pH) \tag{101}$$

The above equations show there are two zero points of charge: for $\sigma_t = 0$, and for $\sigma_v = 0$. The zero point of charge of the mixture is expressed as the zero point of net charge or ZPNC. In this state:

$$\sigma_p + \sigma_v = 0 \tag{102}$$

The net charge density σ_v, in a variable charge system for a 1:1 indifferent electrolyte and for pH values within one unit of zero point of charge, is given by a simplified equation:[31]

$$\sigma_v = 0.135c^{1/2}(pHo - pH) \tag{103}$$

Thus, at pH corresponding to ZPNC:

$$\sigma_p = -0.135c^{1/2}(pH_o - ZPNC) \tag{104}$$

The relationship between the two zero points of charge is as shown:

$$ZPNC = \frac{\sigma_p}{0.135c^{1/2}} + pH_o \tag{105}$$

In Equation 105 pH_o is an intrinsic property of the mixture, whereas ZPNC varies with electrolyte concentration, unless $\sigma_p = 0$

1. Measuring Permanent Charge

From Equations 100 and 101 the permanent charge in a mixed system is measured. When the soil solution pH is adjusted to a value equal to pH_o, the total net charge, σ_t, equals the permanent charge, σ_p. At this pH, equal amounts of cations and anions are adsorbed on the variable charge component so that the permanent charge is obtained by subtracting anions adsorbed from the total cations adsorbed.

From Figure 9, the sign and magnitude of the permanent charge and the pH and concentration dependence of the positive and negative variable surface charge can be determined. The net excess charge at pH_o is due to permanent charge, σ_p. Figure 9 shows the presence of net positive charge.[40] Theory also predicts the ZPNC determined by ion adsorption should be greater than pH_o if σ_p is positive. In summary, the ZPNC falls to the left or the right of pH_o depending on whether the sign of the permanent charge is negative or positive, respectively.

2. The Zero Point of Net Charge (ZPNC)

The pH corresponding to ZPNC can be obtained by ion adsorption measured as a function of pH. The ZPNC corresponds to the pH at which cation and anion adsorption are equal. As predicted by Equation 105, ZPNC moves towards pH_o with increasing electrolyte concentration. However, the Gouy-Chapman theory applies at low, indifferent electrolyte concentrations (<0.01 M) and at pH within one unit of pH_o.

3. Specific Surface Determination

An accurate assessment of the surface charge density also requires an accurate measurement of the surface area of the soil. The problem arises because particles are usually irregular in shape and may "age" on standing. Electron microscopic (direct) measurements of average particle sizes may underestimate the area due to surface roughness and the likelihood of collapse of the surface on a hydrous oxide or a "spongy" organic colloid.[38]

The surface area can also be determined by gas adsorption. However, a serious problem arises when the solid is a hydrate or a hydrous solid. The gas adsorption method requires drying to the

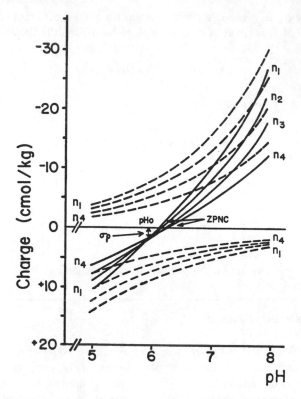

Figure 9   Effect of pH and NH_4Cl concentration on net charge and negative and positive charges (cmol/kg = meq/100 g). (From Wada K. and Okamura, Y., *Soil Sci. Soc. Am. J.*, 47, 902, 1983. With permission.)

point that water does not evolve from the solid during the adsorption measurement and drying is likely to change the surface area and pore structure.[3]

Another procedure, negative adsorption, is an appropriate one for determining specific surface areas. In this case, the specific surface may be determined using the Schofield Equation 84, which is

$$\frac{\Gamma_-}{N} = \frac{q}{(z\beta N)^{1/2}} - \frac{4}{z\beta\Gamma}$$

When the units of Γ_- (now referred to as γ^-) are changed to mol/kg and γ^-/N vs. $(z\beta N)^{1/2}$ is plotted, the slope gives the specific surface in m^2/kg. Thus, the Schofield equation is expressed as

$$\frac{\gamma_a^-}{N} = \frac{qS}{(z\beta N)^{1/2}} - \frac{4S}{z\beta\Gamma} \tag{106}$$

where γ_a^- (mol/kg) = $S\Gamma_-$, S is the specific surface, Γ is in mol/m^2, N is in mol/m^3, β is expressed in m/mol and the subscript a is for anion adsorption. This equation can be used to determine the specific surface of a mixed system provided the selected pH value is sufficiently distant from pH_o (3 to 4 units). At this distance the variable and permanent surface charge are the same sign. Thus, S is the measure of the sum of the specific surfaces of both the permanently and variably charged surfaces.

As discussed earlier, the charge deficit, σ_- of counterions is given by Equation 81. The charge deficit on a variable charge surface and for 1:1 electrolytes is thus:

$$\sigma_- = \left(\frac{c\epsilon RT}{2\pi}\right)^{1/2} \left\{\exp\left[1.15z\left(pH_o - pH\right)\right] - 1\right\}$$ (107)

By experimentally determining the negative adsorption of cations, γ_c^- at 2 or 3 pH units below pH_o for permanently negative charged surfaces and on the acid side of pH_o, the specific surface of the variable charge component S_v can be estimated:[31]

$$\gamma_c^-(mol/kg) = \sigma_-\left(mol/m^2\right) \times S_v\left(m^2/kg\right)$$ (108)

and substituting Equation 107 in the above, we get:

$$S_v = \gamma_c^- \Big/ \left(\frac{c\epsilon RT}{2\pi}\right)^{1/2} \left\{\exp\left[1.15z\left(pH_o - pH\right)\right] - 1\right\}$$ (109)

At 298 K and with $(pH_o - pH) = 2$ in a 0.01 M 1:1 electrolyte

$$S_v\left(m^2/kg\right) = 1.82 \times 10^6 \gamma_c^-$$ (110)

The calculated specific surface values have been reported to be in agreement with the measured ones, especially at lower electrolyte concentration.[41] The agreement, however, is not as good at higher concentrations (0.2 M). This is expected because of likely interaction between electrical double layers.[7] A drawback inherent in the negative adsorption method is that it cannot adequately estimate areas on which double-layer development is impaired (e.g., narrow cracks).[38]

4. Colloidal and Aggregate Stability

The structural and charge characteristics of soils are interrelated, because colloidal stability or flocculation status is dependent on the charge characteristics.[42] Thus, a soil with variable charge or a mixture of variable charge and constant charge surfaces will exhibit colloidal and structural stability different from a constant surface charge soil.

Recalling the surface charge density of a permanent charge system, σ_p, and a variable charge system, σ_v:

$$\sigma_p = \left(\frac{2c\epsilon RT}{\pi}\right)^{1/2} \sinh\frac{zF\psi}{2RT}$$

$$\sigma_v = \left(\frac{2c\epsilon RT}{\pi}\right)^{1/2} \sinh 1.15z\left(pH_o - pH\right)$$

Thus, the colloidal stability in an amphoteric system (capable of mutual flocculation) is dependent on zero point of charge, on concentration of potential determining ions (mainly H^+ and OH^-) and on concentration of supporting electrolyte. Although not shown in the above equations, mean particle size and relative proportions of the constituent colloids are also influential.

On an hydroxylated surface (pH-dependent), the change in surface charge density brought about by changes in pH will act to either enhance the stability of clay suspensions or induce flocculation depending on whether the pH range lies above or below the zero point of charge (pH_o) for the hydroxide, respectively. At pH ranges above pH_o, the predominant negative charge (σ_p) will reinforce repulsion,

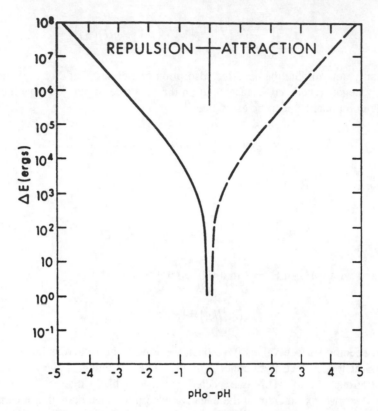

Figure 10 Changes in hydroxide-hydroxide repulsive energy (solid line) and clay-hydroxide attractive energy (broken line) in relation to pH in 0.001 N NaCl. (From E. Swaify, S.A., *Soil Sci. Soc. Am. J.,* 40, 516. 1976. With permission.)

while at pH less than pH_o, the electrostatic energy of repulsion will be reduced due to neutralization of σ_p. Recalling:

$$\sigma_t = \sigma_p + \sigma_v$$

El-Swaify and Emerson[43] showed that the reduction in electrostatic energy of repulsion (inhibition of double-layer swelling) in Na-"illite" and Na-kaolinite/"illite" mixtures was brought about by hydroxide treatments. Low[44] reported that the additional energies "E" enhancing stability or inducing flocculation brought about by a change in the surface charge density may be approximated if a defined interparticle distance, d, is maintained between the particles in suspension.

$$\Delta E = 2\pi \left(\Delta \sigma^2 \right) d / e \tag{111}$$

Figure 10, based on Equation 111, implies that symmetrical colloidal stability is expected for an amphoteric system.[42]

In general, colloidal stability is qualitatively related to the magnitude of charges carried by the colloids. Quantitatively, there appears to be no direct correspondence between the magnitude of such charges and the quantity of <2 μm particles remaining in suspension at a given pH.[37]

The electrostatic contribution to colloidal stability and aggregate breakdown is pH-dependent. From Figure 11 it is evident that the transition from favorable to deteriorated structural conditions is generally abrupt. A practical implication of this is that management practices like liming may

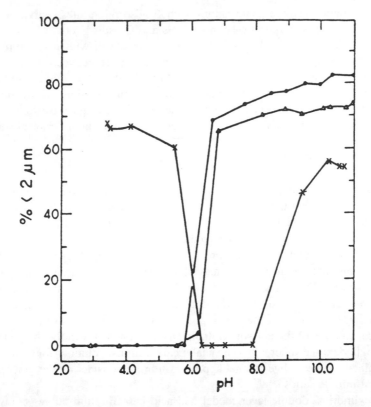

Figure 11 Effect of dispersion pH on the percentage of 2 μm particles recovered from Waikane (•), Hilo (x), and Molokai (Δ) soils. Suspension pH values without pH adjustment were 5.5, 5.9, and 6.8 for the three soils, respectively. (From Tama, K. and El-Swaify, S.A., *Modification of Soil Structure*, John Wiley & Sons, New York, 1978, 41. With permission.)

be detrimental to soil structures.[45] Increasing pH may be detrimental depending upon whether pH elevation causes a shift towards the pH_o (e.g., Hilo-Hydrandept) or away from it (Waikane-Ultisol and Molokai Oxisol).

In one respect the variable surface charge colloids behave in a similar manner to the constant surface charge colloids. Increasing the electrolyte concentration results in increased charge magnitudes; however, in neither of the two systems do the increases result in enhancement of colloidal stability. Thus, the increased electrolyte concentration did not produce repulsive forces strong enough to overcome the flocculating effect of added electrolyte. This phenomenon can be explained by the fact that reduced interparticle repulsion, due to diffuse double-layer collapse, exceeds the simultaneous increase in repulsion due to increased double-layer capacity at the higher electrolyte concentrations.[42]

The coexistence of substantial positive and negative charges in "variable" charge soils at field pHs gives rise to electrostatic bonding between different constituents and between constituents of the same type. Also, the presence of nonelectrostatic bonding gives further aggregate stability to these soils.

5. Phosphate Sorption

Phosphate sorption in variable charged soils — Oxisols and Andepts — is generally high (>700 mg P/kg soil). In soil with mixed charges, P adsorption maxima vary from 100 to 400 mg P/kg soil.[46] Organic matter also inhibits P adsorption by either physically blocking adsorption sites or through the competition of organic anions with P adsorption sites. Fulvic and humic acids effectively compete for P adsorption sites in Andepts and Oxisols.[47]

In soil with mixed charges, silicate-Al was highly correlated with P adsorption. The broken edges of the silicate clays are possible sites of ligand exchange between the open hydroxyl and phosphate groups.[47-49] Further, the effect of silicate-Al on P adsorption might also be related to the surface reactions of clays, including the interaction of phosphate anions with cations in the Stern and diffuse double layers.[46] This is consistent with the role of exchangeable cations on P adsorption, especially Ca^{2+}, the dominant cation on the exchange complex of many tropical soils.

6. Double-Layer Models for Humic Substances

Humic substances are complex and ill-defined polydisperse mixtures of heterogeneous organic polyelectrolytes which are randomly coiled when un-ionized and fully expanded because of intramolecular repulsion when acidic groups are dissociated.[48] In three-dimensional form, extended linear polyelectrolytes are equivalent to charged cylinders or spheres.

At low pH the functional groups are protonated and uncharged, at higher pH the functional groups dissociate and become negatively charged. As expected around charged particles, a diffuse double layer develops. The charge-induced accumulation results in an increase of the specific binding of the positively charged ions. The double-layer model allows for the calculation of the electric potential at the surface of the humic particles and of the pH near the binding sites, pH_s.

An analytical solution of the Poisson–Boltzmann equation is known for charged plane surfaces, while charged cylinders and spheres have no analytical solutions. Solutions for cylinders and spheres have been obtained by the Debye-Huckel approximation by considering the first term only of the one-dimensional infinite series.

Using the cylindrical double-layer model and acid-base titration curves equivalent radii for humic acids were determined.[49] The equivalent radius was dependent on the degree of dissociation, α. For a low degree of dissociation ($\alpha = 0.1$ to 0.3), strong decrease in equivalent radius (1.1 to 1.5 nm) occurred. At $\alpha = 0.3$ to 0.9, the equivalent radius decreased from 0.5 to 0.25 nm. The equivalent radius measured by this method agreed closely with the apparent viscometric radius reported for other humic substances.

Both the cylindrical and spherical double-layer models describe the ionic dependency of humic materials well. The assessed radius is larger for the spherical double-layer model than that for the cylindrical model. The opposite holds for specific surface area.[50] The electrostatic interactions for the spherical model are stronger than those for the cylindrical model. As a consequence pH-titration curves for the spherical model are shifted towards a lower pH_s (pH near binding sites). The spherical double-layer model has the advantage that an average molecular weight is determined directly from the analyses and the number of sites/molecule could be calculated.

Low molecular weight humics are relatively small molecules which can be easily seen as very small rigid particles. Due to cross-linking and association by inorganic bridging ions like Fe^{3+} and Al^{3+}, the high molecular weight humics can be fairly rigid too, with the bulk of the functional groups in the outer shell. This may explain why a description assuming rigid impermeable spheres and cylinders with the double-layer approach gives such satisfactory results.

F. Application of Electrical Double-Layer Theory to Soil Nutrient Dynamics Modeling

Currently, there is very limited direct application of the electrical double-layer theory in agriculture. Likewise, most applications of nutrient dynamics and environmental modeling have been in temperate subtropical regions on soils with predominantly constant surface charge. Computer simulation models provide practical alternatives to address the complex and interactive nature of sustainability issues and environmental impact assessments.[51] Due to the nature and fast rates of chemical and biological transformations in the tropics, the quantification of agricultural sustainability and environmental quality assessment becomes more critical. Nutrient dynamics and soil-and-water quality models must simulate the effects of mixtures of constant and variable charge surfaces on nutrient transformation and solute (or pesticide) movement to be applicable in most tropical soils. Some of the key aspects that need to be considered follow.

In most agricultural soils of the humid tropics the electrolyte concentration rarely exceeds 10^{-2} M, hence the direct effect of concentration on surface charge density is negligible. However, the retention of ions is highly dependent on the magnitude of the net charge as well as the sign. The effects of fertilizers and amendments on soils with variable and mixed charge surfaces differ from those with constant charge surfaces. For example, applications of silicate[34] and phosphate[28,30] have resulted in increased negative surface charge density in soils with low cation retention capacity. A soil phosphorus dynamics model needs to simulate the effect of high-affinity adsorption of added phosphate fertilizer on soils with variable surface charge and its consequences on cation exchange capacity (CEC), buffering capacity, and soil pH.[46] The model must also account for the effects of low-affinity adsorption of SO_4^{2-} and Ca^{2+} on the above soil properties. Desorption of aluminum,[35] when organic residues are applied to soils exhibiting variable charge surfaces (chelation of Al with organic matter), results in lowered pH_o and increased cation retention. As expected in soils with anion retention capacity, addition of phosphates, silicates, organic residues, and sulfates resulted in reduced or even complete elimination of anion retention. Also in such soils, as the organic matter content decreases down the profile, the anion retention capacity increases.[52] Low-affinity specific adsorption of Ca^{2+}, however, could have the opposite effect, increased cation retention capacity.

1. Nitrate Leaching

Simulation of nitrate leaching as described in CERES models[53] assumes complete reservoir mixing of nitrate with soil water in a given layer, and that the flux of nitrate in solution, F_s (cm d^{-1}) is equal to the soil water flux, F_1 (cm d^{-1}). However, in soils with variable charged surfaces and anion retention capacity, $F_s < F_1$. The retention factor, RF ($= F_s/F_1$), is dependent on (1) soil moisture content, $\Theta_{(L)}$, in m^3 m^{-3} where $\Theta_{(L)}$ is greater than the moisture content at the drained upper limit and less than the saturated moisture content; (2) bulk density, ρ_b in kg m^{-3}; and (3) net charge in a given layer, $\sigma_v(L)$. The retention factor has been described by Wild[54] as:

$$RF = F_s/F_t = 1/\left[1+\left(b_{(L)}\rho_{b(L)}/\theta_{(L)}\right)\right]$$ (112)

where $b_{(L)}$ is the retention coefficient of the nitrate ion in the soil layer, L (m^3 kg^{-1}) or ratio of nitrate retained by the soil (kmol kg^{-1}) to nitrate in the soil solution (kmol m^{-3}). Since b is not a readily available input for simulation models, it was estimated for subsoils with variable charge surfaces as:

$$b = \left(1.25\left[Exp\{\Delta pH+0.9\}-1\right]-1.02\ Org\ C\right)\times 10^{-3}$$ (113)

$$R^2 = 0.78$$

The above regression equation applies to b > 0. The ΔpH ($pH_{KCl} - pH_{water}$) is an approximation of surface charge and its sign. Any appreciable anion retention will occur in subsoils with ΔpH of greater than -0.6. The retention coefficient is further reduced with increasing organic carbon content, Org C (%). The typical range of the retention coefficient for nitrate in subsoils with variable charge surfaces is from 0 to 2.4×10^{-3} m^3 kg^{-1}.

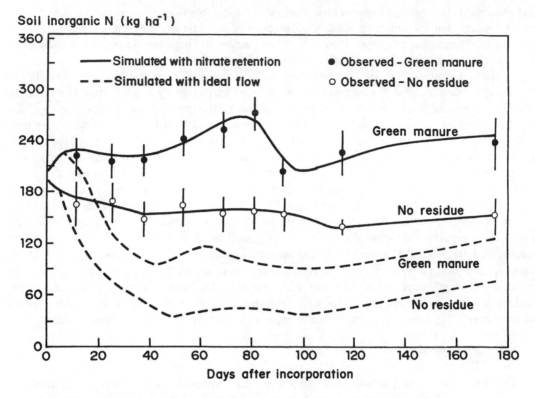

Figure 12   Soil N dynamics as influenced by N leaching and residue management.

The nitrate leaching subroutine of the CERES-Maize model was modified using Equations 112 and 113. Figure 12 shows the performance of the model when compared with the observed data of Bowen and coworkers[55] on an uncropped soil. It is envisaged that the computer simulation models for nutrient dynamics and environmental assessment (nutrient leaching, pesticide movement) would utilize the concepts of electrical double-layer theory as the applications of these models become more widespread in the variable charged soils of the tropics. Due to the complexities of reactions and interactions with biological activities in soils, electrical double-layer theory will only provide a partial but functional solution to modeling the soil system.

2. Transport of Interacting Solutes

The transport of solutes in the unsaturated zone, as evident from the previous section, is a complicated process involving convection and dispersion as well as interactions with the soil matrix (cation exchange, anion exclusion, precipitation, and dissolution). These processes in mixed Na/Ca chloride solutions were quantified using theoretical considerations that were based on the mixed-ion diffuse double-layer theory, the structure of clay particles, the soil's pore size distribution and hydrodynamic principles.[56] In a homogeneous soil, the effect of soil solution-soil matrix interactions on the transport of water and solutes may be significant and generally increases as both the soil water content and the soil solution sodium absorption ratio increase, as the soil solution concentration decreases, and as the clay fraction of soil increases (as soil texture becomes finer).

For a soil whose solid surface charge density is given, both Γ_{ex}^{-} (monovalent anionic charges repulsed from unit area of solid surface in kmol/m^3) and Γ_{ad}^{+} (monovalent cationic charges adsorbed to a unit area of a solid surface) as a function of h (soil water pressure potential head in m^3), R (ratio between molar concentration of monovalent and divalent cations in the soil solution = c_{Na}/c_{Ca})

and sum of the molar concentration of cations, c_o ($= c_{Na} + c_{Ca}$ in kmol/m³) can be estimated using mixed-ion diffuse double-layer theory.[51]

When an anion is repulsed from the negatively charged soil surface, the transport volume is decreased by the fraction θ_{ex}^{Cl}. In other words, for a given water flow regime, the movement of the anion will be accelerated as shown:

$$\theta'(h, R, c_o, c_{Cl}) = \theta(h, R, c_o) - \theta_{ex}^{Cl}(h, R, c_o, c_{Cl})$$ (114)

Similarly, for a cation which is adsorbed to the negatively charged soil surface the transport volume is increased by the fraction θ_{ad}^m, i.e., movement of the cation will be retarded:

$$\theta'(h, R, c_o, c_m) = \theta(h, R, c_o) - \theta_{ad}^m(h, R, c_o, c_m)$$ (115)

for sodium (m = 2) and for calcium (m = 3) cations, respectively, where θ_{ex}^{Cl} and θ_{ad}^m are the exclusion and adsorption water contents.

V. EXAMPLES

These worked examples are amenable to simplifications that yield analytical solutions to electrical double-layer problems. However, mathematical software packages with some coding can be used for numerical solutions to more complicated electrical double-layer problems.

A. Calculating Surface Potential for a Permanent Charged Soil

Using the Gouy-Chapman theory for a 0.01 M (1:1) electrolyte concentration:

$$\sigma_p = \left(\frac{2c\epsilon RT}{\pi}\right)^{1/2} \sinh(zF\psi_o/2RT)$$

given:

$$\sigma_p = -0.135 \, C/m^2$$ (116)

given:
c = 0.01 kmol/m³
ϵ = 8.9 × 10⁻⁹ C²/Jm
z = 1
R = 8.314 × 10³ J/kmol K (8.314 J/mol K)
T = 298 K
F = 9.6487 × 10⁷ C/kmol (9.6487 × 10⁴ C/mol)
ψ_o = the surface potential in volts (V)

Rearranging for ψ_o we get:

$$\psi_o = \frac{2RT}{zF} \sinh^{-1}\left[\sigma_p \Big/ \left(\frac{2c\epsilon RT}{\pi}\right)^{1/2}\right]$$

$$= -5.134 \times 10^{-2} \, J/C \times \sinh^{-1}\left(0.135 \frac{C}{m^2} \Big/ 1.185 \times 10^{-2} \frac{C}{m^2}\right)$$ (117)

$$\psi_o = -1.602 \times 10^{-1} V (or \ 160.2 \ mV) \qquad (118)$$

The surface potential ψ_o for 0.1 *M,* 0.01 *M,* and 0.001 *M* electrolyte concentrations are –102 mV, –160 mV, and –219 mV, respectively. The example illustrates changes in surface potential in a permanent charged soil with changes in electrolyte concentration.

B. Calculating the Surface Potential and the Surface Charge Density for pH-Dependent (Variable Charged) Soil

1. Surface Potential for a Variable Charged Soil

$$\psi_o = \frac{RT}{F} 2.303 (pH_o - pH) \qquad (119)$$

For T = 298 K and ψ_o expressed in V:

$$\psi_o = \frac{8.314 \times 298 \times 2.303}{9.6487 \times 10^7 C} J (pHo - pH) \qquad (120)$$

$$= 0.0592 (pHo - pH) \qquad (121)$$

For $pH_o = 4$ and pH = 6, a variable charged soil has a surface potential, $\psi_o = -0.118$ V or –118 mV,

2. Surface Charge Density for Variable Charged Soil

$$\sigma_v = \left(\frac{2ceRT}{\pi} \right)^{1/2} \sinh z (pH_o - pH)$$

For a 1:1 electrolyte with a concentration of 0.1 *M* (kmol/ml³) and with T, pH_o and pH values as before in the previous example one obtains:

$$\sigma_v = \left(\frac{2 \times 0.1 \times 8.9 \times 10^{-9} \times 8.314 \times 10^3 \times 298}{\pi} \right)^{1/2} \sinh[1.15 \times (-2)] \qquad (122)$$

$$= -1.85 \times 10^{-1} C/m^2$$

For 0.01 *M* and 0.001 *M* electrolyte concentrations, the surface charge densities are -5.88×10^{-2} C/m² and -1.85×10^{-2} C/m². Thus, as expected, the magnitude of surface charge density in a variable charge system decreases with decreasing concentration of the supporting electrolyte.

If in the above case, the calculations were repeated at pH = 2, surface charge densities for 0.1 *M,* 0.01 *M,* and 0.001 *M* electrolyte concentrations would be 1.85×10^{-1} C m², 5.88×10^{-2} C/m², and 1.85×10^{-2} C/m², respectively, meaning that the nature of charge on a pH-dependent soil could be either positive or negative.

C. Calculating S_v

S_v, the specific surface due to the variable charge component, at 298 K and $(pH_o - pH) = 2$ in a 0.01 *M* 1:1 electrolyte, has been shown to be from Equation 110:

$$S_v \left(m^2/kg \right) = 1.82 \times 10^6 \gamma_c^-$$

$$S_v \left(m^2/kg \right) = \gamma_c^- \left(mol/kg \right) / \sigma_- \left(mol/m^2 \right) \tag{123}$$

First σ_- is obtained in C/m² using Equation 107:

$$\sigma_- = \left(\frac{c \varepsilon R T}{2\pi} \right)^{1/2} \left\{ \exp\left[1.15z \left(pH_o - pH \right) \right] - 1 \right\}$$

Substituting in the values for the variables and constants one gets:

$$\sigma_- = \left(\frac{0.01 \times kmol/m^3 \times 8.9 \times 10^{-9} \, C^2/Jm \times 8.314 \times 10^3 \, J \times 298K}{2\pi \quad kmol \, K} \right)^{1/2} \left\{ \exp\left[1.15(2) \right] - 1 \right\} \tag{124}$$

$$= 5.32 \times 10^{-2} \, C/m^2 \tag{125}$$

From F = 9.6487 × 10⁴ C/mol, one gets σ_- in mol/m²:

$$\sigma_- = 5.51 \times 10^{-7} \, mol/m^2 \tag{126}$$

Thus:

$$S_v = \gamma_c^- / 5.51 \times 10^{-7} \tag{127}$$

$$S_v \left(m^2/kg \right) = 1.82 \times 10^6 \gamma_c^- \left(where \ \gamma_c^- \ is \ in \ mol/kg \right) \tag{128}$$

VI. CONCLUSIONS

To conclude this chapter, it may be said that many of the experimentally observed clay colloid properties may be at least qualitatively described in terms of the simple electric double-layer model of Gouy and Chapman. However, the modified Gouy-Chapman-Stern-Grahame model gives a better quantitative picture. The clay colloidal properties which can be successfully described by the double-layer theory include surface charge, cation exchange capacity, buffering capacity, colloidal stability, and the rheological properties. The important solution properties include pH, the electrolyte type (valence), composition, and concentration. Exchange phenomena such as anion adsorption (Schofield's equation) and selectivity of multivalent ions (Erickson's equation) are based on the electric double-layer theory.

The versatility of electric double-layer theory is further evident from its ability to qualitatively explain charge behavior in permanent or variable charge systems and in soils with mixtures of both types of charges. The increasing popularity of electric double-layer theory compared to the thermodynamic approach (Donnan equilibrium) is based on the simplicity and the wide applicability of the theory to soil systems. The versatility of the electric double-layer model is also evident from models that have their derivation from double-layer theory using the Debye-Huckel assumption, e.g., the capacitance model.

REFERENCES

1. Verwey, E.J.W., The electrical double layer and the stability of lyophobic colloids, *Chem. Rev.*, 16, 363, 1935.
2. Grahame, D.C., The electrical double layer and the theory of electrocapillarity, *Chem. Rev.*, 41, 441, 1947.
3. James, R.O. and Parks, G.A., Characterization of aqueous colloids by their electrical double layer and intrinsic surface chemical properties, *Surf. Colloid Sci.*, 29, 119, 1980.
4. Overbeek, J. Th., Electrochemistry of the double layer, *Colloid Sci.*, 1, 115, 1952.
5. Ross, G.E. and Hendricks, S.B., Minerals of the Montmorillonite Group. Their Origin and Relation to Soils and Clays, Paper 205-B, Geological Survey, U.S. Department of the Interior, Washington, D.C. 1945.
6. Gast, R.C., Surface and colloid chemistry, in *Minerals in the Soil Environment*, Dixon, J.B. and Weed, S.B., Eds., Soil Science Society of America, Madison, WI, 1979, 27.
7. van Olphen, H., *Introduction to Clay Colloid Chemistry*, Interscience, New York, 1977, 318.
8. Pauling, L., The structure of micas and related minerals, *Proc. M.S. Natl. Acad. Sci.*, 16, 123, 1949.
9. Marshall, C.E., *The Colloid Chemistry of the Silicate Minerals*, Academic Press, New York, 1949.
10. Parks, G.A. and de Bruyn, P.L., The zero point of charge of oxides, *J. Phys. Chem.*, 66, 967, 1962.
11. Bolt, G.H., The Significance of the Measurement of the Zeta Potential and the Membrane Potential in Soil and Clay Suspensions, Masters thesis, Cornell University, Ithaca, New York, 1952.
12. El-Swaify, S.A., *Soil Physical Chemistry*, booklet for Soil Science 640, University of Hawaii, Honolulu, 1982, 131.
13. Bolt, G.H., Analysis of the validity of the Gouy-Chapman theory of the electric double layer, *J. Colloid Sci.*, 10, 206, 1955.
14. Parks, G.A., Aqueous surface chemistry of oxides and complex oxide minerals. Isoelectric point and zero point of charge, *Adv. Chem.*, 67, 121, 1967.
15. van Raij, G. and Peech, M., Electrochemical properties of some Oxisols and Alfisols of the tropics, *Soil Sci. Soc. Am. Proc.*, 36, 587, 1972.
16. Uehara, G. and Gillman, G.P., *The Mineralogy, Chemistry, and Physics of Tropical Soils with Variable Charge Clays*, Westview Press, Boulder, CO, 1981.
17. Bowden, J.W., Bolland, M.D.A., Posner, A.M., and Quirk, J.P., General model for anion and cation adsorption at oxide surfaces, *Nat. Phys. Sci.*, 245, 81, 1973.
18. Bolt, G.H., Ion adsorption by clay, *Soil Sci.*, 79, 276, 1955.
19. Erickson, E., Cation exchange equilibria on clay minerals, *Soil Sci.*, 74, 103, 1952.
20. Lagerwerff, J.V. and Bolt, G.H., Theoretical and experimental analysis of Gapon's equation for ion exchange, *Soil Sci.*, 87, 217, 1959.
21. Babcock, K.L., Theory of the chemical properties of soil colloidal systems at equilibrium, *Hilgardia*, 34, 417, 1963.
22. Bolt, G.H. and Warkentin, B.P., The negative adsorption of anions by clay suspension, *Kolloid-Z.*, 1, 41, 1958.
23. Bolt, G.H., Stability of hydrophobic colloids and emulsions, *Colloid Sci.*, 1, 302, 1952.
24. Riddick, T.M., *Control of Colloid Stability through Zeta Potential*, Zeta Meter Corp., New York, 1968.
25. Barshad, I., The nature of lattice expansion and its relation to hydration in montmorillonite and vermiculite, *Am. Mineral.*, 34, 675, 1949.
26. Peech, M., Olsen, R.A., and Bolt, G.H., The significance of potentiometric measurements involving liquid junction in clay and soil suspensions, *Soil Sci. Soc. Am. Proc.*, 17, 214, 1953.
27. Warkentin, B.P., Bolt, G.H. and Miller, R.D., Swelling pressures of Montmorillonite, *Soil Sci. Soc. Am. Proc.*, 21, 495, 1957.
28. El-Swaify, S.A. and Sayegh, A.H., Charge characteristics of an Oxisol and an Inceptisol from Hawaii, *Soil Sci.*, 120, 49, 1975.
29. Espinoza, W., Gast, R.G., and Adams, R.S., Jr., Charge characteristics and nitrate retention by two Andepts from South-Central Chile, *Soil Sci. Soc. Am. Proc.*, 39, 842, 1975.
30. Wann, S.S. and Uehara, G., Surface charge manipulation of constant surface potential soil colloids. I. Relation to sorbed phosphorus, *Soil Sci. Soc. Am. J.*, 42, 565, 1978.

31. Uehara, G. and Gillman, G.P., Charge characteristics of soils with variable and permanent charge minerals. I. Theory, *Soil Sci. Soc. Am. J.,* 44, 250, 1980.

32. Bowden, J.W., Posner, A.M., and Quirk, J.P., Adsorption and charging phenomena in variable charge soils, in *Soils with Variable Charge*, Theng, B.K.B., Ed., New Zealand Society of Soil Science, Lower Hutt, New Zealand, 1980, 147.

33. Breeuwsma, A. and Lyklema, J., Interfacial electrochemistry of hematite $(-Fe_2O_3)$, *Discussion Faraday Soc.,* 52, 324, 1971.

34. Hingston, F.J., Atkinson, R.J., and Quirk, J.P., Specific adsorption of anions, *Nature (London),* 215, 1459, 1967.

35. Uehara, G. and Keng, J., Management implications of soil mineralogy in Latin America, in *Soil Management in Tropical America*, Bornemisza, E. and Alvarado, A., Eds., North Carolina State University Press, Raleigh, 1975, 351.

36. Gillman, G.P., The influence of net charge on water dispersible clay and sorbed sulfate, *Aust. J. Soil Res.,* 12, 173, 1974.

37. Tama, K. and El-Swaify, S.A., Charge, colloidal and structural stability relationships in oxide soils, in *Modification of Soil Structure*, Emerson, W.W., Bond, R.D., and Dexter, A.R., Eds., John Wiley & Sons, New York, 1974, 4.

38. Hayes, M.H.B., MacCarthy, P., Malcolm, R.L., and Swift, R.S., Structures of himic substances: the emergence of 'forms', in *Humic Substances, Vol. 2, In Search of Structure*, Hayes, M.H.B., MacCarthy, P., Malcolm, R.L., and Swift, R.S., Eds., John Wiley & Sons, Chichester, 1989, 689.

39. Ayres, A.S. and Hagihara, H.H., Effect of the anion on the sorption of potassium by some humic and hydrol humic Latosols, *Soil Sci.,* 75, 1, 1953.

40. Wada, K. and Okamura, Y., Net charge characteristics of Dystrandepts and theoretical predictions, *Soil Sci. Soc. Am. J.,* 47, 902, 1983.

41. Gillman, G.P. and Uehara, G., Charge characteristics of soils with variable and permanent charge minerals. II. Experimental, *Soil Sci. Soc. Am. J.,* 44, 252, 1980.

42. El-Swaify, S.A., Changes in the physical properties of soil clays due to precipitated aluminum and iron hydroxides. II. Colloidal interactions in the absence of drying, *Soil Sci. Soc. Am. J.,* 40, 516, 1976.

43. El-Swaify, S.A. and Emerson, W.W., Changes in the physical properties of soil clays due to precipitated aluminum and iron hydroxides. I. Swelling and aggregate stability after drying, *Soil Sci. Soc. Am. Proc.,* 39, 1056, 1975.

44. Low, P.F., Mineralogical data requirement for soil physical investigations, in *Mineralogy in Soil Science and Engineering*, Kunze, C.W., White, J.L., and Rust, R.H., Eds., SSSA Special Publication Series No. 3, Soil Science Society of America, Madison, WI, 1968, 1–34.

45. El-Swaify, S.A., Physical and mechanical properties of Oxisols, in *Soils with Variable Charge*, Theng, B.K.G., Ed., New Zealand Society of Soil Science, Lower Hutt, New Zealand, 1980, 303.

46. Agbenin, J.O. and Tiessen, H., The effects of soil properties on the differential phosphate sorption by semiarid soils from Northeast Brazil, *Soil Sci.,* 157, 36, 1994.

47. Sibanda, H.M. and Young, S.D., Competitive adsorption of humic acids and phosphate on goethite, gibbsite and two tropical soils, *J. Soil Sci.,* 37, 197, 1986.

48. Hayes, M.H.B. and Himes F.L., Nature and properties of humus-mineral complexes, in Interactions of Soil Minerals with Natural Organics and Microbes, Huang, P.M. and Schnitzer, M., Eds., *Soil Science Society of America, Madison, WI*, 1986, 103.

49. Barak, P. and Chen, Y., Equivalent radii of humic macromolecules from acid-base titration, *Soil Sci.,* 154, 184, 1992.

50. de Wit, J.C.M., van Riemsdijk, W.H., and Koopal, L.K., Proton binding to humic substances. I. Electrostatic effects, *Environ. Sci. Technol.,* 27, 2005, 1993.

51. Singh, U. and Thornton, P.K., Using crop models for sustainability and environmental quality assessment, *Outlook Agric.,* 21(3), 209, 1992.

52. Wong, M.T.F., Hughes, R., and Rowell, D.L., Retarded leaching of nitrate in acid soils from the tropics: measurement of the effective anion exchange capacity, *J. Soil Sci.,* 41, 655, 1990.

53. Godwin, D.C. and Jones, C.A., Nitrogen dynamics in soil-plant systems in *Modeling Plant and Soil Systems,* Hanks, R.J. and Ritchie, J.T., Eds., *Agronomy Monogr.* 31, American Society of Agronomy, Crop Science Society of America, and Soil Science Society of America, Madison, WI, 1991, 287.

54. Wild, A., Mass flow and diffusion, in *The Chemistry of Soil Processes*, Greenland, D.J. and Hayes, M.H.B., Eds., John Wiley & Sons, Chichester, England, 1981, 37.
55. Bowen, W.T., Jones, J.W., Crasky, R.J., and Quintana, J.O., Evaluation of the nitrogen submodel of CERES-Maize Incorporation, *Agron. J.*, 85, 153, 1993.
56. Russo, D., Numerical analysis of the nonsteady transport of interacting solutes through unsaturated soil. I. Homogeneous systems, *Water Resour. Res.*, 24, 271, 1988.

CHAPTER 2

Chemical Modeling of Ion Adsorption in Soils

John M. Zachara and John C. Westall

CONTENTS

I. INTRODUCTION

Adsorption occurs in soil when aqueous solutes accumulate on mineral surfaces, their coatings, or particulate organic matter. Adsorption is a fundamental process in soil that influences nutrient concentrations, the retardation of microcontaminants, the heterogeneous nucleation of secondary solid phases, and the stability of the colloidal fraction. Proton adsorption controls charge development on soil oxides, hydrous oxides, layer silicate edges, and organic matter, which, in turn, influences the adsorption of other solutes. Weathering, an integral component of the global carbon cycle and soil formation, is driven by adsorption of H^+, $CO_{2(g)}$, organic acids of biotic origin, and reductants/oxidants that destabilize and promote release of structural cations such as Al(III) or Fe(II,III).

The surfaces of soil particles are heterogeneous. This heterogeneity greatly complicates any attempt to develop models for adsorption at these surfaces. Two general approaches exist for modeling ion adsorption to heterogeneous surfaces: mechanistic and semiempirical. In the mechanistic approach (e.g., surface complexation-electric double-layer models), one attempts to represent the soil surface as an assemblage of its components, the properties of which can be determined more or less independently. Since this approach is based on fundamental principles of chemistry and physics, one generally assumes that it is the preferred approach for interpreting experimental data and for performing interpolations and extrapolations based on the data. However, if one analyzes the assumptions on which these models are based, one might reach the conclusion that the assumptions really are not justified, that the models are not so thoroughly grounded in fundamental chemistry and physics, and that the models are not really as "mechanistic" as one generally assumes.

An alternative to this mechanistic approach is a semiempirical one (e.g., affinity [log K] spectrum models). In this latter approach, one accepts the fact that the heterogeneity is too complex to represent in terms of basic principles, and just tries to represent the experimental data as accurately as possible. These models are based on principles such as mass action and mass balance (hence *semi*empirical), such that they are generally consistent with the natural laws that are presumed to be governing the system, but no attempt is made to incorporate explicitly all of the physical processes that are known to occur. These semiempirical models are generally easier to work with than their mechanistic counterparts, and may prove to be as useful in representing, interpolating, and extrapolating soil chemical data. Certainly in the long run, it would be preferable to have purely mechanistic models for adsorption processes, but for the present, the semiempirical models are a viable alternative. This chapter is focused primarily on mechanistic models, but provides an example of the semiempirical approach as well.

In recent years significant scientific advances have been made in understanding adsorption reactions on mineral surfaces (see for example Stumm[1]). These advances have been made possible through the advent of (1) modern surface spectroscopy that allows the direct, albeit at times ambiguous, identification of the chemical nature of surface species,[2] and (2) surface complexation-electric double-layer models (SC-EDL) that allow surface reactions to be modeled as the analogous solution-phase reactions, with an adjustment for electric double-layer effects.[3,4] The success of SC-EDL models in describing the surface reactivity of single-phase mineral suspensions over wide ranges in aqueous chemical conditions has led to their increased use in the interpretation of soil chemical process (see reviews in Sposito[5,6]). However, applications of SC-EDL models to soil have met with difficulty, in part because of the heterogeneous chemical, mineralogical, and physical nature of soil materials.

Site binding models are a general class of models that describe adsorption using mass action expressions between aqueous species and surface sites leading to a surface complex. Site binding models include, but are not limited to, the SC-EDL model. In this chapter we discuss application of site binding models to clay, composite mineral, and soil systems with emphasis on surface charge development and ion adsorption. We focus on models that explicitly include electrostatics (SC-EDL) as well as those which do not (e.g., the surface complexation-nonelectrostatic model [SC-NEM], ion exchange).

II. ADSORPTION MODELS

A. Surface Functional Groups

Mineral surfaces in soil have functional groups capable of complex formation with inorganic and organic ions.[5-9] These functional groups are linked explicitly to surface charge development and include the following:

1. Inorganic hydroxyl groups bound to surface Al, Fe, Mn, or Si on soil oxides or Al and Si exposed on clay mineral edges (SOH)
2. Siloxane cavities on the surfaces of 2:1 layer silicates with a fixed, structural charge arising from isomorphic substitution (X^-) in either the tetrahedral or octahedral layers
3. Organic hydroxyl groups bound to carboxylic acids and phenolate sites on soil organic matter (ROH)
4. Partially coordinated sites on the surfaces of carbonates, sulfates, phosphates, and sulfides

Fixed geometry produces effects that distinguish surface sites from their aqueous phase analogs: (1) they cannot be diluted infinitely, (2) their proximity allows lateral interactions, and (3) their energy reflects the complex stereochemical, structural, and electrostatic conditions of the mineral-water interface.[8]

The acid-base chemistry of these surface functional groups (e.g., SOH, ROH) are often represented by reactions of the type:

$$SOH + H^+ = SOH_2^+ \qquad 1/K_{a1} \quad (K_+) \tag{1}$$

$$SOH = SO^- + H^+ \qquad K_{a2} \quad (K_-) \tag{2}$$

$$ROH = RO^- + H^+ \qquad K_a \tag{3}$$

In this chapter we consistently adopt the 2-pK_a formalism for ionization of inorganic hydroxyls (SOH). The 1-pK_a formalism ($SOH_2^{+1/2} = SOH^{-1/2} + H^+$; see Westall[10] and Bolt and van Riemsdijk[11]) provides an equivalent description of surface charge development on most oxides relevant to soil. Both of these ionization schemes can be readily incorporated into chemical equilibrium models, although the 2-pK_a approach has seen wider application. The use of one over the other is mainly an issue of choice and convenience.

Surface functional groups of the same type can show significant variation in reactive properties. For example, the electronic properties of the ditrigonal cavity on the layer silicate basal plane vary significantly, depending on whether isomorphic substitution is in the octahedral or tetrahedral layers.[5,8] The reactivity of inorganic surface hydroxyls varies both with metal ion center (i.e., SiOH vs. AlOH or FeOH) and with coordination number (single, double, triple). Chemically distinct surface hydroxyls can exist on the same mineral phase as a result of varying coordination and structural environments on differing crystallographic planes (e.g., Hiemstra et al.[12-14]; Barron et al.,[15] White and Zelazny,[16] and Bleam et al.[17]). For example, goethite has three different types of surface hydroxyls (A, B, and C), while gibbsite has at least two (singly coordinated hydroxyls on the edge and doubly coordinated on the basal plane). Such site heterogeneity may influence macroscopic measurements of solute and proton surface binding.[13-15,18] In general, the singly coordinated SOH sites, especially those which function as Lewis acids, are most reactive.[5,8] Acidity constants (pK_a) for carboxylate sites on soil organic matter (ROH) span a large range (e.g., 4 to 10) because of stereochemical and electrostatic effects.

B. Surface Functional Groups and their Interactions with Adsorbates

A great many terms have been used to describe the interaction of adsorbates with surfaces of soil particles — surface complexation, surface coordination, ion exchange, inner-sphere and outer-sphere complex formation, electrostatic binding, specific and nonspecific adsorption, etc. These terms are often used rather loosely, without strong scientific evidence to support the suggested mechanism; rather, they generally reflect an attempt to describe the dependence of extent of adsorption on solution composition variables such as pH and salt concentration. Here we describe some of the nuances associated with these terms; these are well described by Sposito[8].

A surface complex exists when an aqueous species reacts with a surface functional group to form a stable molecular unit (see for example, Schindler and Sposito[9]). Outer- and inner-sphere complexes can be recognized conceptually; the former exhibit a water molecule between the bound ion and the surface functional group while the latter do not retain such intervening water.[5] Evidence for the nature of these surface complexes is derived from spectroscopic, kinetic, and adsorption experiments.[19-22] While the concepts of inner- and outer-sphere complexes provide a useful paradigm, some sorbates (i.e., SO_4^{2-}, CrO_4^{2-}, organic acids, alkali earth cations) may show transitional chemical behavior, or develop a mixed population of surface complexes, depending on surface coverage, pH, and sorbent type.[4,23-25] Mixed complex populations are potentially significant in soil with heterogeneous surface properties and sorbents. Ions bound in the diffuse swarm that provide delocalized charge neutralization are not considered to be surface complexes by the above definitions.

Surface complexes result from the reaction of aqueous species with a surface functional group. Anion (L^{l-}) and cation (M^{m+}) adsorption to hydroxylated surface sites (SOH) on Fe or Al oxides is generally described as follows:

$$SOH + M^{m+} = SOM^{(m-1)} + H^+ \qquad \text{(inner sphere)} \qquad K_M \qquad (4)$$

$$SOH + M^{m+} = SO^- - M^{m+} + H^+ \qquad \text{(outer sphere)} \qquad K_m \qquad (5)$$

$$SOH + L^{l-} = SL^{(l-1)-} + OH^- \qquad \text{(inner sphere)} \qquad K_L \qquad (6)$$

$$SOH + H^+ + L^{l-} = SOH_2^+ - L^{l-} \qquad \text{(outer sphere)} \qquad K_L \qquad (7)$$

Similarly, cation adsorption to fixed-charge sites on layer silicates (X^-) is described with some variant of the following exchange equation:

$$mCX_u + uM^{m+} = uMX_m + mC^{u+} \qquad K_{MX} \qquad (8)$$

The structural identity of surface complexes on mineral surfaces, organic matter, and soil colloidal material is not well established, but it is fundamental to the development of accurate mass-action laws, in terms of reaction stoichiometry and products. The specification of an inner- or outer-sphere complex is also critical for certain types of site binding models, as will be discussed below.[19,22] Definitive chemical information on these species can be obtained only by *in situ* surface spectroscopy (e.g., Hayes et al.,[19] Brown,[2] Chisholm-Brause et al.,[26,27] Manceau et al.,[28] Coombs et al.,[29] Waychunas et al.,[30] and O'Day et al.[25]). Spectroscopic studies have been limited in the past by poor sensitivity and the inability to detect target sorbates at environmentally realistic surface concentrations. Synchrotron light sources have yielded significant improvements in surface sensitivity, however, and high-energy X-ray absorption spectroscopy (XAS) is one method of choice to characterize the structure of inorganic surface complexes.[2] The results of such spectroscopic studies are not without ambiguities, and comparable data sets can support different structural models of surface complexes.[19,31-33] Recent XAS studies have improved understanding of metal ion adsorption by demonstrating the importance of metal ion surface polymer formation at higher sorbate concentrations.[26,27,34,35]

C. Electrostatic Interactions

Since surface functional groups are confined to a surface, the accumulation of surface charge becomes a significant factor in the energetics of adsorption. In contrast, for complexes formed in solution, one generally makes no particular effort to separate electrostatic and chemical components of binding energy, since these are always the same for a particular complex. To account for the variable electrostatic energy of adsorption associated with variably charged surfaces, electric double-layer models have been employed. The term "double layer" as applied to soil particle surfaces can be interpreted as a reference to two components of surface charge: the primary surface charge due to chemically bound ions and fixed charge in layer silicates, and the countercharge, arising from electrostatically bound ions in the aqueous phase, which exactly compensate the primary surface charge.

The primary surface charge arises from surface functional groups and fixed negative charge in layer silicates. Adsorbed proton charge (σ_H) is established through association and dissociation reactions of H^+ as described above (reactions 1 to 3). The intrinsic surface charge of soil (σ_i) is determined primarily by the fixed negative charge (σ_o) in layer silicates (X^-), in combination with the net proton charge (σ_H) carried by inorganic (SOH_2^+, SO^-) and organic hydroxyls (RO^-), i.e., $\sigma_i = \sigma_o + \sigma_H$. The net charge on soil particles (σ_p) results from a combination of σ_i and adsorbed solute charge (σ_{is} and σ_{os}):

$$\sigma_p = \sigma_o + \sigma_H + \sigma_{is} + \sigma_{os} \qquad (9)$$

where σ_{is} and σ_{os} are inner- and outer-sphere complex charge, respectively.[1,5]

According to commonly used electric double-layer models, the countercharge resides in one or more planes or layers that extend out from the surface towards the bulk of solution. Variations on this theme have resulted in the constant capacitance, diffuse-layer, Stern, and triple-layer models for the interface.[36] Other alternatives are possible. In the case of classical ion exchange, the reactions are electroneutral, the counterions are accounted for explicitly in the reaction, and there is no need for an electric double-layer model. In other cases, the surface charge density may be so low or the concentrations of counterions so high that explicit accounting for the countercharge is unnecessary.

Table 1   Electrochemical Equilibrium Equations: Definitions[a]

Type of equations	Scalar[2]	Matrix
Mass action	$\log C_i = \log K_j + \sum_j a_{ij} \log X_i$	$C^{\cdot} = K^{\cdot} + AX^{\cdot}$
Mass balance	$Y_j = \sum_i a_{ij} C_i - T_j$	$Y = AC - T$
Iteration	$z_{jk} = \sum_i \left(\dfrac{a_{ij} a_{ik} C_i}{X_k} \right) = \dfrac{\partial Y_j}{\partial X_k}$	$\Delta X = X_{original} - X_{improved}$ $Z \cdot \Delta X = Y$

Scalar[1] symbols	Description	Matrix or vector symbols	Description[b]
a_{ij}	Stoichiometric coeffient of component j in species i	A	Matrix of a_{ij}
C_i	Free concentration of species i	C	Vector of C_i
		C*	Vector of $\log C_i$
K_i	Stability constant of species i	$K^{\cdot}$	Vector of $\log K_i$
T_j	Total analytical concentration of component j	T	Vector of T_j
X_j	Free concentration of component j	X	Vector of X_j
		$X^{\cdot}$	Vector of $\log X_j$
Y_j	Residual in material-balance equation for component j	Y	Vector of Y_j
zjk	Partial derivative $(\partial Yj/\partial Xk)$	Z	Jacobian of Y with respect to X

[a] After Westall and Hohl.[36]
[b] Indices: i denotes any species; j and k denote components.

D. Models for Adsorption

Ion adsorption can be described with mass-action and material-balance equations that yield a Langmuir isotherm with its attendant assumptions (i.e., discrete sites with uniform adsorption energy). Mass-action and material-balance equations have been incorporated into chemical codes to compute adsorption equilibria.

Mass-action and material-balance equations for adsorption equilibria can be solved by matrix algebra using a computer, as summarized in Table 1.[10,36,37] The procedure involves the definition of species (every chemical entity in the system), components (a minimal set of reactants from which species can form), and a stoichiometry matrix (A), which relates components to species (e.g., Table 2 for surface hydrolysis). The rows of the A matrix are the coefficients in the mass-action equations, while the columns are the coefficients in the material-balance equations. The formulation of these matrices is the basis for adsorption calculations. For conventional chemical components, either the total concentration or free concentration is found analytically, while for electrostatic components the total concentration is found from a series of model-specific equations relating surface charge to surface potential.

The chemical equilibrium problem is expressed in vector notation (Table 1) for numeric solution:

$$C* = K* + A* \qquad \text{mass action} \tag{10}$$

$$Y = {}^tBC - T \qquad \text{material balance} \tag{11}$$

Table 2   Stoichiometry Matrix for Surface Hydrolysis According to the Constant-Capacitance Model (CCM)

Species	Components			
	SOH	$X(\psi_0)$	H^+	
H^+	0	0	1	1
OH^-	0		−1	K_w
SOH	1	0	0	1
SOH_2^+	1	1	1	K_+
SO^-	1	−1	−1	K_-
	T_{SOH}	T_σ	T_H	

$T_{SOH} = N_s \, sa/N_A$
$T_H \quad = C_A - C_B$
$T_\sigma \quad = C\varphi_0 \, sa/F$

where N_s = site density (sites/nm²)
N_A = Avogadro's number (sites/mol)
a = concentration of suspended solid (g/l)
s = surface area (m²/g)
C = capacitance (F/m²)
σ = specific surface charge (C/m²)
φ_0 = potential at surface (V)
F = Faraday's constant (C/mol), C_A and C_B are total concentrations of acid (A) or base (B)
T_H = total hydrogen concentration (mol/l)
T_{SOH} = total surface hydroxyl concentration (mol/l)
T_σ = total surface charge (C/l)

The typical solution of such problems involves specifying the species stoichiometry (A) and stability constants (K), and the total concentrations (T) of components. The free concentrations (X) are computed for the components from mass-action equations that a residual (Y) in the material-balance equations that is less than the convergence criterion. The mass-action equations and solution matrices for the surface ionization example in Table 2 are given in Table 3.

1. Cation Exchange Models

The ideal solution model is used most frequently for computerized cation exchange calculations. It is the simplest form of the regular solution model;[38] the activity coefficients for surface species (f) are unity and the activities of the surface exchange complexes are equal to their mole fractions. For Equation 8, therefore, the following relationships hold, with respect to the thermodynamic (K_{ex}) and conditional (K_c) equilibrium constants:

$$K_{ex} = \left\{ \left[f_{MXm} X_{MXm} \right]^u \left(C^{u+} \right)^m \right\} / \left\{ \left[f_{CXu} X_{CXu} \right]^m \left(M^{m+} \right)^u \right\} \quad \text{and}$$

$$K_c = \left\{ \left[X_{MXm} \right]^u \left(C^{u+} \right)^m \right\} / \left\{ \left[X_{CXu} \right]^m \left(M^{m+} \right)^u \right\} \tag{12}$$

where X represents mole fraction and parentheses denote single ion activities. Under the conditions of ideal exchange, $f_{MXm} = f_{CXu} = 1$ and $K_{ex} = K_c$. The mole fraction-based conditional equilibrium constant (K_c, Equation 12) is the Vanselow selectivity coefficient (K_V; Sposito[39]). Ideal solution behavior is a first assumption in modeling cation exchange in soil because it avoids estimation of activity coefficients for the adsorbed species, which is complex.[38] Unlike the Davies or Debye-Huckel equations for solution species, there are no generalized, simple equations to estimate the activity coefficients for the exchange complex.[6] While the ideal approximation appears valid for selected ion systems with specimen smectites (see Sposito and Mattigod[40], Fletcher and Sposito[41]) and soil,[42] a compositional dependence (i.e., nonideal behavior) in K_c is often observed

Table 3   Surface Hydrolysis — Mass-Action Equations and Solution Matrices

Components:	SOH,	$e^{-F}\psi^{RT}$,	H^+

Species and mass-action equations

H^+				
OH^-			$[H^+]^{-1}$	$K_w = [OH^-]$
SOH_2^+	$[SOH]$	$(e^{-F\psi/RT})$	$[H^+]$	$K_+ = [SOH_2^+]$
SOH				
SO^-	$[SOH]$	$(e^{-F\psi RT})^{-1}$	$[H^+]^{-1}$	$K_- = [SO^-]$

Algebraic description of problem

$$^tT = [T_{SOH} \quad C \cdot \psi \cdot s \cdot a/F \quad T_H]$$
$$^tX = \{[SOH] \quad (e^{-F\psi/RT}) \quad [H^+]\}$$

$$A = \begin{bmatrix} 0 & 0 & 1 \\ 0 & 0 & -1 \\ 1 & 1 & 1 \\ 1 & 0 & 0 \\ 1 & -1 & -1 \end{bmatrix} \qquad C = \begin{bmatrix} [H^+] \\ [OH^-] \\ [SOH_2^+] \\ [SOH] \\ [SO^-] \end{bmatrix} \qquad K = \begin{bmatrix} 1.0 \\ K_w \\ K_+ \\ 1.0 \\ K_- \end{bmatrix}$$

Jacobian

1. For all elements of the Jacobian except $z_{\psi,\psi}$; $z_{jk} = \sum_i (a_{ij}\, a_{ik}\, C_i/X_k)$

2. For $z_{\psi\psi}$; $z_{\psi\psi} = \sum_i (a_{i\psi}a_{i\psi}C_i/X_\psi) + C\,\dfrac{sa}{F}\dfrac{RT}{FX_\psi}$

---

in soil materials as a result of the polyfunctional nature of soil cation exchangers.[43] The ideal approximation is best suited to narrow ranges in exchanger-phase composition and homovalent exchange.

The ideal solution model has been most frequently implemented in speciation codes (e.g., GEOCHEM, MINTEQA2, or MINEQL) using the half-reaction approach.[41,42,44] Half reactions for Equation 8 are defined as

$$C^{u+} + uX^- = CX_u \qquad K'_{CXu} \tag{13}$$

$$M^{m+} + mX^- = MX_m \qquad K'_{MXm} \tag{14}$$

where X^- represents 1 mol of charged surface. The equilibrium constants, K'_{CXu}, and K'_{MXm}, are defined in concentration terms (molality or molarity) as used in speciation codes. The equilibrium constants are also conditional; activity coefficients for the ion exchange complexes (CX_u, MX_m) are unity, consistent with the assumption of ideality:

$$K'_{CXu} = [CX_u]/(C^{u+})[X^-]^u \quad \text{and} \quad K'_{MXm} = [MX_m]/(M^{m+})[X^-]^m \tag{15}$$

where brackets refer to concentration and parentheses refer to single ion activity. As shown by Fletcher and Sposito,[41] K'_{CXu} and K'_{MXm} (in concentration terms of molality) are related to the conditional equilibrium constant (i.e., K_c or K_v, in concentration terms of mole fraction) by the following relationship:

$$K_v = \left(K'_{MXm}\right)^u \left(K'_{CXu}\right)^m \left(m_{MXm} + m_{CXu}\right)^{(m-u)} \tag{16}$$

where m_{MXm} and m_{CXu} are molalities of the individual ion-exchange complexes (MX_m and CX_u).

The half-reaction approach requires a self-consistent, half-reaction database referenced to a common reaction, such as the formation of NaX or CaX_2.[41] The choice of the reference reaction and its log K' is arbitrary. Both Fletcher and Sposito[41] and Stadler and Schindler[45] used high values (log K' >10) to limit computed concentrations of free X^-. It may be simpler just to omit X^- from the list of species, since it does not exist anyway, then a value of 1 can be used for the reference reaction.

The half-reaction approach has been used to model macro-ion exchange in soil,[42,46] and trace metal exchange/adsorption on montmorillonite,[41,45,47,48] smectitic soil clay isolates,[48,49] kaolinite,[50] and kaolinitic soil clay isolates.[51,52] The last three applications incorporated H^+ exchange (K_{HX}) to account for the pH dependency of the cation exchange capacity. The half-reaction approach has great flexibility for describing ion exchange processes. It has, for example, been used to model

1. Cation and anion adsorption on calcite by assuming that the relevant surface reactions were adsorbate exchange for surface Ca and HCO_3, respectively [53,54]
2. Particle interactions between iron oxides and layer silicates[51]
3. Proton-metal exchange on biotic surfaces[55]

2. Models Based on Surface Complexation and Electric Double-Layer Models

Models based on surface complexation reactions with electric double-layer models (SC-EDL models) have been used extensively to describe adsorption to mineral and soil surfaces. Generally the surface complexation models describe the development of primary charge (i.e., σ_H), while the electric double-layer model is used to describe the countercharge. The EDL models include the constant-capacitance (CCM), diffuse double-layer (DLM), triple-layer (TLM), and other select models. Particle surface charge and electrical potential within these are treated as consequences of the surface complexation of protons and other sorbates. Equilibrium constants for the surface reactions [K(int)] differ from those in the aqueous phase, in that they are modified with a correction term for the electrostatic energy of the interface (edl) that effectively represents an activity coefficient ratio for the surface species:[56]

$$K(\text{int}) = K_c \cdot edl \tag{17}$$

where K_c is a conditional equilibrium constant written in terms of activities of aqueous species and concentrations of surface species, and edl is a model-dependent term involving the surface potential or potential within the adsorption plane (see Table 4; Sposito;[5] and Goldberg[56]) for edl terms.

The various EDL models differ in their interfacial structure and stated relationships between surface potential and surface charge. As noted by Westall and Hohl,[36] the models are mathematically degenerate, and conform to a set of comparable, simultaneous equations that are amenable to numeric solution. The reader is directed to reviews for details on their molecular hypotheses, charge potential relationships, and other similarities and differences.[3-6,8,36,56-59] Attributes of these models, including number of adjustable parameters, types of surface complexes considered, and their ability to deal with ionic strength effects, are summarized in Table 5. Of these, the DLM is generally considered to be the least complex and the TLM the most complex.

Table 4   General Formulation of Equilibrium Constants and Electric Double-Layer Corrections for Surface Complexation-Electric Double-Layer (SC-EDL) Models

Reaction	K_c	edl		
		DLM	CCM	TLM
$SOH + H^+ = SOH_2^+$ [a]	$\dfrac{[SOH_2^+]}{[SOH](H^+)}$	$\exp[F\psi/RT]$	$\exp[F\psi/RT]$	
$SOH = SO^- + H^+$ [b]	$\dfrac{[SO^-](H^+)}{[SOH]}$	$\exp[-F\psi/RT]$	$\exp[-F\psi/RT]$	
$SOH + M^{m+} = SOM^{(m-1)} + H^+$	$\dfrac{[SOM^{(m-1)}][H^+]}{[SOH](M^{m+})}$	$\exp[(m-1)F\psi/RT]$	$\exp[(m-1)F\psi/RT]$	
$SOH + L^{l-} = SL^{(l-1)-} + OH^-$	$\dfrac{[SL^{(l-1)-}](OH^-)}{[SOH](L^{l-})}$	$\exp[-(l-1)F\psi/RT]$	$\exp[-(l-1)F\psi/RT]$	
$SOH + M^{m+} = SO^- - M^{m+} + H^+$	$\dfrac{[SO^- - M^{m+}](H^+)}{[SOH](M^{m+})}$ [c]			$\exp[F(m\psi_B - \psi_o)/RT]$
$SOH + H^+ + L^{l-} = SOH_2^+ - L^{l-}$	$\dfrac{[SOH^{2+} - L^{1-}]}{[SOH](H^+)(L^{1-})}$ [c]			$\exp[F(\psi_o - l\,\psi_B)/RT]$

Key: edl, the electrostatic energy of the interface; DLM, diffuse double-layer model; CCM, constant-capacitance model; TLM, triple-layer model.

[a] K(int) for this reaction = K_+
[b] K(int) for this reaction = K_-
[c] Different formulations of K_c have been used.[22,99]

From Goldberg, S., *Adv. Agron.*, 47, 233, 1992. With permission.

SC-EDL models have provided good descriptions of the acid-base properties and adsorption behavior of inorganic and organic cations and anions on many single-phase sorbents that are relevant to soils, including Fe, Al, and Mn oxides and 1:1 and 2:1 phyllosilicates; see reviews by Davis and Kent,[4] Goldberg,[56] and Stadler and Schindler,[45] as well as carbonates,[60] and phosphates.[61] Surface polymer formation has also been adequately modeled.[62] The success of these models results from the conformation of adsorption to mass-action and material-balance laws and flexibility in adjustable parameters, rather than from the mechanistic accuracy of the assumed surface complex or the interfacial structure. The various SC-EDL models are equally capable of fitting acid-base titration data,[36,59] although each model yields different, nonunique values for analogous parameters.[58] Indeed, this ambiguity has led some to caution that surface complexation modeling alone cannot be used to support mechanistic conclusions regarding surface speciation or interfacial structure,[63,64] and that the strength of the SC-EDL model is as a learning tool for chemical insight and in visualizing the effects of complex equilibria.[23,36]

A nonelectrostatic counterpart (the surface complexation-nonelectrostatic model, SC-NEM) exists to the SC-EDL model. It is based on mass-action/material-balance relationships and is analogous to the SC-EDL models except that it does not contain the electrostatic component. The first surface complexation models were of this type.[65,66] The adsorption of some solutes, particularly those forming strong surface complexes, is less sensitive to electrostatics, and, for these, the SC-NEM model provides a comparable fit to the data, as do the SC-EDL models.[24,46] Morel[67] shows the computed effect of the electrostatic component on Pb^{2+} adsorption to alumina. The SC-NEM model has been linked with geochemical transport codes and kinetic geochemical models and is useful in application to soils and geologic materials (i.e., Cowan et al.,[51] Zachara et al.[68]) where electrostatic models based on the assumption of fixed particle geometry are of questionable validity.[69,70]

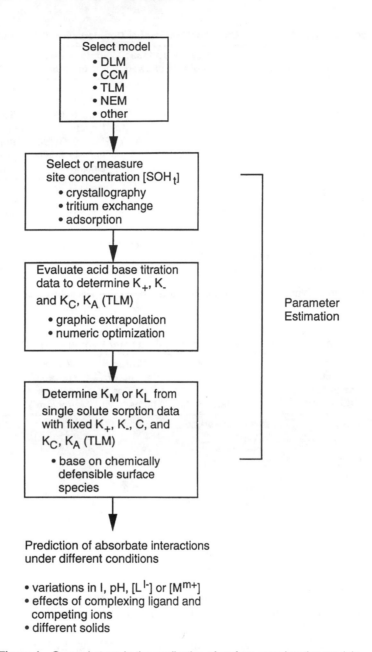

Figure 1 General steps in the application of surface complexation models.

III. PARAMETER DEVELOPMENT FOR SOIL COMPONENTS AND ANALOG PHASES

A. General Considerations

The steps taken in modeling single-phase sorbent/sorbate systems for the SC-EDL and SC-NEM models are shown in Figure 1. The adjustable parameters in these models (Table 5) are weakly constrained and interdependent;[58,59] the challenge in applying the SC-EDL and SC-NEM models derives from estimating/computing parameter values. The surface equilibrium constants (K_+ and

Table 5   Selected Attributes of Surface Complexation-Electric Double-Layer (SC-EDL) Models

Model[a]	Adjustable parameters	Types of surface complexes allowed	Ionic strength effects	Considerations[b]
DLM	$[SOH]_T$, K_+, K_-, K_{si}	Inner sphere	Through GCSG[c]	Restricted to lower I (≤ 0.01 mol/l)
CCM	$[SOH]_T$, K_+, K_-, C_1, K_{si}	Inner sphere	Restricted to fixed I	Restricted to higher I (>0.01 mol/l)
TLM	$[SOH]_T$, K_+, K_-, C_1, C_2, K_C, K_A, K_{si}	Inner and outer	Through GCSG[c] and electrolyte binding	Applicable to variable I; suffers constraint problems

[a] DLM, double-layer model; CCM, constant-capacitance model; TLM, triple-layer model.
[b] Data from Hayes et al.[59]
[c] Gouy-Chapman/Stern-Grahame charge potential relationship $-\sigma_o = \sigma_d = -0.1174 \sqrt{I} \sinh(zF\psi_d/2RT)$

$[SOH]_T$ = total site concentration
K_- = equilibrium constant for $SOH = SO^- + H^+$potential relationship
K_+ = equilibrium constant for $SOH + H^+ = SOH_2^+$
K_{si} = equilibrium constant(s) for cation (K_M) or anion (K_L) adsorbate binding (e.g., reactions 5 to 8)
C_1, C_2 = capacitances[b]
K_C, K_A = TLM equilibrium constants for outer sphere electrolyte binding; C is electrolyte cation, A is electrolyte anion

Table 6   Density (N_s) of Surface Functional Groups on Oxide and Hydrous Oxide Minerals

Mineral phase	Range of site densities (sites/nm²)	Ref.	Proton acceptor groups (sites/nm²)	Proton donor groups (sites/nm²)	Ref.
α-FeOOH	2.6–16.8	James and Parks,[72] Sposito[5]	4.4	6.7	Sposito[5]
α-Fe$_2$O$_3$	5–22	James and Parks[72]			
Ferrihydrite	0.1–0.9 moles per mole of Fe	Dzombak and Morel[3]			
TiO$_2$ (rutile)	12.2	James and Parks[72]	2.6	4.2	James and Parks[72]
TiO$_2$ (rutile and anatase)	2–12	James and Parks[72]			
α-Al(OH)$_3$	2–12	Davis and Hem[71]	2.8	5.6	Sposito[5]
γ-Al$_2$O$_3$	6–9	Davis and Hem[71]			
SiO$_2$ (am)	4.5–12	James and Parks[72]	0	All	James and Parks[72]
Kaolinite	1.3–3.4	Fripiat[179]	0.35	1.0	Sposito[5]

From Davis, J. A. and Kent, D. B., in *Mineral-Water Interface Geochemistry,* Hochella, M. F. and White, A. F., Eds., Mineralogical Society of America, Washington, D.C., 1990, 177. With permission.

K_-, K_M and K_L) are model specific and not interchangeable.[36,58] The hydroxyl site concentration ($[SOH]_T$, generally in moles per liter) in contrast, is a model-independent parameter, unless it is fit as part of the model application. Values of site density (N_s = number of hydroxyl sites per m² of surface) for various common soil minerals are summarized in Table 6. The site density, N_s, is related to $[SOH]_T$ by the relationship $[SOH]_T = (N_s * s * a)/N_a$, where s is specific surface area (m²/g), a is the solids concentration (g/l), and N_a is Avogadro's number. Common methodologies used to determine these adjustable parameters for single-phase sorbents are described below.

For oxides, the site density has been estimated from crystallographic parameters (e.g., Sposito,[5] Davis and Hem[71]) and measured by tritium exchange,[72] acid-base titration,[73] F$^-$ adsorption,[73,74] H$^+$ adsorption,[75] and maximum solute adsorption density (e.g., PO$_4^{3-}$; Goldberg and Sposito[76]). These methods often yield different estimated values of site density, with tritium exchange at the high end and solute adsorption at the low end. Disparities result from the different effective site populations probed by the methods. Only a portion of the hydroxyl sites on oxide surfaces are active in surface complexation reactions;[4] surface hydroxyls coordinated to two or three metal ion centers appear weakly reactive.[6,8]

Hydroxyl site densities have been estimated for kaolinite and smectites (montmorillonite, beidellite) by crystallographic methods.[5,16] Direct measurements for phyllosilicates are difficult because of the multiple-site surface, and the dissolution and readsorption of Al^{3+} at lower pH. Hydroxyl concentrations on the phyllosilicate edge have been estimated from acid-base titration data,[45,50,77] where the following reactions were assumed to control the proton balance:

$$SOH + H^+ = SOH_2^+ \qquad K_{+(SOH)} \qquad (18)$$

$$SOH - H^+ = SO^- \qquad K_{-(SOH)} \qquad (19)$$

$$TOH - H^+ = TO^- \qquad K_{-(TOH)} \qquad (20)$$

$$CX_u + uH^+ = uHX + C^{u+} \qquad K_{HX} \qquad (21)$$

The SOH and TOH groups are consistent with the presence of aluminol (AlOH) and silanol (SiOH) groups exposed at the edges of the octahedral and tetrahedral layers, respectively. Site concentrations were obtained by jointly optimizing $[SOH]_T$, $[TOH]_T$, and the equilibrium constants for Equations 18 through 21 to acid-base titration data over pH ranges where the contributions of Equation 21 (C = Na^+ or Ca^{2+}) were small. The resulting component concentrations and equilibrium constants represent data-fit parameters, rather than measurements of intrinsic properties of the crystallite edge.

Equilibrium constants for surface reactions (K_+, K_-, K_M, K_L) are determined from material balance data where the reactivity of a single component can be isolated (i.e., H^+, M^{m+}, or L^{l-}), beginning first with proton adsorption/desorption reactions and subsequently with single ion adsorption data (all using consistent values of site concentration and the capacitance terms where appropriate). Surface ionization constants, K_+ and K_-, are estimated from alkalimetric titration data (corrected for solid dissolution if significant) using both graphic and numeric procedures. The K_A and K_C are determined simultaneously if the TLM is being used. A comprehensive summary of SC-EDL specific K_+ and K_- for different soil minerals was presented by Goldberg.[56] Nonlinear, least-squares fitting techniques (discussed below) are useful for parameter estimation to (1) avoid the approximations of graphic extrapolation, which bias the ΔpKa of K_+ and K_-,[58,78] and (2) provide error estimates for the optimized constants based on experimental uncertainty.

The numeric procedure is the method of choice for fitting adsorption constants for sorbate species (e.g., K_M, K_L). The number of sorbate surface complexes should be limited to the smallest number that is chemically reasonable and that would describe the experimental data. Surface species for the sorbate must be selected carefully to be consistent with aqueous speciation and spectroscopic data on sorbate chemical bonding and nearest neighbors, if available (see, for example, Hayes et al.,[19] Sposito et al.,[79] Katz and Hayes,[62,80] and Waite et al.[81]). The important surface species need not be those that dominate the aqueous speciation (e.g., Waite et al.[81]). Equilibrium constants may be extracted from single data sets, multiple data sets, and multiple data sets varying in ionic strength, each with differing statistical implications. Dzombak and Morel[3] review parameter estimation, while Hayes et al.[59] demonstrate the sensitivity of the surface equilibrium constants for the DLM, CCM, and the TLM to fixed values of site concentration and the capacitance terms. SC-EDL model equilibrium constants for cationic and anionic sorbates on Fe and Al oxides have been summarized by Goldberg,[56] and predictive regression equations have been developed for select SC-EDL models.[82-84] The Dzombak and Morel[3] compilation is specific to amorphous Fe oxide and the DLM, and all constants have been refit to yield a self-consistent database. Half-reaction ion-exchange constants for montmorillonite were summarized by Fletcher and Sposito.[41]

Ion adsorption on many single-mineral surfaces may not conform to Langmuir adsorption behavior over large concentration ranges.[85-89] Modeling options include the division of the surface site population into subsets, with each subset exhibiting a constant set of parameters and Langmuir behavior (i.e., weak and strong sites on amorphous iron oxyhydroxide),[3] fitting concentration-dependent adsorption constants,[89] or explicitly modeling surface polymerization or precipitation reactions.[62,80,90,91] Some sorbate/sorbent systems, however, do conform to Langmuir behavior over fairly large concentration ranges.[80,91]

B. Numeric Optimization of Equilibrium Constants and Component Concentrations

The FITEQL model, developed by Westall and Morel,[92] Westall,[93,94] Herbelin and Westall[95] is a nonlinear least-squares optimization program that can be used to adjust parameters in a chemical equilibrium model to fit experimental data. The code is based on the Gauss method and is flexible with many potential uses in soil chemistry.[3,96] The adjustable parameters may take various forms, including equilibrium constants for solubility, surface and solution complexation, and ion-exchange reactions; as well as total concentrations of components, such as surface sites (X^-, SOH). FITEQL optimizes values for the adjustable parameters by changing their values from an initial estimate provided by the user to a value that minimizes the sum of the squares of the residuals [$Y_j =$ TOTj(calc) – TOTj(expt)] between the experimental data (expt) and those calculated by the model (calc). The residuals, weighted to reflect uncertainties in the experimental data, are minimized for selected components (j, user identified) where the total and free concentrations of j are known.[3] FITEQL and other such programs (e.g., NONLIN/GMIN; Felmy[97]) offer a bias-free methodology to estimate equilibrium constants and their associated uncertainty.[58] FITEQL may be used to optimize parameters or to perform equilibrium calculations with fixed equilibrium constants and component concentrations.[93]

FITEQL calculations are driven by a chemical equilibrium model of the system (i.e., a set of reactions between components leading to species; see Tables 2 and 3) that is input by the user (see Westall[93,94]). For adsorption calculations, the chemical model includes both aqueous speciation and surface complexation or ion-exchange reactions. The chemical model is developed in matrix format.[37,67,93,94] Hayes et al.,[22] Dzombak and Morel,[3] and Herbelin and Westall[95] provide examples of matrices for adsorption calculations. Other examples are provided in Tables 7 and 8 (discussed below) and in Section IV. C.

FITEQL contains separate matrices for mass-action (A-matrix) and mass-balance (B-matrix) coefficients. The definition of the chemical equilibrium problem includes both A-matrix and B-matrix stoichiometric coefficients. Within FITEQL, the mass-action and mass-balance equations are formulated as follows:

$$\text{mass action} \quad \log C_i = \log K_i + \Sigma_i \left(a_{ij} \log X_j \right) \quad \text{for all species i} \quad (22)$$

$$\text{mass balance} \quad Y_j = \Sigma_i \left(b_{ij} C_i \right) - T_j \quad \text{for all components j} \quad (23)$$

where C_i is the free concentration of species i, K_i is the stability constant of species i, X_j is the free concentration of component j, Y_j is the difference in the material balance equation for component j, T_j is the total analytical concentration of component j, and a and b are stoichiometric coefficients. For most chemical equilibrium problems, the stoichiometric coefficients in the mass-action and mass-balance equations are equal (i.e., $a_{ij} = b_{ij}$), and FITEQL automatically copies the A-matrix mass action coefficients, that were entered as input data, into the B-matrix. There are, however, certain types of soil chemical problems such as those involving multidentate surface

Table 7 FITEQL Stoichiometry Matrix for Zn Sorption to Ferrihydrite Using the Diffuse-Layer Model

Species	Type I X_φ	Type I Zn^{2+}	Type I XOH	Type I YOH	Type II Zn(ads)	Type III H^+	log K
H^+	0	0	0	0	0	1	0.0
OH^-	0	0	0	0	0	−1	−13.78
Zn^{2+}	0	1	0	0	0	0	0.0
$ZnOH^+$	0	1	0	0	0	−1	−9.22
$Zn(OH)_2(aq)$	0	1	0	0	0	−2	−17.12
XOH_2^+	1	0	1	0	0	1	7.18
XOH	0	0	1	0	0	0	0.0
XO^-	−1	0	1	0	0	−1	−8.82
$XOZn^+$	1	1	1	0	1	−1	0.5 (fit)
YOH_2^+	1	0	0	1	0	1	7.18
YOH	0	0	0	1	0	0	0.0
YO^-	−1	0	0	1	0	−1	−8.82
$YOZn^+$	1	1	0	1	1	−1	−2.0 (fit)

Note: According to Dzombak and Morel[3], XOH and YOH are strong and weak sites, respectively.

From Herbelin, A. L. and Westall, J. C., *FITEQL. A Computer Program for Determination of Chemical Equilibrium Constants from Experimental Data, Version 3.1,* Department of Chemistry, Oregon State University, Corvallis, OR, 1994. With permission.

Table 8 Formulation of Ion-Exchange According to Vanselow Convention

Species	A. Mass-action equations components X^-	A. Mass-action equations components X_N	A. Mass-action equations components Na^+	A. Mass-action equations components Ca^{2+}	Constants K	B. Material-balance equations components X^-	B. Material-balance equations components X_N	B. Material-balance equations components Na^+	B. Material-balance equations components Ca^{+2}
N_{Na}	1	0	1	0	1	0	1	0	0
N_{Ca}	2	0	0	1	K_v	0	1	0	0
[NaX]	1	1	1	0	1	1	0	1	0
$[CaX_2]$	2	1	0	1	K_v	2	0	0	1
$[Na^+]$	0	0	1	0	1	0	0	1	0
$[Ca^{+2}]$	0	0	0	1	1	0	0	0	1
						T_{IEC}	$T_N = 1$	T_{Na}	T_{Ca}

Note: The mass-action coefficients constitute the A-matrix and the material-balance coefficients constitute the B-matrix input to FITEQL. Components are listed at the head of columns and the species, for which concentrations are to be calculated, are listed on the left-hand side; [] represents concentration in moles per liter of solution and N_x represents concentration in mole fraction on the exchanger. The stability constant for each species is listed the column labeled K. Aqueous-phase activity coefficients have been omitted for brevity. The total concentrations (T-vector) are listed below the material balance coefficients. X_N is a "dummy component" selected to represent the total number of moles of exchangeable ions associated with the exchanger per liter of solution: $X_N = [NaX] + [CaX_2]$ and T_{IEC} is the total ion exchange capacity in moles/l: $T_{IEC} = [NaX] + 2[CaX_2]$ (see Westall et al.[98] for details).

complexes, coverage of multiple surface sites by large sorbates, ion exchange, and counterion retention where chemical species may require different stoichiometric coefficients in the A- and B-matrices (i.e., $a_{ij} \neq b_{ij}$). FITEQL provides a provision to overwrite the value of b_{ij} copied from a_{ij} with a new value. The B-matrix is also used in the computation of ionic strength and single-ion activity coefficients according to the Davies convention.[95]

FITEQL contains four SC-EDL models (CCM, DLM, TLM, and the Stern layer model) and can be used to perform SC-NEM and ion-exchange calculations using half-reactions or full exchange reactions depending on component specification. The Davies convention is employed for single-ion, aqueous-phase activity coefficients. Activity corrected computations are therefore limited to lower ionic strengths (I < 0.1). The TLM has been incorporated into the NONLIN/GMIN code which uses the Pitzer formulation for higher ionic strength solutions.[97]

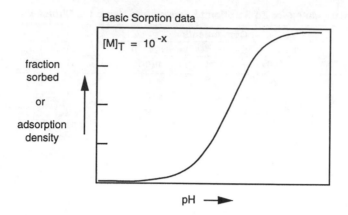

Basic Sorption data

FITEQL Modeling

Data	Error	Fix	Optimize [1]	Output
$[M]_T$ [2]	$[M]_T$ [2]	K_+, K_-, K_M	$[SOH]_T$ [2]	$[SOH]_T$ [2] +/- s [3]
Free M	Free M	$K_+, K_-, [SOH]_T$ [2]	K_M	K_M +/- s [3]
$\log [H^+]$	pH			

1. Basis for optimization (objective function) is the adsorbed concentration of the sorbate (mol/L) which is input as a dummy component (Type II)
2. $[\ ]_T$ = total concentration
3. s = standard deviation

Figure 2 Example application of FITEQL for SC-EDL or SC-NEM models to ion adsorption data on soil.

FITEQL has been used widely for the calculation of K_+/K_- of mineral solids from alkalimetric titration data, and for surface complexation and exchange constants for surface species. A typical application of FITEQL to soil chemical experimental data is shown in Figure 2 for the adsorption of a metal cation (M) by soil material. A FITEQL stoichiometry matrix is shown in Table 7 (from Herbelin and Westall;[95] example no. 10) for the fitting of sorption constants for Zn on the weak and strong sites of ferrihydrite according to the DLM.[3] The example uses a dummy component (the adsorbed Zn concentration) to aid convergence and make best use of the analytical data. The dummy component is a flexible option within FITEQL that has versatile applications in soil chemistry;[95,98] it is not a true chemical component from the thermodynamic sense, but one that fits into the mathematical formulation of the problem as defined in Equations 22 and 23. The dummy component can be used, as noted above, to facilitate numerical convergence or to otherwise modify the mass-action or mass-balance equations for desired effect. Solubility reactions can be incorporated into surface chemical modeling or reference reactions can be fixed for ion-exchange calculations using a dummy component. A stoichiometry matrix for Na-Ca ion exchange according to the Vanselow convention is presented in Table 8. A dummy component is used to aid in computation of exchanger phase mole fractions.

Adjustable parameters, such as component concentrations (e.g., $[SOH]_T$) or adsorption constants for multiple surface species, can be optimized separately or simultaneously in FITEQL. Model

Table 9   Example Applications of FITEQL Relevant to Soil Chemistry

Citation	Sorbent	Solute	Adsorption model[a]	Use of FITEQL
Zachara et al.[180]	$Fe_2O_3 \cdot nH_2O$	CrO_4^{2-}, HCO_3^-, SO_4^{2-} Ca^{2+}, Na^+	TLM	Multicomponent adsorption calculations
Schindler et al.[50]	Kaolinite	Cu^{2+}, Cd^{2+}, Pb^{2+}	CCM and IHRE	Optimized K_{si} for metal binding to edge SOH, and K'_{MeX} for cation exchange[c]
Goldberg and Glaubig[149]	Surface soil	AsO_4^{2-}	CCM	Optimized K_{si} for several As surface species[c]
Zachara et al.[144]	Ultisol subsoil	CrO_4^{2-}	TLM	Optimized SOH_T for hypothesized soil Al-goethite sorbent
Dzombak and Morel[3]	$Fe_2O_3 \cdot nH_2O$	Many metal cations/anions	DLM	K_+ and K_- for $Fe_2O_3 \cdot nH_2O$, extensive self-consistent database for K_{si}
Zachara et al.[54]	$CaCO_3$	Ni^{2+}, Zn^{2+}, Cd^{2+}, Co^{2+}, Ba^{2+}, Sr^{2+}	IHRE	Optimized half-reaction constants (K'_{MeX}) for exchange against surface Ca (K'_{CaX})
Goldberg[64]	FeOOH	Various anionic sorbates	CCM and TLM	Optimized K_{si}, use of V_y to distinguish between potential inner- and outer-sphere complexation mechanisms[c,d]
Hayes et al.[59]	FeOOH	H^+	TLM and CCM	Sensitivity of K_+ and K_- to different input values, error estimates, data groupings
Ronnogren et al.[181]	ZnS	H^+, Zn^{2+}	CCM	Optimized $\log \beta_{101}$, $\log \beta_{2-11}$, $\log \beta_{-101}$[e]
Stadler and Schindler[45]	SWy-1	H^+, Cu^{2+}	IHRE and CCM	Optimized site concentrations and surface equilibrium constants
Zachara and Smith[48]	SWy-1	Na^+, Ca^{2+}, Cd^{2+}	IHRE and TLM	Optimized half-reaction constants for Ca (K'_{CaX}) and Cd (K'_{CdX}) against reference reaction for Na

[a] CCM, constant-capacitance model; TLM, triple-layer model; DLM, diffuse-layer model; IHRE, ideal half-reaction exchange.
[b] NA, not applicable.
[c] K_{si} = equilibrium constant for adsorbate surface complexation.
[d] V_y = SOS/DF = weighted sum of squares of the residuals divided by the degrees of freedom.
[e] Formation constants, β_{pqr}, for the general equilibria $pH^+ + qZn^{2+} + r(-SZn) = HpZnq(-SZn)_r^{(p+Zq)+}$; p, q, and r are stoichiometry coefficients.

convergence becomes less likely when more than two parameters are adjustable. FITEQL will not converge using the TLM if K_+/K_- and K_{Cat}/K_{An} are adjusted simultaneously because the reactions are interdependent. Multiple data sets varying in ionic strength or component concentrations can be used simultaneously within FITEQL through input of serial data. Discussions of the use of FITEQL for extraction of equilibrium constants for surface complexation reactions may be found in Dzombak and Morel,[3] Herbelin and Westall,[95] and Katz and Hayes.[80] Table 9 summarizes representative applications of FITEQL to soil chemical problems.

FITEQL provides output on the goodness of fit of the optimized model parameters to the experimental data and the standard deviation of each adjusted parameter. The overall variance, V_Y, representing the weighted sum of squares of residuals divided by the degrees of freedom (SOS/DF), is the main measure of goodness of fit: $V_Y = SOS/DF$.[3,93-95] An overall variance of approximately one is indicative of good fit, with a range of 0.1 to 20 being generally acceptable.[3,93,94] The variance is dependent on user-input estimates of the absolute [s(abs)] and relative [s(rel)] experimental error associated with each type of input data (e.g., pH, free metal concentration, total metal concentration). Error estimates recommended by Dzombak and Morel[3] for application of FITEQL to acid-base titration data of solids and metal ion sorption experiments are summarized in Table 10.

Table 10   Error Estimates for Acid-Base Titration Data and Sorption Data

Measurement[a]	S[rel][b]	S[abs][c]	Remarks
X_H	0.05	0.0	±0.02 pH unit
X_H	0.10	0.0	±0.04 pH unit
T_H	0.01	0.01·min T_H	
X_M	0.05	0.0	±0.02 pM unit
T_M	0.01	0.01·min T_M	
X_A	0.05	0.0	±0.02 pA unit
T_A	0.01	0.01·min T_A	

[a] H, hydrogen ion (H^+); M, cation; A, anion; X, free concentration; T, total concentration.
[b] S_H[rel] = 0.05 used for pH measurements in sorption experiments; S_H[rel] = 0.10 used for pH measurements in titration experiments.
[c] min, minimum; thus min T_M = the minimum total metal concentration.

From Dzombak, D. A. and Morel, F. M. M., *Surface Complexation Modeling: Hydrous Ferric Oxide*, John Wiley & Sons, New York, 1990. With permission.

IV. APPLICATION TO SOIL SYSTEMS

A. Uses for Adsorption Models in Chemical Studies of Soils

A primary use for adsorption models is in the development of a chemical model of the system. A chemical model is a series of reactions, component concentrations, and equilibrium constants that are consistent with the observed chemical behavior (e.g., pH and ionic strength dependence, stoichiometry, competing ion effects, etc.). A chemical model is a statement of system understanding conveyed in terms of mass-action and material-balance equations, and is a point of departure for understanding the effects of other processes, such as plant uptake, microbiology, or hydrology, and changes in system chemistry, both natural and artificial. The chemical model is fundamental to problem simplification, a need in multispecies reactive transport modeling, because it allows dominant reactions to be identified and insignificant ones to be dismissed.

Adsorption models serve an important role as an extrapolation tool. Once appropriate parameters have been set for regions of chemical space, models provide a means to estimate surface speciation and adsorbed/aqueous concentrations for conditions different than experiment. This estimate is particularly important for metal cations and anions whose sorption, as measured by a concentration or mass distribution ratio (K_d), may vary by orders of magnitude as the pH or other solution variables change.

While the mathematical degeneracy and parameter interdependency of adsorption models, SC-EDL models specifically, limit their direct use to evaluate the mechanisms of surface chemical reactions (see, for example, Goldberg[64]), they have become a valuable interpretive tool to evaluate interfacial processes. Examples of the usefulness of these models include: (1) distinguishing between inner- and outer-sphere complexation reactions,[20,21,99] (2) evaluating rates and plausible reaction suites for mineral dissolution,[77,100,101] surface-mediated oxidation-reduction, and photochemical reactions,[102,103] (3) interpreting particle stability and coagulation phenomena,[104] (4) evaluating the complexation properties of biologic surfaces,[105,106] and (5) determining the contribution of adsorption to solute retardation in groundwater.[107]

Adsorption models are gaining wider acceptance as a performance assessment tool to forecast or predict aqueous concentrations in waste solutions, soil solutions, and groundwater.[108-111] Simplified SC-EDL models have been linked with hydrologic codes to simulate reactive transport.[69,112]

B. Parameters Needed for Adsorption Modeling in Soil

Rationalizing ion sorption behavior with measured soil properties has long been a focus of soil chemistry.[113] Ion sorption in soil occurs by chemical interactions with a heterogeneous collection of materials that vary in surface functional group concentration and affinity for the sorbate. The chemical nature of the adsorption complex on multiple-site, soil-relevant surfaces is emerging only now.[25,114,115] The sorption of trace ions may be controlled by phases present in low mass concentration, and that exist as poorly crystalline particle coatings or phases.[4,5,116] Furthermore, the reactive phases may exhibit complex mineralogic and chemical composition as a result of coprecipitation, overgrowth, oxidation/reduction and precipitation/dissolution cycles, and interaction with organic matter and microorganisms.[117-119] As a result of crystallographic disorder, substituent ions, and coatings by organic matter, these phases may not be easily represented by analog sorbents such as laboratory synthesized Fe or Al oxides, or specimen phyllosilicates. Attempting to reconcile macroscopic measurements of mineralogy and soil chemical properties with the potential contributions of different sites and phases is a central limitation to the chemical modeling of adsorption reactions in soil.

Two general approaches have been taken to model adsorption reactions in natural materials:[4] (1) assume a given sorbent is dominant (e.g., Fe oxide), that analog phase (e.g., goethite) properties apply, and that its site concentration is proportional in some way to its mass content; or (2) assume the adsorption behavior is a soil property that can be fitted, like a single mineral phase, using selected macroscopic characterization measurements of the soil. These approaches are called pseudomechanistic and empirical, respectively, for the sake of discussion. Published examples of these two approaches are summarized in Tables 11 and 12. Each of the modeling studies listed in Tables 11 and 12 used assumptions that should be evaluated critically before attempting to use similar calculations for different materials.

The pseudomechanistic and empirical modeling approaches differ in characterization needs. In the pseudomechanistic approach, one attempts to identify the controlling sorbent, its crystalline structure if inorganic, and its concentration of surface sites by direct or surrogate measurement. Identifying these items may require mineralogic, electrolyte binding, and surface chemical/compositional measurements of significant detail and sophistication. The surface properties of the sorbing phase may also be interrogated, if feasible, through size fractionation experiments or direct-beam/spectroscopic measurement. The empirical approach, in contrast, requires a minimum of characterization information, such as the solute sorption maxima, because the sorption data are fit to yield soil specific reaction parameters.

The major parameters needed for adsorption modeling in soil are summarized in Table 13. The data demands vary depending on whether ion exchange, SC-EDL/SC-NEM, or combinations of these models are to be used, and the number of components [sites (j), solutes (j)] and species to be included in the calculation. The number of sites $[X_j, SOH_{Tj}]$ is a critical parameter, as are the equilibrium constants between these sites and aqueous components leading to surface species (H^+ - K_+, K_-, K_{HX}; and solutes — K_{MXm}, K_{CXm}, K_M, K_L, K_C, K_A). The surface area(s) of the adsorbing phase is required for SC-EDL models (i.e., CCM, TLM).

1. Cation Exchange

a. Site Concentration (X_j)

The concentration of sites associated with each exchanger phase (j) is needed to model cation exchange. The total exchange capacity measured by ion retention, E_T, equals the sum of the individual exchange capacities of the different sorbents (E_j) under the specific conditions of measurement (e.g., pH, ionic strength, surface saturation), where

Table 11   Applications of Surface Complexation-Electric Double-Layer (SC-EDL) Models to Anion Adsorption on Soils

Citation	Sorbate	Solid	Model	SOH_I	S	K±	K_L
Goldberg and Sposito[76]	PO_4^{3-}	44 soils	CCM	q_{max} [a]	Packing area [b]	Al, Fe oxides [c]	Average of individual optimized values
Goldberg and Glaubig[150]	$B(OH)_4^-$	15 soils	CCM	q_{max} [a]	EGME surface area of soil [d]	Optimized with K_{si} [e]	Optimized with K_+/K_-
Sposito et al.[79]	SeO_3^{2-}	Surface soil	CCM	q_{max} (empirical partitioning into two sites)	Packing area	Al, Fe oxides [c]	K_{s1}, K_{s2}, K_{s3} optimized on one soil over different pH ranges
Goldberg and Glaubig[148]	SeO_3^{2-}	Surface soil (0–7.6 cm, Entisol)	CCM	q_{max}	N_2 gas surface area of soil	Optimized with K_{si}	Optimized with K_+/K_-
Goldberg and Glaubig[149]	AsO_4^{2-}	Surface soil (0–7.6 cm, Entisol)	CCM	q_{max}	N_2 gas surface area of soil	Al, Fe oxides [c]	Optimized
Zachara et al.[144]	CrO_4^{2-}	C-horizon material (Ultisols)	TLM	Optimized	ΔS [f]	From Al-goethite	From Al-goethite
Goldberg et al.[182]	$B(OH)_4^-$	Subsoil (25–51 cm, Alfisols)	CCM	q_{max}	EGME surface area of soil	Optimized with K_{si}	Optimized with K_+/K_-

[a] q_{max} = sorbate adsorption maxima.

[b] $S = q_{max} N_A a$ where N_A is Avogadro's number and a is the sorbate packing density.

[c] Average of values reported for Fe and Al oxides $K_+(int) = 10^{7.35}$; $K_-(int) = 10^{-8.95}$.

[d] EGME, ethylene glycol monoethyl ether.

[e] K_{si}, intrinsic equilibrium constant for anion surface complexation.

[f] ΔS, difference in N_2 surface area before and after extraction with citrate-dithionite-bicarbonate.

Table 12 Example Applications of Site Binding Models (SC-EDL, SC-NEM, ion exchange) to Cation Adsorption on Soils and Heterogeneous Natural Sorbents

Citation	Sorbate(s)	Solid	Model[a]	SOH, X_I	S	$K_\pm/K_{HX}$	K_M; K_{MX}
Mouvet and Bourg[183]	Cu^{2+}, Zn^{2+}, Cd^{2+}, Pb^{2+}, Ni^{2+}, Ca^{2+}, Mg^{2+}	River sediment	NEM	Acid-base titration	b	Extrapolation and regression[c]	
Charlet and Sposito[120]	Na^+	Oxisol	TLM	Acid-base titration, ion retention	Not reported	Extrapolation	Extrapolation
Loux et al.[145]	Pb^{2+}, Zn^{2+}, Ni^{2+}, Cd^{2+}, Cu^{2+}	Sandy Wisconsin aquifer sediment	DLM	Extraction[d]	600 m²/g $Fe_2O_3 \cdot nH_2O$	From amorphous[e] $Fe_2O_3 \cdot nH_2O$	From amorphous[e] $Fe_2O_3 \cdot nH_2O$
Osaki et al.[184]	Fe(III), Co^{2+}, Zn^{2+}	Fresh-water and estuarine sediments	CCM	Acid-base titration	Extrapolated from experimental data	Acid-base titration	Extrapolation from experimental data
Payne and Waite[185]	UO_2^{2+}	Weathered schist "Ultisol like"	TLM	Extraction[f]	700 m²/g $Fe_2O_3 \cdot nH_2O$	From amorphous $Fe_2O_3 \cdot nH_2O$	Fit to experimental data
Cowan et al.[51]	Cd^{2+}	Ultisol clay fractions	Multisite NEM and IHRE Al-goethite	a) surface area[g] b) extraction[h]	b	From kaolinite, Al-goethite, K_{HX} from kaolinite	From kaolinite, Al-goethite K_{HX} from kaolinite and montmorillonite
Zachara et al.[147]	Co^{2+}	Ultisol clay fraction	Multisite NEM and IHRE	a) surface area[f] b) extraction	a	K_{HX} from KGa-1	K_{sl} from kaolinite/ Al goethite K_{MX} fitted to soil kaolinite
Zachara and McKinley[49]	Cd^{2+}, UO_2^{2+}	Mollisol clay fraction	TLM and IHRE	Particle size measurements and crystallography	N_2 surface area	Silica and gibbsite	Silica and gibbsite
Zachara and Smith[48]	Cd^{2+}	Mollisol clay fraction	TLM and IHRE	Particle size measurements and crystallography	N_2 surface area	Silica and alumina	Silica and alumina

a NEM, nonelectrostatic model; TLM, triple-layer model; DLM, diffuse-layer model; CCM, constant-capacitance model; IHRE, ideal half-reaction exchange.

b Surface area not required in the NEM.

c Authors noted the following relationship that was used to estimate binding constants for Pb, Ni, Ca, and Mg log $\cdot B_1^{surface}$ = 0.945 log $\cdot B_1^{hydrolysis}$ + 5.6.

d N_t = Fe(ex)[g/kg]·moles sites/g-$Fe_2O_3 \cdot nH_2O$(am) where Fe(ex) is from $NH_2OH \cdot HCl$ extraction.

e Constants from Dzombak and Morel.[3]

f Same as 5 but ammonium oxalate used as extractant.

g $N_{t\text{-kaolinite}}$ = $N_{s\text{-kaolinite}}$ ·N_2 surface area, where N_s is the reported hydroxyl site density on kaolinite.

h Same as 4 except citrate-dithionite-bicarbonate extraction used

Table 13   Parameters Needed for Surface Complexation Modeling in Soil[a]

	Ion exchange	SC-EDL and SC-NEM
Site concentration	X_{i^-}	$SOH_{T, i}$
Reference log K	K_{NaX} [b]	NA[c]
Acid/base	K_{HXi}	$K_{+, i}$; $K_{-, i}$
Electrolyte	$K_{CXu, i}$	$K_{A, i}$; $K_{C, i}$ [d]
Solute	$K_{MXm, i, j}$	$K_{M, i, j, k}$; $K_{L, i, j, k}$
Surface area		s_i [e]

[a] For each site (i), solute/component (j), and surface species (k).
[b] Typically set to high value (log K > 10) to minimize free X^-.
[c] Not applicable.
[d] Generally for the TLM, can also be used to simulate ionic strength effects in the NEM; A, electrolyte anion; C, electrolyte cation.
[e] For each site population (i), needed only for CCM and TLM models.

$$E_T = \Sigma E_j \qquad (24)$$

The exchange capacity represents the concentration of ions bound as outer-sphere complexes and in the diffuse ion swarm.[6] The total site concentration may be established by conventional exchange capacity measurements using various reactive probe ions (recognizing the attendant implications of different solutes, their strength of interaction, and phase specificity). Comprehensive methodologies for, and measurements of, ion retention by various soil types have been reported.[23,120-124]

Cation exchange capacity (CEC) measurements often show pH dependence even when their clay fraction is dominated by smectites containing fixed, structural charge, see for example adsorbed Na, in Figure 3; and Anderson and Sposito.[121] This pH dependence results from Al^{3+} dissolution and adsorption at lower pHs, and ionization of organic material and amphoteric inorganic hydroxyl sites as pH increases. For example, adsorbed Na (Na_s) in Figure 3 reflects the following contributions:

$$E_T = Na_s = \left([X^-] - 3[Al^{3+}{}_s] \right) + [SO^-] + [RO^-] \qquad (25)$$

where X^- represents permanent structural charge in constituents such as layer silicates; Al_s^{3+} is exchangeable Al^{3+}; SO^- are ionized hydroxyls such as SiOH on layer silicate edges; and RO^- is ionized carboxylate or phenolate groups on organic matter (ROH). Organic matter may contribute significantly to the measured CEC in surface soils. Organic matter has a high exchange capacity per unit mass, and the CEC of surface soils generally correlates with its mass content.[6] The measured value of E_T will be at maximum above pH 8,[6] where the ionization of SOH and ROH sites approaches completion.

The modeler must decide whether to subdivide the soil into different ion exchangers (i.e., X^-, SO^-, RO^-), upon which separate mass-action/material-balance equations will be established, or to treat the assemblage as a composite. Each approach can be enacted within a model such as FITEQL, if the appropriate characterization data and equilibrium constants are available. Even though measurement techniques are not sufficiently specific to individually quantify the number of exchange sites on organic matter (RO^-) and clay minerals (X^-) in a soil material, an empirical attempt to segregate sites into these two categories may be required because metal cations show different affinities for these sites that may invalidate a single-site Langmuir approach. Anderson and Sposito[125] reported a Cs/Li exchange method to identify the contributions of structural (i.e., X^-) and pH-dependent charge (i.e., SO^-) in sorbents containing montmorillonite, illite, and nonexpandable vermiculite. This method,

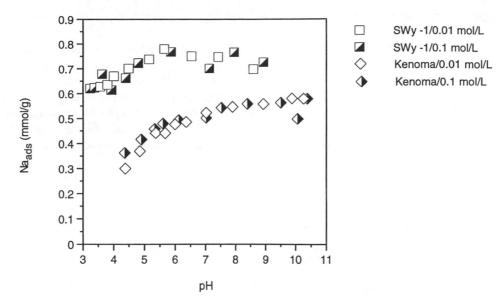

Figure 3 The adsorbed concentration of Na (Na$_{ads}$) on reference smectite (SWy-1) and a Mollisol C horizon smectite (Kenoma) as measured by isotopic dilution with ^{22}Na.

when combined with acid-base titration and ion adsorption measurements, can be used to identify structural and pH variant charge contributions in soil.[121,123] Surface charge distributions defined by such measurements can be used for gross separation of E$_T$ into subclasses of exchanger sites for model calculations.

It must also be decided whether modeling is to be performed at fixed or variable pH. Fixed pH calculations may be driven by an exchange capacity measurement at the target pH. pH-variable calculations require explicit consideration of the equilibria involving H$^+$ that affect E$_T$:

$$Al(OH)_3 - \text{soil} + 3H^+ = Al^{3+} + 3H_2O \qquad (26)$$

$$(3-n)NaX + Al^3 + nH_2O = Al(OH)_n X_{(3-n)} + nH^+ + (3-n)Na^+ \qquad (27)$$

$$SOH = SO^- + H^+ \qquad (28)$$

$$ROH = RO^- + H^+ \qquad (29)$$

The approximation for organic matter (ROH) ionization is a gross one; the proton adsorption of soil organic matter is complex, reflecting molecular heterogeneity of acid sites and electrostatic effects.[126-128] Typically, multiple sites or a site distribution is needed to model the proton chemistry of humus.[98,129,130] Anderson and Sposito[121] suggested, however, that the point of net proton charge for soil organic matter may be below a pH of 3, allowing application of an ion-exchange approach to describe its proton and solute adsorption chemistry (see for example, Westall et al.[98]).

Various options are available to deal with pH-variable CEC in a cation exchange calculation: (1) input analytically measured values as serial data into the calculation, (2) fit a polynomial to the CEC data for empirical extrapolation, or (3) establish a chemical model and parameters that explicitly account for the contributions of different sites/reactions.

b. Equilibrium Constants (K_{HX}, K_{CXu}, K_{MXm})

A standard compendium of equilibrium constants for soil exchange reactions does not exist because of the diverse physicochemical nature of inorganic and organic soil exchangers. Some information, however, does exist. Bruggenwert and Kamphorst[131] summarized exchange constants for layer silicates and soils. Fletcher and Sposito[41] developed a half-reaction database for the exchange of major cations and trace metals on montmorillonite. MINTEQA2 includes a database for complexation/exchange reactions of metal ions on the Suwannee River humic acid.[44] Other more limited databases have been reported for exchange reactions on layer silicates,[45,47-50] calcite,[54] and organic matter.[132-134]

Most modeling studies of minor element exchange on soil have resorted to fitting exchange constants to the experimental data. Subtle differences in mineralogy of the soil exchangers, and the presence of organic material and other difficult-to-characterize minor phases, complicates the direct applicability of equilibrium constants for a reference phase such as montmorillonite. The actual differences in log K, however, between the reference phase constants and those for an analogous soil material may be small (e.g., 0.2 to 0.4 log units).[48,49]

2. Surface Coordination

a. Site Concentration ($SOH_{T,j}$)

The hydroxyl site concentration and its potential distribution between different sorbent phases is a critical measurement for applying SC-EDL models. In soils, the task of measuring hydroxyl site concentrations, which is difficult for oxides and challenging for clays, becomes even harder because of the multiplicity of sites and phases. Davis and Kent[4] recommend the use of a standard value of 3.84 $\mu mol/m^2$, which is an average for reactive mineral phases found in geologic materials and is equal to the value recommended for noncrystalline Fe oxide.[3] Surface hydroxyl concentrations in soil are determined by (1) acid-base titration and/or direct proton adsorption measurements, (2) tritium exchange, (3) ion/solute adsorption, and (4) characterization of the dominant sorbent and assignment of properties based on single-phase analogs. Each of these is described briefly below.

Proton adsorption and adsorbed proton charge (σ_H) derived from acid-base titration can be related directly to the surface hydroxyl concentration under ideal conditions. For example, Charlet and Sposito[120] estimated the $[SOH]_T$ for an Oxisol from the x-intercept of a plot of $\sigma_H/[H^+]$ vs. σ_H. Analysis of titration data for soils containing both structural and pH-variant charge is more complex, requiring additional measurements to quantify proton exchange on structural sites of fixed charge (σ_o).[121-123] Acid-base titrations also suffer from artifacts arising from dissolution of mineral phases and organic material,[135-137] leading to overestimates in proton adsorption and σ_H. Reprecipitation and hydrolysis of dissolution products also affects titration results at higher pH. Poorly crystalline aluminous phases, in particular, can exhibit large solubilities below a pH of five. While such reactions can be explicitly accounted for in the analysis and modeling of the titration data, corrections may exceed proton adsorption density, leading to large uncertainties. Titration data can be corrected for Al^{3+} dissolution through the following relationship:

$$\Gamma_{H+} = \left(C_A + [H^+]\right) - \left(C_B - [OH^-]\right) - 3[Al^{3+}]_T \tag{30}$$

where Al^{3+} (in mol/l) is the total Al(III) released by dissolution. Aqueous phase analyses may underestimate the total extent of dissolution at lower pH in soil materials with structural charge. Al^{3+} may be adsorbed by ion exchange (Equation 25), which promotes continued dissolution of the soluble phase (see example 1 in Section IV.C). A back-titration technique has potential to overcome some of the above noted complications.[138,139]

Table 14 Hydroxyl Site Concentrations for Soil Materials

Sorbent type	N_2^- surface area (m²/g)	Optimized [SOH]$_T$ (mmol/kg)	Measured [SOH]$_T$[a] (mmol/kg)	Ratio[b]
Mollisol smectite	99	425[c]	861	2.02
Goethite-coated sand[d]	6.8	121[e]	110	0.91
Uncoated analog	1.1	27.5[e]	24.2	0.88
Hematite-coated sand[d]	10.5	29.0[e]	266	8.62
Goethite/feryoxhite-coated sand[d]	3.1	44.5[e]	113	2.53
Uncoated analog	0.5	7.8[e]	36.1	4.62
Goethite	32	160[f]	797	5.00

[a] By tritium exchange on Cs⁺-saturated materials.
[b] Measured [SOH]$_T$/optimized [SOH]$_T$.
[c] From ion adsorption measurements, see example 1, Section IV.C.
[d] Atlantic coastal plain sediments from depths of 1.5–2.0 m. The sediments are quartzitic sands with surface precipitates and particle coatings of Fe and Al oxides. See Zachara et al.[68].
[e] Fitted from Co^{2+} pH-edge adsorption data, assuming goethite was the reactive phase. Modeling used a site density for goethite equal to that measured by tritium exchange (15 sites/nm²).
[f] Fitted from a Co^{2+} adsorption isotherm at pH 5.5 assuming the reaction $SOH + Co^{2+} = SOCo^+ + H^+$.

Tritium exchange has not seen widespread application to soils, in spite of its potential to yield a solute-independent measure of the total surface hydroxyl concentration. The method is simple in concept, involving saturation of surface sites with tritiated water, removal of tritiated and physically adsorbed water, and reextraction of tritium bound to surface hydroxyls:

$$SOH + {}^3H_2O = SO^3H + H_2O \qquad (31)$$

$$SO^3H + H_2O = SOH + {}^3H_2O \qquad (32)$$

It is, however, a difficult measurement to perform, requiring a fool-proof vacuum system, and for various reasons, the returned values may not be representative of the solute accessible site population. Balistrieri and Murray[140,141] reported tritium exchange values for marine sediments that correlated with cation exchange capacity, implying tritium retention in the hydration sphere of cationic counterions, rather than exchange with surface hydroxyls. Tritium exchange measurements for Cs-saturated subsurface sediments, clays, and goethite are reported in Table 14. Cesium dehydrates during the removal of physically adsorbed water, eliminating the potential for tritium retention by more strongly hydrated cations such as Ca^{2+}. Surface site concentrations determined by tritium exchange on single-phase sorbents generally exceed those measured by solute adsorption by factors of 3 to over 5.[4,89,142,143] A similar range was observed for the natural sorbents (Table 14) with one exception: the goethite coated sand and its analog. The method holds promise for soil characterization but requires further documentation.

Adsorption measurements are used frequently to estimate hydroxyl site concentrations (Tables 11 and 12). Generally, the Langmuir adsorption maximum is extrapolated from the data and is assumed to equal an effective [SOH]$_T$. Such sorption maxima have been used, along with the sorbate molecular dimensions, to estimate the reactive surface area of the sorbing phase.[76,79] Care must be taken in such extrapolations as the Langmuir sorption maximum may be pH dependent. Linear, least-squares models, such as FITEQL, can be used to incorporate speciation, surface ionization, and competitive ion effects into the fitting, if requisite data are available (i.e., Zachara et al.[144]). Sorption maxima measurements require consideration of (1) the strong pH dependency of the process, (2) competitive ions released by the soil (Al^{3+}, H_4SiO_4, etc.), and (3) precipitation reactions. Precipitation is a major concern for metallic cations and strongly binding anions because sorption maxima measurements require relatively high sorbate concentrations. The adsorptive surface may actually promote solid-phase nucleation or solid-solution behavior at aqueous concentrations below those expected based on the solubility product.[35,62] Aqueous speciation and solubility calculations with a code such as MINTEQA2 should be used in experimental design to avoid concentration ranges where precipitation may occur.[44]

Site concentrations can be estimated from extractable constituents such as poorly crystalline Fe oxides or aluminum-containing phases. The simplistic approach of assuming that hydroxylamine HCl- extractable Fe exhibits a site density and reactivity equal to that of poorly crystalline Fe oxide, as defined by Dzombak and Morel,[3] has led to good predictions of metal ion sorption in select cases.[145,146] As a general approach, however, it fails more often than it succeeds (see, for example, Zachara et al.,[147] Kent et al.[107]). The overall surface area or parts of it can also be assigned a site density, if mineralogic or other data suggest that the surface area is dominated by a certain mineralogic component of known properties.[52,147]

Dividing the total hydroxyl site concentration, $[SOH]_T$, into phase-specific values, $[SOH_j]_T$, may be desirable if characterization indicates the presence of components with different affinities for the sorbate (i.e., silica vs. alumina). Improved surface analytical tools including electron beam spectroscopies (i.e., XPS) and synchrotron radiation tomography hold promise to estimate the mole fractions and concentrations of Fe, Al, and Si in the surface layers of coatings on mineral particles. Such techniques are now beginning to be employed for partitioning surface hydroxyls into different population pools. In practice, however, such assignments are difficult to justify from macroscopic soil characterization data and are often assigned empirically. Sposito et al.[79] proposed the existence of two sorbing sites for SeO_3^{2-} in soil materials based on the shapes of the adsorption edge. Zachara et al.[144] used extraction data to characterize the ratio of Al/Fe in soil goethites, the presumed sorbent for CrO_4^{2-}, and allowed surface equilibria on both of these types of sites.

b. Equilibrium Constants (K_+, K_-, K_M, K_L)

At least six approaches have been used in modeling surface complexation reactions in soils using SC-EDL or SC-NEM models:

1. Global optimization of K_+, K_-, and K_M or K_L, given a measured sorption maximum[148,149]
2. Fixing K_+ and K_- to those of a known oxide and optimizing K_M or K_L
3. Fixing K_+ and K_- to an average value for soil Al and Fe oxides and optimizing K_M or K_L[76]
4. Fixing K_+, K_-, and K_M or K_L to those for a known oxide and optimizing for $[SOH]_T$[144]
5. Predicting adsorption based on K_+, K_-, and K_M or K_L of a known analog phase and an independent measure or estimate of $[SOH]_T$[48,107,145]
6. Predicting adsorption in soil using a set of average constants (K_+, K_-, K_M, K_L) derived from the curve fitting of soil adsorption data.[79,150]

Surface equilibrium constants from analog phases should be applied cautiously to soil materials. The K_+ and K_- (and K_A and K_C for the TLM), K_M, K_L, N_s, and C (the interfacial capacitance) for a single-phase analog sorbent are an interdependent constant set. The ionization constants, K_+ and K_-, depend strongly on the value of N_s and C; and K_M and K_L, in turn, depend on K_+ and K_-. Thus, surface equilibrium constants developed for a specific analog phase may not be appropriate for a comparable soil mineral if its site density is different, if minor substituent ions in its surface layers alter its surface properties, or if the electrostatic environment is different as a result of particle morphology, particle interactions, or site heterogeneity. Sensitivity calculations for various SC-EDL parameters have been performed with sorbate- specific data sets.[48,59,64] McKinley et al.[151] used different sets of TLM constants (K_+, K_-, K_A, K_C, N_s) to describe ionization reactions of AlOH on the smectite edge. It was found that the different constant sets, which provided comparable fits to acid-base titration data for gibbsite, affected both the computed speciation of UO_2^{2+} adsorbed to the smectite edge and the ability to simulate ionic strength effects on UO_2^{2+} adsorption.

3. General Experimental Considerations

Experimental design must be controlled to maximize chances for success in adsorption modeling. Important activities include:

1. Presaturating surface sites with an index cation and anion (i.e., $NaClO_4$, $CaCl_2$)
2. Measuring the concentration of potential competing ions (e.g., Al^{3+} and H_4SiO_4) and other solutes that may affect sorbate speciation in the equilibrated solution, including organic material
3. Maintaining pH and measuring or controlling $CO_2(g)$
4. Employing effective phase separation to eliminate artifacts from dispersed colloids
5. Minimizing microbiologic effects by storing sorbent suspensions under ultraviolet light or using other suitable precautions
6. Documenting whether adsorption is the responsible process for sorption

Adsorption measurements should be performed at lower fractional site occupation (<50%), unless the sorption maximum is being determined, as nonlinear behavior and other complications (surface polymer formation, surface precipitation) may occur as site saturation is approached.[35,62] The implications of pH adjustment should be considered because dissolution, adsorption, and reprecipitation of Al^{3+} and possibly other ions may affect sorbent surface properties and ion retention.[75,117,152-154] Locally high or low pH values after base or acid addition may cause effects (pH shock) to the sorbent or sorbate that are irreversible over the time period of the adsorption measurement. The gas phase composition (i.e., $CO_{2(g)}$, $O_{2(g)}$) can also be highly significant depending on the chemistry of the sorbate and sorbent. Dissolved carbonate species may sorb to mineral surfaces and complex metal cations, both of which can impact the adsorption of the target sorbates.[81,155] Comprehensive solid-phase characterization may provide invaluable insights for modeling. Optional measurements include extractable constituents, clay fraction mineralogy, cation exchange capacity over a range in pH, surface area, particle coating morphology and composition, and organic and inorganic carbon content.

C. Example Calculations

Three examples are presented to illustrate how adsorption modeling can be used to interpret and describe ion retention in soil. The examples include the three types of models experimentalists are likely to apply to soil chemical data: (1) the SC-EDL model, (2) the SC-NEM model, and (3) the log K spectrum model.

1. Surface Charge of a Soil Smectite Using a SC-EDL Model

To describe numerically the adsorption of metal cations or anions to layer silicates in soil, a model is needed of site distribution and energy that is consistent with both proton adsorption and ion retention. The resulting surface charge/site distribution model provides a framework for describing solute adsorption or other related interfacial phenomena. Here we describe the development of such a model for a soil smectite. The experimental and modeling procedures are described in Turner et al. [154]

A low-carbon, beidellite (<2 μm) from a Mollisol C horizon was extracted with citrate-dithionite-bicarbonate, reoxidized with hydrogen peroxide, washed extensively in $NaClO_4$ at pH 6.5, dialyzed against H_2O, and freeze-dried.[68] Samples of the Na-saturated beidellite were suspended in two concentrations of $NaClO_4$, and the adsorbed concentration of Na (Na_{ads}) was determined over a range in pH by isotopic dilution with ^{22}Na (Figure 3). Adsorbed Al^{3+} (Al_{ads}) was determined

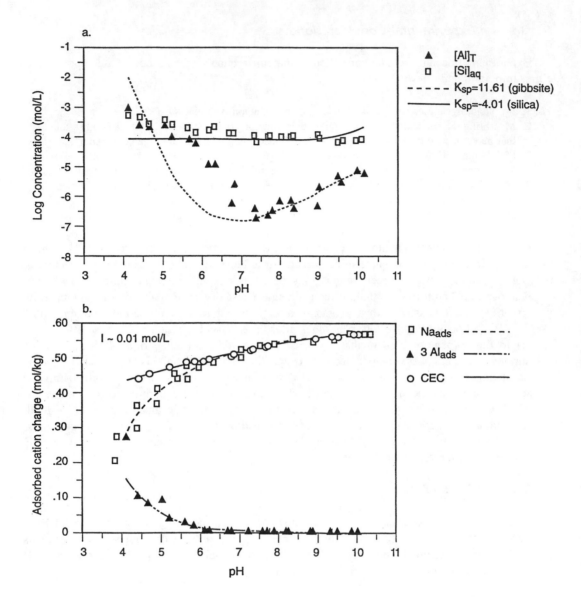

Figure 4   (a) Concentrations of [Al]$_T$ (sorbed plus aqueous) and [Si]$_{aq}$ in an acid-base titration of a Mollisol C horizon smectite (Kenoma). (b) Adsorbed cation charge (Na$^+$, Al^{3+}) and the computed CEC of the Kenoma smectite.

at each level pH by extracting the clay with 0.01 mol/l La^{3+} while maintaining the pH. The CEC was considered to equal the total adsorbed cation charge (Figure 4a):

$$CEC(mol/kg) = \left[Na_{ads}(mol/kg) \right] + 3\left[Al_{ads}(mol/kg) \right] \qquad (33)$$

Adsorbed Al^{3+} was a significant contributor to the CEC below pH 6; aqueous Al^{3+} at these pH values was near analytical detection.

   Acid-base titrations were performed on high-concentration suspensions (10 g/kg) of the Na-saturated clay to minimize dissolution effects. Dissolved Si and Al, and Al$_{ads}$ were measured in a parallel titration. Significant concentrations of both Al^{3+} and H$_4$SiO$_4$ were solubilized from the

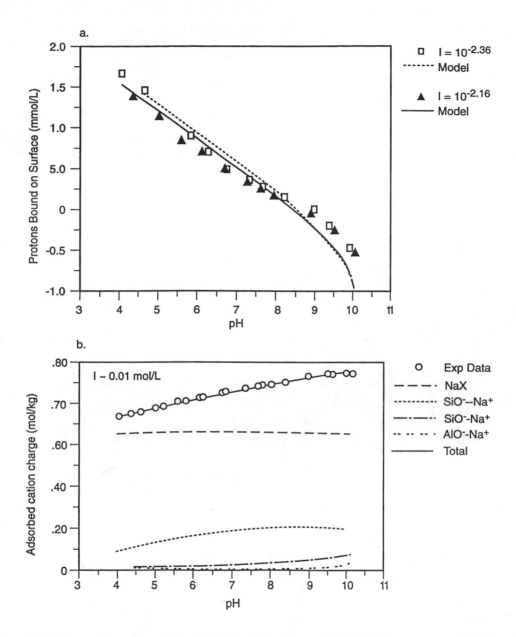

Figure 5 (a) Three-site model prediction of proton adsorption on a Mollisol C horizon smectite (Kenoma). (b) Calibration of a three-site surface charge model of the Kenoma smectite. Site concentrations $[NaX]_T$, $[AlOH]_T$, and $[SiOH]_T$ were optimized.

clay during the titration (Figure 4a); below pH 7 all the Al^{3+} was adsorbed in exchangeable form by the beidellite while aqueous Al^{3+} (as hydroxy complexes) dominated above this pH. The total concentration of Al^{3+} (in mol/l) significantly exceeded the solubility of gibbsite between pH 5 and 7, as a result of the ion exchange of Al^{3+} (Equation 25). The titration data were corrected for Al^{3+} dissolution using Equation 28 (Figure 5a). Aluminum dissolution below pH 6 was high enough (0.1 to 1 mmol/l, Figure 4a) to contribute uncertainty to the computed proton adsorption density, which spanned this same concentration range (Figure 5a).

The CEC was used as a basis for computing the concentrations of fixed-charge (X^-) and ionizable edge sites (SiOH, AlOH) for the hypothetical, fully Na-saturated smectite (i.e., no adsorbed Al^{3+}).

Table 15   Stoichiometry Matrix for Na/H+ Adsorption by Kenoma Soil Smectite[a]

| Species | Type I[b] | | | | | | | Type II[b] | Type III [b,c] | log K |
	Na[+d]	ClO_4^-	SiOH	AlOH	X[-]	ψ_o	ψ_B	Na$_{ads}$	H[+]	
H[+]	0	0	0	0	0	0	0	0	1	
OH[-]	0	0	0	0	0	0	0	0	-1	-13.78
SiOH	0	0	1	0	0	0	0	0	0	
SiO[-]	0	0	1	0	0	-1	0	(1, B)[e]	-1	-6.95
SiO[-]-Na	1	0	1	0	0	-1	1	1	-1	-6.6
AlOH	0	0	0	1	0	0	0	0	0	
$AlOH_2^+$	0	0	0	1	0	1	0	0	1	7.6
$AlOH_2^+$-ClO_4^-	0	1	0	1	0	1	-1	0	1	10.7
AlO[-]	0	0	0	1	0	-1	0	(1, B)[e]	-1	-10.6
AlO[-]-Na	1	0	0	1	0	-1	1	1	-1	-7.3
NaX	1	0	0	0	1	0	0	1	0	15
HX	0	0	0	0	1	0	0	0	1	15

[a] Component concentrations for [SiOH], [AlOH], and X[-] were fit in Figure 5a.
[b] Type I, II, and III components are defined in Westall.[93]
[c] Note Al(OH)$_3$ dissolution may be explicitly included by specifying the K_{sp} as a type III component.[95]
[d] Note the total component concentration equals the electrolyte concentration plus the adsorbed Na concentration [Na] = [Na]$_{aq}$ + [Na]$_{ads}$.
[e] Specified in the B matrix FITEQL input. These species are given a stoichiometry coefficient of 1 for Na$_{ads}$ in the Na$_{ads}$ material-balance equation. This effectively states that adsorbed sodium as measured by isotopic exchange includes Na[+] associated as a counterion with AlO[-] and SiO[-].

The three site concentrations were adjusted numerically based on Table 15 to provide the best fit to the CEC data using the material-balance equation for adsorbed Na (Na$_{ads}$):

$$CEC = Na_{ads} = [NaX] + \left[SiO^- \text{-} Na^+\right] + \left[SiO^- \text{--} Na^+\right] + \left[AlO^- \text{-} Na^+\right] + \left[AlO^- \text{--} Na^+\right] \tag{34}$$

In Equation 32, [NaX], [SiO[-]-Na[+]] and [AlO[-]-Na[+]], and [SiO[-]--Na[+]] and [AlO[-]--Na[+]] represent the concentrations of Na bound by ion exchange, in outer-sphere complexes, and as diffuse layer counterions, respectively. Ionization and outer-sphere complexation reactions for the edge sites (SiOH and AlOH) were computed with the TLM using constants for silica and gibbsite. The ion exchange of Na[+] and H[+] were computed with half-reactions to yield $K_v = 1$. The concentration of Na in the diffuse layer ([SiO[-]--Na[+]] and [AlO[-]--Na[+]]) was determined by assuming that Na[+] associates as a counterion with each SiO[-] and AlO[-]. The SiO[-] and AlO[-] species were included in the Na$_{ads}$ material balance (Equation 32) by appropriate specification in the B-matrix of FITEQL.[95]

The model quantitatively fit the data (Figure 5b), returning values for fixed structural charge (397 mmol/kg) and ionized edge charge at pH 6 (128 mmol/kg) that agreed well with direct measurements of these parameters (348 mmol/kg and 151 mmol/kg, respectively) made with the procedure of Anderson and Sposito.[125] The site concentrations obtained by this fitting, along with the other noted parameters of this model, provided an adequate description of the proton adsorption data (solid lines in Figure 5a), given its uncertainty.

2. Adsorption and Dissociation of Metal-EDTA Complexes Using the SC-NEM Model

Organic chelates of metal ions have been used as micronutrient fertilizers and chelating agents, more generally, have been used as nutrient extractants from soil.[156] The chemical behavior of metal-organic chelates in soil is complex, involving aqueous and interfacial reactions. Adsorption modeling can assist in the interpretation of such complex behavior by numerically integrating the effects of aqueous speciation and surface reactions. Here we describe the chemical behavior of the

Co(II)EDTA^{2-} complex (log K = 17.97) in contact with a sand-textured, low-carbon subsoil that contains grain coatings of crystalline Fe and Al oxides. Experimental details are given in Zachara et al.[118] The modeling objective is to develop a reaction suite consistent with the complex chemical behavior of the chelate.

The adsorption of Co^{2+}, EDTA^{4-}, and Co(II)EDTA^{2-} at 10^{-5} mol/l was measured on the soil (500 g/l) in 0.003 mol/l Ca(ClO$_4$)$_2$. The extent of sorption was determined by radioanalysis (e.g., fractional removal of counts). The behavior of Co(II)EDTA^{2-} was followed by dual-label radioanalysis (^{60}Co, ^{14}C-EDTA), where equality in counts (^{60}Co, ^{14}C) after equilibration was taken as evidence for persistence of the complex, and disparity implied dissociation. Both EDTA^{4-} and Co(II)EDTA^{2-} adsorbed as anions (Figure 6a). The difference in adsorption of Co and EDTA added as the preformed complex indicated that some dissociation had occurred. Aluminum and Fe^{3+} were measured in the aqueous phase after equilibration, and Al^{3+} was observed in significant concentration (Figure 6b).

The SC-NEM was used to describe adsorption because uncertainties in the nature and properties of the sorbing phase did not justify inclusion of the electrostatic component. Furthermore, our goal was to identify first-order effects associated with aqueous speciation and mass action. Low hydroxylamine-HCl-extractable Fe, combined with affirmative X-ray diffraction identification of goethite and other crystalline Fe oxides in the sample, led us to select crystalline goethite as the sorbing phase for model calculations. Surface protonation constants for goethite (K$_+$, K$_-$) were refit to acid-base titration data on this phase using the SC-NEM. An SC-NEM surface complexation constant for Co^{2+} (K$_{Co}$) on goethite in 0.003 mol/l Ca(ClO$_4$)$_2$ was also fit from pH edge data. These SC-NEM goethite constants were used to fit the adsorption data of Co^{2+} on the subsurface material using [SOH]$_T$ as the adjustable parameter. The returned value ([SOH]$_T$ = 10 mmol/kg) provided an excellent fit to the Co^{2+} adsorption data and was assumed, for additional calculations, to equal the concentrations of hydroxyl sites accessible to the anionic sorbates.

The adsorption behavior of EDTA^{4-} was modeled first to provide a basis for the Co(II)EDTA^{2-} system. The adsorption of EDTA^{4-} was described by trial-and-error fitting (optimizing log K) of the adsorption edge data using surface complexes of both major and minor aqueous species of EDTA^{4-}. These calculations used reactions 1 to 14 and 18 and 19 in Table 16, the fitted [SOH]$_T$ for the soil, and aqueous Al^{3+} concentrations at the pH of each sorption measurement supplied as serial data (Figure 6b). The best data fit (Figure 7) was obtained using surface complexes of the dominant aqueous species (AlEDTA$^-$, CaEDTA^{2-}; reactions 20 and 21 in Table 16). Adsorption was presumed to proceed by surface ion-pair formation, consistent with ionic strength effects.[22,157] While qualitative agreement between model and experiment was obtained (Figure 7), the data fit deteriorated below pH 6.5 where Al^{3+} concentrations increased and AlEDTA$^-$ dominated the aqueous speciation of EDTA^{4-}.

The Co(II)EDTA^{2-} adsorption data were modeled using all reactions in Table 15, the fitted [SOH]$_T$, and the Al^{3+} aqueous concentrations. Best-fit equilibrium constants were computed for surface species of the two Co(II)EDTA aqueous species [Co(II)EDTA^{2-}, Co(II)HEDTA] using both the Co and EDTA adsorption data from the Co(II)EDTA^{2-} experiment as the basis for optimization (Co$_{ads}$ and EDTA$_{ads}$ as "dummy" type II components). The value of log K returned for the surface complex of Co(II)EDTA^{2-} (3.27) was equivalent to that computed for the surface complex of CaEDTA^{2-} (3.38) in Figure 7. The model calculation agreed qualitatively with adsorption trends of both Co and EDTA (Figure 8 a, b), indicating that the reaction sequence adopted is a plausible, albeit not unique, one. We conclude that the disparate adsorption behavior of Co and EDTA below a pH of 5.75 results from competitive displacement of Co^{2+} by Al^{3+} from the complex, and the contrasting adsorption behavior of the products, AlEDTA$^-$ and Co^{2+}.

Importantly, the modeling shows weaknesses in understanding and possibly analytical measurements, and the need for further experimentation. The overall variance (V$_Y$ = SOS/DF) of the model calculations was quite high (>50) implying inadequacy in the chemical model. It is clear,

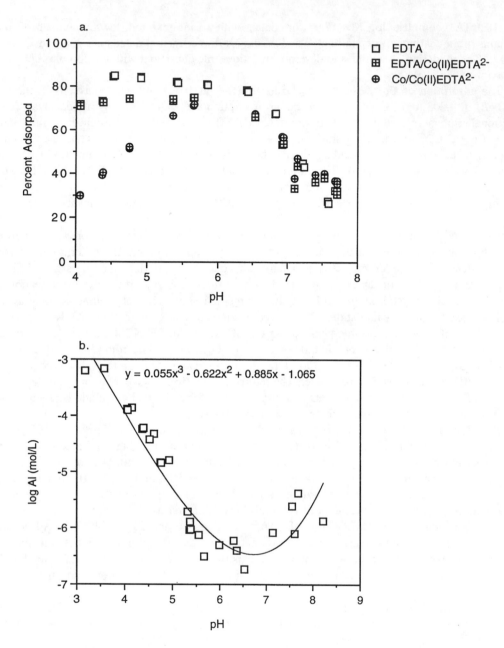

Figure 6   a) Adsorption of free EDTA^{4-} and a Co(II)EDTA^{2-} on a subsurface sand with Fe oxide grain coatings (Oyster 1.5 m). Dual label counting allowed analysis of the sorption behavior of Co and EDTA added as the Co(II)EDTA^{2-} complex. b) Concentrations of aqueous Al in sorption experiments of EDTA^{4-} and Co(II)EDTA^{2-} on the Oyster 1.5-m sediment.

for example, that the disagreement between the model and the Co(II)EDTA^{2-} experiment over the pH range 4.75 to 6.5 (Figure 8) results from our inability to describe the sorption behavior of the free EDTA^{4-} complex below a pH of 7 in Figure 7. Other reactions not considered, such as oxidation of Co(II)EDTA^{2-} to the more stable, weakly sorbing Co(III)EDTA^{-} complex,[119,158] may also have been significant. Thus, the experimentalist is driven to (1) require direct analytical identification of EDTA^{4-} species in the soil suspension (e.g., Co(III)EDTA^{-}, AlEDTA^{-}, FeEDTA^{-}), and (2) perform sorption measurements to define the interaction of these species with the solid phase.

Table 16   Reactions and Equilibrium Constants for Modeling EDTA and Co(II)EDTA^{2-} Adsorption to Subsoil

		log K (I = 0, 25°C)
1.	EDTA^{4-} + H$^+$ = HEDTA^{3-}	11.03
2.	EDTA^{4-} + 2H$^+$ = H$_2$EDTA^{2-}	17.78
3.	EDTA^{4-} + 3H$^+$ = H$_3$EDTA$^-$	20.89
4.	EDTA^{4-} + 4H$^+$ = H$_4$EDTA	23.10
5.	EDTA^{4-} + Ca^{2+} = CaEDTA^{2-}	12.32
6.	EDTA^{4-} + Ca^{2+} + H$^+$ = CaHEDTA$^-$	15.93
7.	EDTA^{4-} + Al^{3+} = AlEDTA$^-$	19.07
8.	EDTA^{4-} + Al^{3+} + H$^+$ = AlHEDTA	21.78
9.	EDTA^{4-} + Al^{3+} + H$_2$O = AlOHEDTA^{2-} + H$^+$	12.81
10.	EDTA^{4-} + Al^{3+} + 2H$_2$O = Al(OH)$_2$EDTA^{3-} + 2H$^+$	2.20
11.	Al^{3+} + H$_2$O = AlOH^{2+} + H$^+$	−4.99
12.	Al^{3+} + 2H$_2$O = Al(OH)$_2^+$ + 2H$^+$	−10.10
13.	Al^{3+} + 3H$_2$O = Al(OH)$_3$ + 3H$^+$	−16.00
14.	Al^{3+} + 4H$_2$O = Al(OH)$_4^-$ + 4H$^+$	−23.00
15.	EDTA^{4-} + Co^{2+} = CoEDTA^{2-}	17.97
16.	EDTA^{4-} + Co^{2+} + H$^+$ = CoHEDTA$^-$	21.40
17.	Co^{2+} + H$_2$O = CoOH$^+$ + H$^+$	−9.67
18.	FeOH + H$^+$ = FeOH$_2^+$	5.6
19.	FeOH = FeO$^-$ + H$^+$	−11.6
20.	FeOH$_2^+$ + CaEDTA^{2-} = FeOH$_2^+$ − CaEDTA^{2-}	3.38 (fit Figure 7)
21.	FeOH$_2^+$ + AlEDTA$^-$ = FeOH$_2^+$ − AlEDTA$^-$	2.52 (fit Figure 7)
22.	FeOH$_2^+$ + Co(II)EDTA^{2-} = FeOH$_2^+$ − Co(II)EDTA^{2-}	3.27 (fit Figure 8)
23.	FeOH$_2^+$ + Co(II)HEDTA$^-$ = FeOH$_2^+$ − Co(II)HEDTA$^-$	3.61 (fit Figure 8)

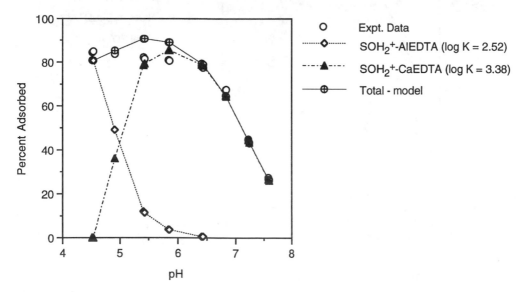

Figure 7   Modeling free EDTA^{4-} (10^{-5} mol/l) sorption to the Oyster 1.5-m sediment. The computed surface speciation approximately followed the aqueous speciation.

3. Metal Ion Sorption by a Kaolinitic Subsoil Using the Log K Spectrum Model

Soil materials contain multiple sorbents and sites that bind metals. Sorption site heterogeneity is difficult to characterize and leads to Freundlich adsorption behavior over ranges in sorbate concentration. Here we describe a semiempirical, log K spectrum approach to model metal ion

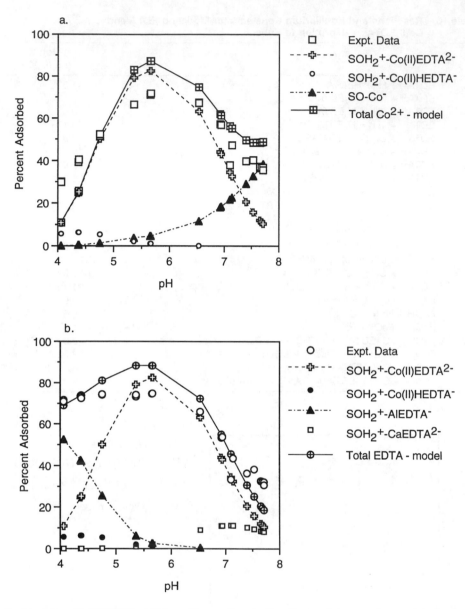

Figure 8   Modeling Co(II)EDTA²⁻ (10⁻⁵ mol/l) sorption to the Oyster 1.5-m sediment. (a) Co added as
Co(II)EDTA²⁻; (b) EDTA added as Co(II)EDTA²⁻. Calculations indicate that increasing Al concentrations
below pH 6 (Figure 6b) induces Co(II)EDTA dissociation. The released Co²⁺ and resulting AlEDTA⁻
sorb independently. The total EDTA sorption reflects contributions of both AlEDTA⁻ and Co(II)EDTA²⁻
adsorption.

binding in soil. The approach embodies implicit recognition of the reaction types involved in metal
ion binding (ion exchange and surface coordination), and is general enough to allow application
to different soil types or heterogeneous environmental complexants. The model is based on mass-
action/material-balance relationships and, like the previous example, does not include electrostatics,
consistent with the belief that heterogeneous particle morphology, surface coatings, and varied
mineralogy preclude defensible and/or realistic computation of such effects in soil. Further details
of the approach and modeling described here may be found in Westall et al.[98] and Wagner et al.[159]
This modeling approach is amenable to large sorbate concentration ranges that defy description by
a single set of Langmuir-type parameters as employed in the previous example.

The modeling objective is to describe Co^{2+} binding to a kaolinitic, clay-sized (<2 μm) separate of an Ultisol subsoil over a range in pH and electrolyte concentration (Figure 9b), and to use the derived parameters to predict Co^{2+} isotherms (Figure 9c) at fixed pH and electrolyte concentration. The Freundlich-type isotherms in Figure 9c preclude a single-site surface complexation approach which would generate a Langmuir isotherm. The experimental procedures for the sorption measurements and mineralogy of the soil are discussed by Zachara et al.[52]

In the log K spectrum model, the sorbent is assumed to contain a series of binding sites that vary systematically in their energy. For this application, four site groups (i) were established (I, II, III, IV). Each site group (I, II, III, IV) contained a surface hydroxyl group (SOH_i) and an ion-exchange site (X_i^-, e.g., structural charge in layer silicates and organic matter) that react with H^+ in the following manner:

$$S_iOH_2^+ = S_iOH + H^+ \qquad K_{a1(i)} \tag{35}$$

$$S_iOH = S_iO^- + H^+ \qquad K_{a2(i)} \tag{36}$$

$$NaX_i + H^+ = HX_i + Na^+ \quad K_{x(i)} \tag{37}$$

These groups were assigned the values of $pK_{a1} = pK_{a2} = -pK_x = 3, 5, 7, 9$ to correspond to the pH range of the data. The site groups are hypothetical, and no attempt is made to justify them from a mineralogic perspective. The material-balance equations for the system were:

$$\left[S_iOH\right]_T = \left[S_iOH\right] + \left[S_iOH_2^+\right] + \left[S_iO^-\right] \tag{38}$$

$$\left[NaX_i\right]_T = \left[NaX_i\right] + \left[HX_i\right] \tag{39}$$

A tableau for this general problem type is given in Westall et al.[98]

To initiate the log K spectrum model, a surface-site/charge model for the kaolinitic separate was established from acid-base titration of KGa-1.[50] KGa-1 was used as a surrogate for the subsoil because Al^{3+} dissolution was less for KGa-1 than for the subsoil, and a more accurate proton balance could be achieved. The total concentrations of the different sites (SOH_i, X_i^-) within each group (I, II, III, IV) were determined by fitting the titration data for KGa-1 in Figure 9a with the equilibrium constants for reactions 33 to 35 fixed as noted above. The model fit the data well, and the resulting site concentrations for the different groups are given in Table 17.

The hydroxyl and ion-exchange type sites were allowed to react with Co^{2+} according to the following reactions:

$$Co^{2+} + 2S_iOH = Co(S_iO^-)_2 + 2H^+ \qquad K_{Co(i)} \tag{40}$$

$$Co^{2+} + 2NaX_i = CoX_2 + 2Na^+ \qquad K_{CoX(i)} \tag{41}$$

The existence of these general reaction types is defensible, given the dependence of Co^{2+} sorption on both pH and Na^+ concentration; Figure 9b and Zachara et al.[52] The equilibrium constants for these reactions on each of the site groups were obtained by fitting the sorption edge data for Co^{2+} on the kaolinitic subsoil at two ionic strengths (Figure 9b). The log K spectrum was used as before, and the site densities of both hydroxyls and ion-exchange sites on the subsoil (i.e., nmol/m²) were

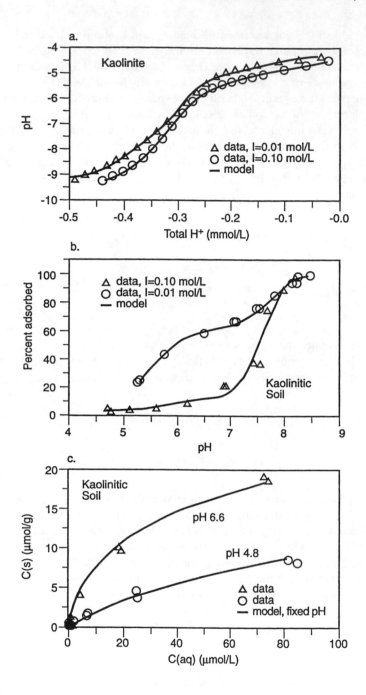

Figure 9 Binding of Co(II) to a kaolinitic subsoil (CP) as a function of pH, [NaClO$_4$], and [Co]$_T$. (a) Acid- base titration of KGa-1. (b) Co(II) adsorption on CP as function of pH ([Co]$_T$ = 1 μmol/l). (c) Co(II) isotherms on CP in 0.01 mol/l NaClO$_4$.

assumed to equal that of kaolinite. The actual site concentrations for the subsoil were computed as the product of the site densities for groups I to IV and the measured N$_2$ surface area of subsoil. The measured Na-CEC of KGa-1 at pH 6.5 was found to scale to the subsoil in a similar way. The resulting adsorption constants are given in Table 17. The model fit to the data was better than that in Figure 8, as a result of the greater number of adjustable parameters.

Table 17   Log K Spectrum Constants for Co Adsorption by Kaolinitic Subsoil in NaClO$_4$

Group	pKa, log K$_x$	[S$_i$OH] mmol/m^2	[NaX$_i$] mmol/m^2	log K$_{Co}$	log K$_{CoX}$
I	3	4,251	870	−0.4	4.8
II	5	281	57	−1.4	6.7
III	7	115	295	−2.8	
IV	9	134	2,534	−1.1	

The model as defined above from Figures 9a and 9b with the parameters in Table 17 was applied to the isotherms in Figure 9c, with no adjustment. The prediction, represented by the solid line, agreed well with the data at both pH values. The agreement is excellent, considering that the total cobalt concentration in the experiments in Figure 9b was 1 μmol/l and that in Figure 9c ranged between 0.1 and 100 μmol/l. We offer this log K spectrum approach for complex environmental complexants, such as soils, that are too complicated to be represented by detailed, sorbent-specific, mechanistic models. Indeed, a semiempirical approach as shown here has both interpolative and extrapolative value, and requires no especially detailed characterization of the solid phase.

D. Alternative Approaches

Throughout this chapter we have emphasized discrete site approaches to modeling adsorption because such calculations are readily performed with chemical models frequently used in soil chemistry (e.g., GEOCHEM, MINTEQ, MINEQL, FITEQL). Differential equilibrium function and site affinity distribution models offer an alternative approach for adsorption calculations with heterogeneous sorbents.[160-165] These models have seen little application to soil materials, in part because the determination of the affinity distribution is not trivial, and is ill-posed (i.e., comparable data fits can be obtained with different distributions).[166]

V. SUCCESSES AND CHALLENGES

A. Merits and Limitations

Site binding models are a primary tool for interpreting ion adsorption observations in single mineral suspensions, with applications ranging from computing surface equilibrium constants to testing hypotheses regarding the interfacial structure of surface complexes. In soil chemical studies, however, site binding models have been used in different ways because of the complex mineralogic and physicochemical properties of soil. They have been used to facilitate understanding of adsorption processes in soil, particularly for computation of the potential contributions of different sorbent types or surface functional groups to sorbate retention. Site binding models have been used to show conformance of adsorption data to a hypothesized reaction of certain stoichiometry and component concentrations, to compute competitive sorbate interactions, to integrate the complex effects of aqueous and surface equilibria, and to evaluate the effects of variable change (e.g., pH, ion composition, ionic strength) on adsorbed and aqueous concentrations. Derived model parameters also provide the basis for comparisons of reactivity between different soil materials, and for both interpolative and extrapolative computations.

Adsorption models have been and will continue to be used descriptively in soil chemical studies. Descriptive applications, which dominate Tables 11 and 12, involve fitting the adsorption data. Curve fitting is a necessity in the analysis of adsorption data for heterogeneous materials whose complexity precludes direct measurement of all required descriptive parameters. The inability to fit a data set given a hypothesized set of mass-action relationships, or the return of chemically unreasonable parameters, are useful outcomes in that they reaffirm that scientific understanding of responsible reactions or the properties of the sorbing phase are inadequate. Important research questions arise from such apparent "failures." It is a fact, however, that adsorption modeling of soil has suffered from the following general inadequacies: (1) computed sorbate adsorption edges are steeper than experimentally observed, and (2) fitted SC-EDL or SC-NEM adsorption constants (co-optimized K_+, K_-, and K_L) are unreasonable given the likelihood of certain dominant sorbent phases (i.e., Fe and Al oxides). On the positive side, the observation that site binding models can generally fit ion adsorption data in soils affirms that anion and cation adsorption can be described by mass-action and material-balance equations.

Calculations where soil property measurements are used with equilibrium constants for analog phases (e.g., Fe oxides, layer silicates) to predict aqueous and adsorbed concentrations over variable space (e.g., pH, sorbate concentration) have been attempted with variable success (see for example, Zachara et al.,[147] Zachara and Smith,[48] and Kent et al.[107]). Two such calculations are shown in Figure 10. In the first (Figure 10a), Cd^{2+} adsorption on a smectitic, clay-sized separate (<2 μm) from a Mollisol C horizon was predicted using ion-exchange/edge complexation constants for the dithionite-citrate-bicarbonate/H_2O_2-extracted soil smectite, its content of extractable Fe(III), and its CEC at the pH of each sorption measurement. The sizable disparity between prediction and experiment may result from the contribution of organic matter (0.73 mass %) that was not included explicitly in the calculation. The second example (Figure 10b) shows the predicted adsorption of SeO_4^{2-} and SeO_3^{2-} on a clay-sized separate (<2 μm) of an Ultisol C horizon, which contains kaolinite, goethite, and a partially chloritized vermiculite. Adsorption reactions on both kaolinite and goethite were considered using the TLM and binding constants (inner sphere for SeO_3^{2-} and outer sphere for SeO_4^{2-}) for the analog phases. The site concentration of kaolinite (mol/g) was estimated from the surface area of the dithionite-citrate-bicarbonate-treated isolate {i.e., $[AlOH]_T = N_{s(AlOH)}$(kaolinite) $* s$ (of the separate in m²/g)}. The site concentration of goethite ($[FeOH]_T$) was estimated as the difference in the measured isotherm sorption maxima (S_{max}, mol/g) for either SeO_4^{2-} or SeO_3^{2-} and the computed site concentration of kaolinite {i.e., $[FeOH]_T = S_{max} - [AlOH]_T$}. The predictions for both Se valence forms were good and demonstrated that AlOH sites associated with kaolinite were plausible locations for outer-sphere SeO_4^{2-} complexation, while the goethite surface was favored for inner-sphere complexation of SeO_3^{2-}.

As evident from Figure 10, the need to characterize the identity as well as concentration of the different reactive sites is the greatest challenge to predictive calculations. The pedogenic process creates reactive phases in soil with a continuum in structure and composition that may defy description by chemical parameters determined for fixed-composition, well-behaved analog sorbents. Unlike thermodynamic constants for solid-phase solubility reactions which apply to diverse soil chemical environments, equilibrium constants for surface reactions on analog sorbents may not directly apply to their soil counterparts as a result of compositional variations, structural disorder, coreacted organic material, and various other factors. For example, the presence of small amounts of coreacted silica in Fe oxides increases the net acidity of the surface and reduces their sorptive affinity for anionic sorbates.[167] Similarly, Al^{3+} may interact with the surface of Fe oxides and change its surface properties.[75,152,153] Soil goethites are widely substituted with Al^{3+} (e.g., Trolard and Tardy,[168] Schwertmann and Carlson[169]) and such substitution may impact their sorption behavior (e.g., Ainsworth et al.[170]). The effects of such substitutions have been implicated in sorption measurements with subsurface materials.[107,117,119,144] The presence of organic matter is a general complication in that small mass quantities may alter mineral surface properties and metal cation reactivity in complex ways.[52,171-173] The power and appeal of prediction, however, continues to drive research on this topic, and the development of sensitive new surface analytical instrumentation may provide requisite information on surface composition and reactivity.

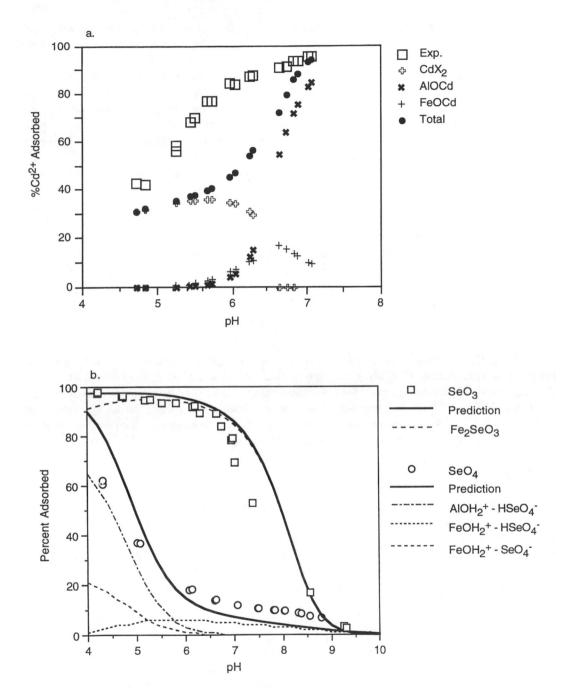

Figure 10 (a) Predicted adsorption of 1 μmol/l Cd^{2+} on a smectitic soil clay separate of a Mollisol C horizon. Computation included ion exchange (CdX_2), coordination to smectite edge sites (AlOCd by the TLM), and binding to Fe-oxide coatings (FeOCd by the SC-NEM). From Zachara, J. M. and Smith, S. C., *Soil Sci. Soc. Am. J.,* 58, 762, 1994. With permission.) (b) Predicted adsorption of 1 μmol/l SeO_3^{2-} and SeO_4^{2-} on an Ultisol C horizon, clay-sized separate. Adsorption allowed to two surface sites parameterized using TLM surface equilibrium constants for kaolinite and goethite. From Zachara, J. M. et al., *Chemical Attenuation Reactions of Selenium,* Electric Power Research Institute, Palo Alto, CA, 1993. With permission.)

B. Mechanistic vs. Empirical Models

Soils represent a great challenge for adsorption modeling because of their heterogeneity and complex composition. Both mechanistic and empirical modeling approaches have merit. Their choice and application is a matter of objective, taste, and intended final application. The level of empiricism, regardless of approach, increases with the complexity of the sorbent. In practicality, the mechanistic approach is not as fundamental as it may appear initially, because it invariably requires assumptions about the identity and properties of the sorbing phase. Also, the electrostatic models often used in a mechanistic approach may not be defensible for soil because the complex surface structure of soil sorbents does not conform to the simple geometry, such as a plane or sphere, embodied in the double-layer corrections of SC-EDL models.

Site binding models are one class of models available to the soil chemist and are the most amenable to solution with existing chemical speciation codes. A variety of other model types, especially for cation exchange computations,[38,43,174] are also available. Isotherm models (e.g., Kinniburgh[87]) are also used extensively, and suites of empirical and mechanistic models have been developed for proton and ion equilibria on humic substances.[127-130,133,160,175] Bolt and Van Riemsdijk[176] proposed that the spatial heterogeneity of soils in the field, combined with the observation that soil pH is typically fairly uniform in a given one-dimensional profile, calls for simple, empirical chemical interaction models, such as the power exchange function for metal ions (M):

$$S_M = K[M]^a [H]^b \tag{42}$$

where S_M is the adsorbed metal concentration, [M] and [H] are the equilibrium concentrations of the metal and the hydrogen ion, and a and b are empirical constants. Mass-action concepts, however, do not have to be discarded to deal with reactive transport and field heterogeneity, as shown by Kent et al.[107,177] Multiple-site Langmuir adsorption combined with limited aqueous speciation has been linked to multidimensional water-flow codes and has been used to describe reactive transport through physically and chemically heterogeneous domains.[178]

C. Applications and Future Directions

Improvements are being made to codes, such as FITEQL, to solve larger chemical equilibrium problems, to facilitate convergence, and to more readily allow analysis of multiple data sets. A generalized kinetic capability has been incorporated into a chemical equilibrium model capable of performing aqueous speciation, solubility, and adsorption computations (SC-NEM and half-reaction exchange).[70] Modern numerical procedures are being tested for their ability to extract information on site distributions and reactivity from adsorption isotherm data on heterogeneous sorbents including soils.[166]

Advances in the understanding of surface chemistry and adsorption reactions in soil will improve prospects for semiquantitative site binding computations. Important needs include investigations of the surface chemical properties of soil particle coatings, the reactivity of complex mineral organic matter associations, the surface speciation of adsorbed constituents, and the phase distribution of sorbed species in composite materials. Evolving information on the chemical structure and identity of surface complexes on reactive soil constituents derived from such techniques as X-ray absorption spectroscopy can be used to constrain model calculations in soil to chemically realistic surface species. While the role of organic matter in cation adsorption/complexation is well acknowledged and models of varying levels of sophistication exist, approaches to deal with its contribution to ion adsorption in composite mineral materials require development.

Increased application of site binding models to soil problems will enhance awareness of the mineralogic, surface chemical, and material-balance data needed for comprehensive modeling, and hasten its development. Additional studies are needed on charge distribution in soil (i.e., σ_H and σ_o; Charlet and Sposito,[120] Anderson and Sposito,[121] and Chorover and Sposito[123]) that are explicitly linked to (1) ion retention measurements, (2) minor cation or anion sorption behavior, and (3) detailed chemical/mineralogic characterization of particle surfaces. While they are time consuming to collect, the importance of such broad spectrum data sets for model formulation, parameter estimation, and calibration/predictive activities cannot be overestimated. Comprehensive studies are needed for testing and evaluating different modeling approaches and for identifying character-ization measurements for successful model application.

Site binding models are ready for broader applications in soil chemistry, with appropriate recognition of their shortcomings. These models are commonly used in experimental design to compute effects of sorbent and sorbate concentrations, pH, and competing solutes, and for both qualitative and quantitative interpretation of data. Continued application to soil chemical data, as shown in the examples, is important to test and refine variants of the mechanistic and empirical approaches, to evaluate their relative merits for different final applications, and to allow systematic comparisons of the reactivity of materials with different properties. Simplified site binding models (i.e., models without electrostatics and a minimal component set) are the algorithms of choice for linkage with water flow codes to calculate reactive transport influenced by adsorption. Indeed, several such codes have taken this approach, and others are under development.[69,112,178] Site binding models, as embodied in speciation codes such as MINTEQA2,[44] are seeing increased use in environmental assessment and cleanup to compute aqueous and sorbed distributions of contami-nants, and to evaluate how they may change with management or manipulation.

VI. CONCLUSIONS

Site-binding models describe adsorption via mass-action and material-balance equations on discrete surface functional groups and have become indispensable tools for describing solute binding to mineral and biotic surfaces. Their application to soils has shown promise, but the models have been challenged by the heterogeneous chemical nature of these materials. Difficulties in character-izing (1) the surface reaction and identity of the adsorbed species, (2) the concentrations of reactive sites, and (3) the distribution of these sites between sorbing phases limits the use of these models for soil materials to descriptive applications. Future advances in surface characterization tools will improve prospects for mechanistic modeling through identification of discrete surface sites involved in adsorption equilibria. Until such advances become reality, however, empirical approaches and assumptions will prevail. Considerable need for innovation exists in the development of appropriate empirical and semiempirical approaches to model adsorption reactions in soil.

ACKNOWLEDGMENTS

This research was supported by the Subsurface Science Program, Office of Health and Envi-ronmental Research, U.S. Department of Energy (DOE) as part of its co-contaminant chemistry research. Pacific Northwest Laboratory is operated for DOE by Battelle Memorial Institute under Contract DE-AC06-76RLO 1830. The authors thank Gary Turner and Julie Wagner for the modeling examples described herein, Sonia Enloe for the text processing, and two anonymous reviewers for helpful comments that improved the manuscript.

REFERENCES

1. Stumm, W., *Chemistry of the Solid-Water Interface*, John Wiley & Sons, New York. 1992.
2. Brown, G. E., Spectroscopic studies of chemisorption reaction mechanisms at oxide-water interfaces, in *Mineral-Water Interface Geochemistry*, Hochella, M. F. and White, A. F., Eds., Mineralogical Society of America, Washington, D.C., 1990, 309.
3. Dzombak, D. A. and Morel, F. M. M., *Surface Complexation Modeling: Hydrous Ferric Oxide*, John Wiley & Sons, New York, 1990.
4. Davis, J. A. and Kent, D. B., Surface complexation modeling in aqueous geochemistry, in *Mineral-Water Interface Geochemistry*, Hochella, M. F. and White, A. F., Eds., Mineralogical Society of America, Washington, D.C., 1990, 177.
5. Sposito, G., *The Surface Chemistry of Soils*, Oxford University Press, New York, 1984.
6. Sposito, G., *The Chemistry of Soils*, Oxford University Press, New York, 1989.
7. Greenland, D. J. and Hayes, M. H. B., *The Chemistry of Soil Constituents*, John Wiley & Sons, New York, 1987.
8. Sposito, G., Surface reactions in natural aqueous colloidal systems, *Chimica*, 43, 169, 1989.
9. Schindler, P. W. and Sposito, G., Surface complexation at (hydr)oxide surfaces, in *Interactions at the Soil Colloid-Soil Solution Interface, NATO ASI Ser.*, Bolt, G. H. et al., Eds., Kluwer Academic, Dordrecht, 1991, 115.
10. Westall, J. C., Adsorption mechanisms in aquatic chemistry, in *Aquatic Surface Chemistry*, Stumm, W., Ed., Wiley Interscience, New York, 1987, 3.
11. Bolt, G. H. and Van Riemsdijk, W. H., The electrified interface of the soil solid phase. A. The electrochemical control system, in *Interactions at the Soil Colloid-Soil Solution Interface, NATO ASI Ser.*, Bolt, G.H. et al., Ed., Kluwer Academic, Dordrecht, 1990, 37.
12. Hiemstra, T., van Riemsdijk, W. H., and Bruggenwert, M. G. M., Proton adsorption mechanism at the gibbsite and aluminum oxide solid/solution interface, *Neth. J. Agric. Sci.*, 35, 281, 1987.
13. Hiemstra, T., van Riemsdijk, W. H., and Bolt, G. H., Multi-site proton adsorption modeling at the solid/solution interface of (hydr)oxides: a new approach. I. Model description and evaluation of intrinsic reaction constants, *J. Colloid Interface Sci.*, 133, 91, 1989.
14. Hiemstra, T., DeWit, J. C. M., and van Riemsdijk, W. H., Multi-site proton adsorption modeling at the solid/solution interface of (hydr)oxides: a new approach. II. Application to various important (hydr)oxides, *J. Colloid Interface Sci.*, 133, 105, 1989.
15. Barron, V., Herrizo, M., and Torrent, J., Phosphate adsorption by aluminous hematites of different shapes, *Soil Sci. Soc. Am. J.*, 52, 547, 1988.
16. White, G. N. and Zelazny, L., Analysis and implications of the edge structure of dioctahedral phyllosilicates, *Clays Clay Miner.*, 36, 141, 1988.
17. Bleam, W. F., Welhouse, G. J., and Janowiak, M. A., The surface coulomb energy and proton coulomb potentials of pyrophyllite {010}, {110}, {100}, and {130} edges, *Clays Clay Miner.*, 41, 305, 1993.
18. Colombo, C., Barron, V., and Torrent, J., Phosphate adsorption and desorption in relation to morphology and crystal properties of synthetic hematites, *Geochem. Cosmochim. Acta*, 58, 1261, 1994.
19. Hayes, K. F., Roe, A. L., Brown, G. E., Hodgson, K. O., Leckie, J. O., and Parks, G. A., In-situ X-ray adsorption study of surface complexes: selenium oxyanions on a-FeOOH, *Science*, 238, 783, 1987.
20. Zhang, P. C. and Sparks, D. L., Kinetics and mechanisms of sulfate adsorption/desorption on goethite using pressure-jump relaxation, *Soil Sci. Soc. Am. J.*, 54, 1266, 1990.
21. Zhang, P. C. and Sparks, D. L., Kinetics of selenate and selenite adsorption/desorption at the goethite/water interface, *Environ. Sci. Technol.*, 24, 1848, 1990.
22. Hayes, K. F., Papelis, C., and Leckie, J. O., Modeling ionic strength effects on anion adsorption at hydrous oxide/solution interfaces, *J. Colloid Interface Sci.*, 125, 717, 1988.
23. Charlet, L. and Sposito, G., Bivalent ion adsorption by an oxisol, *Soil Sci. Soc. Am. J.*, 53, 691, 1989.
24. Cowan, C. E., Zachara, J. M., and Resch, C. T., Cadmium adsorption on iron oxides in the presence of alkaline-earth, *Environ. Sci. Technol.*, 25, 437, 1991.
25. O'Day, P. A., Parks, G. A., and Brown, G. E., Jr., Molecular structure and binding sites of cobalt(II) surface complexes on kaolinite from X-ray absorption spectroscopy, *Clays Clay Miner.*, 42, 337, 1994.
26. Chisholm-Brause, C. J., Day, P. A., Brown, G. E., Jr., and Parks, G. A., Evidence for multinuclear metalion complexes at solid/water interfaces from X-ray adsorption spectroscopy, *Nature*, 348, 528, 1990.

27. Chisholm-Brause, C. J., Hayes, K. F., Roe, A. L., Brown, G. E., Jr., Parks, G. A., and Leckie, J. O., Spectroscopic investigation of Pb(II) complexes at the gamma-Al_2O_3/water interface, *Geochim. Cosmochim. Acta*, 54, 1897, 1990.
28. Manceau, A., Charlet, L., Boisset, M. C., Didier, B., and Spadini, L., Sorption and speciation of heavy metals on Fe and Mn hydrous oxides. From microscopic to macroscopic, *Appl. Clay Sci.*, 7, 201, 1992.
29. Coombs, J. M., Chisholm-Brause, C. J., Brown, G. E., Jr., Parks, G. A., Conradson, S. D., Eller, P. G., Triay, I. R., Hobart, D. E., and Meijer, A., EXAFS spectroscopy study of neptunium(V) sorption at the a-FeOOH/water interface, *Environ. Sci. Technol.*, 26, 376, 1992.
30. Waychunas, G. A., Rea, B. A., Fuller, C. C., and Davis, J. A., Surface chemistry of ferrihydrite. I. EXAFS studies of the geometry of co-precipitated and adsorbed arsenate, *Geochim. Cosmochim. Acta*, 57, 2251, 1993.
31. Manceau, A. and Charlet, L., The mechanism of selenate adsorption on goethite and hydrous ferric oxide, *J. Colloid Interface Sci.*, 168, 87, 1994.
32. Manceau, A., The mechanism of anion adsorption on iron oxides. Evidence for the bonding of arsenate tetrahedra on free $Fe(O,OH)_6$ edges, *Geochim. Cosmochim. Acta*, 59, 3647, 1995.
33. Waychunas, G. A., Davis, J. A., and Fuller, C. C., Geometry of sorbed arsenate on ferrihydrite and crystalline FeOOH: Re-evaluation of EXAFS results and topological factors in predicting sorbate geometry, and evidence for monodentate complexes, *Geochim. Cosmochim. Acta*, 59, 3655, 1995.
34. Roe, A. L., Hayes, K. F., Chisholm-Brause, C., Brown, G. E., Jr., Parks, G. A., Hodgson, K. O., and Leckie, J. O., In situ X-ray adsorption study of lead ion surface complexes at the goethite-water interface, *Langmuir*, 7, 367, 1991.
35. O'Day, P. A., Brown, G. E., Jr., and Parks, G. A., X-ray absorption spectroscopy of cobalt(II) multinuclear surface complexes and surface precipitates on kaolinite, *J. Colloid Interface Sci.*, 165, 269, 1994.
36. Westall, J. C. and Hohl, H., A comparison of electrostatic models for the oxide/solution interface, *Adv. Colloid Interface Sci.*, 12, 265, 1980.
37. Westall, J. C., Chemical equilibrium including adsorption on charged surfaces, in *Particulates in Water*, Kavanaugh, M. C. and Leckie, J. O., Eds., American Chemical Society, Washington, D.C., 1980, 33.
38. Sposito, G., *The Thermodynamics of Soil Solutions*, Oxford University Press, New York, 1981.
39. Sposito, G., The Gapon and Vanselow selectivity coefficients, *Soil Sci. Soc. Am. J.*, 41, 1205, 1977.
40. Sposito, G. and Mattigod, S. V., Ideal behavior in Na-trace metal cation exchange on Camp Bertaux montmorillonite, *Clays Clay Miner.*, 27, 125, 1979.
41. Fletcher, P. and Sposito, G., The chemical modeling of clay/electrolyte interactions for montmorillonite, *Clay Miner.*, 24, 375, 1989.
42. Shaviv, A. and Mattigod, S. V., Cation exchange equilibria in soils expressed as a cation-ligand complex formation, *Soil Sci. Soc. Am. J.*, 49, 569, 1985.
43. Talibudeen, O., Cation exchange in soils, in *The Chemistry of Soil Processes,* Greenland, D. J., and Hayes, M. H. B., Eds., John Wiley & Sons, Chichester, U.K., 1981, 115.
44. Allison, J. D., Brown, D. S., and Novo-Gradac, K. J., *MINTEQA2/PRODEFA2. A Geochemical Assessment Model for Environmental Systems: Version 3.0 User's Manual*, U.S. Environmental Protection Agency, Athens, Georgia, 1991.
45. Stadler, M. and Schindler, P. W., Modeling of H^+ and Cu^{2+} adsorption on calcium-montmorillonite, *Clays Clay Miner.*, 41, 288, 1993.
46. Krupka, K. M., Erikson, R. L., Mattigod, S. V., Schramke, J. A., and Cowan, C. E., *Thermochemical Data Used by the FASTCHEM Package*, Electric Power Research Institute, Palo Alto, CA, 1988.
47. Stadler, M. and Schindler, P. W., The effect of dissolved ligands on the sorption of Cu(II) by Ca-montmorillonite, *Clays Clay Miner.*, 42, 148, 1994.
48. Zachara, J. M. and Smith, S. C., Edge complexation reactions of cadmium on specimen and soil-derived smectite, *Soil Sci. Soc. Am. J.*, 58, 762, 1994.
49. Zachara, J. M. and McKinley, J. P., Influence of hydrolysis on the sorption of metal cations by smectites: importance of edge coordination reactions, *Aquatic Sci.*, 55, 250, 1993.
50. Schindler, P. W., Liechti, P., and Westall, J. C., Adsorption of copper, cadmium, and lead from aqueous solution to the kaolinite/water interface, *Neth. J. Agric. Sci.*, 35, 219, 1987.
51. Cowan, C. E., Zachara, J. M., Smith, S. C., and Resch, C. T., Individual sorbent contributions to cadmium sorption on ultisols of mixed mineralogy, *Soil Sci. Soc. Am. J.*, 56, 1074, 1992.
52. Zachara, J. M., Resch, C. T., and Smith, S. C., Influence of humic substances on Co^{2+} sorption by a subsurface mineral separate and its mineralogic components, *Geochim. Cosmochim. Acta*, 58, 553, 1994.

53. Cowan, C. E., Zachara, J. M., and Resch, C. T., Solution ion effects on the surface exchange of selenite on calcite, *Geochim. Cosmochim. Acta,* 54, 2223, 1990.

54. Zachara, J. M., Cowan, C. E., and Resch, C. T., Sorption of divalent metals on calcite, *Geochim. Cosmochim. Acta,* 55, 1549, 1991.

55. Schiewer, S. and Volesky, B., Modeling of the proton-metal ion exchange in biosorption, *Environ. Sci. Technol.,* 29, 3049, 1995.

56. Goldberg, S., Use of surface complexation models in soil chemical systems, *Adv. Agron.,* 47, 233, 1992.

57. Morel, F. M. M., Yeasted, J. G., and Westall, J. C., Adsorption models: a mathematical analysis in the framework of general equilibrium calculations, in *Adsorption of Inorganics at Solid- Liquid Interfaces,* Anderson, M. A. and Rubin, A. J., Eds., Ann Arbor Science, Ann Arbor, MI, 1981, 263.

58. Dzombak, D. A. and Morel, F. M. M., Adsorption of inorganic pollutants in aquatic systems, *J. Hydraul. Eng.,* 113, 430, 1987.

59. Hayes, K. F., Redden, G., Ela, W., and Leckie, J. O., Surface complexation models: an evaluation of parameter estimation using FITEQL and oxide mineral titration data, *J. Colloid Interface Sci.,* 142, 448, 1991.

60. Van Cappellen, C. L., Stumm, W., and Wersin, P., A surface complexation model of the carbonate mineral-aqueous solution interface, *Geochim. Cosmochim. Acta,* 57, 3505, 1993.

61. Wu, L., Forsling, W., and Schindler, P. W., Surface complexation of calcium minerals in aqueous solution. I. Surface protonation at fluorapatite-water interfaces, *J. Colloid Interface Sci.,* 147, 178, 1991.

62. Katz, L. E. and Hayes, K. F., Surface complexation modeling. II. Strategy for modeling polymer and precipitation reactions at high surface coverage, *J. Colloid Interface Sci.,* 170, 491, 1995.

63. Johnston, C. T. and Sposito, G., Disorder and early sorrow: progress in the chemical speciation of soil surfaces, in *Future Developments in Soil Science Research,* Boersma, L. L., Ed., Soil Science Society of America, Madison, WI, 1987, 89.

64. Goldberg, S., Sensitivity of surface complexation modeling to the surface site density parameter, *J. Colloid Interface Sci.,* 145, 1991.

65. Huang, C. P. and Stumm, W., Specific adsorption of cations on hydrous γ-Al_2O_3, *J. Colloid Interface Sci.,* 43, 409, 1973.

66. Schindler, P. W., Furst, B., Dick, R., and Wolf, P. U., Ligand properties of surface silanol groups. I. Surface complex formation with Fe^{3+}, Cu^{2+}, Cd^{2+}, and Pb^{2+}, *J. Colloid Interface Sci.,* 55, 469, 1976.

67. Morel, F. M. M., *Principles of Aquatic Chemistry,* John Wiley & Sons, New York, 1983.

68. Zachara, J. M., Smith, S. C., McKinley, J. P., and Resch, C. T., Cadmium sorption on specimen and soil smectites in sodium and calcium electrolytes, *Soil Sci. Soc. Am. J.,* 57, 1491, 1993.

69. Yeh, G. T. and Tripathi, V. S., A model for simulating transport of reactive multispecies components: model development and demonstration, *Water Resour. Res.,* 27, 3075, 1991.

70. Yeh, G. T., Iskra, G., Zachara, J. M., Szecsody, J. E., and Streile, G., *KEMOD: A Mixed Chemical Kinetic and Equilibrium Model of Aqueous and Solid Phase Geochemical Reactions,* Pacific Northwest National Laboratory, Richland, WA, 1995.

71. Davis, J. A. and Hem, J. D., The surface chemistry of aluminum oxides and hydroxides, in *The Environmental Chemistry of Aluminum,* Sposito, G.A., Ed., CRC Press, Boca Raton, FL, 1989, 185.

72. James, R. O. and Parks, G. A., Characterization of aqueous colloids by their electrical double-layer and intrinsic surface chemical properties, in *Surface and Colloid Science,* Matijevic, E., Ed., 1982, 119.

73. Sigg, L. and Stumm, W., The interaction of anions and weak acids with the hydrous goethite (a-FeOOH) surface, *Colloids Surfaces,* 2, 101, 1981.

74. Hingston, F. J., Atkinson, R. J., Posner, A. M., and Quirk, J. P., Specific adsorption of anions on goethite, *9th Int. Cong. Soil Sci. Trans.,* 1, 669, 1968.

75. Lovgren, L., Sjoberg, S., and Schindler, P. W., Acid base reactions and Al(III) complexation at the surface of goethite, *Geochim. Cosmochim. Acta,* 54, 1301, 1990.

76. Goldberg, S. and Sposito, G., Chemical model of phosphate adsorption by soils. II. Noncalcareous soils, *Soil Sci. Soc. Am. J.,* 48, 779, 1984.

77. Wieland, E. and Stumm, W., Dissolution kinetics of kaolinite in acidic aqueous solutions at 25°C, *Geochim. Cosmochim. Acta,* 56, 3339, 1992.

78. Koopal, L. K., van Riemsdijk, W. H., and Roffey, M. G., Surface ionization and complexation models: a comparison of methods for determining model parameters, *J. Colloid Interface Sci.,* 118, 117, 1987.

79. Sposito, G., deWit, J. C. M., and Neal, R. H., Selenite adsorption on alluvial soils. III. Chemical modeling, *Soil Sci. Soc. Am. J.*, 52, 947, 1988.

80. Katz, L. E. and Hayes, K. F., Surface complexation modeling. I. Strategy for modeling monomer complex formation at moderate surface coverage, *J. Colloid Interface Sci.*, 170, 477, 1995.

81. Waite, T. D., Davis, J. A., Payne, T. E., Waychunas, G. A., and Xu, N., Uranium(VI) adsorption to ferrihydrite. Application of a surface complexation model, *Geochim. Cosmochim. Acta*, 58, 5465, 1994.

82. Smith, R. W. and Jenne, E. A., Recalculation, evaluation, and prediction of surface complexation constants for metal adsorption on iron and manganese oxides, *Environ. Sci. Technol.*, 25, 525, 1991.

83. Smith, R. W. and Jenne, E. A., Response to comment on "Recalculation, evaluation, and prediction of surface complexation constants for metal adsorption on iron and manganese oxides", *Environ. Sci. Technol.*, 26, 1253, 1992.

84. Dzombak, D. A. and Hayes, K. F., Comment on "Recalculation, evaluation, and prediction of surface complexation constants for metal adsorption on iron and manganese oxides", *Environ. Sci. Technol.*, 26, 1251, 1992.

85. Benjamin, M. M. and Leckie, J. O., Effects of complexation by Cl, SO_4, and S_2O_3 on the adsorption behavior of cadmium on oxide surfaces, *Environ. Sci. Technol.*, 16, 162, 1982.

86. Kinniburgh, D. G., Barker, J. A., and Whitfield, M., A comparison of some simple isotherms for describing divalent cation adsorption by ferrihydrite, *J. Colloid Interface Sci.*, 95, 370, 1983.

87. Kinniburgh, D. G., General purpose adsorption isotherms, *Environ. Sci. Technol.*, 20, 895, 1986.

88. Dzombak, D. A. and Morel, F. M. M., Sorption of cadmium on hydrous ferric oxide at high sorbate/sorbent ratios: equilibrium, kinetics, and modeling, *J. Colloid Interface Sci.*, 112, 588, 1986.

89. Balistrieri, L. S. and Chao, T. T., Adsorption of selenium by amorphous iron oxyhydroxide and manganese dioxide, *Geochim. Cosmochim. Acta*, 54, 739, 1990.

90. Farley, K. J., Dzombak, D. A., and Morel, F. M. M., A surface precipitation model for the sorption of cations on metal oxides, *J. Colloid Interface Sci.*, 106, 226, 1985.

91. Gunneriusson, L., Lovgren, L., and Sjoberg, S., Complexation of Pb(II) at the goethite (alpha-FeOOH)/water interface: the influence of chloride, *Geochim. Cosmochim. Acta*, 58, 4973, 1994.

92. Westall, J. C. and Morel, F. M. M., *FITEQL: A General Algorithm for the Determination of Metal-Ligand Complex Stability Constants from Experimental Data*, Ralph M. Parson Laboratory, Department of Civil Engineering, Massachusetts Institute of Technology, Cambridge, MA, 1977.

93. Westall, J. C., *FITEQL. A Computer Program for Determination of Equilibrium Constants from Experimental Data, Version 1.2*, Department of Chemistry, Oregon State University, Corvallis, OR, 1982.

94. Westall, J. C., *FITEQL. A Computer Program for Determination of Equilibrium Constants from Experimental Data, Version 2.0*, Department of Chemistry, Oregon State University, Corvallis, OR, 1982.

95. Herbelin, A. L. and Westall, J. C., *FITEQL. A Computer Program for Determination of Chemical Equilibrium Constants from Experimental Data, Version 3.1*, Department of Chemistry, Oregon State University, Corvallis, OR, 1994.

96. Gaizer, F., Computer evaluation of complex equilibria, *Coord. Chem. Rev.*, 27, 195, 1979.

97. Felmy, A. R., *GMIN: A Computerized Chemical Equilibrium Model Using a Constrained Minimization of the Gibbs Free Energy*, Pacific Northwest National Laboratory, Richland, WA, 1990.

98. Westall, J. C., Jones, J. D., Turner, G. D., and Zachara, J. M., Models for association of metal ions with heterogeneous environmental sorbents. I. Complexation of Co(II) by leonardite humic acid as a function of pH and $NaClO_4$ concentration, *Environ. Sci. Technol.*, 29, 951, 1995.

99. Hayes, K. F. and Leckie, J. O., Modeling ionic strength effects on cation adsorption at hydrous oxide/solution interfaces, *J. Colloid Interface Sci.*, 115, 564, 1987.

100. Carroll-Webb, S. A. and Walther, J. V., A surface complexation reaction model for the pH- dependence of corundum and kaolinite dissolution rates, *Geochim. Cosmochim. Acta*, 52, 2609, 1988.

101. Amrhein, C. and Suarez, D. L., The use of a surface complexation model to describe the kinetics of ligand-promoted dissolution of anorthite, *Geochim. Cosmochim. Acta*, 52, 2785, 1988.

102. Zinder, B., Furrer, G., and Stumm, W., The coordination chemistry of weathering. II. Dissolution of Fe(III) oxides, *Geochim. Cosmochim. Acta*, 50, 1861, 1986.

103. Siffert, C. and Sulzberger, B., Light-induced dissolution of hematite in the presence of oxalate: a case study, *Langmuir*, 7, 1627, 1991.

104. Liang, L. and Morgan, J. J., Coagulation of iron oxide particles in the presence of organic materials: application of surface chemical model, in *Chemical Modeling of Aqueous Systems,* Vol. 2, Melchior, D. C. and Bassett, R. L., Eds., American Chemical Society, Washington, D.C., 1990, 293.

105. Xue, H. B., Stumm, W., and Sigg, L., The binding of heavy metals to algal surfaces, *Water Res.,* 22, 917, 1988.

106. Xue, H. B. and Sigg, L., Binding of Cu(II) to algae in a metal buffer, *Water Res.,* 24, 1129, 1990.

107. Kent, D. B., Davis, J. A., Anderson, L. C. D., and Rea, B. A., Transport of chromium and selenium in a pristine sand and gravel aquifer: role of adsorption processes, *Water Resour. Res.,* 31, 1041, 1995.

108. Organization for Economic Cooperation and Development, *Radionuclide Sorption from the Safety Evaluation Perspective,* Organization for Economic Cooperation and Development, Paris, France, 1992.

109. Bradbury, M. H. and Baeyens, B., A general application of surface complexation to modeling radio-nuclide sorption in natural systems, *J. Colloid Interface Sci.,* 158, 364, 1993.

110. Turner, D. R., *Mechanistic Approaches to Radionuclide Sorption Modeling,* Center for Nuclear Waste Regulatory Analyses, San Antonio, TX, 1993.

111. Turner, D. R. and Sassman, S. A., Approaches to sorption modeling for high-level waste performance assessment, in *Fourth International Conference on the Chemistry and Migration Behavior of Actinides and Fission Products in the Geosphere,* Charleston, SC, 1994.

112. Cederberg, G. A., Street, R. L., and Leckie, J. O., A groundwater mass transport and equilibrium chemistry model for multicomponent systems, *Water Resour. Res.,* 21, 1095, 1985.

113. Harter, R. D., *Adsorption Phenomena,* van Nostrand Reinhold, New York, 1986.

114. Morris, D. E., Chisholm-Brause, C. J., Barr, M. E., Conradson, S. D., and Eller, P. G., Optical spectroscopic studies of the sorption of UO_2^{2+} species on a reference smectite, *Geochim. Cosmochim. Acta,* 58, 3613, 1994.

115. Chisholm-Brause, C. J., Conradson, S. D., Buscher, C. T., Eller, P. G., and Morris, D. E., Speciation of uranyl sorbed at multiple binding sites on montmorillonite, *Geochim. Cosmochim. Acta,* 58, 3625, 1994.

116. Jenne, E. A., Trace element sorption by sediments and soils — sites and processes, in *Symposium on Molybdenum in the Environment,* Chappel, W. and Peterson, K., Eds., Marcel-Dekker, New York, 1977, 425.

117. Coston, J. A., Fuller, C. C., and Davis, J. A., Pb^{2+} and Zn^{2+} adsorption by a natural aluminum- and iron-bearing surface coating on an aquifer sand, *Geochim. Cosmochim. Acta,* 59, 3535, 1995.

118. Zachara, J. M., Smith, S. C., and Kuzel, L. S., Adsorption and dissociation of Co-EDTA complexes in Fe oxide-containing subsurface sands, *Geochim. Cosmochim. Acta,* 59, 4825, 1995.

119. Zachara, J. M., Gassman, P. L., Smith, S. C., and Taylor, D., Oxidation and adsorption of Co(II)EDTA^{2-} complexes in subsurface materials with iron and manganese oxide grain coatings, *Geochim. Cosmochim. Acta,* 59, 4449, 1995.

120. Charlet, L. and Sposito, G., Monovalent ion adsorption by an oxisol, *Soil Sci. Soc. Am. J.,* 51, 1155, 1987.

121. Anderson, S. J. and Sposito, G., Proton surface-charge density in soils with structural and pH-dependent charge, *Soil Sci. Soc. Am. J.,* 56, 1437, 1992.

122. Chorover, J. and Sposito, G., *Measurement of Surface Charge Components,* University of California Technical Report, 1993.

123. Chorover, J. and Sposito, G., Surface charge characteristics of kaolinitic tropical soils, *Geochim. Cosmochim. Acta,* 59, 875, 1995.

124. Polubesova, T. A., Chorover, J., and Sposito, G., Surface charge properties of a podzolized soil, *Soil Sci. Soc. Am. J.,* 43, 772, 1995.

125. Anderson, S. J. and Sposito, G., Cesium-adsorption method for measuring accessible structural surface charge, *Soil Sci. Soc. Am. J.,* 55, 1569, 1991.

126. Gregor, J. E. and Powell, H. K. J., Protonation reactions of fulvic acids, *J. Soil Sci.,* 39, 243, 1988.

127. Tipping, E., Reddy, M. M., and Hurley, M. A., Modeling electrostatic and heterogeneity effects on proton dissociation from humic substances, *Environ. Sci. Technol.,* 24, 1700, 1990.

128. Bartschat, B. M., Cabaniss, S. E., and Morel, F. M. M., Oligoelectrolyte model for cation binding by humic substances, *Environ. Sci. Technol.,* 26, 284, 1992.

129. Perdue, E. M. and Lytle, C. R., Distribution model for binding of protons and metal ions by humic substances, *Environ. Sci. Technol.*, 17, 654, 1983.

130. Perdue, E. M., Modeling the acid-base chemistry of organic acids in laboratory experiments and in freshwaters, in *Organic Acids in Aquatic Ecosystems,* Perdue, E. M. and Gjessing, E. T., Eds., John Wiley & Sons, New York, 1990, 111.

131. Bruggenwert, M. G. M. and Kamphorst, A., Survey of experimental information on cation exchange in soil systems, in *Soil Chemistry B. Physico-Chemical Models,* Bolt, G.H., Ed., Elsevier Scientific, Amsterdam, 1979, 141.

132. Stevenson, F. J., *Humus Chemistry — Genesis, Composition, Reactions*, John Wiley & Sons, New York, 1982.

133. Sposito, G., Sorption of trace metals by humic materials in soils and natural waters, *Crit. Rev. Environ. Control*, 16, 193, 1986.

134. Baes, A. U. and Bloom, P. R., Exchange of alkaline earth cations in soil organic matter, *Soil Sci.*, 146, 6, 1988.

135. Parker, J. C., Zelazny, L. W., Sampath, S., and Harris, W. G., Critical evaluation of the extension of zero point of charge (ZPC) theory to soil systems, *Soil Sci. Soc. Am. J.*, 43, 668, 1979.

136. Hendershot, W. H. and Lavkulich, L. M., Effect of sesquioxide coatings on surface charge of standard mineral and soil samples, *Soil Sci. Soc. Am. J.*, 47, 1252, 1983.

137. Duquette, M. and Hendershot, W., Contribution of exchangeable aluminum to cation exchange capacity at low pH, *Can. J. Soil Sci.*, 67, 175, 1987.

138. Duquette, M. and Hendershot, W., Soil surface charge evaluation by back-titration. I. Theory and method development, *Soil Sci. Soc. Am. J.*, 57, 1222, 1993.

139. Duquette, M. and Hendershot, W., Soil surface charge evaluation by back-titration. II. Application, *Soil Sci. Soc. Am. J.*, 57, 1228, 1993.

140. Balistrieri, L. S. and Murray, J. W., Metal-solid interactions in the marine environment: estimating apparent equilibrium binding constants, *Geochim. Cosmochim. Acta*, 47, 1091, 1983.

141. Balistrieri, L. S. and Murray, J. W., Marine scavenging: trace metal adsorption by interfacial sediment from MANOP Site H, *Geochim. Cosmochim. Acta*, 48, 921, 1984.

142. Davis, J. A. and Leckie, J. O., Surface ionization and complexation at the oxide/water interface. III. Adsorption of anions, *J. Colloid Interface Sci.*, 74, 32, 1980.

143. Benjamin, M. M. and Bloom, N. S., Effects of strong binding adsorbates on adsorption of trace metals on amorphous iron oxyhydroxide, in *Adsorption from Aqueous Solutions*, Tewari, P.H., Ed., Plenum Press, New York, 1981, 41.

144. Zachara, J. M., Ainsworth, C. C., Cowan, C. E., and Resch, C. T., Adsorption of chromate by subsurface soil horizons, *Soil Sci. Soc. Am. J.*, 53, 418, 1989.

145. Loux, N. T., Brown, D. S., Chafin, C. R., Allison, J. D., and Hassan, S. M., Chemical speciation and competitive cationic partitioning on a sandy aquifer material, *J. Chem. Spec. Bioavail.*, 1, 111, 1989.

146. Smith, K. S., *Factors Influencing Metal Sorption onto Iron-Rich Sediments in Acid-Mine Drainage*, Colorado School of Mines, Golden, CO, 1991.

147. Zachara, J. M., Rai, D., Moore, D. A., Turner, G. D., and Felmy, A. R., *Chemical Attenuation Reactions of Selenium*, Electric Power Research Institute, Palo Alto, CA, 1993.

148. Goldberg, S. and Glaubig, R. A., Anion sorption on a calcareous, montmorillonitic soil — selenium, *Soil Sci. Soc. Am. J.*, 52, 954, 1988.

149. Goldberg, S. and Glaubig, R. A., Anion sorption on a calcareous, montmorillonitic soil — arsenic, *Soil Sci. Soc. Am. J.*, 52, 1297, 1988.

150. Goldberg, S. and Glaubig, R. A., Boron adsorption on California soils, *Soil Sci. Soc. Am. J.*, 50, 1173, 1986.

151. McKinley, J. P., Zachara, J. M., Smith, S. C., and Turner, G. D., The influence of uranyl hydrolysis and multiple site-binding reactions on the adsorption of U(VI) to montmorillonite, *Clays Clay Miner.*, 43, 586, 1995.

152. Anderson, P. R. and Benjamin, M. M., Surface and bulk characteristics of binary oxide suspensions, *Environ. Sci. Technol.*, 24, 692, 1990.

153. Anderson, P. R. and Benjamin, M. M., Modeling adsorption in aluminum-iron binary oxide suspensions, *Environ. Sci. Technol.*, 24, 1586, 1990.

154. Turner, G. D., Zachara, J. M., McKinley, J. P., and Smith, S. C., Surface-charge properties and UO_2^{2+} adsorption of a subsurface smectite, *Geochim. Cosmochim. Acta*, 60, 3399, 1996.

155. van Geen, A., Robertson, A. P., and Leckie, J. O., Complexation of carbonate species at the goethite surface: implications for adsorption of metal ions in natural waters, *Geochim. Cosmochim. Acta*, 58, 2073, 1994.

156. Lindsay, W. L., *Chemical Equilibria in Soils*, John Wiley & Sons, New York, 1979, 449.

157. Girvin, D. C., Gassman, P. L., and Bolton, H., Jr., Adsorption of aqueous cobalt ethylenediaminele-traacetate by δ–Al_2O_3, *Soil Sci. Soc. Am. J.*, 57, 47, 1993.

158. Brooks, S. C., Taylor, D. L., and Jardine, P. M., Reactive transport of EDTA-complexed cobalt in the presence of ferrihydrite, *Geochim. Cosmochim. Acta*, 60, 1899, 1996.

159. Wagner, J., Westall, J. C., and Zachara, J. M., Models for binding of metal ions to heterogeneous materials. II. Adsorption of Co(II) to subsurface materials as a function of pH and electrolyte concentration, *Environ. Sci. Technol.*, (submitted), 1996.

160. Dzombak, D. A., Fish, W., and Morel, F. M. M., Metal-humate interactions. I. Discrete ligand and continuous distribution models, *Environ. Sci. Technol.*, 20, 669, 1986.

161. Fish, W., Dzombak, D. A., and Morel, F. M. M., Metal-humate interactions. II. Application and comparison of models, *Environ. Sci. Technol.*, 20, 676, 1986.

162. Buffle, J., *Complexation Reactions in Aquatic Systems: An Analytical Approach*, John Wiley & Sons, New York, 1988, 692.

163. Altmann, R. S. and Buffle, J., The use of differential equilibrium functions for interpretation of metal binding in complex ligand systems: its relation to site occupation and site affinity distributions, *Geochim. Cosmochim. Acta*, 52, 1505, 1988.

164. Parish, R. S. and Perdue, E. M., Computational methods for fitting statistical distribution models of multi-site binding equilibria, *Mathematical Geology*, 21, 199, 1989.

165. Buffle, J., Altmann, R. S., Fifella, M., and Tessier, A., Complexation by natural heterogeneous compounds: site occupation distribution functions, a normalized description of metal complexation, *Geochim. Cosmochim. Acta*, 54, 1535, 1990.

166. Cernik, M., Borkovec, M., and Westall, J. C., Regularized least-squares methods for the calculation of discrete and continuous affinity distributions for heterogeneous sorbents, *Environ. Sci. Technol.*, 29, 413, 1995.

167. Anderson, P. R. and Benjamin, M. M., Effects of silicon on the crystallization and adsorption properties of ferric oxides, *Environ. Sci. Technol.*, 19, 1048, 1985.

168. Trolard, F. and Tardy, Y., The stabilities of gibbsite, boehmite, aluminous goethites and aluminous hematites in bauxites, ferricretes and laterites as a function of water activity, temperature, and particle size, *Geochim. Cosmochim. Acta*, 51, 945, 1987.

169. Schwertmann, U. and Carlson, L., Aluminum influence on iron oxides. XVII. Unit-cell parameters and aluminum substitution of natural goethites, *Soil Sci. Soc. Am. J.*, 58, 256, 1995.

170. Ainsworth, C. C., Girvin, D. C., Zachara, J. M., and Smith, S. C., Chromate adsorption on goethite: effects of aluminum substitution, *Soil Sci. Soc. Am. J.*, 53, 411, 1989.

171. Tipping, E., Griffith, J. R., and Hilton, J., The effect of adsorbed humic substances on the uptake of copper (II) by goethite, *J. Croat. Chim. Acta*, 56, 613, 1983.

172. Davis, J. A., Complexation of trace metals by adsorbed natural organic matter, *Geochim. Cosmochim. Acta*, 46, 2381, 1984.

173. Murphy, E. M. and Zachara, J. M., The role of sorbed humic substances on the distribution of organic and inorganic contaminants in groundwater, *Geoderma*, 67, 103, 1995.

174. Bond, W. J., On the Rothmun-Kornfeld description of cation exchange, *Soil Sci. Soc. Am. J.*, 59, 436, 1995.

175. Tipping, E., Modeling the competition between alkaline earth cations and trace metal species for binding by humic substances, *Environ. Sci. Technol.*, 27, 520, 1993.

176. Bolt, G. H., and Van Riemsdijk, W. H., Surface chemical processes in soil, in *Aquatic Surface Chemistry*, Stumm, W., Ed., Wiley Interscience, New York, 1987, 127.

177. Kent, D. B., Davis, J. A., Anderson, L. C. D., Rea, B. A., and Waite, T. D., Transport of chromium and selenium in the suboxic zone of a shallow aquifer: influence of redox and adsorption reactions, *Water Resour. Res.*, 30, 1099, 1994.

178. Tompson, A. F. B., Numerical simulation of chemical migration in physically and chemically hetero-geneous porous media, *Water Resour. Res.*, 29, 3709, 1993.

179. Fripiat, J. J., Surface properties of aluminosilicates, *Proc. 12th Natl. Conf., Clays Clay Miner.,* 1964.

180. Zachara, J. M., Girvin, D. C., Schmidt, R. L., and Resch, C. T., Chromate adsorption on amorphous iron oxide in presence of groundwater major ions, *Environ. Sci. Technol.*, 21, 589, 1987.

181. Ronnogren, L., Syoberg, S., Sun, Z., Forsling, W., and Schindler, P. W., Surface reactions in aqueous metal sulfide systems. II. Ion exchange and acid/base reactions at ZnS-H_2O interface, *J. Colloid Interface Sci.*, 145, 396, 1991.

182. Goldberg, S., Forster, H. S., and Heick, E. L., Boron adsorption mechanisms on oxides, clay minerals, and soils inferred from ionic strength effects, *Soil Sci. Soc. Am. J.*, 57, 704, 1993.

183. Mouvet, C. and Bourg, A. C. M., Speciation (including adsorbed species) of copper, lead, and zinc in Meuse River, *Water Res.*, 17, 641, 1983.

184. Osaki, S., Miyoshi, T., Sugihara, S., and Takashima, Y., Adsorption of Fe(III), Co(II) and Zn(II) onto particulates in fresh water on the basis of the surface complexation model. I. Stabilities of metal species adsorbed on particulates, *Sci. Total Environ.*, 99, 105, 1990.

185. Payne, T. E. and Waite, T. D., Surface complexation modeling of uranium sorption data obtained by isotopic exchange techniques, *Radiochim. Acta*, 52/53, 487, 1991.

CHAPTER **3**

Thermodynamics of the Soil Solution

Donald L. Suarez

CONTENTS

I. INTRODUCTION

A. The Concept of the Soil Solution

Soil is traditionally considered to be comprised of three phases, gaseous, solution, and solid. This concept is useful for some purposes but, as will be discussed later, these categories, particularly the solid phase, do not meet the thermodynamic definition of a phase.

The soil solution is the aqueous phase of the soil, which is linked to the gaseous and numerous solid phases via the transport of energy and matter. Since changes in the composition of the solution phase depend in part on the solid and gaseous phases, it is not fruitful to consider solution processes or composition in isolation from the other soil components. Generally the time frame of changes in soil solution composition, at a fixed location in the soil, is intermediate between the rapid changes in the gaseous phase and the slow time frame of changes in the solid phase.

B. Application of Chemical Thermodynamics to Soil Systems

Thermodynamics relates the properties of bulk matter (i.e., macroscopic) to its behavior in chemical processes. Dealing only with observable properties of matter, thermodynamics requires no assumption about molecular level processes or organization. Despite the separation of the thermodynamic relations from mechanistic interpretations or molecular scale information, it is useful to link the thermodynamic data and conclusions to the underlying chemical processes. We also need not assume that all parts of our system are at equilibrium in order to use thermodynamics. In dynamic natural systems, such as soil, it provides the framework to determine what processes are possible, rather than what processes are occurring or are dominant. Our objective in using thermodynamics is to calculate which processes are energetically favorable. We should not assume that our system is at chemical equilibrium and use thermodynamics, in the absence of experimental data, to predict occurrence of solid phases, gas, solution or solid phase composition, nor should we assume that thermodynamically predicted reactions occur.

It is important to remember that the basic criteria for the application of thermodynamics to a system are almost never met in soils. Under field conditions neither the gas nor the liquid phase meet the requirement of homogeneity. The solution phase generally varies spatially both across the landscape and with depth within a profile. In addition there is often disequilibrium among the solution compositions in various pore sizes present within a soil sample. The solid phase, rather than being a pure homogeneous material, is often a mixture of a large number of discrete solid phases which are out of equilibrium with each other, as well as out of equilibrium with the solution phase.

Prediction in the natural system requires that we calculate the thermodynamics of the processes to determine which are possible, and then to consider the kinetics of the reactions to determine which are important. This determination cannot be easily generalized as it is dependent on the time frame of interest in a specific case.

Many solution reactions, such as the formation of some complexes, have rates of the order of nanoseconds to microseconds. Thus, they can be considered instantaneous for the time frames ordinarily of interest to applied soil scientists. Studies on weathering processes associated with mineral transformations are well served by examining thermodynamically feasible reactions, but equilibria calculations are not well suited for predicting present or even future mineralogy. As a result of the relatively slow reaction rates in the mineral-water system at earth surface temperatures, equilibrium models, in general, are essential for descriptions of solution phase speciation, but are more restricted for application to solution-solid interactions.

Lack of equilibrium does not indicate that thermodynamics can be neglected. Kinetic models must consider the concentration of activated complexes, which are in turn dependent on the activity of the species in equilibrium with the complex. Kinetic expressions for reactions far from equilibrium generally consider only forward reaction rates (e.g., rate expressions for dissolution of aluminosilicates). However, complete kinetic descriptions must include back reaction rates as well. If reaction rates are well characterized, then the condition when forward and back reaction rates are equal defines the equilibrium constant as long as the rates are not diffusion controlled. For example, Plummer et al.[1] found good agreement between the equilibrium constant calculated from forward and back reaction rates for calcite and the published equilibrium constant. Ikeda et al.[2] found good correspondence between thermodynamic constants for Na exchange on zeolite determined by equilibrium and kinetic (stopped-flow method) experiments. Also, Zhang and Sparks[3] found good agreement between equilibrium constants determined from static and kinetic (pressure-jump relaxation) experiments on sulfate adsorption on goethite, as did Sparks and Jardine[4] for Ca-K exchange on soil.

Various authors have presented reasons why models based on chemical thermodynamics may not represent natural systems[5] or specifically soil solutions.[6] Among these, the most important is the slow rates of reaction encountered at earth surface temperatures and pressures for many solution-solid reactions. Other important limitations are errors in the model inputs. These include failure to include all pertinent species (particularly important for trace species such as heavy metals which complex strongly with many dissolved organic species) and failure to properly characterize the species (Al may be in dissolved, polymer, or particulate form). Elements may occur in different species or redox states, and *operational* definitions of dissolved (ability to pass through a specified filter size such as 0.1 μm) may be inappropriately substituted for the thermodynamic criteria for a dissolved species. Almost always, it is useful to consider the equilibrium solution composition given a field-determined assemblage of minerals, rather than to also constrain the mineral phase assemblage to be at equilibrium.

C. Selection of Thermodynamic Database

Published data on stability constants or solubility constants usually rely on other previously published thermodynamic data. In many instances the use of data from several sources or the use of data from compilations results in inconsistent thermodynamic databases. A common problem is the updating of solution stability constants without recalculation of the solid phases. For example, the use of revised hydrolysis constants for Al species has not always been accompanied by the recalculation of the solubility constants of the Al solid phases or the stability of other solution species which are in part derived from these constants. In many instances the stability of the solid phase was originally determined from dissolution studies which depend on knowledge of the hydrolysis constants. Similarly, calculations involving solutions with complexes are often incorrectly coupled with older solubility constants which were based on experimental data where these complexes were not considered. Inconsistent databases can lead to erroneous conclusions regarding the relative stability of various phases as well as the saturation status of minerals in solution. Consistency requires that all experimental data be recalculated with the database to be used.

The above examples regarding inconsistent databases have been recently noted by various researchers. A neglected source of inconsistency in databases involves the use of activity coefficient corrections. Many geochemical data for aqueous and solid phases are based, not on classical calorimetry studies, but on solution experiments where total concentrations are measured and activities are calculated. Generally a modified Debye-Hückel equation is used to calculate activities. Soil scientists, in contrast, have generally utilized one of two versions of the Davies equation. In many cases differences in activity among these different calculation methods range from 5 to 10%.

Appendix A provides a list of thermodynamic databases in common usage by soil scientists and hydrochemists. Although primary sources such as the National Bureau of Standards (NBS) series[7] and the more recent Chase et al.[8] are widely accepted and present an internally consistent data set, there may be instances where other data are preferable for our purposes. For example, these databases include ΔH, S, and C_p data for many solids, enabling us to make thermodynamic calculations at high temperatures. Often these data are obtained from calorimetry experiments, which may produce the most precise determination for solids which do not demonstrate reversible reactions at 298 K. Stability of phases such as calcite, aragonite, and gypsum may be more precisely determined using solubility experiments. Other useful general databases include Woods and Garrels[9] and Robie and Hemingway.[10] Nordstrom et al.[11] present a database for major water-mineral reactions.

One limitation to determination of solid phase stability from solution experiments is the need to calculate activities from total solution concentrations — this has resulted in numerous revisions of constants as the activity coefficient and speciation models have been refined. As a result, the databases associated with the thermodynamic computer models are more likely to contain constants developed from solubility or aqueous phase experiments, but may suffer from internal inconsistencies.

II. SOIL SOLUTION EQUILIBRIA

A. Sampling Soil Solution

Analysis of the soil solution requires that we remove it from the soil. Most soil chemists conduct laboratory experiments, where conditions can be more controlled than under field environments. These experiments are almost always performed under conditions of high solution: soil solid ratios for ease of extracting solution. In these cases, solution is obtained by either centrifugation or filtration. These experiments are generally performed with the objective of obtaining thermodynamic data and not necessarily towards predicting chemical processes under field conditions.

Studies of soil chemistry under field conditions are not common, but are required if we are to evaluate the applicability of our laboratory studies to the natural system. In this instance, removal of the soil solution from the soil without changing its chemical composition is not a trivial problem. Most commonly, water is added to a soil sample in sufficient volume so as to allow for a simple extraction of the water phase. This procedure is typically carried out at a fixed or reference water volume to allow comparison among samples. Among the most popular procedures are saturation extracts[12,13] and various extracts of fixed soil:water ratios (e.g., 1:2, 1:5, 1:10).[13]

The saturation extract has the advantage of being the standard water equilibration with the least amount of added water. It is desirable that we add as little water as possible to minimize the disequilibrium between phases. Although not at a fixed soil:water ratio, the saturation extract has the advantage of having a relatively constant ($\approx 2X$) relation between extract water content and water content at field capacity.

Although convenient, none of these diluted extracts can be considered representative of the soil solution. Reaction times are not sufficiently short to allow us to neglect solution-solid or solution-gas reactions caused by the addition of water nor are reaction times sufficiently long to insure solution-soil equilibrium. In many instances, these extracts are adequate for characterization of soil solutions, and the assumption is made that the composition has not been affected by dilution. For some aqueous species (e.g., Cl^-) it is usually reasonable to assume that the solid phases did not release additional Cl^-, and we can calculate the concentration present at the time of sampling by correcting for the dilution process. Although popular for the ease with which they can be obtained, these dilutions and extracts are *not* suitable for evaluating the thermodynamic status of aqueous species or for determining mineral stability with respect to the solution phase.

An additional problem with all laboratory soil extracts is that the removal of a soil sample from the field causes chemical changes, unless special precautions are taken. These changes are caused by solution-gas reactions which are generally neglected. Among these changes are increases in pH due to CO_2 degassing upon removal of either the soils or the soil solution from the *in situ* conditions and their exposure to atmospheric conditions of lower P_{CO_2} than the soil atmosphere. Redox-sensitive species such as Fe, Mn, Cu, Hg, Cr, As, Se, and V are also affected by exposure to atmospheric conditions, especially if the soil is taken from an oxygen-depleted environment.

Procedures to extract undiluted soil solution from the soil have received relatively little attention. Although there are standardized recommended procedures for soluble salts and diluted water extracts,[13] there are no similar recommendations for direct sampling. Published methods of extracting soil solution from a soil sample include displacement methods. Immiscible displacement with an unreactive organic phase such as 1,1,2-trichlorotrifluoroethane,[14] trifluoroethane,[14,15] and ethyl benzylacetate[16] are sometimes utilized. Samples can either be placed in a column and the organic phase placed on the top of the column, thereby displacing the aqueous phase out the bottom of the column, or the organic phase can be added to the soil and centrifuged, with the lighter aqueous phase being subsequently decanted from the top.

Centrifugation methods offer the advantage of being able to extract different fractions of water corresponding to water in different pore sizes (or more precisely, different soil water matric potentials). These procedures are preferred over dilution methods when attempting thermodynamic calculations. Unfortunately there are almost no data available using these methods. Adams et al.,[17] Reynolds,[18] and Elkhatib et al.[19] have described and demonstrated the use of centrifugation to obtain soil solution.

Another technique used to sample soil solution in field studies entails the use of vacuum extractors. Various studies have utilized the method and it also is not without difficulties. As with centrifugation, different pore size distributions are sampled under different suction. Hansen and Harris,[20] among others, have compared results from extractors with measurements of soil solution composition from diluted extracts and found the vacuum extractors not to be representative of the average soil water composition. These results may be unsuitable for mass-balance and mass leaching calculations but may not be a disadvantage in thermodynamic studies, since we do not want to mix nonhomogeneous pore solutions. A review of the use and limitations of extractors was conducted by Litaor.[21] Even *in situ* sampling of soil solutions with an extractor presents errors unless precaution is taken to minimize degassing.[22]

Application of thermodynamics to field studies is limited by the nonhomogeneity of the system. All extraction procedures also suffer from "mixing" of solution from different points in space, as well as from different distributions of pore sizes. It has been well documented that different solution compositions may exist in different size pores, primarily when there is a large compositional change — such as surface application of a soluble fertilizer or change in irrigation water composition.[20] In these instances we obtain average concentrations which may not be appropriate for evaluation of thermodynamic relations because of the existence of more than one solution phase.

B. Variables of State and Thermodynamic Potentials

Application of thermodynamics to a problem requires that we select a system to study. This involves selection of the part of the universe of interest to us, and an attempt to isolate it from all other parts. The system thus defines the region of space to be considered. Factors of importance outside the system cannot be ignored but rather they should be controlled or at least measured (i.e., temperature). Soil scientists often find it useful to select restricted systems which come closer to meeting the thermodynamic requirements. For example, study of ion exchange is usually done with pure or reference minerals rather than on bulk soils. The need to isolate the system also explains why soil scientists usually do experiments under controlled laboratory conditions.

Since thermodynamics is concerned with equilibrium states of a system, it does not deal with rates of chemical processes nor attempt to describe the system while change is occurring. In order to ensure that equilibrium has been achieved, it is necessary to demonstrate that the process is reversible. In the absence of reversibility there is uncertainty as to whether a metastable state has been reached; that is, whether the reaction rates are slow relative to the time scale of measurement. The state of a system is defined by the set of thermodynamic properties which can be measured. State functions, or variables of state are selected properties which have a measurable value for each state and which are not dependent on the path taken to achieve that state. Examples of state functions are temperature, pressure, volume, E, the internal energy of a system, H, the enthalpy, and G, the Gibbs energy. Properties of the system are categorized as extensive or intensive. Extensive properties are properties which depend on mass. Such properties, for example volume and potential energy, are additive for a homogeneous phase. Intensive properties are those which are independent of mass. These properties, such as temperature, pressure, and density, are not additive. A phase is defined as a homogeneous portion of a system.

The first law of thermodynamics is the law of conservation of energy, often expressed as

$$\Delta E = q - w \tag{1}$$

where ΔE is the change in energy in the system, q is the heat added to a system, and w is the work done by the system. When reactions are run in a closed vessel where the volume is constant, w = 0. Thus, changes in E can be determined by measurements of heat released or absorbed.

The state function enthalpy, H, is defined by

$$H = E + PV \tag{2}$$

Changes in enthalpy, ΔH, are expressed by

$$\Delta H = \Delta E + P\Delta V + V\Delta P \tag{3}$$

where P is pressure and V is volume. Commonly, reactions of interest to soil scientists are run at constant pressure rather than constant volume, thus $V\Delta P = 0$. In this case combining Equations 1 and 3 one obtains,

$$\Delta H = q_p \tag{4}$$

Exothermic reactions are defined as those that release heat ($\Delta H < 0$), and endothermic reactions are those that absorb heat ($\Delta H > 0$). An increase in temperature favors a reaction where $\Delta H > 0$.

The enthalpy change when reactants and products are in their standard state is designated by ΔH°. The standard state is taken to be 0.1 MPa (formerly 101.33 kPa or 1 atm) pressure. The term enthalpy (or heat) of formation, ΔH_f^o, is the ΔH of a reaction in which a compound in its standard state is formed from its elements, also in their standard state. From this definition it follows that the enthalpies of formation of the elements are zero. The ΔH_r^o is thus given by

$$\Delta H_r^o = \Delta H_f^o \text{ products } - \Delta H_f^o \text{ reactants} \tag{5}$$

The molar heat capacity is the amount of heat added to a mole of a substance to raise the temperature of the substance 1 K. The term mole is used to denote an amount of a substance which contains 6.022045×10^{23} entities (Avogadro's constant). The heat capacity at constant pressure C_p is

$$C_p = \left(\frac{dq}{dT}\right)_P = \left(\frac{\partial H}{\partial T}\right)_P \tag{6}$$

where T is temperature (K). The heat capacity at constant volume C_v is

$$C_v = \left(\frac{dq}{dT}\right)_v = \left(\frac{\partial E}{\partial T}\right)_v \tag{7}$$

For small changes in temperature (≈ 10 K) we can assume C_p and C_v do not depend on temperature and

$$C_p = \frac{\Delta H}{\Delta T} \qquad C_v = \frac{\Delta E}{\Delta T} \tag{8}$$

Combining these expressions with Equation 3,

$$C_p = C_v + \frac{\Delta(PV)}{\Delta T} \tag{9}$$

For solids and aqueous solutions $\Delta PV/\Delta T$ is relatively small and $C_p \approx C_v$. For 1 mole of a gas

$$PV = RT \tag{10}$$

where R is the gas constant. Substituting RT into Equation 9 one obtains $C_p = C_v + R$.

The second law of thermodynamics defines the change in entropy ΔS of a system taken from state 1 to state 2 by a reversible process

$$\Delta S \equiv \int_1^2 \frac{dq_{rev}}{T} \tag{11}$$

When considering an isothermal process Equation 11 becomes

$$\Delta S = \frac{q_{rev}}{T} \tag{12}$$

At constant pressure

$$\Delta S = \int_{T_1}^{T_2} \frac{n\,C_p}{T}\,\mathrm{d}T \tag{13}$$

where n is the number of moles and C_p is the heat capacity at constant pressure. For relatively small temperature differences, such as the range normally encountered in soils, C_p can be considered constant and thus

$$\Delta S = n\,C_p \ln\frac{T_2}{T_1} \tag{14}$$

This equation provides a means of evaluating changes in ΔS.

The third law of thermodynamics states that the entropy of perfect crystals of pure elements or compounds is zero at absolute zero temperature, thus at any given temperature T,

$$S_T = \int_{o}^{T} \frac{C_p\,\mathrm{d}T}{T} \tag{15}$$

where S_T is the entropy. The Gibbs energy G, defined as

$$G \equiv H - TS \tag{16}$$

is widely used for evaluating the thermodynamics of soil solution reactions. The term ΔG is used to represent changes in the free energy of a species or reaction and can be expressed as

$$\Delta G = \Delta H - T\Delta S \tag{17}$$

At constant temperature, negative values of ΔG indicate that the reaction should proceed spontaneously. Positive values indicate that the reaction cannot occur. By defining the standard pressure (0.1 MPa) and the reference temperature (298.15 K) we obtain the standard free energy of formation ΔG_f^o

$$\Delta G_f^o = \Delta H_f^o - T\Delta S_f^o \tag{18}$$

Under reversible conditions and constant temperature and pressure, the decrease in ΔG represents the maximum useful work that can be obtained by the process. The free energy is an extensive quantity.

The chemical potential μ_i can be defined by[23]

$$\mu_i \equiv \left(\frac{\partial G}{\partial n_i}\right)_{T,P,nj} \tag{19}$$

The chemical potential thus represents the infinitesimal change in the capacity to do work per infinitesimal increase in the amount of the substance. The concept of chemical potential is particularly useful for representing non-ideal systems where activities are calculated. At equilibrium the chemical potential of a component is the same in all parts and in all phases of the system. The differential of the Gibbs energy describes the function

$$dG = -SdT + VdP + \Sigma_i \mu_i dn_i \qquad (20)$$

for a single phase. For more than one phase, Equation 20 is summed for each phase. The free energy of a phase is equal to the sum of the chemical potentials multiplied by the number of moles of the species and summed over all the species in the phase:

$$G = \Sigma_i \mu_i n_i \qquad (21)$$

The derivative of Equation 21 is given by

$$dG = \Sigma_i \mu_i dn_i + \Sigma_i n_i d\mu_i \qquad (22)$$

Substituting Equation 20 into Equation 22 we obtain

$$SdT - VdP + \Sigma_i n_i d\mu_i = 0 \qquad (23)$$

This equation is known as the Gibbs-Duhem equation. For a closed system at constant temperature and pressure,

$$\Sigma_i n_i d\mu_i = 0 \qquad (24)$$

This equation is very useful as it can be used to calculate the thermodynamic properties of one substance based on the measurement of the thermodynamic properties of the other substances.

C. Standard States and Concentration Scales

A standard state is a selected set of conditions (such as defined composition, temperature, and pressure) which serve as a point from which to make comparisons. This is different from the reference state, in which a numerical value (such as zero) is assigned to a property by convention. Three concentration scales are commonly used. Most commonly used for analytical purposes is the molar scale. This scale, with concentration denoted by c or M, is defined by the moles of solute per liter of solution (dm^3). At the specified temperature of 277 K where water is the solvent, a liter is equivalent to 1.0000 kg of pure water. For dilute solutions it is generally acceptable to equate concentrations in the molar scale to the molal scale which is generally utilized for thermodynamic calculations. The molal scale with concentration denoted by m is defined as the moles of solute per kilogram of solvent. This scale has the advantage of being independent of T and P. The thermodynamic databases available for dissolved species and corrections for activity coefficients are in the molal scale. Thus, to the extent that we can assume that the density of the solution = 1.000 g cm^{-3}, then c ≈ m. For most relatively dilute solutions in which the Debye-Hückel or Davies equation is used, M = m is a reasonable assumption. For example, Stumm and Morgan[24] report that a 1.00 m solution is 0.98 M at 298 K. Conversion of molarity into molality can be made by the relation

$$m = M \frac{\text{wt solution}}{(\text{wt solution} - \text{wt solutes})(d)} \qquad (25)$$

where d is the density of the solution.

The mole fraction scale with concentration denoted by N or X is defined as the moles of solute divided by the total number of moles in the system. The mole fraction scale is most commonly used for the solid phase (e.g., for ion exchange). In earlier work on ion exchange, the non-SI term, equivalent fraction, is often used. This concept has been replaced by use of the mole fraction scale where the data are reported as moles of charge, or mol_c.

Although any standard state can be chosen for the individual ionic species in any of the concentration scales selected, the common convention is that the standard state in the molal scale is a "hypothetical one-molal solution".[25] The reference state for the solute is infinite dilution where ideal behavior occurs. By convention the chemical potential of an element in its most stable state = 0 at its standard state pressure (0.1 MPa) and temperature (298.15 K). Similarly the ΔG_f^o of an element = 0 when the element is in its most stable state.

The activity of a component a_i is defined by

$$\mu_i = \mu_i^o + RT \ln a_i/a_i^o \tag{26}$$

where μ_i^o is the chemical potential of the standard state and a_i^o is the activity of the ith ion in the standard state. Using the convention that the standard state activity is unity, then Equation 26 can be written in the more commonly seen form,

$$\mu_i = \mu_i^o + RT \ln a_i \tag{27}$$

where a_i is dimensionless. This convention is convenient in that it allows the activity to approach zero as the solution approaches infinite dilution and $a_i = 1$ for a pure substance. Using this convention, the activity coefficient in the molal scale, γ, is defined by

$$\gamma_i = a_i \frac{m^o}{m_i} \tag{28}$$

Note that the molal activity coefficient is dimensionless, since a_i in Equation 28 is actually the ratio of a_i to a_i^o. As with Equation 26 this equation is more commonly presented with the standard state implicit as

$$\gamma_i = \frac{a_i}{m_i} \tag{29}$$

This ratio of the concentration with that in the standard state accounts for the fact that although the concentrations, activities, and activity coefficients are dimensionless they correspond to a standard state or concentration scale. Thus, the activity of a species in solution is dimensionless but can be assumed to be in reference to the standard state of a 1 m solution, unless specified otherwise.

Although we utilize the molal scale for thermodynamic calculations this is not required. Robinson and Stokes[26] present the equations converting the activity coefficients in the molal scale γ, the molar scale y, and the mole fraction scale f. These are called, respectively, the molal, molar, and rational activity coefficients. In most circumstances it is convenient to choose the standard state such that the activity coefficient approaches unity as the concentration approaches zero. The activity of the species need not be equal to the concentration in the standard state, i.e., the activity of a 1 m solution need not be one.

Table 1   Average Separation of Ions in a Solution of 1:1 Electrolyte

Mole l^{-1}	0.001	0.01	0.1	1.0	10.0
Separation (nm)	9.4	4.4	2.0	0.94	0.44

All values from Robinson, R.A. and Stokes, R.H., *1965.*

The nonideal behavior of an electrolyte increases with increasing ionic strength. This is related to both long-range electrostatic interactions and specific ion interactions. The need to consider ion interactions is evident from examination of the distances between ions as a function of solution composition. Assuming that the ions are arranged in a cubic lattice, Robinson and Stokes[26] calculated the average separation distances for a 1:1 electrolyte. As shown in Table 1 there are only a few molecules of water separating the individual ions for solutions greater than 1.0 *M*.

D. Chemical Equilibria

1. Phase Rule

The Gibbs phase rule defines the number of allowable phases in the system

$$F = C - p + 2 \tag{30}$$

where p is the number of phases, C is the number of components in the system, and F is the degrees of freedom. When F = 0 the system is defined (invariant), meaning that there are no independent intensive variables. When F = 2 there are no independent intensive variables except temperature and pressure, which we almost always specify. At fixed temperature and pressure, F = C – p. The number of independent intensive variables is equal to the degrees of freedom, F, minus the number of independent equations (reactions).

Consider the system halite and water. The number of components = 3 (NaCl solid, NaCl aq, and H_2O), p = 2 (solution + halite solid), thus F = 3. If we specify temperature and pressure and include the equation for the equilibrium reaction NaCl $\rightleftarrows$ Na^+ + Cl^-, then the system is defined and no further information is required (assuming ideal solute behavior). If we consider the above system and add $CaCl_2$, C = 3, p = 2, and F = 4. After considering fixed T and P and the above equilibrium reaction, there is one remaining independent variable required to define the system. This could be either the Ca concentration, Cl concentration, Ca/Na ratio, etc. The imposed equilibrium conditions require that the chemical potentials of each component be the same in every phase.

2. Equilibrium Constants and Concentration Scales

For a generalized reaction

$$aA + bB = cC + dD \tag{31}$$

we represent the ideal concentrations (or activities) by the expression

$$\frac{(C)^c (D)^d}{(A)^a (B)^b} = K \tag{32}$$

where at equilibrium, K is a fixed constant value, and parentheses denote activities. Traditionally soil chemists have used the convention that parentheses represent activities and brackets represent concentrations. In contrast, geochemists and aquatic chemists have used the convention that brackets represent activities and parentheses represent concentrations. The concentration of a pure solid in its most stable form is taken as equal to one. The equilibrium constant is related to the Gibbs energy ΔG by

$$\Delta G = \Delta G_r^o + RT \ln K \tag{33}$$

where ΔG_r^o is

$$\Delta G_r^o = \Delta G_f^o \text{ products} - \Delta G_f^o \text{ reactants} \tag{34}$$

and where ΔG_f^o is the Gibbs energy change for the formation of 1 mole of a compound in its standard state formed from its elements in their standard state. The Gibbs energy of formation of the elements in their standard state is defined as zero. If the reactants and products are at equilibrium then $\Delta G = 0$ and Equation 33 becomes

$$\Delta G_r^o = -RT \ln K \tag{35}$$

Expressing the ΔG_r^o in kJ/mol at the standard state of 0.1 MPa and the reference temperature of 298.15 K and converting to log K yields,

$$\Delta G_r^o = -5.708 \log K \tag{36}$$

3. pe and Eh

Equilibrium reactions in which electron transfer occurs can be treated by considering electron transfer as a mass reaction. Also, the relative oxidative or reductive status in a system can be conceptually characterized by its free electron activity. Thus, pe is defined by

$$pe = -\log\left(e^-\right) \tag{37}$$

and Eh is defined by

$$Eh = \frac{RT \ln 10}{F} pe \tag{38}$$

where F is the Faraday constant. At 298.15 K, and when the electrode potential Eh is expressed in volts, then

$$Eh = 0.05916 \, pe \tag{39}$$

The pe expression is conceptually useful to define the oxidation status of a system. However, natural systems are almost never at thermodynamic equilibrium with respect to oxidation-reduction reactions. Thus, the calculated pe obtained from measuring different redox couples will rarely be the same, or even similar. The most common measure of the redox status of a system is that obtained by a platinum electrode and reference combination.

4. Effects of Temperature and Pressure

Temperature and pressure affect the free energy of the reactants and products and, thus, the equilibrium constants. Many solubility determinations for solids and dissolved species of interest have been calculated for 298 K from experiments at higher temperatures. If the temperature dependence of a reaction is not available, it can be calculated from the changes in enthalpy and entropy. As a first approximation we can assume the van't Hoff relation

$$\log K_{T_1} = \log K_{T_o} - \frac{\Delta H_{T_o}}{2.3\,R}\left(\frac{1}{T_1} - \frac{1}{T_o}\right) \tag{40}$$

where T_o refers to the reference temperature (298.15 K). This equation is useful for relatively small variations from 298 K (25°C), such as under earth surface conditions, where we assume that ΔH at $T_1 = \Delta H$ at T_o. Alternatively it is sometimes possible to evaluate the changes in stability constants by determining the changes in entropy and enthalpy. Similarly, we can represent the change in enthalpy of a reaction ΔH_r to a change in heat capacity:

$$\left(\frac{\partial \Delta H_r}{\partial T}\right)_p = \Delta C_{p_r} \tag{41}$$

where

$$\Delta C_{pr} = \sum \Delta C_p \text{ products} - \sum \Delta C_p \text{ reactants} \tag{42}$$

The heat capacity of a substance at constant pressure, C_p, is given by Equation 6. If we integrate Equation 41 assuming ΔC_{pr} is constant then we obtain:

$$\Delta H_{r_{T_1}} - \Delta H_{r_{T_o}} = \Delta C_{p_r}\left(T_1 - T_o\right) \tag{43}$$

This equation is generally sufficient to represent ΔH changes within the range of interest.

The change in entropy of a system or a specific reaction at constant pressure is given by Equation 13. Again assuming ΔC_{pr} is constant over the temperature range of interest and expressing this relation for the difference in entropy between T and a reference temperature T_o:

$$\Delta S_{T_1} - \Delta S_{T_o} = \Delta C_{p_r} \ln \frac{T_1}{T_o} \tag{44}$$

Combining terms, the Gibbs energy at T_1, given by ΔG_r^1 is

$$\Delta G_r^1 = \Delta G_r^o - \left[\Delta C_{p_r}\left(T_1 - T_o\right) - T_1\left(\Delta C_{p_r}\left(T_1 - T_o\right)\right)\right] \tag{45}$$

and the equilibrium constant at T_1, is

$$\ln K_1 = -\frac{\Delta G_r^1}{RT} \tag{46}$$

III. ACTIVITY COEFFICIENTS

A. Nonideality and the Use of Activity Coefficients

Activity coefficients serve to correct for nonideality, linking the thermodynamic activity functions to the appropriate concentration scales. Activities are linked to concentration via the activity
coefficients (Equation 29). The reference state for the activity coefficient of aqueous species is
generally infinite dilution, but other, constant ionic media reference states may be selected. Examples of the latter include solutions such as 0.1 m often used for modeling adsorption reactions
(Section VII.E) and the historical use of standardized seawater in chemical oceanographic studies.
The infinite dilution reference state is desired in soil studies where variations in ionic strength can
be large. As discussed earlier when describing the phase rule, a dissolved salt such as NaCl is
treated as a component. At equilibrium we know that the chemical potential of a component must
be the same for the solid and aqueous phases, in these instances the activity of the aqueous NaCl
component can be determined. The activity coefficient, called the mean activity coefficient $\gamma\pm$,
expresses the deviation of the solute from ideal behavior. Typically we are interested in the behavior
of an individual chemical species rather than the behavior of a solute.

Single-ion activity calculations and use of single-ion activity coefficients are extra thermodynamic constructs. Using thermodynamics we can measure products and ratios of individual ion-
activity coefficients, but we cannot measure single-ion activity coefficients separately.

Two groups of models have been utilized for calculation of activity coefficients in natural waters.
The ion association models assume that *specific* interactions between ions in solution occur solely
by association. The degree of specific interaction is often measured against alkali metal and alkaline
earth chlorides, most often KCl, which is assumed to be completely unassociated. In contrast ion
interaction models consider that strong electrolytes are completely dissociated and that specific ion
interactions are the result of interactions between free ions.

B. Debye-Hückel Limiting law

The Debye-Hückel limiting law is based on the assumption that deviations of dilute ionic
solutions from ideality are due only to electrostatic interactions among the ions. The extra, free
enthalpy per mole is

$$\mu_i(\text{electric}) = RT\ln\gamma_i \tag{47}$$

and thus the free enthalpy per ion is kT ln γ_i. The work required to increase the charge of an ion
from the uncharged state to a charge z is given by:

$$\Delta G = \frac{-bz^2e^2}{8\pi\varepsilon_o\varepsilon} \tag{48}$$

where e is the electron charge, ϵ_0 is the permittivity of a vacuum (8.854×10^{-12} C$^2 \cdot$ J^{-1} m^{-1}), and
ϵ is the dielectric constant of the solvent. The term b is 1/the Debye length, where the Debye length
is the distance over which an ion exerts its electrostatic field,[27] defined by

$$b = \left(\frac{e^2}{\varepsilon_o\varepsilon kT}\sum c_iz_i^2\right)^{1/2} \tag{49}$$

Substituting the expressions for the free enthalpy per ion ($kT\ln \gamma_i$) into Equation 49 we obtain

$$\ln \gamma_i = \frac{-z_i^2 e_i^2 b}{8\pi \varepsilon_o \varepsilon kT} \tag{50}$$

where c_i is the concentration of i per unit volume. Substituting Equation 49 into Equation 50 and substituting the ionic strength expression

$$I = 1/2 \sum_i m_i z_i^2 \tag{51}$$

we obtain for dilute solutions, the relation:

$$\log \gamma_i = -A z_i^2 I^{0.5} \tag{52}$$

This expression is known as the Debye-Hückel limiting law. The term A is a constant dependent on the dielectric constant, temperature, and density ($1.8248 \times 10^6 \, d^{0.5}/(\varepsilon \, T)^{1.5}$. At 298 K, $\varepsilon = 78.54$, $d = 0.997$ kg dm^{-3}, and we obtain the expression:

$$\log \gamma_\pm = -0.509 \, z_i^2 \, I^{0.5} \tag{53}$$

In addition to the assumptions regarding the extent of the electric field and the representation of ions as point charges, we have also assumed that the dielectric constant of the solvent in the electric field of the ion is equal to the dielectric constant of the pure solvent (which is not likely to be valid for a polar solvent such as water). Despite these limitations the simplified model is able to predict activity coefficients in very dilute solutions ($I < \approx 0.005$) and serves as the basis for all empirical activity coefficient equations. The A parameter increases with temperature, thus the activity coefficients decrease with increasing temperature, at constant pressure.

C. Ion Association Models

1. Associated Species

The ion association model is typically constructed of the undissociated species and additional associated species constructed to fit the model to experimental data. Often these are chemical species which have been identified spectroscopically. In some cases, species and dissociation constants were constructed (ion pairs) which were required to maintain accurate predictions in a mixed electrolyte solution. For example, the prediction of calcite solubility at variable CO_2 partial pressure required the assumption of $CaHCO_3^+$ and $CaCO_3^0$ species in order to maintain the thermodynamic requirement that ΔG_f^0 of calcite is independent of CO_2 partial pressure. These can be considered artificial constructs to account for errors in the empirical activity coefficient equations. The equilibrium constants of these reactions are generally expressed as dissociation constants.

2. Extended Debye-Hückel Equations

Some confusion exists regarding the term "extended Debye-Hückel equation" since the more detailed equation:

$$\log \gamma_\pm = -A z^2 \frac{I^{0.5}}{1 + Ba I^{0.5}} \tag{54}$$

is sometimes called the Debye-Hückel equation[26] in contrast to the Debye-Hückel limiting law (Equation 53). In contrast, others use the term extended Debye-Hückel equation for Equation 54.[24,28] Equation 54 introduces two additional parameters, a, which was originally the hydrated ion size, or distance of closest approach, and the term B, which is a constant depending on density, the dielectric constant of water, and temperature. The equation was traditionally used with the "a" hydrated ion sizes presented by Kielland.[29] This equation is considered to give acceptable results to about I = 0.1.[26]

The extended form of this equation:

$$\log \gamma_\pm = -\frac{Az^2\, I^{0.5}}{1 + Ba\, I^{0.5}} + b\, I \tag{55}$$

is considered the extended Debye-Hückel equation by Robinson and Stokes.[26] It is not presented in most discussions of activity coefficients (e.g., Stumm and Morgan,[24] Pytkowicz and Johnson,[28] Pitzer[30]). This equation has the additional linear term bI, where "b" is an empirical, ion-specific parameter. This term allows for consideration of short-range interactions between ions and solvent and between ions and ions, which are approximately linear with log γ. Both the A and B parameters increase with increasing temperature, with the net result being a decrease in activity coefficients with increasing temperature. The effect of temperature and pressure on these constants is given by Helgeson and Kirkham.[31]

Equation 55 is considered usable to about I = 1.0, but this is very dependent on the mixed electrolyte under consideration.[26] Truesdell and Jones,[32] using Equation 55, optimized values of a and b to fit experimental mean salt single-ion activity coefficients for major ions. The model provided a good fit to the *single* salt data to at least 3.0 *m*. This equation is one of the most frequently used for calculation of activity coefficients and is utilized in the various versions of the chemical equilibrium programs WATEQ (Truesdell and Jones,[32] Ball et al.[33]) and MINTEQ (Brown and Allison,[34] Allison et al.[35]). The a and b parameters for major ions are presented in Table 2.[36] Although the data fit well to high ionic strength this is only for single salt data. In mixed electrolyte systems above 0.1 *m* ionic strength, the model may not always be satisfactory, depending on solution composition. These ion-association models assume that the behavior of an ion in a simple solution can represent its behavior in a more complex system, e.g., the activity of an ion is a universal function of ionic strength. This assumption is clearly not valid at high ionic strength.

Table 2   Parameters of the Extended Debye-Hückel Equation

Selected Ions	a	b
Al^{3+}	6.65	0.19
Ca^{2+}	4.86	0.15
H^+	4.78	0.24
HCO_3^-	5.4	0
K^+	3.71	0.01
Mg^{2+}	5.46	0.22
Na^+	4.32	0.06
Cl^-	3.71	0.01
ClO_4^-	5.30	0.08
OH^-	10.65	0.21
SO_4^{2-}	5.31	−0.07

All values from Parkhurst,[36] except values for HCO_3^- which are from Truesdell and Jones.[32]

The use of empirical parameters to fit experimental data introduces the complication of requiring knowledge of which ion-association equations and which stability constants were used to obtain the parameters. The ion-association constants depend on the ion association model used and the "a" and "b" parameters selected. For example, Truesdell and Jones[32] developed parameter values without including $CaCl^+$ ion pairs. Inclusion of such a species (as done by Parkhurst[36]), is justified when using the ion-association models if it improves the overall fit to the experimental data, but its inclusion requires correction of other constants as well as possible correction to the a and b parameters of Equation 55. Failure to correct the parameters would result in a decreased free Ca^{2+} concentration, meaning that we require a larger value of the activity coefficient to obtain the desired fit to the same experimental data.

3. Davies Equation

The Davies equation is also based on the Debye-Hückel limiting law, but has been modified by an empirical parameter. The equation is often presented as

$$\log \gamma_i = -Az_i^2 \left[\frac{I^{0.5}}{1+I^{0.5}} - 0.3\,I \right] \tag{56}$$

where A is the limiting law parameter, z_i is the valence of the ith species, and 0.3 is the value of the empirical parameter. This equation is extensively used by soil chemists, and is the equation utilized in GEOCHEM (Sposito and Mattigod[37]). While the equation has the advantage of simplicity, in that it contains no ion-specific parameters, this is not of much significance when we perform these calculations in generalized, speciation-chemical equilibria computer programs.

The Davies equation suffers from the simplification that only ion charge is considered, thus we have only one activity coefficient for monovalent ions, one for divalent ions, etc. This approximation, that all species of similar charge behave identically, introduces errors, particularly for divalent and trivalent species. For example, at intermediate ionic strength, activity coefficients for Ca^{2+} and Mg^{2+}can differ by at least 10%. Given these large errors it appears unnecessary to include minor ion pair corrections such as $CaCl^+$ when using the Davies equation. These errors can be minimized if the ion association constants are also determined with the same model in which they are used (in this instance the Davies equation); unfortunately this is almost never done. This inconsistency in use of activity coefficient models generates errors in the application of the thermodynamic constants.

Use of the Davies equation requires specification of the value of the empirical parameter, which is often not stated by those using the equation. For example, some representations present the empirical value as 0.3,[38-40] some as 0.2[24] or 0.24[35]. These different parameter values produce large differences when $I \geq 0.1$. Shown in Table 3 are calculations of the activity coefficients for Ca and Mg as a function of ionic strength using the various constants discussed above. Note that the Davies equation yields very different results depending on which of the recommended constants are used. The calculations shown in Table 3 also show important differences between use of the Davies equation and the extended Debye-Hückel equation. This discrepancy is even larger for trivalent ions. For example, differences in the Al^{3+} activity coefficient calculated using the Davies equation (C = 0.3) and the extended Debye-Hückel equation reach 100% at $\mu = 0.5$! Use of this equation beyond 0.1 ionic strength is generally not recommended, as other equations give better results. Although some compilations list this equation as being suitable for higher ionic strengths than "the extended Debye-Hückel"[24] they are comparing Equation 56, the Davies equation, to the Debye-Hückel equation (54), rather than to the extended form represented by Equation 55.

Since the ion pairs are used as corrections on the single-ion activity coefficients, determination of the ion association constants using the same model as used for determination of the activity

Table 3 Ca^{2+} and Mg^{2+} Activity Coefficients

Equation	μ			
	0.03	0.1	0.3	0.5
Davies, C = 0.2[a]	0.515	0.355	0.252	0.229
Davies, C = 0.24[a]	0.516	0.362	0.265	0.251
Davies, C = 0.3[a]	0.522	0.372	0.290	0.290
D-H extended Ca^{2+} [b]	0.536	0.390	0.289	0.260
D-H extended Mg^{2+} [b]	0.545	0.406	0.315	0.292

[a] C is the adjustable parameter.
[b] Debye-Hückel.

coefficents would serve to reduce the discrepancy among the various equations. This inconsistancy is more prevalent when using the Davies equation since most ion pairs were calculated using the Debye-Hückel equation.

4. Ion Pairs and Complexes

For very dilute solutions the long-range electrostatic model of Debye-Hückel is suitable for representing the activity coefficients of major ions in water. Short-range interactions between oppositely charged ions can be treated by regarding the linked ion as a distinct entity, called an ion pair. These species differ from complexes but can operationally be treated in a similar manner, represented by reactions with thermodynamic equilibrium constants. The term ion pair is used for ions with primarily coulombic attraction, and the ions are separated by water of hydration. The term complex is usually reserved for entities which have coordinate covalent bonds, and the ions are not separated by water of hydration. Coordinate covalent bonds differ from ordinary covalent bonds in that the shared electrons are donated by only one of the atoms.[41] Typically complexes can be observed spectroscopically (such as by examination of infrared spectra).

The ion association model has been used to account for deviations between predicted and experimentally determined activities using various modifications of the Debye-Hückel model. As discussed earlier, the determination of thermodynamic constants from solution experiments depends on the activity coefficient model utilized. In most cases these may be considered as empirical constructs to fit the models to experimental data. The ion interaction models include specific ion interaction parameters and utilize only a very few ion pair entities.

An additional conceptual difficulty when comparing the ion association models with the ion interaction model is the different ionic strengths calculated by each model. Since the ion association model considers the ion pairs to be thermodynamic entities, species such as $CaSO_4^0$ do not contribute to the ionic strength. In contrast models based on the Pitzer equations consider that the ion interactions reduce the activity of the dissolved species, but in the above case this does not diminish the contribution of Ca^{2+} and SO_4^{2-} ions to the ionic strength.

As the attractive forces between ions in an ion pair are electrostatic (coulombic), the most important associations are between ions of higher valence and with cations of higher electronegativity. Thus, for divalent cations the stability of ion pairs with hydroxyls and carbonate ions generally follows the relation Cu > Fe > Zn > Mg > Ca = Sr > Ba. Other factors affecting the degree of association are the extent to which the cation can dehydrate, and the ionic radius. In contrast, the divalent-sulfate pairs all have similar K_d values of $\approx 10^{-2.3}$, suggesting to Garrels and Christ[42] that the association is that of a hydrated ion (all are of comparable radii) and the sulfate ion.

Increasing temperature generally results in an increase in the stability of the complex or ion pair, meaning a greater degree of ion association or an increase in the ion interaction parameters. This result of increasing ion pair stability is interpreted by Garrels and Christ[42] as being the result of the expansion of water with increasing temperature. As the solution expands, the proximity of water molecules around the dissolved ion diminishes and the polar effects of the water molecules on the dissolved ion are reduced. This polarity of the water molecules is the force causing the high degree of ionization of ionic salts in water. Chloride ion pairs are significant for trace metal elements but minor for alkaline earth metals and generally not given for alkali metals. Listings of ion pair equilibrium constants are available in various equilibrium models (WATEQ, MINTEQ, GEOCHEM), as well as in compilations such as Woods and Garrels,[9] Parkhurst,[36] and Nordstrom et al.[11]

5. Ion Interaction Model: the Pitzer Equation

The ion interaction model differs from the above models in that it does not attempt to represent the activity or activity coefficients of individual chemical species. Rather it represents the activity coefficient of the individual ion which is present in the salt. With the exception of a very few species the model does not include ion pairs. In this sense it is closer to a traditional thermodynamic approach, dealing with macroscopic properties. The Pitzer equations consist of a Debye-Hückel expression for long-range electrostatic interactions coupled with a detailed ion interaction model with virial coefficients. At high ionic strength, the activity coefficients are not universal functions of ionic strength but depend on the relative concentrations of the various ions present in solution. The activity coefficients are expressed in an expansion of the form[43]

$$\ln \gamma_i = \ln \gamma_i^{DH} + \sum_j B_{ij}(I) m_j + \sum_j \sum_k C_{ijk} m_j m_k + \ldots \tag{57}$$

where γ_i^{DH} is a modified Debye-Hückel activity coefficient and B_{ij} and C_{ij} are specific coefficients for each ion interaction.[43] The activity coefficient expression[44] for cations is given by

$$\ln \gamma_M = z_m^2 F + \sum_a m_a (2 B_{Ma} + Z C_{Ma}) + \sum_c m_c \left(2 \Phi_{Mc} + \sum_a m_a \Psi_{Mca} \right) + \sum_{a < a'} \sum m_a m_{a'} \Psi_{aa'M} +$$

$$|z_M| \sum_c \sum_a m_c m_a C_{ca} + \sum_n m_n (2 \lambda_{nM}) + \sum_n \sum_a m_n m_a \xi_{naM} \tag{58}$$

where the subscript M refers to a cation and the terms m_c, m_a, and m_n refer to the molalities of cation c, anion a, and neutral species n, z is the ion charge. The term F is given by

$$F = -A^\phi \left(\frac{I^{1/2}}{1 + b I^{1/2}} + \frac{2}{b} \ln(1 + b I^{1/2}) \right) +$$

$$\sum_c \sum_a m_c m_a B'_{ca} + \sum_{c < c'} \sum m_c m_{c'} \Phi'_{cc'} + \sum_{a < a'} \sum m_a m_{a'} \Phi'_{aa'} \tag{59}$$

The term A^ϕ is equal to one third the Debye-Hückel limiting slope and is equal to 0.39 at 298 K. The term C_{MX} is given by

$$C_{MX} = \frac{C_{MX}^\phi}{2 |Z_M Z_X|^{1/2}} \tag{60}$$

where

$$Z = \sum_i |z_i| m_i \qquad (61)$$

and C_{MX}^{ϕ} is the third virial coefficient. The second virial coefficients B are given by the following ionic strength-dependent terms.

$$B_{MX}^{\phi} = \beta_{MX}^{(0)} + \beta_{MX}^{(1)} \exp\left(-\alpha_1 \sqrt{I}\right) + \beta_{MX}^{(2)} \exp\left(-\alpha_2 \sqrt{I}\right)$$

$$B_{MX} = \beta_{MX}^{(0)} + \beta_{MX}^{(1)} g\left(\alpha_1 \sqrt{I}\right) + \beta_{MX}^{(2)} g\left(\alpha_2 \sqrt{I}\right) \qquad (62)$$

$$B_{MX}' = \beta_{MX}^{(1)} \frac{g'\left(\alpha_1 \sqrt{I}\right)}{I} + \beta_{MX}^{(2)} \frac{g'\left(\alpha_2 \sqrt{I}\right)}{I}$$

The functions g and g′ are defined by

$$g(x) = 2 \frac{\left(1 - (1+x)e^{-x}\right)}{x^2}$$

$$\qquad (63)$$

$$g'(x) = -2 \frac{\left(1 - \left(1 + x + \frac{x^2}{2}\right)e^{-x}\right)}{x^2}$$

with $x = \alpha_1 I^{0.5}$ or $\alpha_2 I^{0.5}$. When either the cation M or the anion X is univalent then $\alpha_1 = 2$. When the valence of the ions is 2-2 or higher then $\alpha_1 = 1.4$. For all electrolytes $\alpha_2 = 12$ and $b = 1.2$. The second virial coefficients Φ, which also depend on ionic strength, are given by

$$\Phi_{ij}^{\phi} = \theta_{ij} + \theta_{ij}^E(I) + I\theta_{ij}^{E'}(I)$$

$$\Phi_{ij} = \theta_{ij} + \theta_{ij}^E(I) \qquad (64)$$

$$\Phi_{ij}' = \theta_{ij}^{E'}(I)$$

The expressions $\theta_{ij}^{E'}(I)$ and $\theta_{ij}^E(I)$ account for the unsymmetrical mixing effects and are functions of ionic strength and ion charge of the electrolyte pair. The second and third virial coefficients λ_{ni} and ξ_{nij} (Equation 58) represent the interactions between ions and neutral species and are assumed constant, as are the third virial coefficients C_{MX}^{ϕ} and ψ_{ijk}. Thus, in order to calculate the ion interaction parameters we need the B^0, B^1, B^2, and C^{ϕ} values for each pair of ions. A listing of the parameters is provided by Pitzer[30] for a large number of 1-1, 2-1, 3-1, and 2-2 electrolyte pairs.

Treating individual ion interactions, the *Pitzer* equations do not require the large number of ion pairs which are required by the other models described earlier. For example, for the major ion chemistry, the GMIN model requires only consideration of the ion pairs $CaCO_3^0$, $MgCO_3^0$, and $MgOH^+$.[45] The model is considered accurate to 20.0 m and can be used down to infinite dilution. Because the model considers ion interaction parameters, with the exception of a few ions such as $MgCO_3^0$ and $CaCO_3^0$, it does not require the generation of distinct ion pair entities. The *Pitzer* equations are clearly to be preferred over the other activity coefficient models at high ionic strength, but are not likely to quickly replace the other models at low to intermediate ionic strength. This is in part due to the complexity of the ion interaction parameters and until recently the limited data on the temperature dependence of these values. More importantly use of the model is limited by the lack of discussion of these equations in soil and environmental chemistry texts and the consideration that the present models are satisfactory for application to most environments.

The ion interaction models in unmodified form do not calculate the activity of the individual chemical species in a solution. At present they are not useful for describing adsorption phenomena, such as adsorption of trace metals on oxides or soils, where evidence exists that the ion pairs and complex species are important. This restriction also extends to studies of plant ion uptake which is very dependent on the chemical species of the element and not just the free metal activity. However, the *Pitzer* equations can be modified by addition of ion pairs where such ion associations are known to exist.[46] Harvie et al.[47] developed the necessary parameters for use of this model for the major dissolved species in natural waters.

The chemical equilibrium model GMIN[45] utilizes the *Pitzer* equations with the Harvie et al.[47] parameter values for thermodynamic calculations and equilibria predictions in mineral-solution systems. This model is utilized for prediction of brine chemistry with consideration of major ions and corresponding mineral phases. The PHRQPITZ model[48] includes *Pitzer* equations for a larger group of chemical components. The *Pitzer* equations have also been included as an option in the unsaturated water flow-chemical transport model UNSATCHEM.[49]

D. Stoichiometric Activity Coefficients

1. Mean Salt Theory

Experimental data to high ionic strength is available for mean molal activity coefficients $\gamma_{\pm}$ of many salts. Individual ion activity coefficients cannot be directly determined but can be obtained from these data with certain assumptions. The MacInnes convention[50] assumes that the single-ion activity coefficients of K^+ and Cl^- are equal to each other and thus equal to the mean activity coefficient of KCl in water. With this assumption that $\gamma \pm KCl = \gamma_{K^+} = \gamma_{Cl^-}$, then we can determine the activity coefficients of other cations at the same ionic strength in MCl_x. For example

$$\gamma_{Na^+} = \frac{\gamma^2 \pm Na\ Cl}{\gamma \pm K\ Cl} \tag{65}$$

and

$$\gamma_{Ca^{2+}} = \frac{\gamma^3 \pm NaCl}{\gamma^2 \pm KCl} \tag{66}$$

These formulations are subject to some of the same assumptions as the extended Debye-Hückel equation discussed above. We are not explicitly accounting for ion interactions but rather considering an overall stoichiometric coefficient to be multiplied by the total molal concentration to obtain activities. In the above example, we have assumed that the Cl^- ion activity coefficient in KCl solutions is the same as that of Cl^- ions in a $CaCl_2$ solution of the same ionic strength. These assumptions require that the salts used do not form substantial complexes or ion pairs. This approach is best suited for brines where one salt predominates (e.g., NaCl brines). Although extensively utilized in the past, the extended Debye-Hückel formulation and parameters used by Truesdell and Jones[32] incorporate this information at high ionic strength. In addition, the Pitzer equations provide greater accuracy within a more generalized framework.

IV. OSMOTIC PRESSURE

A. Concepts

The osmotic pressure of a solution is defined as the external pressure that must be applied to the solution to raise the vapor pressure of the solvent A to that of pure A.[27] In aqueous soil solutions this osmotic pressure is related to the salt content of the solution, although the concept is also applicable to mixtures of solvents. The concept is most useful to plant scientists interested in water uptake. Osmotic and matric potential both result in a lowering of vapor pressure in the soil solution. Since plants act as a semipermeable membrane, they must perform work to extract relatively dilute water from saline solutions.

B. Calculation of Osmotic Pressure

In dilute solution, van't Hoff noted that the osmotic pressure π could be predicted by the universal gas law

$$PV = nRT \tag{67}$$

Relating $P = \pi$ and since $c = nM$, where c is molarity, then

$$\pi = cRT \tag{68}$$

This relation breaks down rapidly for solutions above about 0.2 m.

The osmotic pressure can be measured experimentally by the pressure exerted on a semipermeable membrane separating a salt solution from pure water. Since the chemical potential of the solution is equal on both sides, the lowering of potential due to dilution is given by:

$$\Delta\mu = RT \ln \frac{P_a}{P_a{}^0} \tag{69}$$

The increase in potential due to the osmotic pressure is

$$d\mu_a = V_a dP \tag{70}$$

thus,

$$\int_o^\pi V_a dP = -RT \ln \frac{P_a}{P_a^o} \tag{71}$$

For aqueous solutions we can consider that the solution is incompressible, where

$$V_a \pi = RT \ln \frac{P_a^o}{P_a} \tag{72}$$

If we assume that the partial molar volume V_a equals V_a^o, the molar volume of pure water, then for an ideal solution:

$$\pi \, V_a^o = RT \ln X_a \tag{73}$$

where X_a is the mole fraction of solvent in the solution. If we assume that the mol fraction of solute $X_b = n_b/n_a$ (dilute solution) then:

$$\pi = mRT \tag{74}$$

This equation, while not quite as accurate as Equation 73, is reasonably accurate to at least 0.5 m for nonelectrolyte salts and is preferred to the van't Hoff expression. For electrolytes, osmotic pressure is preferably expressed by a modification of Equation 73 given by Robinson and Stokes,[26] which accounts for nonideality as:

$$\pi = \frac{RTv \, W_a}{1000 \, V_a} \phi m \tag{75}$$

where V_a again is the partial molal volume of the solvent, v is the number of moles of ions given by a mole of electrolyte, W_a is the molecular weight of the solvent. Note $W_a/1000$ equals X_a of Equation 73. The term ϕ represents the molal osmotic coefficient.

C. Osmotic Coefficients

The molal osmotic coefficient is defined by:

$$\ln a_{H_2O} = -\frac{vm \, W_{H_2O}}{1,000} \phi \tag{76}$$

where a_{H_2O} is the activity of water.

If we consider a mixed electrolyte solution and let m_i represent the molality of the i ion then:

$$\ln a_{H_2O} = \frac{W\phi}{1,000} \left(\sum_{i=1} m_i \right) \tag{77}$$

V. GAS PHASE AND SOLUBILITY OF SOIL GASES

A. Gas Phase

Under earth surface conditions of temperature and pressure, the behavior of most pure gases can be represented by the universal gas law, given above (Equation 67). The activity of a gas can thus be represented by its pressure. Also, mixtures of gases, within the temperature range and pressure of interest to soil scientists, usually behave as ideal gas mixtures.

For ideal gases, the activity of each gas in a mixture is equal to its partial pressure. The partial pressure is defined as the pressure that a gas present in a mixture would exert if it were the only gas present in that volume. Thus, the partial pressure is equal to the total pressure multiplied by the mole fraction of the gas in the mixture:

$$a_i = P_i = P_T \cdot X_i \tag{78}$$

and the total pressure is equal to the sum of the partial pressures. These relations assume no interaction between the gases in a mixture.

Under nonideal conditions various equations have been developed based on the ideal gas law. Among these equations is the van der Waals' equation, given by

$$\left(P + \frac{an^2}{V^2}\right) \times (V - nb) = nRT \tag{79}$$

where a and b are constants specific to each gas. The volume term, V – nb, is linearly related to the amount of gas present. Other equations are also used to represent nonideality, such as a virial expansion of the form

$$\frac{PV}{nRT} = 1 + \frac{nB}{V} + \frac{n^2C}{V^2} + \frac{n^3D}{V^3} \tag{80}$$

where B, C, and D are the temperature-dependent virial coefficients. Alternatively the Redlich-Kwong equation[51] may be used:

$$P(V - nb) + \frac{na(V - nb)}{T^{0.5}V(V + nb)} = nRT \tag{81}$$

where a and b are gas-specific parameters.

B. Gas-Solution Equilibria

Solution theory is applicable to both liquid and solid systems. In this section we develop the relations for gas-solution systems. Nonelectrolyte solution theory is applicable to both uncharged molecules in liquids, as well as solid solutions in which the particles may be charged or uncharged. The simple expression given by Raoult's law was observed to be useful for conditions of constant temperature and pressure. Raoult's law considers that the partial pressure of a component in a mixture divided by the vapor pressure of the pure substance is equal to the mole fraction of A in the mixture.

$$\frac{P_A}{P_A^o} = X_A \tag{82}$$

The degree to which Raoult's law is obeyed depends on the extent to which the component physical properties are similar. Solutions in which components only vary isotopically almost exactly follow Raoult's law.

This relation is valid for a vapor phase which behaves as an ideal gas. For ideal gases or solutions, upon mixing components there is no change in volume ($\Delta V = 0$) or enthalpy ($\Delta H = 0$). The change in entropy caused by mixing n_A moles of A with n_B moles of B is given by:

$$\Delta S = -n_A R \ln X_A - n_B R \ln X_B \tag{83}$$

For dissolution of gases into a liquid or for liquid mixtures, ideal behavior is not observed. For dilute solutions the partial pressure of gas A is not equal to its mole fraction X_A in the solution multiplied by its pure vapor pressure.

James Cameron Gifford Library - Issue Receipt

Customer name: Rushby, Helena

Title: Soil physics.
ID: 1003946147
Due: 14/01/2015 23:59

Title: Soil physical chemistry / edited by Donald
L. Sparks.
ID: 1001569497
Due: 14/01/2015 23:59

Total items: 2
19/11/2014 12:02

All items must be returned before the due date
and time.

The Loan period may be shortened if the item is
requested.

The idealized pressure represented by Raoult's law is defined as the fugacity of a gas. Thus, the fugacity of an ideal gas is defined by:

$$f_i = f_i^o X_i \tag{84}$$

where f_i is the fugacity of gas "i" in a solution. Most solutions have low vapor pressures, and the vapor phase fugacity can be represented by the partial pressure, thus Raoult's law can be applied. When there is equilibrium between gas, liquid, or solid phases then the fugacity of a component in a vapor phase is equal to the fugacity of the component in the liquid or solid.

The activity of a component for any solution is defined as:

$$a_i = \frac{f_i}{f_i^o} \tag{85}$$

where f_i is the fugacity of the dissolved component "i" and f_i^o is its fugacity in the standard state. The standard state is usually the pure component at 1 atm total pressure and specified temperature. Thus, for a pure substance $a_i = 1$, and for an ideal solution the activity of component $a_i = X_i$.

For the dilute component in a mixture (particularly gases dissolved into water), the fugacity of the gas A is not equal to its mole fraction X_A multiplied by its pure vapor fugacity at the same total pressure, but rather is often linearly proportional to its mole fraction. This is Henry's law, given by

$$f_A = k\, X_A \tag{86}$$

This relation is satisfactory for many gases in the pressure range of interest to soil scientists, including the major atmospheric gases. As discussed earlier, we can also assume, for our conditions, that $f_a = P_a$. In the above relation we have assumed that there are no other components in the system. In the presence of electrolytes, P_A above can be designated as P_o, the partial pressure derived from the dissolution of the gas into pure water. The partial pressure in a salt solution, P_s, is given by:

$$P_s = \frac{\gamma_s m_s}{K} \tag{87}$$

where γ_s is the activity coefficient of the gas in the salt solution.

VI. SOLID PHASE STABILITY

A. Solubility Products and Ion Activity Products

Solubility products are the equilibrium constants relating the free energy of the solid phase to the dissolved products in solution. The reaction $CaCO_{3(s)} = Ca^{2+} + CO_3^{2-}$ is represented by the expression:

$$\frac{(Ca^{2+})(CO_3^{2-})}{(CaCO_3)_s} = K_{sp} \tag{88}$$

Since the activity of the pure solid phase under standard conditions is by convention $= 1$, the solid phase term $(CaCO_3)_s$ is usually not presented. Reactions considering minerals with ion substitution or minerals of high surface area (considered not to be in the standard state), which have an excess surface free energy, need to explicitly account for the activity of the solid phase in the equilibrium expression. Although not frequently utilized, this convention, i.e., the solution activities are equal to $K_{sp} \cdot a_{solid}$, seems preferable to the more commonly used alternative of altering the equilibrium constant for the phase (such as different constants for poorly crystallized phases).

It is often useful to determine if the solution phase is in equilibrium with a solid phase. This can be done by calculating the free energy of the solution species and comparing this value with the free energy of the solid. More conveniently, comparison is made between the ion activities in solution with the solubility product. The term ion activity product (IAP) is the product of the activities in solution corresponding to the equilibrium expression considered. For the case above, the $CaCO_3$ ion activity product is given by the product of the determined Ca^{2+} and CO_3^{2-} activities. Values of the ion activity product greater than the solubility product denote that the solution is supersaturated with respect to the solid phase considered. The term Ω is defined by

$$\Omega = \frac{IAP}{K_{sp}} \tag{89}$$

At equilibrium, $\Omega = 1$; for $\Omega < 1$ the solution is undersaturated with respect to the solid phase; and $\Omega > 1$ denotes supersaturation. Alternatively the term saturation index (SI) is utilized, defined by

$$SI = \log \frac{IAP}{K_{sp}} \tag{90}$$

$SI = 0$ denotes equilibrium and $SI > 0$ denotes supersaturation.

Note that the above solubility product relation is typically written in terms of the species in the solid, rather than a more realistic reaction describing the predominant entities in the reaction. In the above case, (HCO_3^-) is almost always the predominant carbonate species in solution, thus substituting:

$$\frac{\left(CO_3^{2-}\right)\left(H^+\right)}{\left(HCO_3^-\right)} = K_2 \tag{91}$$

where K_2 is the second dissociation constant of carbonic acid, we obtain:

$$\frac{\left(Ca^{2+}\right)\left(HCO_3^-\right)}{\left(H^+\right)} = \frac{K_{sp}}{K_2} = K' \tag{92}$$

This reaction with constant K' is conventionally distinguished from the solubility expression to avoid confusion. Nonetheless *numerical* solution of equilibria problems generally requires that we consider reactions with the predominant species of the chemical components.

B. Phase Diagrams

Phase diagrams provide a means of representing the stability fields of various phases as a function of variables such as temperature, pressure, or chemical composition. These diagrams were apparently first applied to natural systems by Pourbaix,[52] who developed a large series of Eh-pH diagrams of dissolved species for various elements. These diagrams have been further popularized by Garrels and Christ[42] and Lindsay,[40] among others. The diagrams are excellent for providing a visualization of the relative stability of phases or species, as well as the location of the stability field relative to other phases or species. Nonetheless the diagrams suffer from the simplifications inherent in representing complex systems in two dimensions. These can lead to misleading representations of natural systems. For example, plots of Ca^{2+} vs. pH can be made showing the solubility of calcite with different lines drawn for various CO_2 pressures. This is not of use if we wish to see how the three variables interact, i.e., how pH decreases as P_{CO_2} increases. A cursory look at such a diagram could lead to the assumption that increased CO_2 *decreases* calcite solubility (if pH is held constant). Data interpretation is also made difficult when variables are combined, especially the same variable on both axes (i.e., plots of log H_2PO_4-pH vs. log Al^{3+} + 3 pH or log pe + pH). In many instances such plots may give the impression that the concentration of a metal is decreasing with increasing pH, suggesting a solid phase control, while in fact the data may indicate that the total metal concentration is constant and that most of the changes in the variable are related to pH or metal hydrolysis. In these cases, direct evaluation of Al or phosphate changes with pH and the comparison of ion activity products to K_{sp} provides a clearer representation of possible solid phase control. The diagrams are best suited for simple systems where two variable representations are suitable (e.g., pe-pH diagrams of dissolved Se species in a simple electrolyte or at infinite dilution).

Stability fields for the diagrams are constructed by writing a reaction involving the relevant species or solid phases, obtaining the ΔG_f^0 for the species, calculating the ΔG_r^0, determining the equilibrium reaction constant, K, and subsequently solving for the variables defining the axes (for example pH and Ca) by specifying some fixed value for all other variables (such as ionic strength and CO_2 partial pressure. Often other variables must be assumed constant. For example, in order to construct a stability diagram for Ca solid phases plotted with the pH-P_{CO_2} axes, we write the reaction

$$\left(CaCO_3\right)_s + H_2O = Ca(OH)_2 + CO_2 \qquad (93)$$

The standard-state free energy of the reaction is given by

$$\Delta G_r^o = \Delta G_f^o Ca\left(OH_2\right) + \Delta G_f^o CO_2 - \Delta G_f^o\left(CaCO_3\right) - \Delta G_f^o\, H_2O \qquad (94)$$

At equilibrium $\Delta G = 0$ and we solve for log K using Equation 36. Assuming that the activity of water, calcite, and $Ca(OH)_{2s} = 1$, then PCO_2 at equilibrium with both phases is equal to K. In this example this line on the pH-PCO_2 diagram represents the stability boundary of these two phases. Similarly for an aqueous reaction the stability line between the two species is drawn for equal activities of the species.

C. Solid Substitution and Solid Solutions

As discussed earlier, for pure solid phases the activity is equal to one. Most naturally occurring minerals are not pure phases but have some degree of ion substitution. The extent to which substitution can take place, as well as the thermodynamic stability of the solid solution, depends on the mineral structure and the similarity of the two ions. The charge of the ion, as well as its ionic radius, are most important. Also important is the size of the site and the coordination number. For example, Al can be tetrahedrally or octahedrally coordinated. Clearly substitution for Al is markedly different for those sites. With increasing temperature, thermal motion increases, thus increasing the effective ionic radius. This increase is greater for the smaller ionic radii. Thus, the differences among ions become less, and greater amounts of substitution are possible at elevated temperatures. With decreasing temperature the solid solution may become thermodynamically unstable and may transform into two discrete phases or, as is often the case, remain as a homogeneous but now metastable phase.

The activity of a component in a solid solution can be treated in the same manner as for gases or liquids. For an ideal solid solution:

$$a_i = X_i \tag{95}$$

Thus, if the $MgCO_3$ behaved as an ideal solid solution with $CaCO_3$ then for $X_{MgCO_3} = 0.2$, $a_{MgCO_3} = 0.2$. Many binary nonelectrolyte solutions (including mineral phases) can be represented by regular solution theory, whereby the solid phase activity coefficient λ is represented by:

$$\ln \gamma_1 = \frac{B}{RT} X_2^2 \tag{96}$$

where B is a constant independent of mole fraction, and λ is the solid phase mole fraction activity coefficient. Equation 96 can be rewritten for a binary system:

$$\log \gamma_1 = B^1 (1 - X_1)^2 \tag{97}$$

such that as $X_1 \to 1$, $a_1 \to 1$. The standard states of the components are the pure components.

Consider the following example, assuming 5% Mg substitution into calcite, $X_{CaCO_3} = 0.95$ and $X_{MgCO_3} = 0.05$. If we assume ideal solution, then $\lambda_{CaCO_3} = 1$ and $\lambda_{MgCO_3} = 1$ and $a_{CaCO_3} = 0.95$. The equilibrium representation of calcite solubility under such conditions is given by:

$$\frac{(Ca^{2+})(CO_3^{2-})}{0.95} = 10^{-8.47} \tag{98}$$

and

$$\frac{(Mg^{2+})(CO_3^{2-})}{0.05} = K_{MgCO_3} \tag{99}$$

From the above we see that unless $\lambda_{CaCO_3} > (X_{CaCO_3})^{-1}$, ideal substitution of an ion into calcite *reduces* the apparent solubility of Ca in the solution. In the above example if we assume magnesite, $MgCO_3$, has a $K_{sp} = 10^{-8.03}$, then we can calculate the Mg^{2-}/Ca^{2+} ratio in equilibrium with this solid. Combining Equations 98 and 99 we obtain:

$$\frac{a_{Mg^{2+}}}{a_{Ca^{2+}}} = \frac{10^{-8.03} \cdot 0.05}{10^{-8.47} \cdot 0.95} \tag{100}$$

Assuming that:

$$\frac{a_{Mg^{2+}}}{a_{Ca^{2+}}} \simeq \frac{m_{Mg}}{m_{Ca}} \tag{101}$$

then

$$\frac{m_{Mg}}{m_{Ca}} \approx 0.144 \qquad (102)$$

or X_{Mg} in the solution phase = 0.125. In this example, we have assumed that λ_{CaCO_3} and λ_{MgCO_3} = 1, which is not likely valid. The complexity of determining solid phase equilibria and solubility can be seen by the following example.

Dissolving a solid solution, such as that given above $(Mg_{.05})$ $(Ca_{.95})$ (CO_3) in deionized water, with stoichiometric dissolution, would result in a Mg/Ca ratio in solution which is equal to that in the solid phase. Such a system is *not* in equilibrium, but may be in a metastable equilibrium state. This condition has been termed stoichiometric saturation.[53,54] In the above case, the stoichiometric solubility constant K_{ss} is given by

$$K_{ss} = \left(Ca^{2+}\right)^{0.95} \left(Mg^{2+}\right)^{0.05} \left(CO_3^{2-}\right) \qquad (103)$$

where K_{ss} is experimentally determined (in this case assumed to equal the ion activity product). The excess molar free energy is given by the expression[53]

$$G_{ex} = RT\left[\ln K_{ss} - X\ln\left(X K_{CaCO3} - (1-X)\ln(1-X)K_{MgCO3}\right)\right] \qquad (104)$$

where in our case X = 0.95. Solid phase free energies and solid phase activity coefficients can be determined from fitted parameters and the end-member solubility products. This concept may be useful for minerals where stoichiometric dissolution occurs and precipitation of a secondary solid solution, recrystallization of the solid phase, and diffusion in the solid phase *does not* occur (or is exceedingly slow). In this case the equilibrium condition, that the chemical potential of a component in the solid and liquid phases are equal, has not been met. If the kinetics of the reaction are sufficiently fast (or we wait sufficiently long) then, upon further reaction the solid phase continues dissolving and a magnesian calcite with a lower Mg substitution starts to precipitate. Such a system would have a continually changing composition in the solid phase as the solution phase continues to change (increasing Mg/Ca ratio). The final equilibrium solid phase would be intermediate in composition between the initial solid placed in distilled water and the initially formed precipitate. Similarly the final equilibrium Mg/Ca ratio would be intermediate between the initial solid phase cation ratio and the predicted composition based on solid solution theory using the initial solid phase composition. In the above example:

$$0.144 < \frac{m_{Mg}}{m_{Ca}} > 0.05 \qquad (105)$$

Determination of Mg-carbonate "stability" has, not surprisingly, generated considerable discussion. See, for example, the enlightening discussion by Thorstenson and Plummer,[55,56] Garrels and Wollast,[57] and Lafon[58] concerning binary solid solution (Mg-calcite) and equilibrium with an aqueous phase.

D. Surface Area

The effects of surface area on stability are generally neglected but are nonetheless important in soil systems, where many fine grained particles exist. The increase in solubility with increasing surface area (decrease in particle size) is related to the excess free energy of the surface, and is thus readily treated by the thermodynamic relations discussed above. Langmuir and Whittemore[59] presented data on the change in stability of ferric oxyhydroxides as a function of particle size. For cubes whose edge dimension is greater than 76 nm, goethite is more stable than hematite. They reported surface enthalpies of hematite (expressed as $1/2 \, \alpha \, Fe_2O_3$) and goethite ($\alpha \, FeOOH$) as 770 and 1250 ergs/cm^2, respectively, at 343.15 K. In the absence of additional data, Langmuir and Whittemore[59] assumed that $H_s^{343.15} = H_s^{298.15}$ and that $\Delta G_s = \Delta H_s$. In this case the excess free energy for hematite and goethite equals 7.7×10^{-5} J/cm^2 and 1.25×10^{-4} J/cm^2, respectively. Expressing this in kilojoules per mole,

$$\Delta G_s \text{ hematite } = \frac{\left(7.70 \times 10^{-5} J\right)}{cm^2} \cdot \frac{79.84g}{0.5 \, mole} \cdot \frac{cm^3}{5.26g} \cdot$$

$$\frac{particle}{10^{-18} cm^3} \cdot \frac{6 \times 10^{-12} cm^2}{particle} = 7.013 \, kJ/0.5 \, mole \tag{106}$$

Similarly, the excess surface free energy of a 10 nm crystal of goethite equals 15.53 kJ/mol. The changes in solubility given by $\Delta \log K$ (actually $Ka_{solid} - K$) for the reactions

$$0.5\left(\alpha Fe_2O_3\right) + 1.5H_2O = Fe^{3+} + 3OH^- \tag{107}$$

$$\alpha \, FeOOH + H_2O = Fe^{3+} + 3OH^- \tag{108}$$

are given by the relation $\Delta \log K = \Delta G / -5.708$, which equals 1.23 and 2.72, respectively, for hematite and goethite for particles of 10 nm. Since the excess surface free energy of goethite exceeds that of hematite, the stability of goethite decreases relative to hematite with decreasing particle size. In the above cases the equilibrium constant remains the same and the activity of the solid phases are 16.98 and 524.8, respectively, for hematite and goethite.

Chave and Schmaltz[60] determined the change in solubility of calcite as related to particle size. From these data they calculated surface energies of 1.88×10^{-9} J/m^2 and 3.01×10^{-9} J/m^2 for particles of 1 and 0.2 μm, respectively. The corresponding solid phase activities are 1.02 and 1.15. For many minerals, BET surface areas (determined by N_2 adsorption) are at least 10 times greater than surface areas calculated from particle size. Thus, surface area effects on stability may be more important than are presently considered.

VII. EXCHANGE EQUILIBRIA

A. Binary Exchange

For binary exchange between monovalent ions we can write the reaction:

$$AX + B^+ = BX + A^+ \tag{109}$$

where AX and BX are the respective exchange phase concentrations of A and B and the equilibrium reaction is given by:

$$\frac{\left(A^+\right)(BX)}{\left(B^+\right)(AX)} = K_{AB} \tag{110}$$

where parentheses denote activity and K_{AB} is the equilibrium constant. Expressing this relation in terms of concentration and activity coefficients instead of activity we obtain:

$$\frac{[m_{A^+} \cdot \gamma_{A^+}][\lambda_{BX}][X_B]}{[m_{B^+} \cdot \gamma_{B^+}][\lambda_{AX}][X_A]} = K_{AB} \tag{111}$$

where λ_{BX} and λ_{AX} are the exchanger phase rational activity coefficients and X_A and X_B are the mole fraction of exchanger A and B. If the exchanger forms an ideal solution, then Equation 111 reduces to:

$$\frac{\gamma_{A^+}}{\gamma_{B^+}} \cdot \frac{m_{A^+}}{m_{B^+}} = K_{AB} \cdot \frac{X_A}{X_B} \tag{112}$$

In a similar fashion for monovalent-divalent exchange and an ideal solid solution:

$$A_2X + C^{2+} = CX + 2A^+ \tag{113}$$

$$\frac{\gamma_A^2 \cdot [m_A]^2}{\gamma_C \cdot [m_C]} = K_{AC} \cdot \frac{X_{A_2}}{X_C} \tag{114}$$

As discussed earlier it is not often realistic to consider the solid phase solution in terms of ideal solution theory. More realistically, we can apply regular solution theory, discussed earlier for solid solution minerals. Substituting Equation 96 into Equation 110 we obtain:

$$\frac{(A^+)e^{\frac{B}{RT}X_B^2} \cdot X_B}{(B^+)e^{\frac{B}{RT}X_A^2} \cdot X_A} = K_{AB} \tag{115}$$

Simplifying and taking logarithms of both sides:

$$\ln\frac{(A^+)}{(B^+)} = \ln K_{AB} + \ln\frac{X_A}{X_B} - \frac{B}{RT}\left[X_A^2 - X_B^2\right] \tag{116}$$

Since $[X_A^2 - X_B^2] = [X_A + X_B][X_A - X_B]$ and $X_A + X_B = 1$, then,

$$\ln\frac{(A^+)}{(B^+)} = \ln K_{AB} + \ln\frac{X_A}{X_B}\frac{B}{RT}\left[X_A - X_B\right] \tag{117}$$

As an approximation for $0.9 > X_i > 0.1$

$$\ln\frac{X_A}{X_B} \approx 2\left[X_A - X_B\right] \tag{118}$$

Substituting Equation 117 into Equation 116 and taking antilogarithms yields:

$$\frac{\left(A^+\right)}{\left(B^+\right)} \cong K_{AB}\left(\frac{X_A}{X_B}\right)^{1-(B/2RT)} \tag{119}$$

If we consider that n = 1 − B/2RT, then application of regular solution theory results in the same form of equation as:

$$\frac{\left(A^+\right)}{\left(B^+\right)} = K_{AB}\left(\frac{X_A}{X_B}\right)^{n} \tag{120}$$

When B = 0, n = 1 and this equation reduces to the ideal solid solution Equation 112. In a similar manner for monodivalent exchange, using regular solution theory,

$$\frac{\left(A^+\right)^2}{\left(C^{2+}\right)} \cong K_{AC}\left(\frac{X_{A_2}}{X_C}\right)^{n} \tag{121}$$

Again when B = 0, n = 0 and this reduces to the ideal solid solution Equation 114. As written, Equation 113 indicates that 1 mole of exchanger has 2 moles of charge. For soil cation exchange the origin of the charge is usually substitution of an Al for a Si ion or substitution of a divalent ion (Mg^{2+}) for Al, thus the exchange site can be considered to have one unit charge.

$$2AX + C^2 = CX_2 + 2A^+ \tag{122}$$

then, using regular solution theory we obtain:

$$\frac{\left(A^+\right)^2}{\left(C^{2+}\right)} = K_{AC}\left(\frac{X_A^2}{X_C}\right)^{n} \tag{123}$$

For an ideal solid solution n = 1 and Equation 123 reduces to the Vanselow equation:

$$K_{ex} = \frac{\left(A^+\right)^2 X_C}{\left[X_A\right]^2 \left(C^{2+}\right)} \tag{124}$$

Alternatively, we can write the reaction:

$$2AX + C^{2+} = 2C_{0.5}X + 2A^+ \tag{125}$$

and we obtain for an ideal solid solution,

$$K = \frac{\left[C_{0.5}X\right]^2 \left(A^+\right)^2}{\left[AX\right]^2 \left(C^{2+}\right)} \tag{126}$$

This equation is related to the Gapon equation:

$$K_G = \frac{X_B A^+}{X_A \sqrt{C^{2+}}}$$ (127)

by the relation $K_G^2 = K$, assuming the activities of A^+ and C^{2+} in solution are equal to their concentrations in solution. These expressions all differ in their representation of the mechanism of exchange, but all can be considered to have a thermodynamic basis with differing conventions. The Gapon equation represents the convention that 1 mole of exchanger charge reacts with z^{-1} moles of cation i, meaning 1 mole of cation charge.

B. Exchange Isotherms

Exchange isotherms are used to present a visual representation of exchange data. They consist of plots of charge (or equivalent) fraction of a species in the solution phase plotted against the charge fraction on the exchanger phase. The thermodynamic nonpreference isotherm follows from the concepts that the exchanger forms an ideal solution and that the $\Delta G_r^o = 0$ (K = 1). Plots of experimental data allow comparison with the nonpreference isotherm. Others provide information about chemical information such as ion preference and site heterogeneity. For a binary exchange system:

$$E_1 = \frac{z_1 X_1}{z_1 X_1 + z_2 X_2}$$ (128)

where E_1 is the equivalent fraction of ion 1 on the exchanger phase (replaced by the term charge fraction in SI). The equivalent fraction of ion 1 in the solution phase (charge fraction), $\tilde{E}_1$, is given by

$$\tilde{E}_1 = \frac{z_1 m_1}{z_1 m_1 + z_2 m_2}$$ (129)

For homovalent exchange, the nonpreference isotherm is a straight line, where $\tilde{E}_1 = E_1$. The assumption of an ideal solution in the aqueous and solid phases is not realistic for soil systems and numerous empirical equations have been proposed. The above discussion has assumed that the exchanger phase contains a fixed charge. Soils contain variable charge as well as permanent charge. A detailed discussion of exchange properties is presented in Chapters 1 and 2.

C. Exchange Models

As developed and used by soil scientists, exchange models do not meet the thermodynamic rigor needed to regard the constants as equilibrium constants. Derivations of the regular solution relations, Equation 120 and 121, are also based on the assumption that $0.1 < X_i < 0.9$, which is not valid for many soil exchange reactions. Because soil is a mixture of various minerals and organic matter the assumption that the exchanger is a simple phase (homogeneous portion of a system) is also not valid. Another important assumption regards the use of binary expressions for multi-ion reactions. Rather than a thermodynamic equilibrium constant, we utilize the term K_{ex}, the exchange selectivity coefficient.

Despite the empirical nature of the expressions, it is still desirable to formulate equations which have stable or constant selectivity coefficients which allow application over a wide range of conditions, particularly the exchange phase composition. Among these empirical models, the Vanslow and Gapon equations are the most frequently used. As with solid substitution into mineral phases, the degree to which the exchanger phase approximates an ideal solution depends on the similarity of the properties of the ions involved. For specific minerals, such as smectite or illite, Ca-Mg exchange can be represented by a constant exchange selectivity coefficient.[61,62]

Multiphase systems, such as soils, are generally not suited to the use of a single exchanger selectivity coefficient, even for homovalent binary exchange. It can be shown that a mixture of two phases, both of which exhibit ideal solid solution exchange (invariant exchange coefficient), exhibits a variable selectivity coefficient whenever K_{ij} of phase one is not equal to K_{ij} of phase 2.

Since the empirical equations are related to the thermodynamic equilibrium constants when one assumes ideal or regular solution theory, deviations in the selectivity coefficient are directly related to the solid phase activity coefficients. For example, the changes in the Vanslow selectivity coefficient can be used to calculate the activities of the exchanger phase for Ca-Na exchange $a_{xi} \to 1$ as $X_i \to 1$ by the reference state $a_i = 1$ for a pure solid (exchanger) phase. Thus, as $X_i \to 1$, $K_v \to K$; changes in K_v are related to changes in the solid phase activity coefficients by[6]

$$d \ln K_v = z_2 d \ln f_i - z_i d \ln f_2 \tag{130}$$

$$\ln K = \ln K_v - z_2 \ln f_i + z_1 \ln f_2 \tag{131}$$

where f_i is the rational activity coefficient of $M_i X_{zi}$ at any X_i.

D. Multi-ion Exchange

Until relatively recently the cation exchange selectivity of clays and soils has been examined almost exclusively in binary systems. The implicit assumption that binary systems can be combined and used to represent exchange in multi-ion systems has not been extensively evaluated.

Using the law of mass action for ternary monovalent cation exchange and by combining Equation 120 and a similar equation for A-D exchange, we obtain

$$\frac{\left(A^+\right)}{\left(B^+\right)\left(D^+\right)} = k_{AB} \cdot k_{AD} \frac{\left(X_A\right)^{n+m}}{\left(X_B\right)^n \left(X_D\right)^m} \tag{132}$$

Use of binary exchange reactions to construct ternary systems has been successfully applied to several specimen minerals (e.g., Sposito and Le Vesque[61]).

E. Adsorption

Adsorption models can be considered as variations of exchange models. They differ from the thermodynamic exchange models presented above in that they include specific assumptions or mechanisms. The models are generally extended for use beyond systems for which the assumptions may be valid so that they often become empirical models. The application of the Langmuir absorption model to soils with heterogeneous adsorption sites is a case in point. These adsorption and exchange models are treated in detail in Chapters 2 and 1, respectively.

Surface complexation models and their use for adsorption studies on soils have been reviewed by Goldberg.[63] The surface complexation models consist of mechanistic models which employ thermodynamic concepts. These models contain at least one coulombic correction factor to account for the effect of surface charge on surface complexation. The coulombic correction factors can be considered as solid phase activity coefficients. The constant-capacitance model utilizes constant ionic strength as the reference state for the dissolved species, thus solution activity coefficient calculations are not required. The standard state for the surface species is an uncharged environment.

It is assumed that all adsorption is by inner-sphere complexation. The intrinsic equilibrium constants are thus obtained by extrapolating the conditional constants obtained under measured or assumed charge conditions to zero net surface charge.

Subsequent to development of the constant-capacitance model, more detailed models, termed the triple-layer model and modified triple-layer model, have been described.[64] The triple-layer model considers all adsorption except for protons and hydroxyls to be outer-sphere complexation. The reference state for the dissolved species is infinite dilution; activities are calculated using one of the single-ion activity coefficient models described earlier. The intrinsic equilibrium constants are again extrapolated to zero net charge as this is the reference state for the surface species. The modified triple-layer model considers inner- as well as outer-sphere complexation. The reference state for the surface species is now infinite dilution as well as zero net charge. By consideration of solution and surface activity coefficients, this model includes intrinsic conditional equilibrium constants which are numerically equivalent to thermodynamic equilibrium constants.

Readers desiring more comprehensive information on the application of thermodyamics to soil and geological systems are referred to two excellent texts by Fletcher[51] and Anderson and Crerar.[65]

REFERENCES

1. Plummer, L. N., Parkhurst, D. L., and Wigley, T. M. L., Critical review of the kinetics of calcite dissolution and precipitation, in *Chemical Modeling in Aqueous Systems*, Jenne, E. A., Ed., Am. Chem. Soc. Symp. Ser. 93., American Chemical Society, Washington, D.C., 1979, 537.

2. Ikeda, T., Nakahara, J., Sasaki, M., and Yasunaga, T., Kinetic behavior of alkali metal ion on zeolite 4A surface using the stopped-flow method, *J. Colloid Interface Sci.*, 97, 278, 1984.

3. Zhang, P. and Sparks, D. L., Kinetics and mechanisms of sulfate adsorption/desorption on goethite using pressure-jump relaxation, *Soil Sci. Soc. Am. J.*, 54, 1266, 1990.

4. Sparks, D. L. and Jardine, P. M., Thermodynamics of potassium exchange using a kinetics approach, *Soil Sci. Soc. Am. J.*, 45, 1094, 1981.

5. Morgan, J. J., Applications and limitations of chemical thermodynamics in natural water systems, in *Equilibrium Concepts in Natural Water Systems*, Stumm, W., Ed., American Chemical Society, Washington, D.C., 1967, 1–29.

6. Sposito, G., Thermodynamics of the soil solution, in *Soil Physical Chemistry*, Sparks, D. L., Ed., CRC Press, Boca Raton, FL, 1986, 147.

7. Wagman, D. D., Evans, W. H., Parker, V. B., Schumm, R. H., Halow, I., Bailey, S. M., Churney, K. L., and Nuttall, R. L., The NBS tables of chemical thermodynamic properties, *J. Phys. Chem. Ref. Data*, Suppl. 2, 11, 1982.

8. Chase, M.W., Davies, C.A., Downey, J.R., Jr., Frurip, D.J., McDonald, R.A., and Syverud, A.N., *JANAF Thermochemical Tables*, 3rd ed., American Institute of Physics, New York, 1985.

9. Woods, T.L. and Garrels, R. M., *Thermodynamic Values at Low Temperature for Natural Inorganic Materials: An Uncritical Summary*, Oxford University Press, New York, 1987.

10. Robie, R. A. and Hemingway, B. S., Thermodynamic properties of minerals and related substances at 298.15 K and 1 bar (10^5 Pa) pressure and at higher temperatures, Geological Survey Bull., 2131, U.S. Geological Survey, Washington, D.C., 1995.

11. Nordstrom, D. K., Plummer, L. N., Langmuir, D., Busenburg, E., May, H. M., Jones, B. F., and Parkhurst, D. L., Revised chemical equilibrium data for major water-mineral reactions and their limitations, in *Chemical Modeling of Aqueous Systems, Vol. 2*, Melchior, D. C. and Bassett, R. L., Eds., ACS Ser. 416, American Chemical Society, Washington, D.C., 1990, 398.

12. U.S. Salinity Laboratory Staff, *Diagnosis and Improvement of Saline and Alkali Soils*, USDA Handb. 60. U.S. Government Printing Office, Washington, D.C., 1954.

13. Rhoades, J .D., *Soluble Salts in Methods of Soil Analysis*, Part 2, Am. Soc. Agron. Monogr. No. 9, American Society of Agronomy, Madison, WI, 1982.

14. Kinniburgh, D. G. and Miles, D. L., Extraction and chemical analysis of interstitial waters from soils and rocks, *Environ. Sci. Technol.*, 17, 362, 1983.

15. Philips, I. R. and Bond, W. J., Extraction procedure for determining solution and exchangeable ions in the same sample, *Soil Sci. Soc. Am. J.,* 53, 1294, 1989.
16. Elkhatib, E. A., Bennett, O. L., Baligar, V. C., and Wright, R. J., A centrifuge method for obtaining soil solution using an immiscible liquid, *Soil Sci. Soc. Am. J.,* 50, 297, 1986.
17. Adams, F. C., Burmester, N., Hue, V., and Long, F. L., A comparison of column-displacement and centrifuge methods for obtaining soil solutions, *Soil Sci. Soc. Am. J.,* 44, 733, 1980.
18. Reynolds, B., A simple method for the extraction of soil solution by high speed centrifugation, *Plant Soil,* 78, 437, 1984.
19. Elkhatib, E. A., Hern, J. L., and Staley, T. E., A rapid centrifuge method for obtaining soil solution, *Soil Sci. Soc. Am. J.,* 51, 578, 1987.
20. Hansen, E. A. and Harris, A. R., Validity of soil water samples collected with porous ceramic cups, *Soil Sci. Soc. Am. Proc.,* 39, 528, 1975.
21. Litaor, M. I., Review of soil solution samplers, *Water Resour. Res.,* 24, 727, 1988.
22. Suarez, D. L., Prediction of pH errors in soil-water extractors due to degassing, *Soil Sci. Soc. Am. J.,* 51, 64, 1987.
23. Denbigh, K., *The Principles of Chemical Equilibrium,* Cambridge University Press, New York, 1981.
24. Stumm, W. and Morgan, J., *Aquatic Chemistry,* 3rd ed., John Wiley & Sons, New York, 1996.
25. Stokes, R., Thermodynamics of solutions, in *Activity Coefficients in Electrolyte Solutions,* Vol. 1, Pytkowicz, R. M., Ed., CRC Press, Boca Raton, FL, 1979.
26. Robinson, R. A. and Stokes, R. H., *Electrolyte Solutions,* Butterworths, London, 1965.
27. Moore, W. J., *Physical Chemistry,* Prentice-Hall, Englewood Cliffs, NJ, 1972.
28. Pytkowicz, R. M. and Johnson, K. S., Lattice theories and a new lattice concept for ionic solutions, in *Activity Coefficients in Electrolyte Solutions,* Vol. 1, Pytkowicz, R. M. Ed., CRC Press, Boca Raton, FL, 1979, 209.
29. Kielland, J., Individual activity coefficients of ions in aqueous solutions, *J. Am. Chem. Soc.,* 59, 1675, 1937.
30. Pitzer, K. S., Ion interaction approach: theory and data correlation, in *Activity Coefficients in Electrolyte Solutions,* 2nd ed., Pitzer, K.S., Ed., CRC Press, Boca Raton, FL, 1991, chap. 3.
31. Helgeson, H. C. and Kirkham, D. H., Theoretical prediction of the thermodynamic behaviour of aqueous electrolytes at high pressures and temperature. II. Debye-Huckel parameters for activity coefficients and relative partial molal properties, *Am. J. Sci.,* 274, 1199, 1974.
32. Truesdell, A. H. and Jones, B. F., WATEQ, a computer program for calculating chemical equilibria of natural waters, *J. Res. U.S. Geol. Surv.,* 2, 233, 1974.
33. Ball, J. W., Jenne, E. A., and Cantrell, M. W., WATEQ3: A Geochemical Model with Uranium Added, U.S. Geol. Surv. Open-File Rep. 81-1183, U.S. Geological Survey, Menlo Park, CA, 1981.
34. Brown, D. S. and Allison, J. A., MINTEQA1, an Equilibrium Metal Speciation Model, EPA-600/3-87-012, Office of Research and Development, U.S. Environmental Protection Agency, Athens, GA, 1987.
35. Allison, J. D., Brown, D. S., and Novo-Gradac, K. J., *MINTEQA2/PRODEFA2, a Geochemical Assessment Model for Environmental Systems: Version 3.0 User's Manual,* Environmental Research Laboratory, U.S. Environmental Protection Agency, Athens, GA, 1991.
36. Parkhurst, D. L., Ion-association models and mean activity coefficients of various salts, in *Chemical Modeling of Aqueous Systems,* Vol. 2, Melchior, D. C. and Bassett, R. L., Eds., ACS Ser. 416, American Chemical Society, Washington, D.C., 1990, 30.
37. Sposito, G. and Mattigod, S., *GEOCHEM: A Computer Program for the Calculation of Chemical Equilibria in Soil Solutions and Other Natural Water Systems,* Department of Soil and Environmental Science, University of California, Riverside, 1977, 110.
38. Davies, C. W., *Ion Association,* Butterworths, London, 1962.
39. Sposito, G., *The Chemistry of Soils,* Oxford University Press, New York, 1989.
40. Lindsay, W., *Chemical Equilibria in Soils,* John Wiley & Sons, New York, 1979.
41. Pytkowicz, R. M., *Equilibria, Nonequilibria, and Natural Waters,* Vol. 1, John Wiley & Sons, New York, 1983.
42. Garrels, R. M. and Christ, C. L., *Solutions, Mineral, and Equilibria,* Harper and Row, New York, 1965, 450.

43. Pitzer, K. S., Theory: ion interaction approach, in *Activity Coefficients in Electrolyte Solutions*, Vol. 2, Pytkowicz, R. M., Ed., CRC Press, Boca Raton, FL, 1979, 157.
44. Felmy, A. R. and Weare, J. H., The prediction of borate mineral equilibria in natural waters: application to Searles Lake, California, *Geochim. Cosmochim. Acta*, 50, 2771, 1986.
45. Felmy, A. R., GMIN: *A Computerized Chemical Equilibrium Model Using a Constrained Minimization of the Gibbs Free Energy*, Pacific Northwest Laboratories, Richland, WA, 1990.
46. Clegg, S. L. and Whitfield, M., Activity coefficients in natural waters, in *Activity Coefficients in Electrolyte Solutions*, 2nd ed., Pitzer, K. S., Ed. CRC Press, Boca Raton, FL, 1991, chap. 6.
47. Harvie, Ch. E., Moller, N., and Weare, J. H., The prediction of mineral solubilities in natural waters: the Na-K-Mg-Ca-H-Cl-SO_4-OH-HCO_3-CO_3-H_2O system to high ionic strengths at 25°C, *Geochim. Cosmochim. Acta*, 48, 723, 1984.
48. Plummer, L.N., Parkhurst, D.L, Flemming, G.W., and Dunkle, S.A., A computer program incorporating Pitzer's equation for calculation of geochemical reactions, in *Brines. U.S. Geological Survey, Water Resources Investigations Report*, 88-4153, U.S. Geological Survey, Reston, VA, 1988.
49. Suarez, D. L. and Simunek, J., UNSATCHEM: Unsaturated water and solute transport model with equilibrium and kinetic chemistry, *Soil Sci. Soc. Am. J.*, 61, 1633, 1997.
50. MacInnes, D. A., *The Principles of Electrochemistry*, Reinhold, New York, 1939.
51. Fletcher, P., *Chemical Thermodynamics for Earth Scientists*, Longman Scientific, Essex, U.K. 1993.
52. Pourbaix, M. J. N., *Thermodynamics of Dilute Aqueous Solutions*, E. Arnold, London, 1949.
53. Glynn, P. and Reardon, E. J., Solid-solution aqueous-solution equilibria: thermodynamic theory and representation, *Am. J. Sci.*, 290, 164, 1990.
54. Konigsberger, E. and Gamsjager, H., Comment, solid-solution aqueous-solution equilibria: thermodynamic theory and representation, *Am. J. Sci.*, 292, 199, 1992.
55. Thorstenson, D. C. and Plummer, L. N., Equilibrium criteria for two-component solids reacting with fixed composition in an aqueous phase — example: the magnesian calcites, *Am. J. Sci.*, 277, 1203, 1977.
56. Thorstenson, D. C. and Plummer, L. N., Reply, *Am. J. Sci.*, 278, 1478, 1978.
57. Garrels, R. and Wollast, R., Discussion, *Am. J. Sci.*, 278, 1469, 1978.
58. Lafon, G. M., Discussion. Equilbrium criteria for two-component solids reacting with fixed composition in an aqueous phase — example: the magnesium calcites, *Am. J. Sci.*, 278, 1455, 1978.
59. Langmuir, D. and Whittemore, D., Variations in the stability of precipitated ferric oxyhydroxides, in *Nonequilibrium Systems in Natural Water Chemistry*, Adv. in Chem. Ser. No. 106, American Chemical Society, Washington, D.C., 1971, 209.
60. Chave, K. E. and Schmaltz, R. F., Carbonate-seawater interactions, *Geochim. Cosmochim. Acta.*, 30, 1037, 1966.
61. Sposito, G. and Le Vesque, C. S., Sodium-calcium-magnesium exchange on Silver Hill illite, *Soil Sci. Soc. Am. J.*, 49, 1153, 1985.
62. Suarez, D. L. and Zahow, M., Calcium-magnesium exchange selectivity of Wyoming montmorillonite in chloride, sulfate, and perchlorate solutions, *Soil Sci. Soc. Am. J.*, 53, 52, 1989.
63. Goldberg, S., Use of surface complexation models in soil chemical systems, in *Advances in Agronomy*, Vol. 47, Sparks, D. L., Ed., Academic Press, New York, 1994, 234.
64. Hayes, K. F. and Leckie, J. O., Modeling ionic strength effects on cation adsorption at hydrous oxide/solution interfaces, *J. Colloid Interface Sci.*, 115, 564, 1987.
65. Anderson, G. M. and Crerar, D. A., *Thermodynamics in Geochemistry. The Equilibrium Model*, Oxford University Press, New York, 1993.
66. Kharaka, Y. K., Gunter, W. D., Aggarwal, P. K., Perkins, E. H., and DeBraal, J. D., *SOLMNEQ.88: A Computer Program Code for Geochemical Modeling of Water-Rock Interactions*, U.S. Geol. Surv. Water Resour. Invest. Rep. 88-4227. U.S. Government Printing Office, Washington, D.C., 1988.
67. Martell, A. E. and Smith, R., *Critical Stability Constants*, 6 Vols., Plenum Press, New York, 1974-1989.
68. Sadiq, M. and Lindsay, W. L., Selection of standard free energies of formation for use in soil chemistry, *Colorado State Univ. Tech. Bull.*, 1979, 134.
69. Barin, I., *Thermochemical Data of Pure Substances*, Part 1-3, VCH Publishers, Weinheim, West Germany, 1993.

APPENDIX A

Allison et al., 1991[35]	ΔH_r^o log K
Ball et al. 1981[33]	ΔH_r^o log K
Brown and Allison, 1987[34]	ΔH_r^o log K
Kharaka et al., 1988[66]	K as f (T,P)
Martell and Smith, 1974-1989[67]	$\Delta H_r^o \Delta S_r^o$ log K
Robie and Hemingway, 1995[10]	$\Delta G_f^o \Delta H_f^o$ log K
Sadiq and Lindsay, 1979[68]	$\Delta G_f^o \Delta H_f^o$ log K
Wagman et al., 1982[7]	$\Delta G_f^o, \Delta H_f^o S^o$
Woods and Garrels, 1987[9]	$\Delta G_f^o, \Delta H_f^o S^o$
Barin, 1993[69]	$C_p \Delta H_f^o \Delta G_f^o$ log K

Kinetics and Mechanisms of Chemical Reactions at the Soil Mineral/Water Interface

Donald L. Sparks

CONTENTS

0-87371-883-6/99/$0.00+$.50
© 1999 by CRC Press LLC

I. INTRODUCTION

Without question, one of the important paradigms in our society is preservation of the environment. Worldwide, concerns have been voiced about numerous soil and water contaminants. These include plant nutrients (e.g., nitrate and phosphate), heavy metals, radionuclides, pesticides, and other organic chemicals.

The reactions that these contaminants undergo with natural particles, such as sediments and soils, involving sorption, desorption, precipitation, complexation, redox, and dissolution phenomena, are critical in determining their fate and mobility in the subsurface environment.

Much of the research on migration and retention of contaminants on natural materials has been studied from a macroscopic, equilibrium approach. The focus of many of these investigations has been on the determination of distribution coefficients (determined primarily on a 24-hour basis), and the use of equilibrium-based models such as the Freundlich, Langmuir, and the various surface complexation models, e.g., constant-capacitance and triple-layer, to determine numerous sorption parameters, information on the physical description of the electric double layer, and data conformance over a wide range of experimental conditions such as varying pH and ionic strength. While the surface complexation models are predicated on molecular descriptions of the electric double layer, equilibrium-derived data are employed and, thus, no direct molecular information is provided.

The above criticism of equilibrium approaches is not meant to imply that they are not useful, since they provide important data on the final state of a reaction. However, they provide no information on reaction rates or mechanisms. Moreover, such equilibrium studies are usually not relevant to field settings, since reactions involving subsurface materials are seldom, if ever, at equilibrium. To understand the rates of chemical reactions on particle surfaces, one must study the kinetics of the reactions. Kinetic studies can assist in revealing reaction mechanisms. Of course, to determine mechanisms directly, one must use microscopic and spectroscopic surface techniques. Ideally, one should follow reaction rates microscopically and/or spectroscopically and couple these with macroscopically observed processes. Such approaches will be discussed later. In short, time-dependent reactions are important factors in controlling the fate and transport of contaminants in the subsurface environment.

Early studies by Way[1] on ion exchange in soils showed that reaction rates were often instantaneous. Similar conclusions were reached by Gedroiz[2] and Hissink.[3] The finding that ion-exchange kinetics were diffusion-controlled was discovered by Boyd et al.[4] In their seminal paper they also elucidated rate-limiting steps for ion exchange. Kelley[5] correctly hypothesized that rates of ion exchange should be highly dependent on the adsorbent. For example, reaction rates on kaolinite, which has only external surface sites, should be higher than on vermiculite, which has planar and edge external as well as internal sites. Reactions on internal sites, depending on their geometry, could be quite slow. Unfortunately, Kelley's[5] astute observations went unnoticed and little research on reaction rates on soils and soil components appeared over the next three decades. There were notable contributions by Helfferich[6] on ion-exchange kinetics and by Mortland and coworkers[7,8] and Scott and coworkers[9] on the kinetics of potassium release from vermiculites and mica.

In the late 1970s, and certainly in the 1980s and 1990s, the kinetics of environmentally important reactions at the soil mineral/water interface has become a major leitmotif in the soil and environmental sciences and in environmental engineering. This intense interest is in large part due to the recognition that reactions in natural settings are usually time dependent and thus, to predict accurately the fate of contaminants in the subsurface environment, a knowledge of the reaction kinetics is imperative. While major advances have been made in understanding time-dependent reactions on natural materials such as soils and sediments, there are still many unknowns and needs that are complicated by the complex, heterogeneous nature of natural materials. Research needs include models that accurately describe both chemical kinetics and transport processes in multiple-site, heterogeneous systems; better kinetic methods; more extensive studies on the effect of residence time ("aging") on contaminant retention/release; and mechanistic studies that employ time-resolved *in situ* microscopic and spectroscopic techniques.

In this chapter, I shall discuss the application of chemical kinetics to heterogeneous systems such as soils and soil components (clay minerals, organic matter, and humic substances), with emphasis on sorption/release processes. A critical review of kinetic models that can be used to describe reaction rates on heterogeneous surfaces will be covered. The chapter will also cover the kinetics of important inorganic and organic sorption/desorption and dissolution reactions at the soil mineral/water interface. Additionally, there are discussions on the use of *in situ* spectroscopic and microscopic techniques to confirm reaction mechanisms at the soil mineral/water interface. For additional details on these topics and other aspects of kinetics of soil chemical and geochemical processes the reader should consult a number of recent books.[10-14]

II. TIME SCALES OF SOIL CHEMICAL PROCESSES

A variety of chemical reactions occur in soils and often in combination with one another. Reaction time scales can vary from microseconds for many ion association reactions to microseconds and milliseconds for some ion exchange and sorption reactions, to years for many mineral-solution and mineral crystallization phenomena and for some sorption/desorption reactions (Figure 1). Ion association reactions include ion pairing, inner- and outer-sphere complexation, and chelation in solution. Gas-water reactions involve gaseous exchange across the air-liquid interface. Ion exchange reactions occur when cations and anions are adsorbed and desorbed from soil surfaces via electrostatic attractive forces. Ion exchange reactions are reversible and stoichiometric. Sorption reactions can involve physical sorption, outer-sphere complexation, inner-sphere complexation, and surface precipitation. Mineral-solution reactions include precipitation/dissolution of minerals, and coprecipitation reactions, whereby small constituents become a part of mineral structures.[10,11]

The type of soil component can drastically affect the reaction rate. For example, sorption reactions are often more rapid on clay minerals such as kaolinite and smectites than on vermiculitic and micaceous minerals. This is in large part due to the availability of sites for sorption. For example, kaolinite has readily available planar external sites and smectites have primarily internal sites that are also quite available for retention of sorptives. Thus, sorption reactions on these soil constituents are often quite rapid, even occurring on time scales of seconds and milliseconds.[10]

Metal and metalloid sorption reactions on oxides, hydroxides, and humic substances depend on the type of surface and metal being studied, but the chemical reaction rate appears to be rapid. For example, chemical reaction rates of molybdate, sulfate, selenate, and selenite on goethite occurred on millisecond time scales.[15-17] Half-times for divalent Pb, Cu, Zn sorption on peat ranged from 5 to 15 s.[18] Some studies have shown that heavy metal sorption on oxides[19-20] and clay minerals[20] increases with longer equilibration times. The mechanism for these lower reaction rates is not well understood and has been ascribed to diffusion phenomena, sites of lower reactivity, and surface nucleation/precipitation.[19-21] Recent findings on slow metal retention mechanisms at the mineral/water interface will be discussed later.

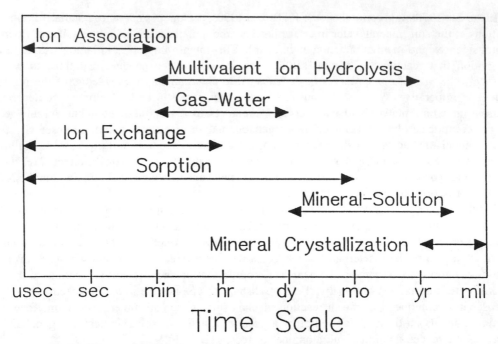

Figure 1   Time ranges required to attain equilibrium by different types of reactions in soil environments. (From Amacher, M. C., in *Rates of Soil Chemical Processes,* SSSA Spec. Publ. No. 27, Sparks, D. L. and Suarez, D. L., Eds., Soil Science Society of America, Madison, WI, 1991, 19. With permission.)

On the other hand, vermiculite and micas have multiple sites for retention of metals and organics, including planar, edge, and interlayer sites, with some of the latter sites being partially to totally collapsed. Consequently, sorption and desorption reactions on these sites can be slow, tortuous, mass transfer-controlled. Often, an apparent equilibrium may not be reached even after several days or weeks. Thus, with vermiculite and mica, sorption can involve two to three different reaction rates — high rates on external sites, intermediate rates on edge sites, and low rates on interlayer sites.[22-24]

Sorption/desorption of metals and organic chemicals on soils is often very slow, which has been attributed to diffusion into micropores of inorganic minerals and into humic substances, retention on sites of lower reactivity, and surface nucleation/precipitation. These reactions will be discussed in more detail later.

III. APPLICATION OF CHEMICAL KINETICS TO HETEROGENEOUS SURFACES

The study of chemical kinetics, even in homogeneous systems, is complex and often arduous. When one attempts to study the kinetics of reactions in heterogeneous systems such as soils, sediments, and even of soil components such as clay minerals, hydrous oxides, and humic substances, the difficulties are greatly magnified. This is largely due to the complexity of soils, which are made up of a mixture of inorganic and organic components. These components often interact with each other and display different types of sites with various reactivities for inorganic and organic sorptives. Moreover, the variety of particle sizes and porosities in soils and sediments further adds to their heterogeneity. In most cases, both *chemical kinetics* and multiple transport processes are occurring simultaneously. Thus, the determination of *chemical kinetics,* which can be defined as "the investigation of rates of chemical reactions and of the molecular processes by which reactions

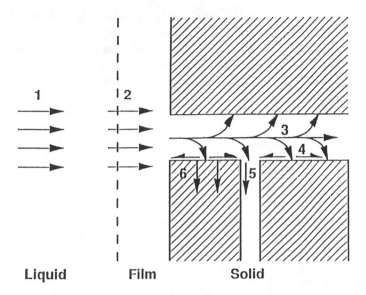

Figure 2   Transport processes in solid-liquid soil reactions. Nonactivated processes: **1**, transport in the soil solution; **2**, transport across a liquid film at the solid-liquid interface; **3**, transport in a liquid-filled macropore. Activated processes: **4**, diffusion of a sorbate at the surface of the solid; **5**, diffusion of a sorbate occluded in a micropore; **6**, diffusion in the bulk of the solid. (From Aharoni, C. and Sparks, D. L., in *Rates of Soil Chemical Precesses,* SSSA Spec. Publ. No. 27, Sparks, D. L. and Suarez, D. L., Eds., Soil Science Society of America, Madison, WI, 1991, 1. With permission.)

occur where transport is not limiting,"[25] is extremely difficult, if not impossible, in heterogeneous systems. In these systems, one is studying *kinetics*, which is a generic term referring to time-dependent or nonequilibrium processes. Thus, apparent and not mechanistic rate laws and rate parameters are determined.[10,26]

A. Rate-Limiting Steps

Both transport and chemical reaction processes can affect the reaction rates in the subsurface environment. Transport processes include:[27] transport in the solution phase, transport across a liquid film at the particle/liquid interface (film diffusion, FD), transport in liquid-filled macropores, all of which are nonactivated diffusion processes and occur in mobile regions, and particle diffusion (PD) processes, which include diffusion of sorbate occluded in micropores (pore diffusion) and along pore-wall surfaces (surface diffusion) and diffusion processes in the bulk of the solid, all of which are activated diffusion processes (Figure 2). Pore and surface diffusion within the immediate region can be referred to as intra-aggregate (intraparticle) diffusion and diffusion in the solid can be called interparticle diffusion. The actual chemical reaction (CR) at the surface, e.g., adsorption, is usually instantaneous. The slowest of the CR and transport processes is ratelimiting.

B. Rate Laws

There are two important reasons for investigating the rates of soil chemical processes:[10,28] (1) to determine how rapidly reactions attain equilibrium, and (2) to infer information on reaction mechanisms. One of the most important aspects of chemical kinetics is the establishment of a rate equation or law. By definition, a rate law is a differential equation. For the following reaction:[28,29]

$$aA + bB \rightarrow yY + zZ \qquad (1)$$

the rate is proportional to some power of the concentrations of reactants A and B and/or other species (C, D, etc.) in the system and a, b, y, and z are stoichiometric coefficients and are assumed to be equal to one in the discussion that follows on rate laws. The power to which the concentration is raised may equal zero (i.e., the rate is independent of that concentration), even for reactant A or B. Rates are expressed as a decrease in reactant concentration or an increase in product concentration per unit time. Thus, the rate of conversion of reactant A above, which has a concentration [A] at any time t, is ($-d[A]/(dt)$) while the rate with regard to product Y having a concentration [Y] at time t is ($d[Y]/(dt)$).

The rate expression for Equation 1 is therefore

$$\left| d[Y]/dt \right| = \left| -d[A]/dt \right| = k[A]^{\alpha}[B]^{\beta} \tag{2}$$

where k is the rate constant, α is the order of the reaction with respect to reactant A and can be referred to as a partial order for the total reaction, and β is the order with respect to reactant B. These orders are experimentally determined and not necessarily integral numbers. The sum of all the partial orders, α, β is the overall order (n) of the total reaction and may be expressed as

$$n = \alpha + \beta + \dots \tag{3}$$

Once the values of α, β, etc. are determined experimentally, the rate law is defined. Reaction order provides only information about the manner in which rate depends on concentration. Order does not mean the same as "molecularity," which concerns the number of reactant particles (atoms, molecules, free radicals, or ions) entering into an elementary reaction. One can define an elementary reaction as one in which no reaction intermediates have been detected or need to be postulated to describe the chemical reaction on a molecular scale. An elementary reaction is assumed to occur in a single step and to pass through a single transition state.[29]

To demonstrate that a reaction is elementary, one can use experimental conditions that are different from those employed in determining the reaction rate law. For example, if one conducted kinetic studies using a flow technique with set steady-state flow rates, one could see if reaction rate and rate constants changed with flow rate. If they did, one would not be determining mechanistic rate laws (see definition below).

Rate laws serve three purposes: they assist one in predicting the reaction rate, mechanisms can be proposed, and reaction orders can be ascertained. There are four types of rate laws that can be determined for soil chemical processes:[26] mechanistic, apparent, transport with apparent, and transport with mechanistic. Mechanistic rate laws assume that only chemical kinetics are operational and transport phenomena are not occurring. Consequently, it is difficult to determine mechanistic rate laws for most soil chemical systems due to the heterogeneity of the system caused by different particle sizes, porosities, and types of retention sites. There is evidence that with some kinetic studies, using chemical relaxation techniques and pure systems (e.g., clay minerals, oxides), that mechanistic rate laws are determined or closely approximated, since the agreement between equilibrium constants calculated from both kinetic and equilibrium studies is comparable.[30,31]

The heterogeneity of natural materials would indicate that in most cases transport processes affect the reaction rate. Thus, soil structure, stirring, mixing, and flow rate all would affect the kinetics. Transport with apparent rate laws emphasize transport-limited phenomena. One often assumes first-order or zero-order reactions (see discussion below on reaction order). In determining transport with mechanistic rate laws, one attempts to describe *simultaneously* transport-controlled and chemical kinetics phenomena. One is thus trying to explain accurately both the chemistry and physics of the system.

C. Determination of Reaction Order and Rate Constants/Coefficients

There are three basic ways to determine rate laws and rate constants/coefficients:[10,26,28,29] (1) initial rates, (2) directly using integrated equations and graphing the data, and (3) using non-linear least square analysis.

Let us assume the following elementary reaction between species A, B, and Y:

$$A + B \underset{k_{-1}}{\overset{k_1}{\rightleftharpoons}} Y \tag{4}$$

A forward reaction rate law can be written as

$$d[A]/dt = -k_1[A][B] \tag{5}$$

where k_1 is the forward rate constant and α and β (see Equation 2) are each assumed to be 1.

The reverse reaction rate law for Equation 4 is

$$d[A]/dt = +k_{-1}[Y] \tag{6}$$

where k_{-1} is the reverse rate constant.

Equations 5 and 6 are only applicable far from equilibrium where back or reverse reactions are insignificant. If both forward and reverse reactions are occurring, Equations 5 and 6 must be combined, such that

$$d[A]/dt = -k_1[A][B] + k_{-1}[Y] \tag{7}$$

Equation 7 applies the principle that the net reaction rate is the difference between the sum of all reverse reaction rates and the sum of all forward reaction rates.

One way to ensure that back reactions are not important is to measure initial rates. The initial rate is the limit of the reaction rate as time reaches zero. With an initial rate method, one plots the concentration of a reactant or product over a short reaction time period during which the concentrations of the reactants change so little that the instantaneous rate is hardly affected. Thus, by measuring initial rates, one could assume that only the forward reaction in Equation 4 predominates. This would simplify the rate law to that given in Equation 5, which, as written, would be a second-order reaction, first-order in reactant A and first-order in reactant B. Equation 5 under these conditions would represent a second-order irreversible elementary reaction. To measure initial rates, one must have available a technique that can measure rapid reactions such as a chemical relaxation method and an accurate analytical detection system to determine product concentrations.

Integrated rate equations can also be used to determine rate constants/coefficients. If one assumes that reactant B in Equation 5 is in large excess of reactant A, which is an example of the "method of isolation" to analyze kinetic data, and $Y_0 = 0$, where Y_0 is the initial concentration of product Y, Equation 5 can be simplified to:

$$d[A]/dt = -k_1'[A] \tag{8}$$

where $k_1' = k_1[B]$.

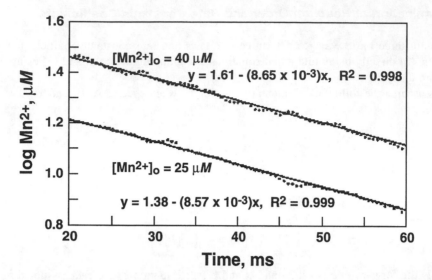

Figure 3 Initial reaction rates depicting the first-order dependence of Mn^{2+} sorption as a function of time for initial Mn^{2+} concentrations ($[Mn^{2+}]_o$) of 25 and 40 μ*M*. (From Fendorf, S. E. et al., *Soil Sci. Soc. Am. J.*, 57, 57, 1993. With permission.)

The first-order dependence of [A] can be evaluated using the integrated form of Equation 8, using the initial conditions at t = 0, A = A$_o$,

$$\log[A]_t = \log[A]_o - \frac{k_1' t}{2.303} \qquad (9)$$

The half time ($t_{1/2}$) for the above reaction is equal to $0.693/k_1'$ and is the time required for half of reactant A to be consumed.

If a reaction is first-order a plot of log $[A]_t$ vs. t should result in a straight line with a slope = $-k_1'/2.303$ and an intercept of log $[A]_o$. An example of first-order plots for Mn^{2+} sorption on δ-MnO_2 at two initial Mn^{2+} concentrations, $[Mn^{2+}]_o$, 25 and 40 μ*M*, is shown in Figure 3. One sees that the plots are linear at both concentrations, which would indicate that the sorption process is first order. The $[Mn^{2+}]_o$ values, obtained from the intercept of Figure 3, were 24 and 41 μ*M*, which are in good agreement with the two $[Mn^{2+}]_o$ values. The rate constants were 3.73×10^{-3} and 3.75×10^{-3} s^{-1} at $[Mn^{2+}]_o$ of 25 and 40 μ*M*, respectively. The findings that the rate constants are not significantly changed with concentration is a good indication that the reaction in Equation 8 is first order under the experimental conditions that were imposed.

It is dangerous to conclude that a particular reaction order is correct, based simply on the conformity of data to an integrated equation. As illustrated above, multiple initial concentrations that vary considerably should be employed to see if the rate is independent of concentration. One should also test multiple integrated equations. It may also be useful to show that reaction rate is not affected by a species whose concentration does not change considerably during an experiment; they may be substances not consumed in the reaction (i.e., catalysts) or present in large excess.[10,23,29]

Least squares analysis can also be used to determine rate constants/coefficients. With this method, one fits the best straight line to a set of points that are linearly related as y = mx + b, where y is the ordinate and x is the abscissa datum point, respectively. The slope, m, and the intercept, b, can be calculated by least squares analysis.

Curvature may result when kinetic data are plotted. This may be due to an incorrect assumption of reaction order. If first-order kinetics is assumed and the reaction is really second-order, downward curvature is observed. If second-order kinetics is assumed but the reaction is first order, upward curvature is observed. Curvature can also be due to fractional, third, higher, or mixed reaction order. Nonattainment of equilibrium often results in downward curvature. Temperature changes during the study can also cause curvature; thus, it is important that temperature be accurately controlled during a kinetic experiment.

IV. KINETIC MODELS

A. Ordered Models

First-order kinetics models often describe reactions at the particle/solution interface. Both single first-order and multiple first-order reactions have been described by many investigators (for example, see References 10, 23, 28, and 33).

It is not uncommon to observe biphasic kinetics, i.e., a rapid reaction rate followed by a much slower reaction rate. Such data can often be described by two first-order reactions. Some investigators have interpreted such biphasic kinetics to mean reactions on two types of sites, e.g., external, readily accessible sites (slope 1) and internal, difficultly accessible sites (slope 2).[22,24]

However, it is unsound to conclude anything about mechanisms based solely on multiple rate constants that are calculated from multiple slopes of kinetic plots. There are other ways to ascertain reaction mechanisms more definitively, such as calculating energies of activation, elucidating rate-limiting steps through stopped-flow and interruption approaches, using independent or direct methods to determine mechanisms such as spectroscopic techniques, and employing blocking agents that are specific for certain reaction sites. An example of the latter approach is found in the research of Jardine and Sparks,[22] who studied potassium-calcium exchange on a Delaware (U.S.) soil at three temperatures and observed two apparent simultaneous first-order reactions at 283 and 298 K (Figure 4). They hypothesized that the first, more rapid reaction, was predominantly due to adsorption on external planar sites of the organic matter and kaolinite in the soil. The slower reaction was ascribed to vermiculitic clay sites that promoted slow pore and surface diffusion. These hypotheses were seemingly validated by using a large organic polymer, cetyltrimethylammonium bromide (CTAB), which because of its size, is sterically hindered from internal sites. Thus, CTAB should only block external planar sites. When CTAB was applied to the soil, the first slope was eliminated, while the second slope was still present, indicating multireactive sites.

While first-order models have been used widely to describe the kinetics of chemical reactions on natural materials, a number of other simple kinetic models also have been employed. These include various ordered equations, such as zero-order, second-order, and fractional-order, and Elovich, power function or fractional power, and parabolic diffusion models. A brief discussion of some of these will be given; the final forms of the equations are given in Table 1. For more complete details and applications of these models one may consult Sparks.[10,28]

B. Elovich Equation

The Elovich equation was originally developed to describe the kinetics of heterogeneous chemisorption of gases on solid surfaces.[34] It seems to describe a number of reaction mechanisms, including bulk and surface diffusion and activation and deactivation of catalytic surfaces.

In soil chemistry, the Elovich equation has been used to describe the kinetics of sorption and desorption of various inorganic materials on soils.[10,28] It can be expressed as:[35]

$$q = (1/\beta)\ln(\alpha\beta) + (1/\beta)\ln t \qquad (10)$$

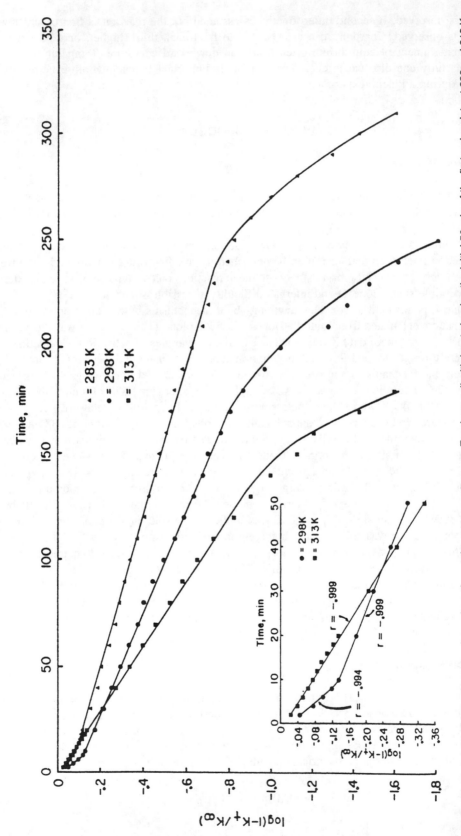

Figure 4   First-order kinetics for potassium adsorption at three temperatures on Evesboro soil with inset showing the initial 50 min of the first-order plots at 298 and 313 K. (From Jardine, P. M. and Sparks, D. L., *Soil Sci. Soc. Am. J.*, 48, 39, 1984. With permission.)

Table 1 Linear Forms of Kinetic Equations Commonly Used in Environmental Soil Chemistry[a]

Zero-order[b]

$$[A]_t = [A]_0 - k_1't$$

First-order[b]

$$\log [A]_t = \log[A]_0 - \frac{k_1't}{2.303^c}$$

Second-order[b]

$$\frac{1}{[A]_t} = \frac{1}{[A]_0} + k$$

Elovich

$$q = (1/\beta) \ln (\alpha\beta) + (1/\beta) \ln t$$

Parabolic diffusion

$$\left(\frac{1}{t}\right)\left(\frac{Q_t}{Q_\infty}\right) = \frac{4}{\pi^{1/2}}\left(\frac{D}{r^2}\right)^{1/2}\frac{1}{t^{1/2}} - \frac{D}{r^2}$$

Power function

$$\ln q_t = \ln k + v \ln t$$

[a] From Sparks;[28] terms in equations are defined in the text of the chapter.
[b] Describing the reaction A → Y.
[c] ln x = 2.303 log x is the conversion from natural logarithms (ln) to base 10 logarithms (log).

From Sparks, D. L., *Environmental Soil Chemistry*, Academic Press, San Diego, 1995. With permission.

where q is the amount of sorbate per unit mass of sorbent at time t and α and β are constants during any one experiment. A plot of q vs. ln t should give a linear relationship if the Elovich equation is applicable with a slope of $(1/\beta)$ and an intercept of $(1/\beta) \ln (\alpha\beta)$. An application of Equation 10 to phosphate sorption on soils is shown in Figure 5.

Some investigators have used the α and β parameters from the Elovich equation to estimate reaction rates. For example, it has been suggested that a decrease in β and/or an increase in α would increase reaction rate. However, this is questionable. The slope of plots using Equation 10 changes with the concentration of the adsorptive and with the solution-to-soil ratio.[36] Therefore, the slopes are not always characteristic of the soil but may depend on various experimental conditions.

Some researchers also have suggested that "breaks" or multiple linear segments in Elovich plots could indicate a changeover from one type of binding site to another.[37] However, such mechanistic suggestions may not be correct.[10,28]

C. Parabolic Diffusion Equation

The parabolic diffusion equation is often used to indicate that diffusion-controlled phenomena are rate limiting. It was originally derived based on radial diffusion in a cylinder where the ion concentration on the cylindrical surface is constant, and initially the ion concentration throughout the cylinder is uniform. It is also assumed that ion diffusion through the upper and lower faces of the cylinder is negligible. Following Crank,[38] the parabolic diffusion equation, as applied to soils, can be expressed as:

$$(Q_t/Q_\infty) = \frac{4}{\pi^{1/2}}\left(\frac{Dt}{r^2}\right)^{1/2} - \frac{Dt}{r^2} - \frac{1}{3\pi^{1/2}}\left(\frac{Dt}{r^2}\right)^{3/2} \tag{11}$$

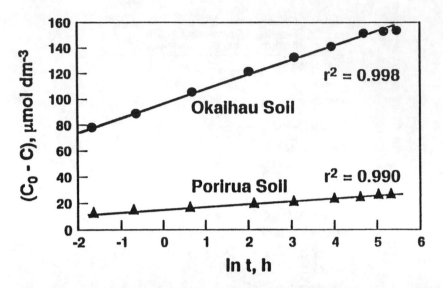

Figure 5 Plot of Elovich equation for phosphate sorption on two soils where C_0 is the initial phosphorus concentration added at time 0 and C is the phosphorus concentration in the soil solution at time t. The quantity $(C_0 - C)$ can be equated to q, the amount sorbed at time t. (From Chien, S. H. and Clayton, W. R., *Soil Sci. Soc. Am. J.*, 44, 265, 1980. With permission.)

where r is the radius of the cylinder, Q_t is the quantity of diffusing substance that has left the cylinder at time t, Q_∞ is the corresponding quantity after infinite time, and D is an "apparent" diffusion coefficient.

For the relatively short times in most experiments, the third and subsequent terms may be ignored, and thus

$$\frac{Q_t}{Q_\infty} = \frac{4}{\pi^{1/2}}\left(\frac{Dt}{r^2}\right)^{1/2} - \frac{Dt}{r^2}$$

or

$$\frac{1}{t}\left(\frac{Q_t}{Q_\infty}\right) = \frac{4}{\pi^{1/2}}\left(\frac{D}{r^2}\right)^{1/2}\frac{1}{t^{1/2}} - \frac{D}{r^2} \tag{12}$$

and thus a plot of $\dfrac{(Q_t/Q_\infty)}{t}$ vs. $1/t^{1/2}$ should give a straight line with a slope

$$\frac{4}{\pi^{1/2}}\left(\frac{D}{r^2}\right)^{1/2}$$

and intercept $(-D/r^2)$. Thus, if r is known, D may be calculated from both the slope and intercept.

The parabolic diffusion equation has successfully described metal reactions on soils and soil constituents,[22,39] feldspar weathering,[40] and pesticide reactions.[41]

D. Fractional-Power or Power-Function Equation

This equation can be expressed as

$$q = kt^v \tag{13}$$

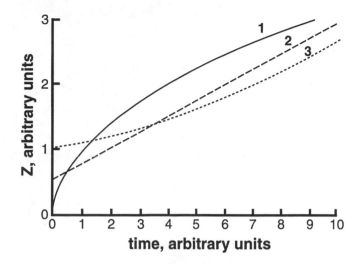

Figure 6 Plot of Z vs. time implied by (1) the power-function model, (2) the Elovich model, and (3) the first-order model. The equations for the models were differentiated and expressed as explicit functions of the reciprocal of the rate, Z. (From Aharoni, C. and Sparks, D. L., *Rates of Soil Chemical Processes,* SSSA Spec. Publ. No. 27, Sparks, D. L. and Suarez, D. L., Eds., Soil Science Society of America, Madison, WI, 1991, 1. With permission.)

where q is amount of sorbate per unit mass of sorbent at time t, k and v are constants, and v is positive and <1. Equation 13 is empirical, except for the case where v = 0.5, when it is similar to the parabolic diffusion equation.

Equation 13 and various modified forms have been used by a number of researchers to describe the kinetics of reactions on natural materials.[42,43]

E. Z(t) and Diffusion Models

In a number of studies it has been shown that several simple kinetic models, as described previously, describe rate data well, based on correlation coefficients and standard errors of the estimate.[35,44,45] Despite this, there is often not a consistent relation between the equation that gives the best fit and the physicochemical and mineralogical properties of the adsorbent(s) being studied. Another problem with some of the kinetic models is that they are empirical and no meaningful rate parameters can be obtained.

Aharoni and Ungarish[46] and Aharoni[47] noted that some simple kinetic models are approximations to which more general expressions reduce in certain limited time ranges. They suggested a generalized empirical equation by examining the applicability of power-function, Elovich, and first-order equations to experimental data. By writing these as the explicit functions of the reciprocal of the rate, Z, which is $(dq/dt)^{-1}$, one can show that a plot of Z vs. t should be convex if the power-function equation is operational (1 in Figure 6), linear if the Elovich equation is appropriate (2 in Figure 6), and concave if the first-order equation is appropriate (3 in Figure 6). However, Z vs. t plots for soil systems (Figure 7) are usually S-shaped, convex at small t, concave at large t, and linear at some intermediate t. These findings suggest that the reaction rate can best be described by the power-function equation at small t, by the Elovich equation at an intermediate t, and by a first-order equation at large t. Thus, the S-shaped curve indicates that the above equations may be applicable, each at some limited time range.

One of the reasons a particular kinetic model appears to be applicable may be that the study is conducted during the time range when the model is most appropriate. While sorption, for example, decreases over many orders of magnitude before equilibrium is approached, with most methods and experiments, only a portion of the entire reaction is measured and over this time range the

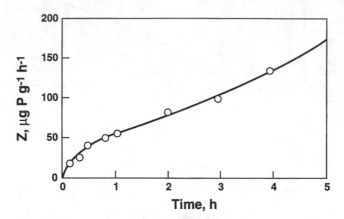

Figure 7 Sorption of phosphate by a typic Dystrochrept soil plotted as Z vs. time. The circles represent the experimental data of Polyzopoulos et al.[48] The solid line is a curve calculated according to a homogeneous diffusion model. (From Aharoni, C. and Sparks, D. L., in *Rates of Soil Chemical Processes*, SSSA Spec. Publ. No. 27, Sparks, D. L. and Suarez, D. L., Eds., Soil Science Society of America, Madison, WI, 1991, 1. With permission.)

assumptions associated with a simple kinetic model (power-function, Elovich, and first-order) are valid. Aharoni and Suzin[49,50] showed that the S-shaped curves could be well described using homogeneous and heterogeneous diffusion models. In homogeneous diffusion situations, the final and initial portions of the S-shaped curves (conforming to the power-function and first-order equations, respectively) predominated, whereas, in instances where the heterogeneous diffusion model was operational, the linear portion of the S-shaped curve, that conformed to the Elovich equation, predominated. Derivations of homogeneous and heterogeneous diffusion models can be found in Aharoni and Sparks.[27]

F. Implications of Diffusion Models

The finding that slower reactions at the soil particle/liquid interface can be described by diffusional models indicates that the kinetics of chemical processes cannot be considered separately from physically limited transport phenomena. Thus, such a combination of processes cannot be treated using first-order or other-order chemical kinetics equations. When one states that a reaction between the molecular species A and B is of first order with respect to A, one assumes that the molecules of A have equal chances of participating in the reaction and therefore the rate is proportional to the concentration, C_A. This reasoning can be extended to a reaction between an adsorbing surface and an adsorptive solute. The concentration C_A, in this case, refers to the number of reactive sites per unit area, which corresponds to the number of unoccupied sites per unit area $(1 - \theta_A)$. However, by using first-order kinetics (or other-order kinetics) one tacitly assumes that all of the surface sites are potential reactants at any time, and they have an opportunity of participating in the sorption process. If one assumes that there are sites that cannot be reached directly from the fluid phase, but can be reached after the sorbate has undergone sorption and desorption at other sites, one cannot separate chemical kinetics from diffusion-limited kinetics. The overall kinetic process obeys a diffusion equation since diffusion is the rate-limiting process. However, the diffusion coefficient, which reflects the rate at which the sorbate jumps from one site to another, is determined by the rate of the chemical reactions by which the sorbent-sorbate bonds are created and destroyed. Additionally, the activation energy for diffusion is equivalent to the activation energy of the chemical reaction.

G. Multiple-Site Models

Based on the previous discussion, it is evident that simple chemical kinetics models, such as ordered reaction models and the power-function and Elovich models, may not be appropriate to describe reactions in heterogeneous systems such as soils, sediments, and soil components. In these systems where there is a range of particle sizes and multiple retention sites, both chemical kinetics and transport phenomena are occurring simultaneously, and a fast reaction is often followed by a slower reaction(s). In such systems, nonequilibrium models that describe both chemical and physical nonequilibrium and that consider multiple components and sites are more appropriate. Physical nonequilibrium is ascribed to some rate-limiting transport mechanism such as FD or PD while chemical nonequilibrium is due to a rate-limiting mechanism at the particle surface (CR). Nonequilibrium models include two-site, multiple-site, radial diffusion (pore diffusion), surface diffusion, and multiprocess models (Table 2). Emphasis in this chapter will be on use of these models to describe sorption phenomena.

The term "sites" can have a number of meanings:[57] (1) specific, molecular scale reaction sites; (2) sites of differing degrees of accessibility (external, internal); (3) sites of differing sorbent type (organic matter and inorganic mineral surfaces); and (4) sites with different sorption mechanisms. With chemical nonequilibrium sorption processes, the sorbate may undergo two or more types of sorption reactions, one of which is rate limiting. For example, a metal cation may sorb to organic matter by one mechanism and to mineral surfaces by another mechanism, with one of the mechanisms being time dependent.

1. Chemical Nonequilibrium Models

Chemical nonequilibrium models describe time-dependent reactions at sorbent surfaces. The one-site model is a first-order approach that assumes that the reaction rate is limited by only one process or mechanism on a single class of sorbing site and that all sites are of the time-dependent type. In many cases this model appears to describe soil chemical reactions quite well. However, often it does not. This model would seem not appropriate for most heterogeneous systems, since multiple sorption sites exist.

The two-site (two compartment, two box) or bicontinuum model has been widely used to describe chemical nonequilibrium[52,58-62] and physical nonequilibrium[63-65] (Table 2). This model assumes that there are two reactions occurring, one that is fast and reaches equilibrium quickly and a slower reaction that can continue for long time periods. The reactions can occur either in series or in parallel.[57]

In describing chemical nonequilibrium with the two-site model it is assumed that the sorbent has two types of sites. One site involves an instantaneous equilibrium reaction and the other site involves the time-dependent reaction. The instantaneous equilibrium reaction is described by an equilibrium isotherm equation, while a first-order equation is usually employed to describe the time-dependent reaction.

Jardine et al.[66] modeled the transport of Al through Ca-saturated kaolinite columns using a two-site nonequilibrium transport process (Figure 8a). They assumed and showed that type 1 sites were in local equilibrium with the solution phase and involved an instantaneous Ca-Al exchange mechanism. Type 2 sites involved a time-dependent Al polymerization reaction mechanism and were described by first-order kinetics.

The polymerization mechanism was indirectly confirmed by investigating the effect of influent pH on Al transport. When influent pH was lowered, the slower kinetic reaction was eliminated and the Al breakthrough curve was described with a one-site equilibrium model (Figure 8b).

With the two-site model there are two adjustable or fitting parameters, the fraction of sites at local equilibrium (X_1) and the rate constant (k). A distribution (K_d) or partition coefficient (K_p) is determined independently from a sorption/desorption isotherm.

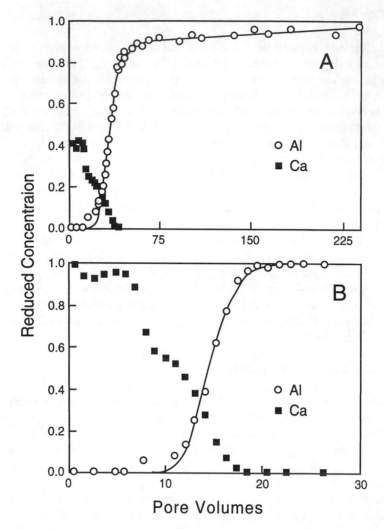

Figure 8   (a) Breakthrough curve for 0.73 mg l⁻¹ Al at pH 4.29 on kaolinite with corresponding desorbed Ca. Solid line is the fitted curve from a two-site, nonequilibrium model. (b) Breakthrough curve for 1.50 mg l⁻¹ Al at pH 3.97 on kaolinite with corresponding desorbed Ca. Solid line is the fitted curve from the one-site, equilibrium model. (From Jardine, P. M. et al., *Soil Sci. Soc. Am. J.*, 49, 867, 1985. With permission.)

Connaughton et al.[51] used a two-site model to describe naphthalene desorption from contaminated soil (Figure 9a). The model did not describe the data well and fitted and estimated desorption rate coefficients (k_d) did not agree, with the estimated k_d values being higher than the values obtained from fitting. The estimated k_d values were based on the relation log $k_d = 0.301 - 0.688$ log K_p.[57]

Connaughton et al.[51] related this discrepancy to the greater desorption times of their experiments and to the use of the two-site model to describe the entire desorption process. The two-site model, that the $K_p - k_d$ relationship was based on, assumes that the initial desorption is instantaneous, which is not the case for naphthalene desorption. One major disadvantage in using the two-site model to describe heterogeneous systems such as soil is the assumption that only two sorptive sites are present. Thus, the fitting parameters in the two-site model probably do not conform to actual reaction rates on multiple sites in soils. Moreover, it is difficult to relate the fitting parameters to known properties of the sorbent. For example, Wu and Gschwend[53] found two different sets of fitting parameters described tetrachlorobenzene sorption on a Charles River sediment if different

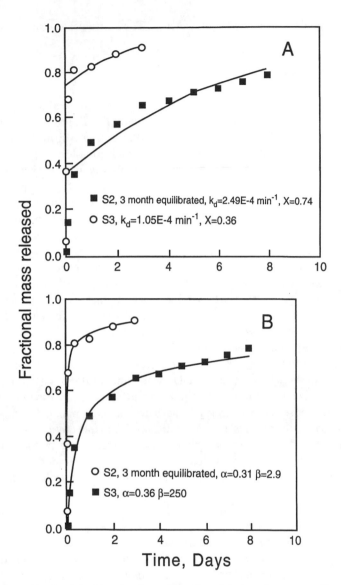

Figure 9   (a) Fitted two-site model with release profiles of naphthalene from two soils, S2 and S3 (S2 is a freshly contaminated soil, reacted with naphthalene for 3 months, and S3 is a field [aged] contaminated soil); R^2 = 0.88, 0.91 for S2 and S3 (regression fits, respectively, where X is the fraction of sites that reach an equilibrium in release instantaneously and k_d is the desorption rate coefficient). (b) Mass fractional release of naphthalene from S2 and S3 soils fitted with a multisite continuum compartment (Γ) model. (From Connaughton, D. F. et al., *Environ. Sci. Technol.*, 27, 2397, 1993. With permission.)

sediment mean aggregate sizes were used. Consequently, the parameters had to be experimentally determined for each sediment size.

To account for the multiple sites that may exist in heterogeneous systems Connaughton et al.[51] developed a multisite compartment model (Γ) that incorporates a continuum of sites or compartments with a distribution of rate coefficients that can be described by a gamma density function. A fraction of the sorbed mass in each compartment is at equilibrium and there is a desorption rate coefficient or distribution coefficient for each compartment or site (Table 2). The multisite model has two fitting parameters, α, a shape parameter, and $1/\beta$, which is a scale parameter that determines the mean standard deviation of the rate coefficients. Figure 9b shows application of the Γ model to desorption of naphthalene from contaminated soils. The entire desorption process was described well with this model.

2. Physical Nonequilibrium Models

There are a number of models that can be used to describe physical nonequilibrium reactions. Since transport processes in the mobile phase are not usually rate limiting, physical nonequilibrium models focus on diffusion in the immobile phase or intra-aggregate-diffusion processes (e.g., pore and/or surface diffusion). The transport between mobile and immobile regions is accounted for in physical nonequilibrium models in three ways:[57] (1) explicitly with Fick's law to describe the physical mechanism of diffusive transfer; (2) explicitly by using an empirical first-order mass-transfer expression to approximate solute transfer; and (3) implicitly by using an effective or lumped dispersion coefficient that includes the effects of sink/source differences and hydrodynamic dispersion and axial diffusion.

A pore diffusion model (Table 2) has been used by a number of investigators to study sorption processes using batch systems.[53,67-70] Wu and Gschwend[53] successfully used the pore diffusion model to describe chlorobenzene congener sorption/desorption on soils and sediments. Figure 10 shows experimental and model fits for tetrachlorobenzene and pentachlorobenzene sorption on soils. The sole fitting parameter in this model is the effective diffusion coefficient (D_e). This parameter may be estimated *a priori* from chemical and colloidal properties. However, this estimation is only valid if the sorbent material has a narrow particle size distribution such that an accurate, average particle size can be defined. Moreover, in the pore diffusion model it is assumed that there is an average representative D_e, which means there is a continuum in properties across an entire pore size spectrum. This is not a valid assumption for micropores (<2.0 nm), since there are higher adsorption energies of sorbates in micropores, which causes increased sorption. The increased sorption causes reduced diffusive transport rates and nonlinear isotherms for sorbents with pores < several sorbate diameters in size. Other factors can cause reduced transport rates in micropores including steric hindrance which increases as the pore size approaches the solute size and greatly increased surface area to pore volume ratios (which occurs as pore size decreases).

Another problem with the pore diffusion model is that sorption and desorption kinetics may have been measured over a narrow concentration range. This is a problem since a sorption/desorption mechanism in micropores at one concentration may be insignificant at another concentration.

Fuller et al.[55] used a pore space diffusion model (Table 2) to describe arsenate adsorption on ferrihydrite that included a subset of sites whereby sorption was at equilibrium. A Freundlich model was used to describe sorption on these sites. Diffusion into the particle was described by Fick's second law of diffusion; homogeneous, spherical aggregates and diffusion only in the aqueous phase were assumed.

Figure 11 shows the fit of the model when sorption at all sites was controlled by intra-aggregate diffusion. The fit was better when sites that had attained sorption equilibrium were included (Figure 11). The latter model assumed that there was an initial rapid sorption on external surface sites before intra-aggregate diffusion.

Pedit and Miller[56] have developed a general multiple-particle class pore diffusion model that accounts for differences in physical and sorptive properties for each particle class (Table 2). The model includes both instantaneous equilibrium sorption and time-dependent pore diffusion for each particle class. The pore diffusion portion of the model assumes that solute transfer between the intraparticle fluid and the solid phases is fast vis-à-vis intraparticle pore diffusion processes.

Surface diffusion models, assuming a constant surface diffusion coefficient, have been used by a number of researchers.[54,71] The dual resistance model (Table 2) combines both pore and surface diffusion.

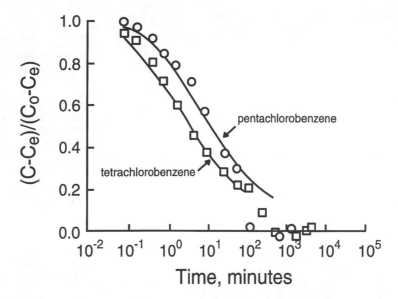

Figure 10 Experimental and model-fitting results for pentachlorobenzene and tetrachlorobenzene sorption on Iowa soils where C is the dissolved concentration of organic chemical in the bulk solution, C_0 is the initial concentration, and C_e is the equilibrium concentration. The points represent experimental data and the solid lines represent fit of the data to the radial diffusion (pore diffusion) model. (From Wu, S. and Gschwend, P. M., *Environ. Sci. Technol.,* 20, 717, 1986. With permission.)

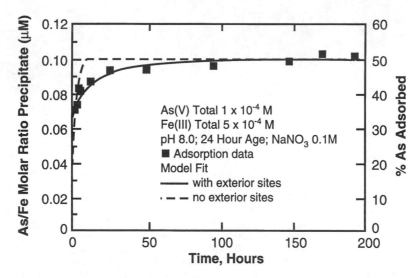

Figure 11 Comparison of pore space diffusion model fits of As(V) sorption with experimental data (dashed curve represents sorption where all surface sites are diffusion-limited and the solid curve represents sorption on equilibrium sites plus diffusion-limited sites). From Fuller, C. C. et al., *Geochim. Cosmochim. Acta,* 57, 2271, 1993. With permission.)

V. KINETIC METHODOLOGIES

A number of methodologies can be used to study the rates of soil chemical processes. These can be broadly classified as methods for slower reactions (>15 s), which include batch and flow techniques, and rapid techniques that can measure reactions on millisecond and microsecond time scales. It should be recognized that none of these methods is a panacea for kinetic analyses. They

Table 2   Comparison of Sorption Kinetic Models

Conceptual model	Fitting parameter(s)	Model limitations
One-site model $S \xrightarrow{k_d} C$	k_d	Cannot describe biphasic sorption/desorption
Two-site model $S_1 \overset{X_1 K_p\ k_d}{\leftrightarrow} C \leftarrow S_2$	k_d,* K_p, X_1	May not describe the "bleeding" or slow, reversible, nonequilibrium desorption for residual sorbed compounds[52]
Radial diffusion penetration retardation (pore diffusion) model[53] $S' \overset{K_p}{\leftrightarrow} C^{\cdot} \xrightarrow{D_{eff}} C$	**$D_{eff} = f(n,t)D_m n/(1-n)\rho_s K_p$	Cannot describe instantaneous uptake without additional correction factor; did not describe kinetic data for times greater than 10^3 min[53]
Dual-resistance surface diffusion model[54] $S' \xrightarrow{D_s} C'_s \xrightarrow{k_b} C$	D_s, k_b	Model calibrated with sorption data predicted more desorption than occurred in the desorption experiments[54]
Multisite continuum compartment model[51] $F(t) = 1 - \dfrac{M(t)}{M} = 1 - \left(\dfrac{\beta}{\beta+t}\right)^{\alpha}$	α, β	Assumption of homogeneous, spherical particles and diffusion only in aqueous phase
Pore space diffusion model[55] $\left(e + \dfrac{S_a}{n}K_s C(r)^{(1-1/n)}\right)\dfrac{\partial C(r)}{\partial t} = D_e\left(\dfrac{\partial^2 C(r)}{\partial r^2} + \dfrac{2\partial C(r)}{r\partial r}\right)$	D_e, ϵ, K_s, $1/n$, F_{eq}	
Multiple particle class pore diffusion model[56] $\left(\dfrac{\theta_p^i + \rho_a^i}{\partial C_p^i}\right)\dfrac{\partial q_r^i(r,t)}{(r,t)}\dfrac{\partial C_p^i(r,t)}{\partial t} = \dfrac{\theta_p^i D_p^i}{r^2}\dfrac{\partial}{\partial_r}\left(\dfrac{r^2 \partial C_p^i(r,t)}{\partial r}\right)$ $\quad -\theta_p^i \lambda_p^i C_p^i(r,t) - \rho_a^i \lambda_r^i q_r^i(r,t)$	θ_p^i, ρ_a^i, D_p^i, λ_p^i, λ_r^i	Multiple fitting parameters; variations in sorption equilibrium and rates that might occur within a particle class or an individual particle grain are not addressed

Partially adapted from Connaughton et al.[51]

Abbreviations used are as follows: S, concentration of the bulk sorbed contaminant (g g^{-1}); C, concentration of the bulk aqueous-phase contaminant (g ml^{-1}); k_d, first-order desorption rate coefficient (min^{-1}); S_2, concentration of the sorbed contaminant that is rate limited (g g^{-1}); S_1, concentration of the contaminant that is in equilibrium with the bulk aqueous concentration (g g^{-1}); X_1, fraction of the bulk sorbed contaminant that is in equilibrium with the aqueous concentration; K_p, sorption equilibrium partition coefficient (ml g^{-1}); D_{eff}, effective diffusivity of sorbate molecules or ions in the particles (cm^2 s^{-1}); S'·, concentration of contaminant in immobile bound state (mol g^{-1}); C', concentration of contaminant free in the pore fluid (mol cm^{-3}); n, porosity of the sorbent (cm^3 of fluid cm^{-3}); D_m, pore fluid diffusivity of the sorbate (cm^2 s^{-1}); ρ_s, specific gravity of the sorbent (g cm^{-3}); f (n,t), pore geometry factor; k_b, boundary layer mass transfer coefficient (m s^{-1}); r, radius of the spherical solid particle, assumed constant (m); ρ, macroscopic particle density of the solid phase (g m^{-3}); C'_s, solution-phase solute concentration corresponding to an equilibrium with the solid-phase solute concentration at the exterior of the particle (g l^{-1}); D_s, surface diffusion coefficient (m s^{-1}).

*K_p can be determined independently.

**K_p, D_m, and ρ_s can be determined independently; F(t), fraction of mass released through time t; M(t), mass remaining after time t; M, total initial mass; β, scale parameter necessary for determination of mean and standard deviation of k_s; α, shape parameter; ϵ, internal porosity of sorbent; C(r), concentration of sorptive in the aqueous phase in the pore fluid at radial distance r; S_a is the surface of sorbent per unit volume of solid; 1/n, the adsorption isotherm slope; K_s, adsorption isotherm intercept; D_e, effective diffusion coefficient; a, radius of the aggregate; F_{eq}, equilibrium fraction of adsorption sites; θ_p^i, intraparticle porosity of particle class i; ρ_a^i, apparent particle density of particle class i; r, radial distance; C_p^i(r, t), intraparticle fluid-phase solute concentration of the particle class i; D_p^i, pure diffusion coefficient for particle class i; λ_p^i, intraparticle fluid-phase first-order reaction rate coefficient for particle class i; λ_r^i, intraparticle solid-phase first-order reaction rate coefficient for particle class i; q_r^i (r, t), intraparticle solid-phase solute concentration of particle class i.

all have advantages and disadvantages. For comprehensive discussions on kinetic methodologies one should consult Sparks,[10] Amacher,[11] Sparks and Zhang,[72] and Sparks et al.[73]

A. Batch Methods

Batch methods have been the most widely used kinetic techniques. In the simplest traditional batch technique, an adsorbent is placed in a series of vessels such as centrifuge tubes with a particular volume of adsorptive. The tubes are then mixed by shaking or stirring. At various times a tube is sacrificed for analysis, i.e., the suspension is either centrifuged or filtered to obtain a clear supernatant for analysis. A number of variations of batch methods exist and these are discussed in Amacher.[11]

There are a number of disadvantages to traditional batch methods. Often the reaction is complete before a measurement can be made, particularly if centrifugation is necessary, and the solid:solution ratio may be altered as the experiment proceeds. Too much mixing may cause abrasion of the adsorbent, altering the surface area, while too little mixing may enhance mass transfer and transport processes. Another major problem with all batch techniques, unless a resin or chelate material such as Na-tetraphenylboron is used, is that products are not removed. This can cause inhibition in further adsorbate release and promotion of secondary precipitation in dissolution studies. Moreover, reverse reactions are not controlled, which makes the calculation of rate coefficients difficult and perhaps inaccurate.

Many of the disadvantages listed above for traditional batch techniques can be eliminated by using a method like that of Zasoski and Burau,[74] shown in Figure 12. In this method an absorbent is placed in a vessel containing the adsorptive, pH and suspension volume are adjusted, and the suspension is vigorously mixed with a magnetic stirrer. At various times, suspension aliquots are withdrawn using a syringe containing N_2 gas. The N_2 gas prevents CO_2 and O_2 from entering the reaction vessel. The suspension is rapidly filtered and the filtrates are then weighed and analyzed. With this apparatus a constant pH can be maintained, reactions can be measured at 15-s intervals, excellent mixing occurs, and a constant solid-to-solution ratio is maintained.

B. Flow Methods

Flow methods can range from continuous flow techniques (Figure 13), which are similar to liquid-phase chromatography, to stirred-flow methods (Figure 14) that combine aspects of both batch and flow methods. Important attributes of flow techniques are that one can conduct studies at realistic soil-to-solution ratios that better simulate field conditions, the adsorbent is exposed to a greater mass of ions than in a static batch system, and the flowing solution removes desorbed and detached species.

With continuous flow methods, samples can be injected as suspensions or spread dry on a membrane filter. The filter is attached to its holder by securely capping it, and the filter holder is connected to a fraction collector and peristaltic pump, the latter maintaining a constant flow rate. Influent solution then passes through the filter, reacts with the adsorbent, and at various times, effluents are collected for analysis. Depending on flow rate and the amount of effluent needed for analysis, samples can be collected about every 30 to 60 s. One of the major problems with this method is that the colloidal particles may not be dispersed, i.e., the time necessary for an adsorptive to travel through a thin layer of colloidal particles is not equal at all locations of the layer. This plus minimal mixing promotes significant transport effects. Thus, apparent rate laws and rate coefficients are measured, with the rate coefficients changing with flow rate. There can also be dilution of the incoming adsorptive solution by the liquid used to load the adsorbent on the filter, particularly if the adsorbent is placed on the filter as a suspension, or if there is washing out of remaining adsorptive solution during desorption. This can cause concentration changes not due to adsorption or desorption.

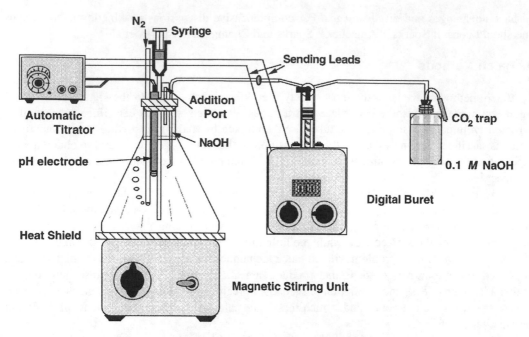

Figure 12 Schematic diagram of equipment used in batch technique of Zasoski and Burau.[74] (From Zasoski, R. G. and Burau, R. G., *Soil Sci. Soc. Am. J.*, 47, 372, 1978. With permission.)

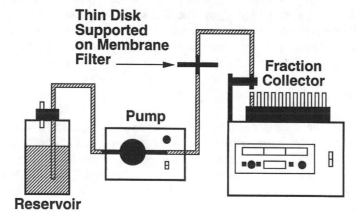

Figure 13 Continuous flow method experimental setup. Background solution and solute are pumped from the reservoir through the thin disk and are collected as aliquots by the fraction collector. (From Amacher, M. C., in *Rates of Soil Chemical Processes*, SSSA Spec. Publ. No. 27, Sparks, D. L. and Suarez, D. L., Eds., Soil Science Society of America, Madison, WI, 1991, 19. With permission.)

A more preferred method for measuring soil chemical reaction rates is the stirred-flow method. The experimental setup is similar to the continuous flow method (Figure 13), except there is a stirred-flow reaction chamber rather than a membrane filter. A schematic of this method is shown in Figure 14. The sorbent is placed into the reaction chamber, where a magnetic stir bar or an overhead stirrer (Figure 14) keeps it suspended during the experiment. There is a filter placed in the top of the chamber which keeps the solids in the reaction chamber. A peristaltic pump maintains a constant flow rate and a fraction collector is used to collect the leachates. The stirrer effects perfect mixing, i.e., the concentration of the adsorptive in the chamber is equal to the effluent concentration.

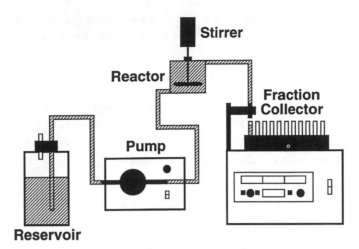

Figure 14 Stirred-flow reactor experimental setup. Background solution and solute are pumped from the reservoir through the stirred reactor containing the solid phase and are collected as aliquots by the fraction collector. Separation of solid and liquid phases is accomplished by a membrane filter at the outlet end of the stirred reactor. (From Amacher,[11] (From Amacher, M. C., in *Rates of Soil Chemical Processes,* SSSA Spec. Publ. No. 27, Sparks, D. L. and Suarez, D. L., Eds., Soil Science Society of America, Madison, WI, 1991, 19. With permission.)

This method has several advantages over the continuous flow technique and other kinetic methods. Reaction rates are independent of the physical properties of the porous media, the same apparatus can be used for adsorption and desorption experiments, desorbed species are removed, continuous measurements allow for monitoring reaction progress, experimental factors such as flow rate and adsorbent mass can be easily altered, a variety of solids can be used (however, sometimes fine particles can clog the filter, causing a buildup in pressure which results in a nonconstant flow rate) with the technique, the adsorbent is dispersed, and dilution errors can be measured. With this method, one can also use stopped-flow tests and vary influent concentrations and flow rates to elucidate possible reaction mechanisms.[75]

C. Relaxation Techniques

As noted earlier, many soil chemical reactions are very rapid, occurring on millisecond and microsecond time scales. These include metal and organic sorption-desorption reactions, ion exchange processes, and ion association reactions. Batch and flow techniques, which measure reaction rates of >15 s, cannot be employed to measure these reactions. Chemical relaxation methods must be used to measure very rapid reactions. These include pressure-jump (p-jump), electric field pulse, temperature-jump (t-jump), and concentration jump (c-jump) methods. These methods are fully outlined and described in other sources.[10,72] Only a brief discussion of the theory of chemical relaxation and a description of p-jump methods, which have been the most widely used chemical relaxation method to describe reaction rates at the mineral/water interface, will be given here. The theory of chemical relaxation can be found in a number of sources.[76-78] It should be noted that relaxation techniques are best used with soil components such as oxides and clay minerals and not with soils. Soils are heterogeneous, which complicates the analyses of the relaxation data.

All chemical relaxation methods are based on the theory that the equilibrium of a system can be rapidly perturbed by some external factor such as pressure, temperature, or electric field strength. Rate information can then be obtained by measuring the approach from the perturbed equilibrium to the final equilibrium by measuring the relaxation time, τ (the time that it takes for the system to relax from one equilibrium state to another, after the perturbation pulse) by using a detection system such as conductivity. The relaxation time is related to the specific rates of the elementary reactions involved. Since the perturbation is small, all rate expressions reduce to first-order equations regardless of reaction order or molecularity.[78] The rate equations are then linearized such that:

$$\tau^{-1} = k_1 \left(C_A + C_B \right) + k_{-1} \tag{14}$$

where k_1 and k_{-1} are the forward and backward rate constants and C_A and C_B are the concentrations of reactants A and B at equilibrium. From a linear plot of τ^{-1} vs. $(C_A + C_B)$ one could calculate k_1 and k_{-1} from the slope and intercept, respectively. Pressure-jump relaxation is based on the principle that chemical equilibria depend on pressure, as shown below:[78]

$$\left(\frac{\partial \ln K^\circ}{\partial \ln p} \right)_T = -\Delta V / RT \tag{15}$$

where K° is the equilibrium constant, ΔV is the standard molar volume change of the reaction, p is pressure, and R and T were defined earlier. For a small perturbation,

$$\frac{\Delta K^\circ}{K^\circ} = \frac{-\Delta V \Delta p}{RT} \tag{16}$$

Details on the experimental protocol for a p-jump study can be found in several sources.[10,15,79]

Fendorf et al.[32] used an electron paramagnetic resonance stopped-flow (EPR-SF) method (an example of a c-jump method) to study reactions in colloidal suspensions *in situ* on millisecond time scales. If one is studying an EPR active species (paramagnetic) such as Mn, this technique has several advantages over other chemical relaxation methods. With many relaxation methods, the reactions must be reversible and reactant species are not directly measured. Moreover, in some relaxation studies, the rate constants are calculated from linearized rate equations that are dependent on equilibrium parameters. Thus, the rate parameters are not directly measured.

With the EPR-SF method of Fendorf et al.[32] the mixing can be done in <10 msec and EPR digitized within a few microseconds. A diagram of the EPR-SF instrument is shown in Figure 15. Dual 2-ml in-port syringes feed a mixing cell that is located in the EPR spectrometer. This allows for EPR detection of the cell contents. A single outflow port is fitted with a 2-ml effluent collection syringe equipped with a triggering switch. The switch activates the data acquisition system. Each run consists of filling the in-port syringes with the desired reactants, flushing the system with the reactants several times, and initiating and monitoring the reaction. Fendorf et al.[32] used this system to study the kinetics of Mn^{2+} sorption on γ-MnO_2 (see Figure 3). The sorption reaction was complete in 200 msec. Data were taken every 50 μsec and 100 points were averaged to give the time-dependent sorption of Mn(II).

D. Choice of Kinetic Method

The method that one chooses to study the kinetics of soil chemical reactions depends on several factors. The reaction rate will certainly dictate the choice of method. With batch and flow methods, the most rapid measurements one can make require about 15 s. For more rapid reactions, one must use relaxation techniques where millisecond and microsecond time scales can be measured.

Another factor in deciding on a kinetic method is the objective of one's experiment(s). If one wishes to measure the chemical kinetics of a reaction where transport is minimal, most batch and flow techniques are unsuitable and a relaxation technique should be employed. On the other hand, if one wants to simulate time-dependent reactions in the field, perhaps a flow technique would be more realistic than a batch method.

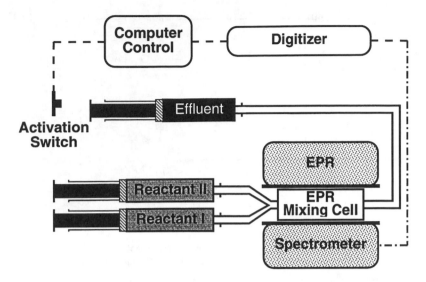

Figure 15   Schematic diagram of the electron paramagnetic resonance monitored stopped-flow kinetic apparatus. (From Fendorf, S. E. et al., *Soil Sci. Soc. Am. J.*, 57, 57, 1993. With permission.)

VI. KINETICS OF IMPORTANT REACTIONS ON NATURAL PARTICLES

In the past two decades numerous studies have been conducted on the kinetics of metal, radionuclide, plant nutrient, and organic chemical reactions on natural materials.

In this section emphasis will be placed on the kinetics of sorption-desorption and precipitation/dissolution reactions on soils and soil constituents.

A. Sorption/Desorption Reaction Rates

1. Heavy Metals and Metalloids

The chemical reaction of heavy metals on soil components is rapid, occurring on millisecond time scales. For such rapid reactions, chemical relaxation techniques, i.e., pressure-jump relaxation must be employed.[10,72,79,80]

The use of p-jump relaxation to measure the kinetics of ion sorption/desorption on metal oxide surfaces was pioneered by several Japanese chemists. Their research includes some of the following adsorption/desorption kinetic studies: divalent metal ion,[30] phosphate,[81] chromate,[82] and uranyl[83] adsorption reactions on γ-Al$_2$O$_3$. Hayes and Leckie[80] were the first to use p-jump relaxation to study adsorption/desorption kinetics of a metal ion contaminant (Pb^{2+}) on goethite (α-FeOOH). Other successive studies monitored the rapid adsorption/desorption kinetics of molybdate,[15] sulfate,[16] selenate and selenite,[17] Cu^{2+},[84] and arsenate and chromate[85] on goethite. Additionally, studies have investigated borate adsorption/desorption kinetics on pyrophyllite[86] and on γ-Al$_2$O$_3$.[87]

Details of many of these studies are summarized in Hayes and Leckie,[80] Sparks,[10] and Sparks and Zhang[72] and will not be detailed here. A recent study of Grossl et al.[85] will be summarized to illustrate rapid chemical reaction rates of chromate and arsenate on goethite. A double relaxation was observed for both arsenate and chromate adsorption/desorption over a pH range of 6.5 to 7.5 for arsenate and 5.5 to 6.5 for chromate, respectively (Figures 16 and 17). Based on the double relaxations, a two-step process, resulting in the formation of an inner-sphere bidentate surface complex (Figure 18), was proposed. The first step involves an initial ligand exchange reaction of the aqueous oxyanion (H$_2$AsO$_4^-$ or HCrO$_4^-$) with goethite, forming an inner-sphere monodentate

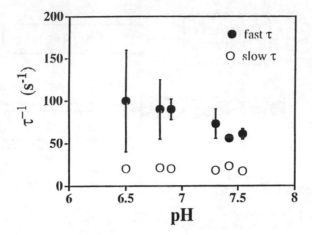

Figure 16 τ^{-1} values determined from p-jump experiments for arsenate adsorption/desorption on goethite, as a function of pH. (From Grossl, P. R. et al., *Environ. Sci. Technol.*, 31, 321, 1997. With permission.)

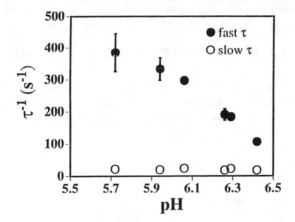

Figure 17 τ^{-1} values determined from p-jump experiments for chromate adsorption/desorption on goethite, as a function of pH. (From Grossl, P. R., et al., *Environ. Sci. Technol.*, 31, 321, 1997. With permission.)

surface complex. This first step produces the signals associated with the fast τ values. The succeeding step involves a second ligand exchange reaction, resulting in the formation of an inner-sphere bidentate surface complex. This step produces the signal associated with the slow τ values.

To determine if the mechanism displayed in Figure 18 was plausible and consistent with the kinetic data, the following linearized rate equations relating reciprocal relaxation time values (τ^{-1}) to the concentrations of reactive species were used:

$$\tau_{fast}^{-1} + \tau_{slow}^{-1} = k_1\big([XOH]+[ion\ species]\big)+k_{-1}+k_2+k_{-2} \tag{17}$$

$$\tau_{fast}^{-1} \cdot \tau_{slow}^{-1} = k_1\big[k_2+k_{-2}\big]\big([XOH]+[ion\ species]\big)+k_{-1}k_{-2} \tag{18}$$

where the ion species are $H_2AsO_4^-$ or $HCrO_4^-$. The derivation of these equations was obtained from Bernasconi[78] and is based on the two-step reaction system (A + B $\leftrightarrow$ C $\leftrightarrow$ D). If the mechanism portrayed in Figure 18 is accurate then a plot of $\tau_{fast}^{-1} + \tau_{slow}^{-1}$ and $\tau_{fast}^{-1} \cdot \tau_{slow}^{-1}$ as a function of the concentration term ([XOH] + [ion species]) should be linear. Plots of Equations 17 and 18 were

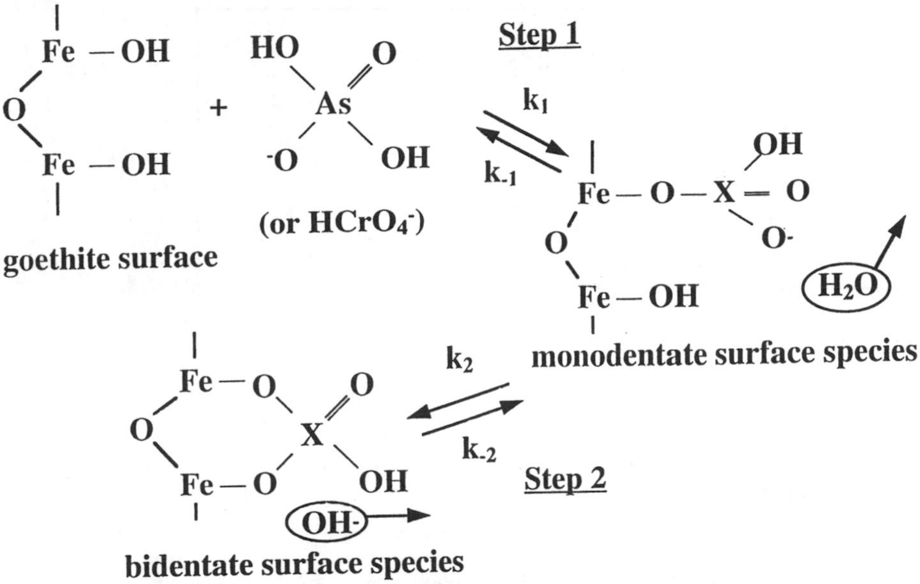

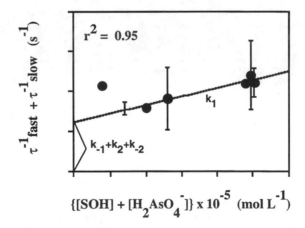

Figure 18   Proposed mechanism for oxyanion adsorption/desorption on goethite. The X represents either As(V) or Cr(VI). (From Grossl, P. R. et al., *Environ. Sci. Technol.,* 31, 321, 1997. With permission.)

$$\{[SOH] + [H_2AsO_4^-]\} \times 10^{-5} \ (mol \ L^{-1})$$

Figure 19   Evaluation of the linearized rate Equation 17 for the mechanism displayed in Figure 18 for arsenate. (From Grossl, P. R. et al., *Environ. Sci. Technol.,* 31, 321, 1997. With permission.)

linear for both arsenate and chromate, suggesting that the proposed mechanism was plausible. Figures 19 and 20 show the linear relationships for Equations 17 and 18, respectively, for arsenate.

From the plots in Figures 19 and 20 forward and reverse rate constants were obtained for the adsorption and desorption reactions of both the monodentate and bidentate steps, where k_1 = slope (Figure 19); k_{-1} = intercept (Figure 19) – slope (Figure 20)/slope (Figure 19); k_2 = intercept (Figure 19) – k_{-1} – k_{-2}; and k_{-2} = intercept (Figure 20)/k_{-1} . The calculated rate constants for both chromate and arsenate adsorption/desorption on goethite are listed in Table 3.

For both oxyanions, the rate constants for the reverse reactions (associated with the breaking of arsenate or chromate/goethite bonds) were lower than the rate constants for the forward reactions (formation of the inner-sphere oxyanion/goethite surface complexes). Therefore, the rate-limiting steps were the reverse reactions. The equilibrium constants listed in Table 3 were calculated using

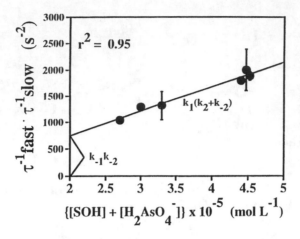

Figure 20 Evaluation of the linearized rate Equation 18 for the mechanism displayed in Figure 18 for arsenate. (From Grossl, P. R. et al., *Environ. Sci. Technol.*, 31, 321, 1997. With permission.)

Table 3   Calculated Rate Constants for Chromate and Arsenate Adsorption/Desorption on Goethite

	Step I	Step II
Arsenate	$k_1 = 10^{6.3}$ l/mol s	$k_2 = 15\ s^{-1}$
	$k_{-1} = 8\ s^{-1}$	$k_{-2} = 8\ s^{-1}$
	$K_{eq} = 10^{5.35}$ l/mol s	$K_{eq} = 10^{0.26}$
Chromate	$k_1 = 10^{5.8}$ l/mol s	$k_2 = 16\ s^{-1}$
	$k^{-1} = 129\ s^{-1}$	$k_{-2} = 38\ s^{-1}$
	$K_{eq} = 10^{3.7}$ l/mol s	$K_{eq} = 10^{-0.4}$

From Grossl, P.R. et al, *Environ. Sci. Technol.*, 31, 321, 1997. With permission.

the rate constants for each reaction step in our proposed mechanism (Figure 18) from the following relationship:

$$K_{eq} = k_f/k_r \tag{19}$$

The calculated equilibrium constant for step 1 for arsenate was $10^{5.35}$ and for step 2 was $10^{0.26}$, while the calculated K_{eq} for step 1 for chromate was $10^{3.7}$ and for step 2 was $10^{-0.4}$. The adsorption of both oxyanions and subsequent formation of inner-sphere surface complexes are thermodynamically favorable, with the exception of the equilibrium constant for the second step associated with chromate adsorption (slightly less than 1). Thus, the bidentate chromate/goethite surface complex is less likely to form than the monodentate surface complex. This is in agreement with spectroscopic data obtained from X-ray absorption fine structure (XAFS) analyses which indicate a mixture of both monodentate and bidentate chromate surface complexes, but at low surface coverage a greater proportion of chromate is associated with the monodentate complex than the bidentate complex. The results from both kinetic and XAFS experiments suggest that arsenate is more likely to form an inner-sphere surface complex with goethite than chromate.

While the initial sorption of heavy metals is rapid, with the chemical reaction step occurring on millisecond time scales, further sorption is usually quite slow (Figure 21) occurring over time scales of days and longer. This slow sorption has been ascribed to interparticle or intraparticle diffusion in pores, sites of low reactivity, and surface precipitation.[88-90] Examples from the literature of each hypothesis will be discussed. However, these hypotheses have primarily been based on

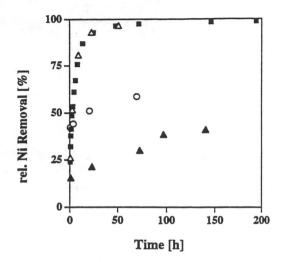

Figure 21   Kinetics of Ni sorption (%) on pyrophyllite (■), kaolinite (△), gibbsite (▲), and montmorillonite (○) from a 3 mM Ni solution at pH = 7.5 and an ionic strength I = 0.1 M (NaNO₃). (From Scheidegger, A. M. et al., *J. Colloid Interface Sci.*, 186, 118, 1997. With permission.)

macroscopic observations. To confirm the mechanism(s), *in situ* surface spectroscopic/microscopic techniques should be employed. These will be discussed later.

Obviously, an important factor affecting the degree of slow sorption of metals (and for that matter also of organic chemicals) is the time period the sorbate has been in contact with the sorbent (residence time). Bruemmer et al.[19] studied Ni^{2+}, Zn^{2+}, and Cd^{2+} adsorption on goethite, a porous Fe-oxide that has defect structures in which metals can be incorporated to satisfy charge imbalances. Bruemmer et al.[19] found at pH 6 that as reaction time increased from 2 h to 42 d (at 293 K), adsorbed Ni^{2+} increased from 12% to 70% of total adsorption, and total increases in Zn^{2+} and Cd^{2+} adsorption over this time increased 33% and 21%, respectively. The kinetic reactions could be well described using a Fickian diffusion model. Metal uptake was hypothesized to occur via a three-step mechanism: (1) adsorption of metals on external surfaces; (2) solid-state diffusion of metals from external to internal sites; and (3) metal binding and fixation at positions inside the goethite particle.

Fuller et al.[55] combined kinetic sorption and desorption experiments with spectroscopic observations[90] to study As sorption on ferrihydrite. Using X-ray absorption fine structure (XAFS) spectroscopy, they found that As was sorbed predominantly as inner-sphere bidentate complexes, regardless of whether the As was adsorbed postmineralization of the ferrihydrite, or it was present during precipitation. No As surface precipitates were observed. Slow As sorption and desorption were explained as slow diffusion of the As to or from interior surface complexation sites that exist within disordered aggregates of crystallites. The kinetic reactions could be described using a Fickian diffusion model.

Slow metal sorption has also been ascribed to conversion of the metal sorbate from a high-energy state to a low-energy state. For example, Lehmann and Harter[91] measured the kinetics of chelate-promoted Cu^{2+} release from a soil to assess the strength of the bond formed. Sorption/desorption was biphasic, which was attributed to high- and low-energy bonding sites. With increased residence time from 30 min to 24 h, Lehmann and Harter[91] speculated that there was a transition of Cu from low-energy sites to higher-energy sites (as evaluated by release kinetics). Incubations for up to 4 d showed a continued uptake of Cu and a decrease in the fraction released within the first 3 min, which was referred to as the low-energy adsorbed fraction.

Ainsworth et al.[92] studied the adsorption/desorption of Co^{2+}, Cd^{2+}, and Pb^{2b+} on hydrous ferric oxide (HFO) as a function of oxide aging and metal-oxide residence time. Oxide aging did not cause hysteresis of metal cation sorption-desorption. Aging the oxide with the metal cations resulted

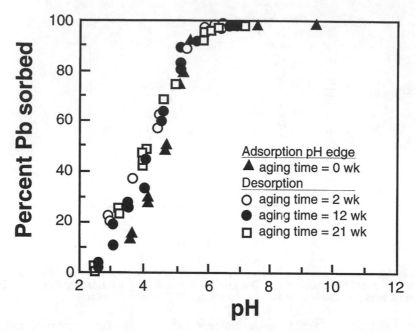

Figure 22   Fractional sorption-desorption of Pb^{2+} to hydrous Fe-oxide (HFO) as a function of pH and HFO-Pb^{2+} aging time. (From Ainsworth, C. C. et al., *Soil Sci. Soc. Am. J.*, 58, 1615, 1994. With permission.)

in hysteresis with Cd^{2+} and Co^{2+} but little hysteresis was observed with Pb^{2+}. With Pb^{2+}, between pH 3 and 5.5 there was slight hysteresis over a 21 week aging process (hysteresis varied from <2% difference between sorption and desorption to ≈10%). At pH 2.5 Pb^{2+} desorption was complete within a 16 h desorption period and was not affected by aging time (Figure 22). However, with Cd^{2+} and Co^{2+}, extensive hysteresis was observed over a 16 week aging period and the hysteresis increased with aging time (Figures 23 and 24). After 16 weeks of aging 20% of the Cd^{2+} and 53% of the Co^{2+} was not desorbed, and even at pH 2.5, hysteresis was observed. The extent of reversibility with aging for Co^{2+}, Cd^{2+}, and Pb^{2+} was inversely proportional to the ionic radius of the ions, i.e., $Co^{2+} < Cd^{2+} < Pb^{2+}$. Ainsworth et al.[92] attributed the hysteresis to Co and Cd incorporation into a recrystallizing solid (probably goethite) via isomorphic substitution and not to micropore diffusion.

Scheidegger et al.[93] investigated Ni sorption on pyrophyllite, kaolinite, gibbsite, and montmorillonite. Nickel sorption was characterized by a rapid reaction in which 15% to 40% of the initial Ni was removed within the first hour. Following the rapid reaction, the rate of sorption decreased significantly. Using XAFS, the slow sorption was directly ascribed to the formation of mixed Ni-Al hydroxides on the sorbents. The "surface" precipitates had a structure similar to a Ni-bearing mineral, tacovite ($Ni_6Al_2(OH)_{16} CO_3 \cdot H_2O$). Scheidegger et al.,[93] hypothesized that the release of Al from the sorbents into solution was the rate-determining step for the formation of the mixed Ni/Al hydroxide-like phases.

2. Organic Contaminants

There have been a number of studies on the kinetics of organic chemical sorption/desorption with soils and soil components. Many of these investigations have shown that sorption/desorption is characterized by a rapid, reversible stage followed by a much slower, nonreversible stage[60,94-95] or biphasic kinetics. The rapid phase has been ascribed to retention of the organic chemical in a labile form that is easily desorbed. However, the much slower reaction phase involves the entrapment of the chemical in a nonlabile form that is difficult to desorb. This slower sorption/desorption

Table 4   Sorption Distribution Coefficients for Herbicides in "Freshly Aged" and "Aged" Soils

Herbicide	Soil	$K_d{}^a$	$K_{app}{}^b$
Metolachlor	CVa	2.96	39
	CVb	1.46	27
	W1	1.28	49
	W2	0.77	33
Atrazine	CVa	2.17	28
	CVb	1.32	29
	W3	1.75	4

[a] Sorption distribution coefficient (l kg^{-1}) of "freshly aged" soil based on a 24 h equilibration period.
[b] Apparent sorption distribution coefficient (l kg^{-1}) in contaminated soil ("aged" soil) determined using a 24 h equilibration period.

Adapted from Pignatello and Huang.[99]

A number of studies have also shown that with "aging" the nonlabile portion of the organic chemical in the soil/sediment becomes more resistant to release.[61,67,97-99] However, Connaughton et al.[51] did not observe the nonlabile fraction increasing with age for naphthalene contaminated soils.

One way to gauge the effect of time on organic contaminant retention in soils is to compare K_d (sorption distribution coefficient) values for "freshly aged" and "aged" soil samples. In most studies, K_d values are measured based on a 24-h equilibration between the soil and the organic chemical. When these values are compared to K_d values for field soils previously reacted with the organic chemical ("aged" samples) the latter have much higher K_d values, indicating that much more of the organic chemical is in a sorbed state. For example, Pignatello and Huang[99] measured K_d values in "freshly aged" (K_a) and "aged" soils (K_{app}, apparent sorption distribution coefficient) reacted with atrazine and metolachlor, two widely used herbicides. The "aged" soils had been treated with the herbicides 15 to 62 months before sampling. The K_{app} values ranged from 2.3 to 42 times higher than the K_d values (Table 4).

Scribner et al.,[98] studying simazine (a widely used triazine herbicide for broadleaf and grass control in crops) desorption and bioavailability in aged soils, found that K_{app} values were 15 times higher than K_d values. Scribner et al.[98] also showed that 48% of the simazine added to the freshly aged soils was biodegradable over a 34-d incubation period while none of the simazine in the aged soil was biodegraded.

One of the implications of these results is that while many transport and degradation models for organic contaminants in soils and waters assume that the sorption process is an equilibrium process, the above studies clearly show that kinetic reactions must be considered when making predictions about the mobility and fate of organic chemicals. Moreover, calculation of K_d values based on a 24-h equilibration period, which is commonly used in fate and risk assessment models, can be inaccurate, since 24-h K_d values often overestimate the amount of organic chemical in the solution phase.

The finding that many organic chemicals are quite persistent in the soil environment has both good and bad features. The beneficial aspect is that the organic chemicals are less mobile and may not be readily transported in groundwater supplies. The negative aspect is that their persistence and inaccessibility to microbes may make decontamination more difficult, particularly if *in situ* remediation techniques such as biodegradation are employed.

B. Kinetics of Mineral Dissolution

1. Rate-Limiting Steps

Dissolution of minerals involves several steps (Stumm and Wollast[100]): (1) mass transfer of dissolved reactants from the bulk solution to the mineral surface, (2) adsorption of solutes, (3) interlattice transfer of reaction species, (4) surface chemical reactions (CR), (5) removal of reactants from the surface, and (6) mass transfer of products into the bulk solution. Under field conditions mineral dissolution is slow and mass transfer of reactants or products in the aqueous phase (steps 1 and 6) are not rate-limiting. Thus, the rate-limiting steps are either transport of reactants and products in the solid phase (step 3) or surface chemical reactions (step 4) and removal of reactants from the surface (step 5).

Transport-controlled dissolution reactions or those controlled by mass transfer or diffusion can be described using a parabolic rate law given below:[100]

$$r = \frac{dC}{dt} = kt^{-1/2} \tag{20}$$

where r is the reaction rate, C is the concentration in solution, t is time, and k is the reaction rate constant. Integrating, C increases with $t^{1/2}$,

$$C = C_0 + 2kt^{1/2} \tag{21}$$

where C_0 is the initial concentration in solution.

If the surface reactions are slow compared to the transport reactions, dissolution is surface-controlled which is the case for most dissolution reactions of silicates and oxides. In surface-controlled reactions the concentrations of solutes next to the surface are equal to the bulk solution concentrations and the dissolution kinetics are zero-order if steady-state conditions are operational on the surface. Thus, the dissolution rate, r is

$$r = \frac{dC}{dt} = kA \tag{22}$$

and r is proportional to the surface area, A, of the mineral. Thus, for a surface-controlled reaction the relationship between time and C should be linear. Figure 26 compares transport- and surface-controlled dissolution mechanisms.

Intense arguments have ensued over the years concerning the mechanism for mineral dissolution. Those who supported a transport-controlled mechanism believed that a leached layer formed as mineral dissolution proceeded and that subsequent dissolution took place via diffusion through the leached layer.[101] Advocates of this theory found that dissolution was described by the parabolic rate law (Equation 20). However, the "apparent" transport-controlled kinetics may be an artifact caused by dissolution of hyperfine particles, formed on the mineral surfaces after grinding that are highly reactive sites, or by use of batch methods that cause reaction products to accumulate causing precipitation of secondary minerals. These experimental artifacts can cause incongruent reactions and pseudoparabolic kinetics. Recent studies employing surface spectroscopies, such as X-ray photoelectron spectroscopy (XPS) and nuclear resonance profiling,[102,103] have demonstrated that although some incongruency may occur in the initial dissolution process, which may be diffusion-controlled, the overall reaction is surface-controlled. An illustration of the surface-controlled dissolution of γ-Al_2O_3 resulting in a linear release of Al^{3+} with time is shown in Figure 27. The dissolution rate, r, can be obtained from the slope of Figure 27.

2. Surface-Controlled Dissolution Mechanisms

Dissolution of oxide minerals via a surface-controlled reaction by ligand-promoted and proton-promoted processes has been described by Stumm and coworkers[104-106] using a surface coordination

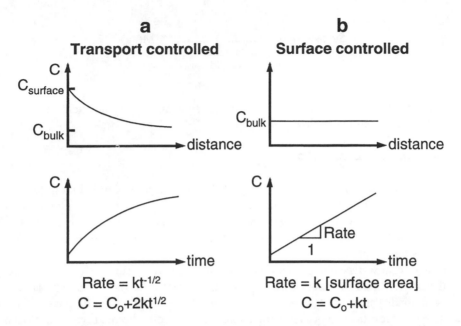

a
Transport controlled

b
Surface controlled

Rate = $kt^{-1/2}$
$C = C_o + 2kt^{1/2}$

Rate = k [surface area]
$C = C_o + kt$

Figure 26 Transport vs. surface-controlled dissolution. Schematic representation of concentration in solution, C, as a function of distance from the surface of the dissolving mineral. In the lower part of the figure, the change in concentration is given as a function of time. From Stumm, W., *Chemistry of the Solid-Water Interface,* John Wiley & Sons, New York, 1992. With permission.)

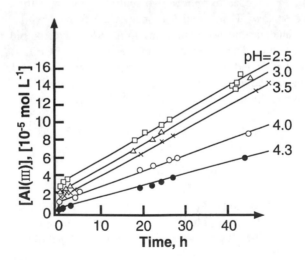

Figure 27 Linear dissolution kinetics observed for the dissolution of γ-Al_2O_3. Representative of processes whose rates are controlled by a surface reaction and not by transport. (From Furrer, G. and Stumm, W., *Geochim. Cosmochim. Acta,* 50, 1847, 1986. With permission.)

approach. The important reactants in these processes are H_2O, H^+, OH^-, ligands, and reductants and oxidants. The reaction mechanism occurs in two steps:[100]

$$\text{Surface sites + Reactants (} H^+, OH^-, \text{ or ligands)} \xrightarrow{\quad \text{Fast} \quad} \text{Surface species} \quad (23)$$

$$\text{Surface sites} \xrightarrow[\substack{\text{detachment of} \\ \text{metal (M)}}]{\text{slow}} \text{M(aq)} \tag{24}$$

Thus, the attachment of the reactants to the surface sites is fast and detachment of metal species from the surface into solution is slow and rate-limiting.

3. Ligand-Promoted Dissolution

Figure 28 shows how the surface chemistry of the mineral affects dissolution. One sees that surface protonation of the surface ligand increases dissolution by polarizing interatomic bonds close to the central surface ions that promotes the release of a cation surface group into solution. Hydroxyls that bind to surface groups at higher pHs can ease the release of an anionic surface group into the solution phase.

Ligands that form surface complexes via ligand exchange with a surface hydroxyl add negative charge to the Lewis acid center coordination sphere, and lower the Lewis acid acidity. This polarizes the M-oxygen bonds causing detachment of the metal cation into the solution phase. Thus, inner-sphere surface complexation plays an important role in mineral dissolution. Ligands such as oxalate, salicylate, F^-, EDTA, and NTA increase dissolution but others, e.g., SO_4^{2-}, CrO_4^{2-} and benzoate inhibit dissolution. Phosphate and arsenate enhance dissolution at low pH and dissolution is inhibited at pH>4.[12]

The reason for these differences may be that bidentate species that are mononuclear promote dissolution while binuclear bidentate species inhibit dissolution. With binuclear bidentate complexes, more energy may be needed to remove two central atoms from the crystal structure. With phosphate and arsenate, at low pH mononuclear species are formed while at higher pH (around pH = 7) binuclear or trinuclear surface complexes form. Mononuclear bidentate complexes are formed with oxalate while binuclear bidentate complexes form with CrO_4^{2-}. Additionally, the electron donor properties of CrO_4^{2-} and oxalate are also different. With CrO_4^{2-} a high redox potential is maintained at the oxide surface which restricts reductive dissolution.[12,100]

Dissolution can also be inhibited by cations such as VO^{2+}, Cr(III), and Al(III) that block surface functional groups.

One can express the rate of the ligand-promoted dissolution, R_L, as

$$R_L = k'_L(\equiv ML) = k'_L \, C_L^s \tag{25}$$

where k'_L is the rate constant for ligand-promoted dissolution (time^{-1}), $\equiv ML$ is the metal-ligand complex, and C_L^s is the surface concentration of the ligand (mol m^{-2}). Figure 29 shows that Equation 25 adequately described ligand promoted dissolution of γ-Al$_2$O$_3$.

4. Proton-Promoted Dissolution

Under acid conditions, protons can promote mineral dissolution by binding to surface oxide ions, causing bonds to weaken. This is followed by detachment of metal species into solution. The proton-promoted dissolution rate, R_H, can be expressed as:[12]

$$R_H = k'_H(\equiv MOH_2^+)^j = k'_H(C_H^s)^j \tag{26}$$

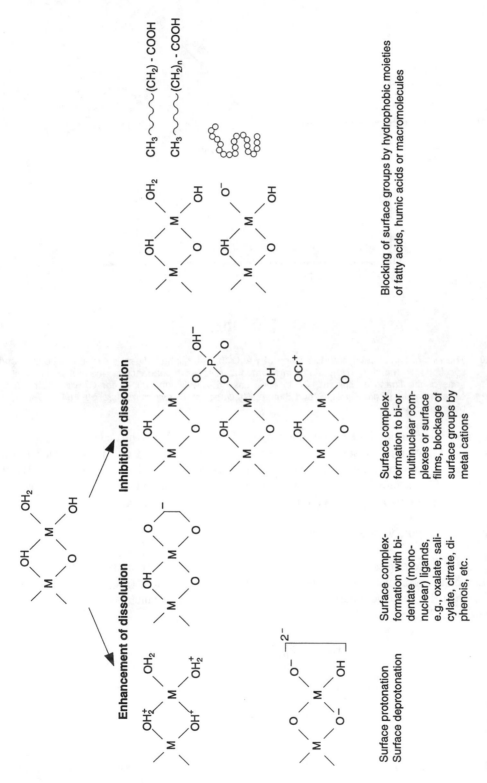

Figure 28   The dependence of surface reactivity and of kinetic mechanisms on the coordinative environment of the surface groups. (From Stumm, W. and Wollast, R., *Rev. Geophys.*, 28, 53, 1990. With permission.)

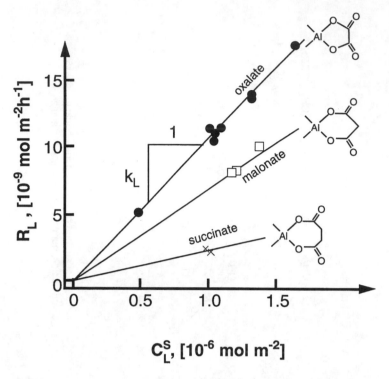

Figure 29   The rate of ligand-catalyzed dissolution of γ-Al$_2$O$_3$ by the aliphatic ligands oxalate, malonate, and succinate, R_L (nmol m^{-2} h^{-1}), can be interpreted as a linear dependence on the surface concentrations of the ligand complexes, C_L^s. In each case the individual values for C_L^s were determined experimentally. (From Furrer, G. and Stumm, W., *Geochim. Cosmochim. Acta,* 50, 1847, 1986. With permission.)

where k_H' is the rate constant for proton-promoted dissolution, $\equiv$MOH$_2^+$ is the metal-proton complex, C_H^s is the concentration of the surface adsorbed proton complex (mol^{-2}), and j corresponds to the oxidation state of the central metal ion in the oxide structure (i.e., j = 3 for Al(III) and Fe(III) in simples cases). If dissolution occurs by only one mechanism j is an integer. Figure 30 shows an application of Equation 26 for the proton-promoted dissolution of γ-Al$_2$O$_3$.

5. Overall Dissolution Mechanisms

The rate of mineral dissolution, which is the sum of the ligand-promoted, proton-promoted, and deprotonation-promoted (or bonding of OH$^-$ ligands) dissociation ($R_{OH} = k_{OH}'$ (C_{OH}^s)^i) rates along with the pH-independent portion of the dissolution rate (k_{H_2O}') which is due to hydration, can be expressed as:[12]

$$R = +k_L'\left(C_L^s\right) + k_H'\left(C_H^s\right)^j + k_{OH}'\left(C_{OH}^s\right)^i + k_{H_2O}' \tag{27}$$

Equation 27 is valid if dissolution occurs in parallel at varying metal centers.[104]

6. Dissolution Kinetics of Polynuclear Surface Species

Recent studies using surface spectroscopic and microscopic techniques such as XAFS, electron paramagnetic resonance spectroscopy (EPR), X-ray photoelectron spectroscopy (XPS), auger electron

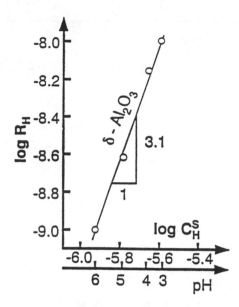

Figure 30   The dependence of the rate of proton-promoted dissolution of γ-Al$_2$O$_3$, R$_H$ (mol m^{-2} h^{-1}), on the surface concentration of the proton complexes, C$_H^S$ (mol m^{-2}). (From Furrer, G. and Stumm, W., *Geochim. Cosmochim. Acta*, 50, 1847, 1986. With permission.)

spectroscopy (AES), scanning electron microscopy (SEM), and atomic force microscopy (AFM) have shown that the formation of polynuclear surface species (e.g., surface precipitates) on natural materials is an important phenomenon.[107-111]

Multinuclear metal hydroxides of Pb, Ni, Co, Cu, and Cr(III) on oxides and aluminosilicates have been discerned.[93,108,110,112-121] Such surface precipitates have been observed at metal surface loadings far below a theoretical monolayer coverage, and in a pH range well below the pH where the formation of metal hydroxide precipitates would be expected according to the thermodynamic solubility product.

While metal surface precipitates could have important ramifications with respect to environmental quality (bioavailability, mobility, and fate of metals in soils and waters) via dissolution, little information is available in the literature on the dissolution kinetics of surface precipitates. Scheidegger and Sparks.[21] studied the dissolution of mixed Ni-Al polynuclear surface precipitates from pyrophyllite. Nickel detachment was slow and depended strongly on the pH and the experimental method (Figure 31). Under steady-state conditions, a constant Ni detachment rate was observed which was much slower than the dissolution of a crystalline Ni(OH)$_2$ reference compound.

The mixed Ni/Al hydroxide-like phases explain the finding of Scheidegger et al.[93] that the dissolution rates (Si-release) are strongly enhanced (relative to the dissolution rates of the clays alone) as long as Ni sorption is pronounced. This suggests that the surface complexes of Ni destabilize surface metal ions (Al and Si) relative to the bulk solution, and therefore lead to an enhanced dissolution of the clay. The association of Ni with Al could explain why the enhanced dissolution rate is only observable where Ni sorption is pronounced (Figure 32).

The structure of the mixed Ni-Al polynuclear surface precipitates is also not changed after extensive dissolution. EXAFS analysis shows that a tacovite-like structure remains after dissolution that is identical to the precipitate before commencement of dissolution.[21,93] It is obvious that metal surface precipitates must be carefully considered in metal surface complexation modeling, metal speciation, and risk assessments for the migration of contaminants in polluted sites.

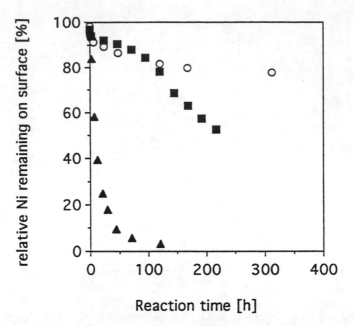

Figure 31   Kinetics of Ni detachment from surface precipitates at pH = 4. Relative Ni remaining on the surface
(%) is shown for the *conventional method* (○) and the *replenishment method* (■) as a function of
the reaction time; 98% of the initial Ni was sorbed in the beginning of the detachment experiment.
The dissolution of an equivalent amount of crystalline $Ni(OH)_2$ (in mol) at pH = 4 is given for
comparison (▲). (From Scheidegger, A. M. and Sparks, D. L., *Chem. Geol.*, 132, 157, 1996. With
permission.)

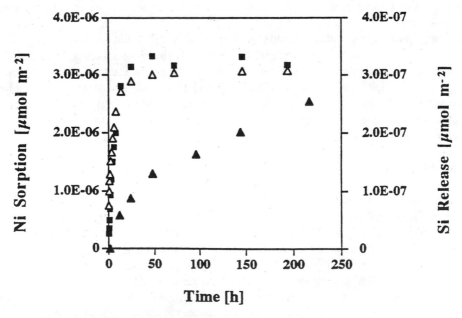

Figure 32   The kinetics of Ni sorption on pyrophyllite from a 3×10^{-3} *M* Ni solution at pH = 7.5. (■) denotes
the amount of sorbed Ni (μmol m^{-2}) and (△) the amount of simultaneous dissolved Si (μmol m^{-2}).
The dissolution of untreated pyrophyllite at pH = 7.5 is shown for comparison (▲). (From Scheidegger, A. M. et al., *J. Colloid Interface Sci.*, 186, 118, 1997. With permission.)

VII. CONFIRMATION OF REACTION MECHANISMS USING SPECTROSCOPIC AND MICROSCOPIC TECHNIQUES

In spite of many decades of intensive efforts by soil chemists to understand sorption processes, our understanding of the mechanisms of chemical reactions at the solid/liquid interface is still not definitive. One of the main reasons for this is that studies of the reactions between environmental particle surfaces and aqueous solutions have been limited to macroscopic studies, until quite recently. While macroscopic equilibrium studies and models reveal some important information about sorption/desorption phenomena, no mechanistic or molecular information is revealed. Kinetic studies can reveal something about reaction mechanisms at the soil particle/solution interface, particularly if energies of activation are calculated and stopped-flow or interruption techniques are employed. However, molecular and/or atomic resolution surface techniques should be employed to corroborate the proposed mechanism hypothesized from equilibrium and kinetic studies. These techniques can be used either separately or, preferably, simultaneously with kinetic investigations. While the latter approach is preferable, only limited studies have been reported in the literature. Examples of both approaches will be cited in the following discussions. Additionally, an overview will be presented of contemporary spectroscopic and microscopic techniques that are important for studying sorption processes on soils and soil minerals. Without question, the application of molecular and atomic resolution techniques to elucidation of sorption mechanisms should be a major research thrust in soil and environmental chemistry for decades to come.

There are two principal subdivisions in molecular spectroscopy: *in situ* and non-*in situ* methods.[23,122] The principal invasive non-*in situ* techniques used for soil and aquatic systems are XPS, AES, and secondary mass spectroscopy (SIMS). Each of these techniques yields detailed information about the structure and bonding of minerals, and the chemical species present on the mineral surfaces. The disadvantage of invasive techniques is that they often must be performed under adverse experimental conditions, e.g., desiccation, high vacuum, heating, or particle bombardment. Such conditions may yield data that are misleading as a result of experimental artifacts.[123,124] XPS is the most widely used non-*in situ* surface-sensitive technique. It has been used to study sorption mechanisms of inorganic cations and anions such as Cu^{2+}, Co^{2+}, Ni^{2+}, Cd^{2+}, Cr^{3+}, Fe^{3+}, selenite, and uranyl in soil and aquatic systems.[109,111,125-132]

In situ methods require little or no alteration of the sample from its natural state.[129] They can be applied to aqueous solutions or suspensions; most involve the input and detection of photons. Examples of *in situ* techniques are EPR, Fourier-transform infrared (FTIR), nuclear magnetic resonance (NMR), X-ray absorption (XAS), and Mössbauer spectroscopies. However, many other techniques are available.[23,133]

EPR spectroscopy is a technique for detecting paramagnetism. Electron paramagnetism occurs in all atoms, ions, organic free radicals and molecules with an odd number of electrons. EPR is based upon the resonant absorption of microwaves by paramagnetic substances and describes the interaction between an electronic spin subjected to the influence of a crystal field and an external magnetic field.[23,134] The method is applicable to transition metals of Fe^{3+}, Cu^{2+}, Mn^{2+}, V^{4+}, and molybdenum (V) and has been widely used to study metal ion sorption on soil mineral components[135-138] and soil organic matter.[139-144] Several review articles on EPR are available.[134,145-146]

The Mössbauer effect is based on the recoil-free emission and resonant absorption of γ rays by specific atomic nuclei in a solid. γ rays are used as a probe of nuclear energy levels which, in turn, are sensitive to the details of both local electron configuration and the electric and magnetic fields of the solid.[147] In natural systems, Mössbauer effects specifically relate to iron.[145] Mössbauer spectroscopy is able to distinguish between high spin Fe^{2+} and Fe^{3+} without interference from any other element.[23] It also provides information on the chemical nature of chemical entities bound to the iron. Application of Mössbauer spectroscopy in soil science is not common.[141,148-151] There are some review articles available on application of Mössbauer spectroscopy to soil materials.[145,147,152]

Application of infrared (IR) spectroscopy to the study of sorbed species has a long history. The introduction of Fourier transform techniques has made a significant contribution to the development of new investigation techniques such as diffuse reflectance infrared Fourier transform (DRIFT) and attenuated total reflectance (ATR) spectroscopy. IR spectroscopy now extends far beyond classical chemical analysis and is successfully applied to study sorption processes of inorganic and organic soil components. These techniques, and other vibrational spectroscopies such as Raman, are the subject of numerous reviews.[123,153-156]

The use of NMR spectroscopy to study surfaces has a shorter history, and fewer applications than vibrational spectroscopies. The primary reason is that the sensitivity of NMR is much lower than IR. Properties that might be exploited are the chemical shift, NMR relaxation times, and magnetic couplings to nuclei that are characteristic of a surface.[157] Most NMR studies in the field of soil science concentrate on the characterization of soil organic matter and soil humification processes and therefore involve ^1H, ^{13}C, and ^{15}N NMR. Reviews on these and related topics are available.[122,157-159] In the past few years ^3Li, ^3Na, ^{39}K, ^{111}Cd, and ^{133}Cs NMR spectroscopy have been increasingly used as tools to elucidate cation exchange sites on clay mineral surfaces.[160-163] Since it is virtually impossible to obtain any useful molecular information by observing the nucleus of a paramagnetic metal directly, studies of cation exchange have focused on diamagnetic metals, such as Cd^{2+}, which have a spin of 1/2 and an acceptable natural abundance (e.g., 12% and 13%, respectively, for the two NMR-active isotopes, ^{113}Cd and ^{111}Cd). NMR is essentially a bulk spectroscopic technique.[23] The advent of high-resolution, solid-state NMR techniques, such as magic angle sample spinning (MAS) and cross-polarization (CP), along with more sensitive, high-magnetic-field, user-friendly, pulsed NMR spectrometers, has brought increased applications to heterogeneous aqueous systems.[23,122] In particular, ^{27}Al and ^{29}Si NMR in zeolites and other minerals have proven valuable for the structural elucidation of samples whose disorder has prevented diffraction techniques from being very useful.[164-166]

One of the most powerful noninvasive surface sensitive techniques is XAS. XAS is a powerful, element-specific, *in situ* technique that can be used to determine the local structure (bond distance, number and type of nearest neighbors) around a sorbing element, even when the element is at low concentration levels (depending on element and matrix as low as 0.03% to 0.05% per weight[107]). XAS can be used to probe most types of phases (crystalline or amorphous solids, liquids, gases) and at structural sites ranging from those in crystals and glasses to those at interfaces, such as the mineral/water interface. Like other spectroscopic methods, XAS is not without limitations. It requires intense X-rays, and these are generated by electrons/positrons that circulate in a storage ring of a synchrotron facility at energies of 1 to 6 GeV in paths curved by a magnetic field. Synchrotron facilities are not readily accessible. In addition, in most cases studies are primarily limited to elements heavier than Sc.[23] There are several reviews on XAS that the reader can consult. Particularly relevant to soil science are those by Brown et al.,[167] Brown,[133] Charlet and Manceau,[107] Fendorf et al.,[124] and Schulze and Bertsch.[168]

By convention, XAS spectra (one scans near the X-ray absorption edge K, L, M of the element of concern) can be divided into two energy regions: (1) the X-ray absorption near edge structure (XANES) region and (2) the extended edge X-ray absorption structure (EXAFS) region. The XANES region runs to about 50 eV above the absorption edge.[168] It is usually characterized by intense resonance features arising from electron transitions to unoccupied bound state and continuum levels, and from multiple scattering of the emitted photoelectrons by atoms surrounding the absorber. For many first row transition elements the pre-edge resonances provide information on the site geometry of the absorber, which is commonly related to the oxidation state.[23] For example, Cr oxidation states can be deduced from the presence or absence of a predominant pre-edge feature that is characteristic of Cr(VI), which is tetrahedrally coordinated, but is nearly absent for Cr(III), which is octahedrally coordinated.[167-169] XANES was also used to determine oxidation states of Mn,[170-172] Se,[173] Ce,[174] Ti,[169] Tl,[175] and U[176] in natural environments. Although interatomic distances can be deduced from the XANES region,[177-179] the main application of XANES is "fingerprinting". By comparison of known spectra with unknown samples important qualitative information on bonding environments can be deduced. This method has proven to be beneficial for investigating unknown heterogeneous samples such as soils.[180]

The EXAFS region which follows the XANES region in the XAS spectra runs up to about 800 eV beyond the edge.[107] The frequency oscillations in this region arise from constructive and destructive interference patterns between the outgoing and the returning photoelectronic wave that has been backscattered from first and sometimes second shell neighboring atoms. The frequency of the oscillations is inversely related to the bond distance between the absorber and neighboring atoms. The amplitude of these oscillations is related to both the identity and number of atoms surrounding the central absorber.[168] EXAFS has been applied to problems in physics, chemistry, biochemistry, and materials science for quite some time.[181] It is also well established in research on earth and marine material.[177,182,183] One of the earliest uses of EXAFS to determine sorption mechanisms of ions on natural surfaces was in the research of Hayes et al.[184] who studied selenate and selenite adsorption on goethite. They showed that selenate was absorbed as an outer-sphere complex and selenite was adsorbed as an inner-sphere complex.[184] This interpretation was later questioned by Manceau and Charlet,[185] who found that selenate ions form binuclear bidentate surface complexes on goethite. Recently, numerous studies have demonstrated the usefulness of EXAFS for providing specific chemical speciation information on contaminants associated with sorptive phases, including soils.[23]

While XAS provides local chemical information, it provides no information on spatial resolution of surface species. Such information can only be obtained by microscopic methods. Scanning electron microscopy (SEM) and transmission electron microscopy (TEM or HRTEM, high-resolution TEM) are well-established methods for acquiring both chemical and micromorphological data on soils and soil materials. TEM can provide spatial resolution of surface alterations and the amorphous nature or degree of crystallinity of sorbed species (ordering). It can also be combined with electron spectroscopies to determine elemental analysis.[23] In the last two decades several books and review articles on microcharacterization of soils using electron microscopy have been published. A partial list of these includes the books edited by Smart and Tovey,[186] Bullock and Murphy,[187] and Douglas,[188] as well as review articles by Whalley,[189] Bisdom et al.,[190] Tovey et al.,[191] and Chen.[192]

From the very inception of the scanning tunneling microscope (STM) in 1981 it was apparent that the technique would revolutionize the study of mineral surfaces and surface-related phenomena. Indeed, by the end of the 1980s, applications of STM were beginning to appear in the earth sciences literature.[193,194] However, the major event for the environmental science community came with the development of scanning force microscopy (SFM; also known as atomic force microscopy, AFM). SFM allows imaging of mineral surfaces in air or immersed in solution, and at subnanometer scale resolution.[195] Applications to date include determining the molecular to atomic scale structure of mineral surfaces,[196,197] probing forces at the mineral/water interface,[198,199] visualizing sorption of hemimicelles and macromolecular organic substances such as humic and fulvic acid,[195,200] determining clay particle thicknesses and morphology of clay-sized particles,[201-205] imaging soil bacteria,[206] and measuring directly the kinetics of growth, dissolution, heterogeneous nucleation, and redox processes (see later discussion).

Nevertheless, one must realize that SFM imaging of soil samples presents a challenge for the microscopist because natural samples tend to be heterogeneous. SFM does not provide chemical data; hence, minerals must be identified primarily based on morphology. Yet, morphology alone can lead to ambiguous results. Even atomic-scale imaging may not be conclusive because the crystal structures of many soil minerals are similar.[23] To work with soil samples, Maurice[195] suggested the following approach: (1) it is essential to compile as many images as possible of well-characterized minerals to use as a catalog for particle identification, (2) the soil samples should be characterized as fully as possible by XRD and other chemical and physical analytical techniques, and (3) techniques for isolating different components and fractions can greatly simplify image interpretation. Several review articles on SFM are available; particularly relevant to soil scientists are those by Hochella,[207] Vempati and Cocke,[208] and Maurice.[195]

VIII. USE OF KINETIC AND SPECTROSCOPIC APPROACHES
TO ELUCIDATE SORPTION MECHANISMS

An ideal way to confirm sorption/desorption mechanisms is to combine kinetic investigations with surface spectroscopic/microscopic experiments. There are a few examples in the literature of studies where mechanisms of metal sorption reactions on soil components have been hypothesized via kinetic experiments and verified in separate spectroscopic investigations.[55,85,90,93,209] An example of this approach can be found in the recent research of Fuller et al.[55] and Waychunas et al.,[90] who studied the kinetics and mechanisms of As(V) sorption on ferrihydrite. Adsorption was investigated during coprecipitation, in which As(V) was present in solution during the hydrolysis and precipitation of Fe, and after coprecipitation (postsynthesis adsorption). In the postsynthesis adsorption studies, As(V) uptake was initially rapid and then slowly increased for up to 8 d. The rapid uptake was ascribed to adsorption on surface sites near the outside of aggregates, while the slower adsorption was attributed to diffusion of As(V) to adsorption sites on ferrihydrite surfaces within aggregates of colloidal particles. The latter were caused by coagulation and crystallite growth processes. These processes resulted in a decrease in the number of adsorption sites, and as aggregates formed, adsorption sites were buried in large clusters of the particles.[55] In the coprecipitation studies, initial As(V) uptake was much greater than observed for the postsynthesis adsorption studies, and the uptake rate was not diffusion controlled as As(V) was coordinated by surface sites before crystallite growth.

The mechanistic hypotheses, based on the kinetic studies, were verified with companion EXAFS studies.[90] Analyses of the EXAFS data provided no evidence for surface precipitation, one possible mechanism that has been proposed for slow metal sorption processes. Arsenate retention in both the coprecipitation and postsynthesis adsorption studies involved primarily inner-sphere bidentate and monodentate binding on sites initially adsorbing arsenate. Waychunas et al.[90] hypothesized that these defect sites probably adsorb As(V) as a bidentate complex first, and then sorb as a monodentate complex. Monodentate complexes accounted for about 30% of all As-Fe correlations and occurred at only low As loading levels.

Scheidegger et al.[210] studied the kinetics of Ni surface precipitate formation on clay minerals and gibbsite using XAFS. Reaction times of minutes to months were investigated. Figure 33 shows the formation of multinuclear Ni complexes on pyrophyllite (pH = 7.5) within minutes. There was a small peak at $R \approx 2.8$ Å in the radial structure function (RSF) of pyrophyllite treated with Ni for 15 min (Figure 33f). This peak represented the second Ni shell and reflected backscattering among multinuclear Ni complexes. The presence of multinuclear surface complexes after a sorption time of only 15 min (28% of the initial Ni sorbed) is a surprising finding. Traditionally, adsorption (strictly a two-dimensional process) is considered to be the predominant sorption mode responsible for metal uptake on mineral surfaces within the first few hours,[28] while surface precipitation is considered to be a much slower process occurring on time scales of hours to days.[211]

As reaction time progressed (15 min to 24 h) and relative Ni removal in solution increased from 28% to 97% the peak at $R \approx 2.8$ Å in the RSFs increased in intensity (Figure 33b, c, d, and e). As time increased from 15 min to 3 months, the number of Ni second-neighbor (N) atoms at a distance of 2.99 to 3.03 Å increased from N = 1.4 to 4.5. These findings indicate the formation of surface precipitates that increase in size with progressing reaction time. Moreover, adsorption and the onset of surface precipitation can occur simultaneously.

Ideally, one would prefer to study sorption reaction mechanisms by following reaction rates spectroscopically (time-resolved or real-time studies) using *in situ* approaches. Such studies are

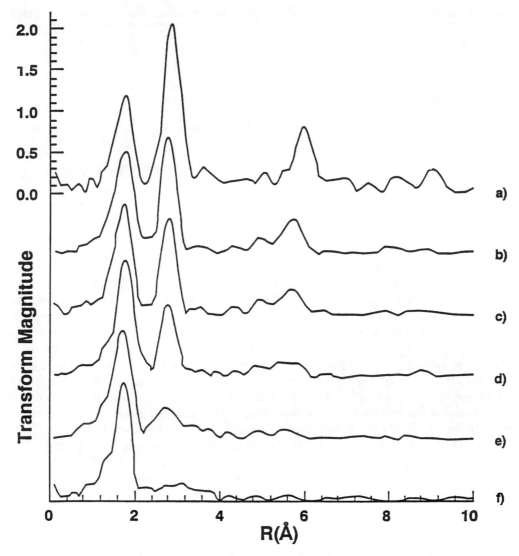

Figure 33   Radial structure functions (RSFs) of pyrophyllite samples reacted with Ni for (a) 3 months, (b) 24 h, (c) 12 h, (d) 3 h, (e) 75 min, and (f) 15 min. The spectra are uncorrected for phase shift. Note the appearance of a peak at R of about 2.8 Å with increasing reaction time. (From Scheidegger, A. M. et al., *J. Phys. IV*, 7, C2-773, 1997. With permission.)

scarce in the literature. The quick EXAFS technique, abbreviated (QEXAFS), depends on a constant monochromator scan rate and fast array defectors to obtain a full EXAFS spectrum in a fraction of a second, compared to tens of minutes in a traditional EXAFS method. Thus, milli- or micro-second time scales can be spectroscopically monitored.[168,212-214] Such studies would be very useful for many soil particle/solution reactions that are rapid. Energy dispersive EXAFS, abbreviated (DEXAS), can also be used to determine a full EXAFS spectrum in a fraction of a second.[215] However, detection can only be determined in the transmission mode.

A recent example of time-resolved *in situ* spectroscopic analyses is the research of Hunter and Bertsch.[216] They employed attenuated total reflectance Fourier transform infrared spectroscopy (ATR-FTIR) to quantitatively measure the degradation kinetics of tetraphenylboron (TPB) on clay minerals. The mechanisms of degradation were ascribed to surface-facilitated oxidation at Lewis acid and Brönsted acid sites. First-order models, based on these mechanisms, described the time-dependent data quite well.

Table 5   Studies on the Kinetics of Mineral Reactions Using Scanning Force Microscopy (SFM)

Dove et al.[222]	Calcite precipitation
Hellman et al.[223]	Albite dissolution
Hillner et al.[224,225]	Calcite growth and dissolution
Johnsson et al.[226]	Muscovite dissolution
Dove and Hochella[217]	Calcite precipitation mechanisms and inhibition by orthophosphate
Gratz and Hillner[218]	Step dynamics and spiral growth on calcite
Bosbach and Rammensee[219]	Gypsum growth and dissolution
Junta and Hochella[109]	Mn(II) oxidation on hematite, goethite, and albite
Stipp et al.[220]	Calcite surface structure
Maurice et al.[205]	Dissolution of hematite in organic acids
Fendorf et al.[221]	Precipitation kinetics of chromium hydroxide on goethite and silica

Note: Studies are listed in chronological order.

IX. USE OF KINETIC AND MICROSCOPIC APPROACHES

Scanning force microscopy (SFM) has been used increasingly as an *in situ* technique for imaging mineral surfaces immersed in aqueous solution, over the course of dissolution, precipitation, and heterogeneous nucleation reactions.[109,205,217-221] SFM permits a direct measure of surface-controlled growth and dissolution rates by providing three-dimensional data on changes in microtopography. *In situ* SFM has the perhaps unique ability to detect separate processes, such as dissolution and secondary phase formation, occurring simultaneously on a mineral surface.[195] Some of the studies in which SFM have been used to study the kinetics of mineral reactions are reported in Table 5.

Recently, Fendorf et al.[221] studied the kinetics of Cr(III) sorption reactions on single goethite and silica particles using a flow-cell mounted in a SFM. This procedure enabled one to study the reactions *in situ* and to react the surface while imaging (real-time measurements). Figure 34a shows an image of the unreacted silica in an aqueous environment. The surface is mostly flat and smooth with no island outcroppings. One hour after a 1 mM Cr(III) solution at pH 6 was introduced into the flow cell one sees that the surface morphology of silica has changed dramatically (Figure 34b). Surface clusters have formed on the surface and within 2 h (figure not shown) the clusters have expanded in width and girth. The precipitates form as discrete surface clusters on the silica surface rather than distributing across the surface.[221]

X. CONCLUSIONS AND FUTURE RESEARCH NEEDS

Research on the kinetics and mechanisms of sorption/release reactions at the soil particle-solution interface will be a common theme in soil and environmental sciences for decades to come. This research emphasis is in large part due to the need to more accurately control the long-term fate and transport of contaminants in the subsurface environment. Without such data, economically sound decisions about soil remediation cannot be made and risk assessment models are not complete. For further advancement to occur in the area of sorption/desorption the following research is needed: long-term sorption/desorption rate studies, a better understanding of residence time effects on nutrient, radionuclide, metal, and organic retention/release mechanisms on soils and other natural materials, and increased use of time-resolved, *in situ* spectroscopic and microscopic techniques to confirm sorption/desorption reaction mechanisms.

In just the last decade major advances in surface spectroscopic and microscopic techniques have greatly enhanced one's ability to study sorption phenomena at the solid/solution interface. One of the most important findings is the direct verification that the chemical properties of natural interfaces are highly heterogeneous.[23,227] The microtopography of the natural mineral/water interface is invariably complex (microtopography includes surface features on atomic and molecular scales

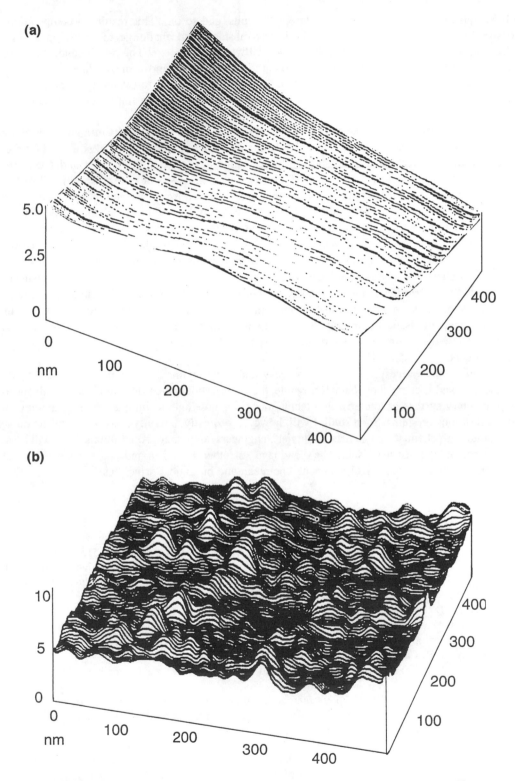

Figure 34   Using a flow cell, a single particle of silica was imaged in an aqueous environment. The unreacted silica (a) is relatively flat and smooth over the 500 × 500 nm scan region; no pronounced outcroppings from the surface are observed. After reacting with 1 m*M* Cr(III) for 1 h at pH 6.0 (b) a different surface morphology is apparent: distinct surface clusters have formed which protrude away from the silica surface. (From Fendorf, S. E. et al., *Soil Sci. Soc. Am. J.,* 60, 99, 1996. With permission.)

such as steps, kinks, defect outcrops, microcracks, pits, and so on). This results in compositional and structural heterogeneity which is the largest obstacle and challenge to obtaining a better understanding of sorption processes at the mineral/water interface.[227] The heterogeneity complicates evaluation of spectroscopic data and theoretical modeling, and it makes interface thermodynamic observations much harder to rationalize without resorting to oversimplification. Ironically, it is this very feature of natural surfaces which makes them so important geochemically and environmentally.[23]

We must realize that our understanding of sorption processes is still in its infancy. One reason for this is that the physical and chemical characteristics of the first few layers of solid and fluid on either side of the interface are just starting to be explored. For example, Knight and Dvorkin[228] obtained experimental evidence that the first three or four monolayers of water on a mineral surface are distinctly different from bulk water. With solids, it is known for a number of nonmineral crystalline solids that the first few atomic layers are structurally (and electronically) modified due to the termination of the bulk solution.[229,230] Therefore, the properties of the near surface of the mineral and the first few atomic layers of the fluid will be modified relative to their bulk phases.[23]

Hochella[227] clearly demonstrated our lack of understanding of sorption phenomena by posing fundamental questions. For example, during the time that a species is attached to a specific surface site, how is the sorbed molecule and the site on which it sits electronically modified? Due to this, what is the molecular orientation of the attachment, and what are the relative probabilities that this sorbed species will dissociate, migrate to another site, diffuse into the near-surface, or desorb back into solution? How and at what rate do surface species diffuse and combine with other species to form products?[23]

Answering these questions will not be easy and will require detailed macroscopic, kinetic, microscopic, and spectroscopic data. If possible, the studies should include a multitude of advanced, complimentary surface characterization techniques since no single technique provides a complete depiction of most systems. Such studies will, however, be costly and only possible if soil scientists collaborate with chemists, physicists, biologists, engineers, and other professionals. This will mean that soil scientists must understand the "language" of other scientists and engineers and be well versed in fundamental analytical, physical, chemical, and biological principles.[23]

REFERENCES

1. Way, J. T., On the power of soils to absorb manure, *J. R. Agric. Soc. Engl.*, 11, 313, 1850.
2. Gedroiz, K. K., Colloidal chemistry as related to soil science. II. Rapidity of reaction exchange in the soil, colloidal condition of the soil saturated with various bases and the indicator method of determining the colloidal content of the soil, *Zh. Opytn. Agron.*, 15, 181, 1914.
3. Hissink, D. J., Base exchange in soils, *Trans. Faraday Soc.*, 20, 551, 1924.
4. Boyd, G. E., Adamson, A. W., Meyers, L. S., Jr., The exchange adsorption of ions from aqueous solutions by organic zeolites. II. Kinetics. *J. Am. Chem. Soc.*, 69, 2836, 1947.
5. Kelley, W. P., Cation exchange in soils, in *ACS Monogr.* 1948.
6. Helfferich, F., *Ion Exchange,* Vol. No. 2003414, University Microfilms International, Ann Arbor, MI, 1962.
7. Mortland, M. M., Kinetics of potassium release from biotite, *Soil Sci. Soc. Am. Proc.*, 22, 503, 1958.
8. Mortland, M. M. and Ellis, B. G., Release of fixed potassium as a diffusion-controlled process, *Soil Sci. Soc. Am. Proc.*, 23, 363, 1959.
9. Scott, A. D. and Reed, M. G., Chemical extraction of potassium from soils and micaceous minerals with solution containing sodium tetraphenylboron. II. Biotite, *Soil Sci. Soc. Am. Proc.*, 26, 41, 1962.
10. Sparks, D. L., *Kinetics of Soil Chemical Processes*, Academic Press, San Diego, CA, 1989.
11. Amacher, M. C., Methods of obtaining and analyzing kinetic data, in *Rates of Soil Chemical Processes,* SSSA Spec. Publ. No. 27, Sparks, D. L. and Suarez, D. L., Eds., Soil Science Society of America, Madison, WI, 1991, 19.

12. Stumm, W., *Chemistry of the Solid-Water Interface*, John Wiley & Sons, New York, 1992.
13. Schwarzenbach, R. T., Gschwend, P. M., and Boden, D. M., *Environmental Organic Chemistry*, John Wiley & Sons, New York, 1993.
14. Sposito, G., *Chemical Equilibria and Kinetics in Soils*, John Wiley & Sons, New York, 1994.
15. Zhang, P. C. and Sparks, D. L., Kinetics and mechanisms of molybdate adsorption/desorption at the goethite/water interface using pressure-jump relaxation, *Soil Sci. Soc. Am. J.*, 53, 1028, 1989.
16. Zhang, P. C. and Sparks, D. L., Kinetics and mechanisms of sulfate adsorption/desorption on goethite using pressure-jump relaxation, *Soil Sci. Soc. Am. J.*, 54, 1266, 1990.
17. Zhang, P. C. and Sparks D. L., Kinetics of selenate and selenite adsorption/desorption at the goethite/water interface, *Environ. Sci. Technol.*, 24, 1848, 1990.
18. Bunzl, K., Schmidt, W., and Sansoni, B., Kinetics of ion exchange in soil organic matter. IV. Adsorption and desorption of Pb^{2+}, CU^{2+}, ZN^{2+}, and CA^{2+} by peat, *J. Soil Sci.*, 27, 32, 1976.
19. Bruemmer, G. W., Gerth, J., and Tiller, K. G., Reaction kinetics of the adsorption and desorption of nickel, zinc and cadmium by goethite. I. Adsorption and diffusion of metals, *J. Soil Sci.*, 39, 37, 1988.
20. Lövgren, L., Sjöberg, S., and Schindler, P. W., Acid/base reactions and Al(III) complexation at the surface of goethite, *Geochim. Cosmochim. Acta*, 54, 1301, 1990.
21. Scheidegger, A. M. and Sparks, D. L., Kinetics of the formation and the dissolution of nickel surface precipitates on pyrohyllite, *Chem. Geol.*, 132, 157, 1996.
22. Jardine, P. M. and Sparks, D. L., Potassium-calcium exchange in a multireactive soil system. I. Kinetics, *Soil Sci. Soc. Am. J.*, 48, 39, 1984.
23. Scheidegger, A. M., and Sparks, D. L., A critical assessment of sorption-desorption mechanisms at the soil mineral/water interface, *Soil Sci.,* 161, 813, 1996.
24. Comans, R. N. J. and Hockley, D. E., Kinetics of cesium sorption on illite, *Geochim. Cosmochim. Acta*, 56, 1157, 1992.
25. Gardiner, W. C., Jr., *Rates and Mechanisms of Chemical Reactions*, Benjamin, New York, 1969.
26. Skopp, J., Analysis of time dependent chemical processes in soils, *J. Environ. Qual.*, 15, 205, 1986.
27. Aharoni, C. and Sparks, D. L., Kinetics of soil chemical reactions: A theoretical treatment, in *Rates of Soil Chemical Processes*, SSSA Spec. Publ. No. 27, Sparks, D. L. and Suarez, D. L, Eds., Soil Science Society of America, Madison, WI, 1991, 1.
28. Sparks. D. L., *Environmental Soil Chemistry*, Academic Press, San Diego, 1995.
29. Bunnett, J. F., Kinetics in solution, in *Investigations of Rates and Mechanisms of Reactions*, Bernasconi, C. F., Ed., John Wiley & Sons, New York, 1986, 171.
30. Hachiya, K., Sasaki, M., Ikeda, I., Mikami, N., and Yasunaga, T., Static and kinetic studies of adsorption-desorption of metal ions on a γ-Al_2O_3 surface. II. Kinetic studies by means of pressure-jump technique, *J. Phys. Chem.*, 88, 27, 1984.
31. Tang, L. and Sparks, D L., Cation exchange kinetics on montmorillonite using pressure-jump relaxation, *Soil Sci. Soc. Am. J.*, 57, 42, 1993.
32. Fendorf, S. E., Sparks, D. L., Franz, J. A., and Camaioni, D. M., Electron paramagnetic resonance stopped-flow kinetic study of manganese (II) sorption-desorption on birnessite, *Soil Sci. Soc. Am. J.*, 57, 57, 1993.
33. Sparks, D. L., Fendorf, S. E., Zhang, P. C., and Tang, L., Kinetics and mechanisms of environmentally important reactions on soil colloidal surfaces, in *Migration and Fate of Pollutants in Soils and Subsoils*, Petruzzelli, D. and Helfferich, F. G., Eds., Springer-Verlag, Berlin, 1993, 141.
34. Low. M. J. D., Kinetics of chemisorption of gases on solids, *Chem. Rev.*, 60, 267, 1960.
35. Chien, S. H. and Clayton, W. R., Application of Elovich equation to the kinetics of phosphate release and sorption in soils, *Soil Sci. Soc. Am. J.*, 44, 265, 1980.
36. Sharpley, A. N., Effect of soil properties on the kinetics of phosphorus desorption, *Soil Sci. Soc. Am. J.*, 47, 462, 1983.
37. Atkinson, R. J., Hingston, F. J., Posner, A. M., and Quirk, J. P., Elovich equation for the kinetics of isotope exchange reactions at solid-liquid interfaces, *Nature (London)*, 226, 148, 1970.
38. Crank, J., *The Mathematics of Diffusion*, 2nd ed., Oxford University Press (Clarendon), London, 1976.
39. Chute, J. H. and Quirk, J. P., Diffusion of potassium from mica-like materials, *Nature (London)*, 213, 1156, 1967.
40. Wollast, R., Kinetics of the alteration of K-feldspar in buffered solutions at low temperature, *Geochim. Cosmochim. Acta*, 31, 635, 1967.

41. Weber, W. J., Jr. and Gould, J. P., Sorption of organic pesticides from aqueous solution, *Adv. Chem. Ser.*, 60, 280, 1966.

42. Kuo, S. and Lotse, E. G., Kinetics of phosphate adsorption and desorption by lake sediments, *Soil Sci. Soc. Am. Proc.*, 38, 50, 1974.

43. Havlin, J. L. and Westfall, D. G., Potassium release kinetics and plant response in calcareous soils, *Soil Sci. Soc. Am. J.*, 49, 366, 1985.

44. Onken, A. B. and Matheson, R. L., Dissolution rate of EDTA-extractable phosphate from soils, *Soil Sci. Soc. Am. J.*, 46, 276, 1982.

45. Sparks, D. L. and Jardine, P. M., Comparison of kinetic equations to describe K-Ca exchange in pure and in mixed systems, *Soil Sci.*, 138, 115, 1984.

46. Aharoni, C. and Ungarish, M., Kinetics of activated chemisorption. I. The non-Elovichian part of the isotherm, *J. Chem. Soc. Faraday Trans.*, 172, 400, 1976.

47. Aharoni, C., Kinetics of adsoprtion: the S-shaped Z(t) plot, *Adsorpt. Sci. Technol.*, 1, 1, 1984.

48. Polyzopoulos, N. A., Keramidas, V. Z., and Pavlatou, A., On the limitations of the simplified Elovich equation in describing the kinetics of phosphate sorption and release from soils, *J. Soil Sci.*, 37, 81, 1986.

49. Aharoni, C. and Suzin, Y., Application of the Elovich equation to the kinetics of occlusion. I. Homogenous microporosity, *J. Chem. Soc. Faraday Trans. 1*, 78, 2313, 1982.

50. Aharoni, C. and Suzin, Y., Application of the Elovich equation to the kinetics of occlusion. III. Heterogenous microporosity, *J. Chem. Soc. Faraday Trans. 1*, 78, 2329, 1982.

51. Connaughton, D. F., Stedinger, J. R., Lion, L. W., and Shuler, M. L., Description of time-varying desorption kinetics: release of naphthalene from contaminated soils, *Environ. Sci. Technol.*, 27, 2397, 1993.

52. Karickhoff, S. W., Sorption kinetics of hydrophobic pollutants in natural sediments, in *Contaminants and Sediments*, Vol. 2, Baker, R. A., Ed., Ann Arbor Science, Ann Arbor, MI, 1980, 193.

53. Wu, S. and Gschwend, P. M., Sorption kinetics of hydrophobic organic compounds to natural sediments and soils, *Environ. Sci. Technol.*, 20, 717, 1986.

54. Miller, C. T. and Pedit, J., Use of a reactive surface-diffusion model to describe apparent sorption-desorption hysteresis and abiotic degradation of lindane in a subsurface material, *Environ. Sci. Technol.*, 26, 1417, 1992.

55. Fuller, C. C., Davis, J. A., and Waychunas, G. A., Surface chemistry of ferrihydride. II. Kinetics of arsenate adsorption and coprecipitation, *Geochim. Cosmochim. Acta*, 57, 2271, 1993.

56. Pedit, J. A. and Miller, C. T., Heterogenous sorption processes in subsurface systems. II. Diffusion modeling approaches, *Environ. Sci. Technol.*, 29, 1766, 1995.

57. Brusseau, M. L. and Rao, P. S. C., Sorption nonideality during organic contaminant transport in porous media, *CRC Crit. Rev. Environ. Control*, 19, 33, 1989.

58. Leenheer, J. A. and Ahlrichs, J. L., A kinetic and equilibrium study of the adsorption of carbaryl and parathion upon soil organic matter surfaces, *Soil Sci. Soc. Am. Proc.*, 35, 700, 1971.

59. Hamaker, J. W. and Thompson, J. M., Adsorption, in *Organic Chemicals in the Environment*, Goring, C. A. I. and Hamaker, J. W., Eds., Marcel Dekker, New York, 1972, 39.

60. Karickhoff, S. W. and Morris, K. R., Sorption dynamics of hydrophobic pollutants in sediment suspensions, *Environ, Toxicol. Chem.*, 4, 469, 1985.

61. McCall, P. J. and Agin, G. L., Desorption kinetics of picloram as affected by residence time in the soil, *Environ. Toxicol. Chem.*, 4, 37, 1985.

62. Jardine, P. M., Dunnivant, F. M., Selim, H. M., and McCarthy, J. F., Comparison of models for describing the transport of dissolved organic carbon in aquifer columns, *Soil Sci. Soc. Am. J.*, 56, 393, 1992.

63. Nkedi-Kizza, P., Biggar, J. W., Selim, H. M., van Genuchten, M. T., Wierenga, P. J., Davison, J. M., and Nielsen, D. R., On the equivalence of two conceptual models for describing ion exchange during transport through an aggregated Oxisol, *Water Resour. Res.*, 20, 1123, 1984.

64. Lee, L. S., Rao. P. S. C., Brusseau, M. L., and Ogwada, R. A., Nonequilibrium sorption of organic contaminants during flow through columns of aquifer materials, *Environ. Toxicol. Chem.*, 7, 779, 1988.

65. van Genuchten, M. T. and Wagenet, R. J., Two-site/two-region models for pesticide transport and degradation: theoretical development and analytical solutions, *Soil Sci. Soc. Am. J.*, 53, 1303, 1989.

66. Jardine, P. M., Parker, J. C., and Zelazny, L. W., Kinetics and mechanisms of aluminum adsorption on kaolinite using a two-site nonequilibrium transport model, *Soil Sci. Soc. Am. J.*, 49, 867, 1985.

67. Steinberg, S. M., Pignatello, J. J., and Sawhney, B. L., Persistence of 1,2 dibromoethane in soils: entrapment in intra particle micropores, *Environ. Sci. Technol.*, 21, 1201, 1987.

68. Ball, W. P. and Roberts, P. V., Long-term sorption of halogenated organic chemiclas by aquifer materials. I. Equilibrium, *Environ. Sci. Technol.*, 25, 1223, 1991.

69. Harmon, T. C., Semprini, L., and Roberts, P. V., Simulating solute transport using laboratory-based sorption parameters, *J. Environ. Eng.*, 118, 666, 1992.

70. Pignatello, J. J., Ferrandino, F. J., and Huang, L. Q., Elution of aged and freshly added herbicides from a soil, *Environ. Sci. Technol.*, 27, 1563, 1993.

71. Weber, W. J., Jr. and Miller, C.T., Modeling the sorption of hydrophobic contaminants by aquifer materials. I. Rates and equilibria, *Water Res.*, 22, 457, 1988.

72. Sparks, D. L. and Zhang, P. C., Relaxation methods for studying kinetics of soil chemical phenomena, in *Rates of Soil Chemical Processes*, Soil Sci. Soc. Am. Spec. Publ. 27, Sparks, D. L. and Suarez, D. L., Eds., Soil Science Society of America, Madison, WI, 1991, 61.

73. Sparks, D L., Fendorf, S. E., Toner, C. V., IV, and Carski, T. H., Kinetic methods and measurements, in *Methods of Soil Analysis, Pt. 3: Chemical Methods*, Soil Sci. Soc. Am. Book Series No. 5., Sparks, D. L., Ed., Soil Science Society of America, Madison, WI, 1996, 1275.

74. Zasoski, R. G. and Burau, R. G., A technique for studying the kinetics of adsorption in suspensions, *Soil Sci. Soc. Am. J.*, 42, 372, 1978.

75. Bar-Tal, A., Sparks, D. L., Pesek, J. D., and Feigenbaum, S., Analysis of adsorption kinetics using a stirred-flow chamber. I. Theory and critical tests, *Soil Sci. Soc. Am. J.*, 54, 1273, 1990.

76. Eigen, M., Ionic reactions in aqueous solutions with half-times as short as 10^{-9} second. Applications to neutralization and hydrolysis reactions, *Discuss. Faraday Soc.*, 17, 194, 1954.

77. Takahashi, M. T., and Alberty, R. A., The pressure-jump methods, in *Methods in Enzymology*, Vol. 16, Kustin, K., Ed., Academic Press, New York, 1969, 31.

78. Bernasconi, C. F., *Relaxation Kinetics*, Academic Press, New York, 1976.

79. Grossl, P. R. and Sparks, D. L., Evaluation of contaminant ion adsorption/desorption on goethite using pressure-jump relaxation kinetics, *Geoderma*, 67, 87, 1995.

80. Hayes, K. F. and Leckie, J. O., Mechanism of lead ion adsorption at the goethite-water interface, *ACS Symp. Ser.*, 323, 114, 1986.

81. Mikami, N., Sasaki, M., Hachlya, K., Ikeda, R. D., and Yasunaga, T., Kinetics of the adsorption of PO_4 on the γ-Al_2O_3 surface using the pressure-jump technique, *J. Phys. Chem.*, 87, 1454, 1983.

82. Mikami, N., Sasaki, M., Kikuchi, T., and Yasunaga, T., Kinetics of the adsorption-desorption of chromate on γ-Al_2O_3 surfaces using the pressure-jump technique, *J. Phys. Chem.*, 87, 5245, 1983.

83. Mikami, N., Sasaki, M., Hachiya, K., and Yasunaga, T., Kinetic study of the adsorption-desorption of the uranyl ion on a γ-Al_2O_3 surface using the pressure-jump technique, *J. Phys. Chem.*, 87, 5478, 1983.

84. Grossl, P. R., Sparks, D. L., and Ainsworth, C. C., Rapid kinetics of Cu (II) adsorption/desorption on goethite, *Environ. Sci. Technol.*, 28, 1422, 1994.

85. Grossl, P. R, Eick, M. J., Sparks, D. L., Goldberg, S., and Ainsworth, C. C., Arsenate and chromate retention on goethite. II. Kinetic evaluation using a p-jump relaxation technique, *Environ. Sci. Technol.*, 31, 321, 1997.

86. Keren, R., Grossl, P. R., and Sparks, D. L., Equilibrium and kinetics of borate adsorption-desorption on pyrophyllite in aqueous suspensions, *Soil Sci. Soc. Am. J.*, 58, 1116, 1994.

87. Toner, C. V., IV, and Sparks, D. L., Chemical relaxation and double layer model analysis of boron adsorption on alumina, *Soil Sci. Soc. Am. J.*, 59, 395, 1995.

88. Sparks, D. L., Kinetics of sorption/release reactions on natural particles, in *Structure and Surface Reactions of Soil Particles*, Huang, P. M., Senesi, N., and Buffle, J., Eds., John Wiley & Sons, New York, 1998, 413.

89. Sparks, D. L., Kinetic processes at the soil particle-solution interface, SSSAJ Spec. Publ., Soil Science Society of America, Madison, WI, 1998, in press.

90. Waychunas, G. A., Rea, B. A., Fuller, C. C., and Davis, J. A., Surface chemistry of ferrihydrite. I. EXAFS studies of the geometry of coprecipitated and adsorbed arsenate, *Geochim. Cosmochim. Acta*, 57, 2251, 1993.

91. Lehmann, R. G., and Harter, R. D., Assessment of copper-soil bond strength by desorption kinetics, *Soil Sci. Soc. Am. J.*, 48, 769, 1984.

92. Ainsworth, C. C., Pilou, J. L., Gassman, P. L., and Sluys, W. G. V. D., Cobalt, cadmium, and lead sorption to hydrous iron oxide: residence time effect, *Soil Sci. Soc. Am. J.*, 58, 1615, 1994.

93. Scheidegger, A. M., Lamble, G. M., and Sparks, D. L., Spectroscopic evidence for the formation of mixed-cation hydroxide phases upon metal sorption on clays and aluminum oxides, *J. Colloid Interface Sci.*, 186, 118, 1997.

94. Karickhoff, S. W., Brown, D. S., and Scott, T. A., Sorption of hydrophobic pollutants on natural sediments, *Water Res.*, 13, 241, 1979.

95. DiToro, D. M. and Horzempa, L. M., Reversible and resistant components of PCB adsorption-desorption: isotherms, *Environ. Sci. Technol.*, 16, 594, 1982.

96. Carroll, K. M., Harkness, M. R., Bracco, A. A., and Balcarcel, R. B., Application of a permeant/polymer diffusional model to the desorption of polychlorinated biphenyls from Hudson River sediments, *Environs. Sci. Technol.*, 28, 253, 1994.

97. Pavlostathis, S. G. and Mathavan, G. N., Desorption kinetics of selected volatile organic compounds from field contaminated soils, *Environ. Sci. Technol.*, 26, 532, 1992.

98. Scribner, S. L., Benzing, T. R., Sun, S., and Boyd, S. A., Desorption and bioavailability of aged simazine residues in soil from a continuous corn field, *J. Environ. Qual.*, 21, 115, 1992.

99. Pignatello, J. J. and Huang, L. Q., Sorptive reversibility of atrazine and metolachlor residues in field soil samples, *J. Environ. Qual.*, 20, 222, 1991.

100. Stumm, W. and Wollast, R., Coordination chemistry of weathering. Kinetics of the surface-controlled dissolution of oxide minerals, *Rev. Geophys.*, 28, 53, 1990.

101. Petrovic, R., Berner, R. A., and Goldhaber, M. B., Rate control in dissolution of alkali feldspars. I. Study of residual feldspar grains by x-ray photoelectron spectroscopy, *Geochim. Cosmochim, Acta*, 40, 537, 1976.

102. Schott, J. and Petit, J. C., New evidence for the mechanisms of dissolution of silicate minerals, in *Aquatic Surface Chemistry*, Stumm, W., Ed., Wiley Interscience, New York, 1987, 293.

103. Casey, W. H., Westrich, H. R., Arnold, G. W., and Banfield, J. F., The surface chemistry of dissolving labradorite feldspar, *Geochim. Cosmochim. Acta*, 53, 821, 1989.

104. Furrer, G. and Stumm, W., The coordination chemistry of weathering. I. Dissolution kinetics of γ-Al_2O_3 and BeO, *Geochim. Cosmochim. Acta*, 50, 1847, 1986.

105. Zinder, B., Furrer, G., and Stumm, W., The coordination chemistry of weathering. II. Dissolution of Fe(III) oxides, *Geochim. Cosmochim. Acta*, 50, 1861, 1986.

106. Stumm, W. and Furrer, G., The dissolution of oxides and aluminum silicates: examples of surface-coordination-controlled kinetics, in *Aquatic Surface Chemistry*, Stumm, W., Ed., Wiley Interscience, New York, 1987, 197.

107. Charlet, L. and Manceau, A., Structure, formation, and reactivity of hydrous oxide particles: insights from x-ray absorption spectroscopy, in *Environmental Particles*, Buffle, J. and van Leeuwen, H. P., Eds., Lewis Publishers, Boca Raton, Fl, 1993, 117.

108. Fendorf, S. E., Lamble, G. M., Stapelton, M. G., Kelley, M. J., and Sparks, D. L., Mechanisms of chromium (III) sorption on silica. I. Cr(III) surface structure derived by extended x-ray absorption fine structure spectroscopy, *Environ. Sci. Technol.*, 28, 284, 1994.

109. Junta, J. L. and Hochella, M. F., Jr., Manganese (II) oxidation at mineral surfaces: a microscopic and spectroscopic study, *Geochim. Cosmochim. Acta*, 58, 4985, 1994.

110. O'Day, P. A., Parks, G. A., and Brown, G. E., Jr., Molecular structure and binding sites of cobalt(II) surface complexes on kaolinite from X-ray absorption spectroscopy, *Clays Clay Miner.*, 42, 337, 1994.

111. Wersin, P., Hochella, M. F., Jr., Persson, P., Redden, G., Leckie, J. O., and Harris, D. W., Interaction between aqueous uranium (VI) and sulfide minerals: spectroscopic evidence for sorption and reduction, *Geochim. Cosmochim. Acta*, 58, 2829, 1994.

112. Chisholm-Brause, C. J., O'Day, P. A., Brown, G. E., Jr., and Parks, G. A., Evidence for multinuclear metal-ion complexes at solid/water interfaces from X-ray absorption spectroscopy, *Nature*, 348, 528, 1990.

113. Chisholm-Brause, C. J., Roe, A. L., Hayes, K. F., Brown, G. E., Jr., Parks, G. A., and Leckie, J. O., Spectroscopic investigation of Pb(II) complexes at the γ-Al_2O_3/water interface, *Geochim. Cosmochim. Acta*, 54, 1897, 1990b.

114. Roe, A. L., Hayes, K. F., Chisholm-Brause, C. J., Brown, G. E., Jr., Parks, G. A., Hodgson, K. O., and Leckie J. O., In situ X-ray absorption study of lead ion surface complexes at the goethite/water interface, *Langmuir*, 7, 367, 1991.

115. Charlet, L. and Manceau, A., X-ray absorption spectroscopic study of the sorption of Cr(III) at the oxide-water interface. II. Adsorption, co-precipitation and surface preciptitation on ferric hydrous oxides, *J. Colloid Interface Sci.*, 148, 443, 1992.

116. Fendorf, S. E. and Sparks, D. L., Mechanisms of chromium (III) sorption on silica. II. Effect of reaction conditions, *Environ. Sci. Technol.*, 28, 290, 1994.

117. O'Day, P. A., Brown, G. E., Jr., and Parks, G. A., X-ray absorption spectroscopy of cobalt (II) multinuclear surface complexes and surface precipitates on kaolinite, *J. Colloid Interface Sci.*, 165, 269, 1994.

118. Bargar, J. R., Brown, G. E., Jr., and Parks, G. A., XAFS study of lead (II) chemisorption at the α-Al_2O_3-water interface, in *209th American Chemical Society National Meeting*, Abstract, ENVR 152, 1995.

119. Papelis, C. and Hayes, K. F., Distinguishing between interlayer and external sorption sites of clay minerals using X-ray absorption spectroscopy, *Colloid Surfaces*, 107, 89, 1996.

120. Scheidegger, A. M., Lamble, G. M., and Sparks, D. L., Investigation of Ni sorption on pyrophyllite: an XAFS study, *Environ. Sci. Technol.*, 30, 548, 1996.

121. O'Day, P. A., Chisholm-Brause, C. J., Towle, S. N., Parks, G. A., and Brown, G. E., Jr., X-ray absorption spectroscopy of Co(II) sorption complexes on quartz (α-SiO_2) and rutile (TiO_2), *Geochim. Cosmochim. Acta*, 60, 2515, 1996.

122. Johnston, C. T., Sposito, G., and Earl, W. L., Surface spectroscopy of environmental particles by Fourier-transform infrared and nuclear magnetic resonance spectroscopy, in *Environmental Particles*, Buffle, J. and van Leeuwen, H. P., Eds., Lewis Publishers, Boca Raton, FL, 1993, 1.

123. Perry, D. L., Taylor, J. A., and Wagner, C. D., X-ray-induced photoelectron and Auger spectroscopy, in *Instrumental Surface Analysis of Geologic Materials*, Perry, D. L, Ed., VCH Publishers, New York, 1990, 45.

124. Fendorf, S. E., Sparks, D. L., Lamble, G. M., and Kelley, M. J., Applications of X-ray absorption fine structure spectroscopy to soils, *Soil Sci. Soc. Am.*, 58, 1583, 1994

125. Koppelmann, M. H., Emerson, A. B., and Dillard, J. G., Adsorbed Cr(III) on chlorite, illite, and kaolinite: an x-ray photoelectron spectroscopic study, *Clays Clay Miner.*, 28, 119, 1980.

126. Dillard, J. G. and Koppelman, M. H., X-ray photoelectron spectroscopic (XPS) surface characterization of cobalt on the surface of kaolinite, *J. Colloid Interface Sci.*, 95, 298, 1982.

127. Schenk, C. V. and Dillard, J. G., Surface analysis and the adsorption of Co(II) on goethite, *J. Colloid Interface Sci.*, 95, 398, 1983.

128. Hochella, M. F. and Carim, A. H., A reassessment of electron escape depths in silicon and thermally grown silicon dioxide thin films, *Surf. Sci.*, 197, 260, 1988.

129. Davison, N. and Whinnie, W. R., X-ray photoelectron spectroscopic study of cobalt(II) and nickel(II) on hectorite and montmorillonite, *Clays Clay Miner.*, 39, 22, 1991.

130. Stipp, S. L. and Hochella, M. F., Structure and bonding environments at the calcite surface as observed with X-ray photoelectron spectroscopy (XPS) and low energy electron diffraction (LEED), *Geochim. Cosmochim. Acta*, 55, 1723, 1991.

131. Scheidegger, A. M., Borkovec, M., and Sticher, H., Coating of silica sand with goethite: preparation and analytical identification, *Geoderma*, 58, 43, 1993.

132. Papelis, C., X-ray photoelectron spectroscopic studies of cadmium and selenite adsorption on aluminum oxide, *Environ. Sci. Technol.*, 29, 1526, 1995.

133. Brown, G. E., Jr., Spectroscopic studies of chemisorption reaction mechanisms at oxide-water interfaces, in *Mineral-Water Interface Geochemistry*, Hochella, M. F. and White, A. F., Eds., Mineralogical Society of America, Washington, D.C., 1990, 309.

134. Calas, G., Electron paramagnetic resonance, in *Reviews in Mineralogy*, Vol. 18, *Spectroscopic Methods in Mineralogy and Geology*, Hawthorne, F. C., Ed., Mineralogical Society of America, Washington, D.C., 1988, 513.

135. McBride, M. B., Cu^{2+} adsorption characteristics of aluminum hydroxide and oxyhydroxides, *Clays Clay Miner.*, 30, 21, 1982.

136. McBride, M. B., Fraser, A. R., and McHardy, W. J., Cu^{2+} interaction with microcrystalline gibbsite. Evidence for oriented chemisorbed copper ions, *Clays Clay Miner.*, 32, 12, 1984.

137. Bleam, W. F. and McBride, M.B., The chemistry of adsorbed Cu(II) and Mn(II) in aqueous titanium dioxide suspensions, *J. Colloid Interface Sci.*, 110, 335, 1986.

138. Wersin, P., Charlet, L., Karthein, R., and Stumm, W., From adsorption to precipitation: sorption of Mn^{2+} on $FeCO_3$(s), *Geochim. Cosmochim. Acta*, 53, 2787, 1989.

139. Senesi, N. and Sposito, G., Residual copper(II) complexes in purified soil and sewage sludge fulvic acids: an electron spin resonance study, *Soil Sci. Soc. Am. J.*, 48, 1247, 1984.

140. Senesi, N., Bocian, D. F., and Sposito, G., Electron spin resonance investigation of copper(II) complexation by soil fulvic acid, *Soil Sci. Soc. Am. J.*, 49, 114, 1985.

141. Goodman, B. A. and Cheshire, M. V., Characterization of iron-fulvic acid complexes using Mossbauer and EPR spectroscopy, *Sci. Total Environ.*, 62, 229, 1987.

142. Senesi, N. and Calderoni, G., Structural and chemical characterization of copper, iron and manganese complexes formed by paleosol humic acids, *Organic Geochem.*, 13, 1145, 1988.

143. Senesi, N., Garrison, S., Holtzclaw, K., and Bradford, G. R., Chemical properties of metal-humic fractions of a sewage sludge-amended aridisol, *J. Environ. Qual.*, 18, 186, 1989.

144. Neto, L. M., Nascimento, O. R., Talamoni, J., and Poppi, N. R., EPR of micronutrients — humic substances complexes extracted from a Brazil soil, *Soil Sci.*, 151, 363, 1991.

145. Cheshire, M. V., ESR and Mossbauer spectroscopy applied to soil matrices, in *15th World Congr. Soil Science*, ISSS, Acapulco, Mexico. Transactions, Vol. 3a, Commission II, 1994.

146. Senesi, N., Spectroscopic studies of metal ion — humic substance complexation in soil, in *15th World Congr. Soil Science,* ISSS, Acapulco, Mexico, Transactions, Vol. 3a, Commission II, 1994.

147. Hawthorne, F. C., Mossbauer spectroscopy, in *Reviews in Mineralogy, Vol. 18, Spectroscopic Methods in Mineralogy and Geology*, Hawthorne, F. C., Ed., Mineralogical Society of America, Washington, D.C., 1988, 573.

148. Goodman, B. A. and Cheshire, M. V., A Mossbauer spectroscopic study of the effect of pH on the reaction between iron and humic acid in aqueous media, *J. Soil Sci.*, 30, 85, 1979.

149. Kallianou, C. S. and Yassoglou, N. J., Bonding and oxidation state of iron-fulvic acid in humic complexes extracted from some Greek soils, *Geoderma*, 35, 209, 1985.

150. Kodama, H., Schnitzer, M., and Mirai, E., An investigation of iron(II)-fulvic acid reaction by Mossbauer spectroscopy and chemical methods, *Soil Sci. Soc. Am. J.*, 52, 994, 1988.

151. Goodman, B. A., Cheshire, M. V., and Chadwick, J., Characterization of the Fe(III)-fulvic acid reaction by Mossbauer spectroscopy, *J. Soil Sci.*, 42, 25, 1991.

152. Goodman, B. A., The use of Mossbauer spectroscopy in the study of soil colloidal materials, in *Soil Colloids and their Association in Aggregates*, Vol. 215, DeBoodt, M. F., Hayes, H. B., and Herbillon, A., Eds., Plenum Press, New York, 1990, 119.

153. Hair, M. I., *Infrared Spectroscopy in Surface Chemistry*, Marcel Dekker, New York, 1967.

154. Bell, A. T., Applications of fourier transform infrared spectroscopy to studies of adsorbed species, in *ACS Symp. Ser. No. 137*, American Chemical Society, Washington, D.C., 1980.

155. McMillan, P. F. and Hofmeister, A. M., Infrared and Raman Spectroscopy, in *Reviews in Mineralogy, Vol. 18, Spectroscopic Methods in Mineralogy and Geology*, Hawthorne, F. C., Ed., Mineralogical Society of America, Washington, D.C., 1988, 573.

156. Piccolo, A., Advanced infrared techniques (FT-IR, DRIFT, and ATR) applied to organic and inorganic soil materials, in *15th World Congr. Soil Science*, ISSS, Acapulco, Mexico, Transactions, Vol. 3a, Commission II, 1994.

157. Wilson, M. A., *NMR Techniques and Applications in Geochemistry and Soil Chemistry*, Pergamon Press, Oxford, 1987.

158. Wershaw, R. L. and Mikita, M. A., *NMR of Humic Substances and Coal*, Lewis Publishers, Chelsea, MI, 1987.

159. Hatcher, P. G., Bortiatynski, J. M., and Knicker, H., NMR techniques (C, H, and N) in soil chemistry, in *15th World Congr. Soil Science*, ISSS, Acapulco, Mexico, Transactions, Vol. 3a, Commission II, 1994.

160. Bank, S., Bank, J. F., and Ellis, P. D., Solid-state ^{113}Cd nuclear magnetic resonance study of exchanged montorillonites, *J. Phys. Chem.*, 93, 4847, 1989.

161. Luca, V., Cardile, C. M., and Meingold, R. H., High-resolution multinuclear NMR study of cation migration in montmorillonite, *Clay Miner.*, 24, 115, 1989.

162. Laperche, V., Lambert, J. F., Prost, R., and Fripiat, J. J., High resolution solid-state NMR of exchangeable cations in the interlayer surface of a swelling mica: ^{23}Na, ^{111}Cd, and ^{133}Cs vermiculites, *J. Phys. Chem.*, 94, 8821, 1990.

163. Weiss, C. A., Kirkpatrick, R. J., and Altaner, S. P., The structural environments of cations adsorbed onto clays: cesium-133 variable temperature MAS NMR spectroscopy of hectorite, *Geochim. Cosmochim. Acta*, 54, 1655, 1990.

164. Altaner, S. P., Weiss, C. A., and Kirkpatrick, R. J., Evidence from ^{29}Si NMR for the structure of mixed-layer illite/smectite clay minerals, *Nature*, 331, 669, 1988.

165. Herrero, C. P., Sanz, J., and Serratosa, J. M., Dispersion of charge deficits in the tetrahedral sheet of phyllosilicates. Analysis from ^{29}Si NMR spectra, *J. Phys. Chem.*, 93, 4311, 1989.

166 Woessner, D. E., Characterization of clay minerals by [27]Al nuclear magnetic resonance spectroscopy, *Am. Mineral.*, 74, 203, 1989.

167. Brown, G. E., Jr., Parks, G. A., and Chisholm-Brause, C. J., In situ x-ray absorption spectroscopic studies of ions at oxide-water interfaces, *Chimia*, 43, 248, 1989.

168. Schulze, D. G. and Bertsch, P. M., Synchrotron x-ray techniques in soil, plant, and environmental research, *Adv. Agron.*, 55, 1, 1995.

169. Bidoglio, G., Gibson, P. N., O'Gorman, M. O., and Roberts, K. S., X-ray absorption spectroscopy investigation of surface redox transformation of thallium and chromium on colloidal mineral oxides, *Geochim. Cosmochim. Acta*, 57, 2389, 1993.

170. Manceau, A., Gorshkov, A. I., and Drits, V., Structural chemistry of Mn, Fe, Co and Ni in Mn hydrous oxides. I. Inforamtion from XANES spectroscopy, *Am. Mineral.*, 77, 1133, 1992.

171. Manceau, A. Gorshkov, A. I., and Drits, V., Structural chemistry of Mn, Fe, Co and Ni in Mn hydrous oxides. II. Information from XANES spectroscopy, electron and x-ray diffraction, *Am. Mineral.*, 77, 1144, 1992.

172. Lytle, C. M., Lytle, F. W., and Smith, B. N., Use of XAS to determine the chemical speciation of bioaccumulated manganese in Ptamogeton pectinatus, *J. Environ. Qual.*, 25, 311, 1996.

173. Pickering, I. J., Brown, G. E., Jr., and Tokunaga, T. K., Quantitative speciation of selenium in soils using x-ray absorption spectroscopy, *Environ. Sci. Technol.*, 29, 2456, 1995.

174. Bidoglio, G., Gibson, P. N., Haltier, E., Omenetto, N., and Lipponen, M., XANES and laser fluorescence spectroscopy for rare earth speciation at mineral water interfaces, *Radiochim. Acta*, 59, 191, 1992.

175. Bajt, S., Sutton, S. R., and Delaney, J. S., X-ray microprobe analysis of iron oxidation states in silicates and oxides using X-ray absorption near edge structure (XANES), *Geochim. Cosmochim. Acta*, 58, 5209, 1994.

176. Bertsch, P. M., Hunter, D. B., Sutton, S. R., Bajt, S., and Rivers, M. L., In situ chemical speciation of uranium in soils and sediments by micro x-ray absorption spectroscopy, *Environ. Sci. Technol.*, 28, 980, 1994.

177. Waychunas, G. A., Apted, M. J., and Brown, G. E., Jr., X-ray K-edge absorption spectra of Fe minerals and model compounds: near-edge structure, *Phys. Chem. Miner.*, 10, 1, 1983.

178. Petiau, J., Calas, G., and Sainctavit, P., Recent developments in the experimental studies of XANES, *J. Phys. Colloq.*, 48, 1987.

179. Waychunas, G. A., Synchotron radiation XANES spectroscopy of Ti minerals: effects of Ti bonding distances, Ti valence, and site geometry on absorption edge structure, *Am. Mineral.*, 72, 89, 1987.

180. Manceau, A., Boisset, M. C., Sarbet, G., Hazemann, J., Mench, M., Cambier, P., and Prost, R., Direct determination of lead speciation in contaminated soils by EXAFS spectroscopy, *Environ. Sci. Technol.*, 30, 1540, 1996.

181. Sayers, D. E., Stern, E. A., and Lytle, F. W., New technique for investigating noncrystalline structures: Fourier analysis of the extended x-ray absorption fine structure, *Phys. Rev. B: Solid State*, 27, 1204, 1971.

182. Calas, G. and Petiau, J., Structure of oxide glasses: spectroscopic studies of local order and crystallochemistry: geo-chemical implications, *Bull. Mineral.*, 196, 33, 1983.

183. Petiau, J. and Calas, G., EXAFS and edge structure; application to nucleation in oxide glasses, *J. Phys. Colloq.*, 46, 41, 1985.

184. Hayes, K. F., Roe, A. L., Brown, G. E., Jr., Hodgson, K. O., Leckie, J. O., and Parks, G. A., In situ x-ray absorption study of surface complexes: selenium oxyanions on α-FeOOH, *Science*, 238, 783, 1987.

185. Manceau, A. and Charlet, L., The mechanism of selenate adsorption on goethite and hydrous ferric oxide, *J. Colloid Interface Sci.*, 164, 87, 1994.

186. Smart, P. and Tovey, N. K., *Electron Microscopy of Soils and Sediments*, Oxford University Press, New York, 1981.

187. Bullock, P. and Murphy, C. P., *Soil Micromorphology*, Vol. 1, A. B. Academic, Berkhamsted, UK, 1983.

188. Douglas, L. A., *Soil Micromorphology. A Basic and Applied Science*, Elsevier, Amsterdam, 1990.

189. Whalley, W. B., Scanning electron microscopy and the sedimentological characterization of soils, in *Geomorphology of Soils*, Richards, K. S., Arnett, R. R., and Ellis, S., Eds., Elsevier, Amssterdam, 1985, 183.

190. Bisdom, E. B. A., Tessier, D., and Schoute, J. F. T., Micromorphological techniques in research and teaching (submicroscopy), in *Soil Micromorphology: A Basic and Applied Science*, Douglas, L. A., Ed., Elsevier, Amsterdam, 1990, 581.

191. Tovey, N. K., Krinsley, D. H., Dent, D. L, and Corbett, W. M., Techniques to quantitatively study the microfabric of soils, *Geoderma*, 53, 217, 1992.

192. Chen, Y., Electron microscopy techniques applied to soil organic matter and soil structure studies, in *15th World Congr. Soil Science*, ISSS, Acapulco, Mexico. Transactions, Vol. 3a, Commission II, 1994.

193. Hochella, M. F., Eggleston, C. M., Elings, V. B., Parks, G. A., Brown, G. E., Jr., Wu, C. M., and Kjoller, K. K., Mineralogy in two dimensions: scanning tunneling microscopy of semiconducting minerals with implications for geochemical reactivity, *Am. Mineral.*, 74, 1233, 1989.

194. Eggleston, C. M. and Hochella, M. F., Scanning tunneling microscopy of sulfide surfaces, *Geochim. Cosmochim. Acta*, 54, 1511, 1990.

195. Maurice, P. A. Scanning probe microscopy of mineral surfaces in *Structure and Surface Reactions of Soil Particles*, Huang, P. M., Senesi, N., and Buffle, J., Eds., John Wiley & Sons, New York, 1998.

196. Johnsson, P. A., Eggleston, C. M., and Hochella, M. F., Imaging molecular-scale structure and micro-topography of hematite with atomic force microscope, *Am. Mineral.*, 76, 1442, 1991.

197. Ohnesorge, F. and Binnig, G., True atomic resolution by atomic force microscopy through repulsive and attractive forces, *Science*, 260, 1451, 1993.

198. Ducker, W. A., Senden, T. J., and Pashley, R. M., Direct measurement of forces using an atomic force microscope, *Nature*, 353, 239, 1991.

199. Ducker, W. A., Senden, T. J., and Pashley, R. M., Measurement of forces in liquids using a force microscope, *Langmuir*, 8, 1831, 1992.

200. Manne, S., Cleveland, J. P., Gaub, H. E., Stucky, G. D., and Hansma, P. K., Direct visualization of surfactant hemimicelles by force microscopy of the electrical double layer, *Langmiur*, 10, 4409, 1994.

201. Hartman, H. G. S., Yang, A., Manne, S., Gould, A. A. C., and Hansma, P. K., Molecular-scale imaging of clay mineral surfaces with atomic force microscope, *Clays Clay Miner.*, 38, 337, 1990.

202. Friedbacher, G., Hansma, P. K., Ramli, E., and Stucky, G. D., Imaging powders with atomic force microscope: from biominerals to commercial materials, *Science*, 253, 1261, 1991.

203. Lindgreen, H., Garnaes, J., Hansen, P. L., Besenbach, F., Laegsgaard, E.,. Stensgaard, I., Gould, S. A., and Hansma, P. K., Ultrafine particles of North Sea illite/smectite clay minerals investigated by STM and AFM, *Am. Mineral.*, 76, 1218, 1991.

204. Blum, A. E. and Eberl, D. D., Determination of clay thicknesses and morphology using scanning force microscopy, in *7th Int. Symp. Water-Rock Interaction*, Rotterdam, Park City, Utah, 1992.

205. Maurice, P. A., Hochella, M. F., Jr., Parks, G. A., Sposito, G., and Schwertmann, U., Evolution of hematite surface microtopography upon dissolution by simple organic acids, *Clays Clay Miner.*, 43, 29, 1995.

206. Grantham, H. C. and Dove, P. M., Investigation of bacterial-mineral interactions using fluid tapping mode atomic force microscopy, *Geochim. Cosmochim. Acta*, 60, 2473, 1996.

207. Hochella, M. F., Atomic structure, microtopography, composition, and reactivity of mineral surfaces in *Reviews in Mineralogy, Vol. 23, Mineral-Water Interface Geochemistry*, Hochella, M. F. and White, A. F., Eds., Mineralogical Society of America, Washington, D.C., 1990, 87.

208. Vempati, R. K. and Cocke, D. L., Applications of scanning force microscopy to soil minerals, in *15th World Congr. Soil Science*, ISSS, Acapulco, Mexico, Transactions, Vol. 3a, Commission II, 1994.

209. Fendorf, S. E., Eick, M. J., Grossl, P. R., and Sparks, D. L., Arsensate and chromate retention mechanisms on goethite. I. Surface structure, *Environ. Sci. Technol.*, 31, 315, 1997.

210. Scheidegger, A. M., Lamble, G. M., and Sparks, D. L., The kinetics of nickel sorption on pyrophyllite as monitored by x-ray absorption fine structure (XAFS) spectroscopy, *J. Phys. IV*, 7, C2-773, 1997.

211. McBride, M. B., *Environmental Chemistry of Soils*, Oxford University Press, New York, 1994.

212. Frahm, R., Quick XAFS: potentials and practical applications in materials science, in *X-ray Absorption Fine Structure*, Hasnain, S. S., Ed., Ellis Harwood, New York, 1991, 731.

213. Lytle, F. W. and Greegar, R. B., New developments in XAS experiments in *X-ray Absorption Fine Structure*, Hasnain, S. S., Ed., Ellis Harwood, New York, 1991, 625.

214. Dobson, B. R., Quick scanning EXAFS facilities at Daresbury, *SRS Synchrotron Radiat. News*, 7 (1), 21, 1994.
215. Baker, G., Dent, A. J., Derbyshire, G., Greaves, G. N., Catlow, C. R. A., Couves, J. W., and Thomas, J. M., Time resolved structural studies of nickel exchanged zeolite and nickel oxide using energy dispersive EXAFS, in *X-ray Absorption Fine Structure*, Hasnain, S. S., Ed., Ellis Harwood, New York, 1991, 738.
216. Hunter, D. B. and Bertsch, P. M., *In situ* measurements of tetraphenylboron degradation kinetics on clay mineral surfaces by FTIR, *Environ. Sci. Technol.*, 28, 686, 1994.
217. Dove, P. M. and Hochella, M. F. Jr., Calcite precipitation mechanisms and inhibition by orthophosphate: *in situ* observations by scanning force microscopy, *Geochim. Cosmochim. Acta*, 57, 705, 1993.
218. Gratz, A. J. and Hillner, P. E., Poisoning of calcite growth viewed in the atomic force microscope (AFM), *J. Cryst. Growth*, 129, 789, 1993.
219. Bosbach, D. and Rammensee, W., In situ investigation of growth and dissolution on the (010) surface of gypsum by scanning force microscopy, *Geochim. Cosmochim. Acta*, 58, 843, 1994.
220. Stipp, S. L., Eggleston, C. M., and Nielsen, B. S., Calcite surface structure observed at microtopographic and molecular scales with atomic force microscopy (AFM), *Geochim. Cosmochim, Acta*, 58, 3032, 1994.
221. Fendorf, S. E., Li, G., and Gunter, M.E., Micromorphologies and stabilities of chromium (III) surface precipitates elucidated by scanning force microscopy, *Soil Sci. Soc. Am. J.*, 60, 99, 1996.
222. Dove, P. M., Hochella, M.F., Jr., and Reeder, R. J., In situ investigation of near-equilibrium calcite precipitation by atomic force microscopy, in *Water-Rock Interaction, Vol. 7*, Kharaka, Y. K. and Maest, A. S., Eds., A. A. Balkema, Rotterdam, 1992, 141.
223. Hellman, R., Drake, B., and Kjoller, K., Using atomic force microscopy to study the structure, topography and dissolution of albite surfaces, in *Water-Rock Interaction, Vol. 7*, Kharaka, Y. K. and Maest, A. S., Eds., A. A. Balkema, Rotterdam, 1992, 149.
224. Hillner, P. E., Gratz, A. J., Manne, S., and Hansma, P. K., Atomic-scale imaging of calcite growth and dissolution in real-time, *Geology*, 20, 359, 1992.
225. Hillner, P. E., Manne, S., Gratz, A. J., and Hansma, P. K., AFM images of dissolution and growth on a calcite crystal, *Ultramicroscopy*, 44, 1387, 1992.
226. Johnsson, P. A., Hochella, M. F., Jr., Parks. G. A., Blum, A. E., and Sposito, G., Direct observation of muscovite basal-plane dissolution and secondary phase formation: an XPS, LEED, and SFM study, in *Water-Rock Interaction, Vol. 7*, Kharaka, Y. K. and Maest, A. S., Eds., A. A. Balkema, Rotterdam, 1992, 159.
227. Hochella, M. F., Jr., The changing face of mineral-fluid interface chemistry, in *Water-Rock Interaction, Vol. 7*, Kharaka, Y. K. and Maest, A. S., Eds., A. A. Balkema, Rotterdam, 1992. 7.
228. Knight, R. and Dvorkin, J., Seismic and electrical properties of sandstones at low saturation, *J. Geophys. Res.*, 12, 17425, 1992.
229. Somorjai, G. A. and van Hove, M. A., Adsorbate-induced restructuring of surfaces, *Prog. Surface Sci.*, 30, 201, 1989.
230. Somorjai, G. A., Modern concepts in surface science and heterogeneous catalysis, *J. Phys. Chem.*, 94, 1013, 1990.

CHAPTER 5

Precipitation/Dissolution Reactions in Soils

Wayne P. Robarge

CONTENTS

I. INTRODUCTION

Precipitation and dissolution reactions of primary and secondary minerals are an integral part of the chemistry of soils and have ramifications for chemical weathering, nutrient availability, soil genesis, global change, and environmental concerns. Chemical weathering is one of the major processes controlling the global hydrogeochemical cycle of elements.[1] As a dissolution reaction, chemical weathering represents the release of plant nutrients such as Mg, Ca, K, P, Fe, Mn, and B from primary minerals for plant uptake and recycling in plant ecosystems. Work by Sparks[2] has demonstrated the importance of K release from minerals on crop growth in agronomic systems, while nutrient release and recycling has long been of interest for sustainable productivity in forested ecosystems.[4-6] As essentially a H^+ consuming process, mineral dissolution is also important in countering the long-term effects of acidic deposition on watersheds.[7-10]

Because of the differing environments of formation for the primary and secondary minerals found in soils, several attempts have been made to relate the relative degree of soil development to the type of minerals present in the different size fractions in soils.[11] Minerals do vary greatly in their stability in pedogenic environments, which in turn has a direct effect on soil genesis.[12] Additional examples of precipitation and dissolution reactions of importance in soil formation include eluviation (formation of spodic horizons by ligand enhanced dissolution), absence of quartz in the fine clay fraction (influence of particle size on dissolution rate), formation of secondary minerals like iron oxide and kaolinite *in situ* due to weathering of soil parent material (Figure 1), and accumulation of carbonates and sulfates in soils from subhumid to arid regions (calcic or gypsic horizons which form as material is dissolved in surface horizons and precipitated in the lower subsoil).

Perhaps the best example of precipitation and dissolution reactions arising from anthropogenic inputs in agronomic soils are those known to occur adjacent to concentrated phosphate fertilizer granules.[13] The presence of monocalcium phosphate monohydrate in superphosphate fertilizer granules produces a succession of hydrolytic reactions resulting in the movement of an acidic concentrated phosphate solution into the surrounding soil crumbs. The nature of the resulting aluminum, iron, or calcium secondary solid phases formed by the reaction of this concentrated solution with the soil has direct implications for P availability for the current and successive crops. Indeed, it might be said that study of the reactions of the phosphate anion in soils has contributed much to our current understanding about sorption and surface-induced precipitation on mineral surfaces. Pioneering work by van Riemsdijk et al.[14,15] with aluminum oxide surfaces, and by Griffin and Jurinak[16] with calcium carbonate, as well as subsequent studies by others, has demonstrated that the sorption of phosphate by these common soil mineral surfaces can lead to a solid-phase transition that results in the formation of a new solid-phase and renewal of the original sorbing surface, without the need for excessive concentrations of phosphate in solution. The possibility that such surface-induced heterogeneous nucleation reactions can occur on mineral surfaces has profound environmental implications for both anions with chemistries similar to phosphate, and for cations as well. Fendorf[17] and others have demonstrated that chromium hydroxide precipitation can occur on the surfaces of silica, goethite, and smectites. The formation of the precipitate is promoted by the presence of the mineral surface and directly affects the stability of chromium in the soil environment. However, the resulting morphology of the surface-induced precipitate differed dramatically between the original three mineral surfaces. Thus, it is apparent that one must not only consider whether surface-induced precipitation of a potentially toxic ion has occurred, but also the resulting surface morphological features of the precipitate, in order to predict possible future reactions with changes in the soil environment.

Mineral weathering reactions have been studied for decades, but most information generated from such studies is empirical in nature and has limited application for making quantitative predictions.[11] The approach taken in this chapter, therefore, is that proposed by Lasaga,[18] in that major advances in understanding precipitation and dissolution reactions in soils can only arise from a sound understanding of the chemical and physical laws that govern these processes. Furthermore, it is acknowledged that these processes span a range in physical scale and it is necessary to develop the proper points of view for the atomic, submicron, micron, laboratory, watershed, and continental scales of these processes.

Figure 1   Secondary mineral formation in saprolite from the Piedmont Region of North Carolina: (a) edge of weathering biotite grain (with cleavage; bar = 10 μm); (b) growth of tubular halloysite on the biotite edge (bar = 2 μm). (Source: Ruben Kretzschmar, Institute of Terrestrial Ecology, Schlieren, Switzerland, personal communication.)

This chapter begins with a review of traditional precepts based on thermodynamics, and concepts about the nature of mineral surfaces that have been deduced either from pure solution studies or physical inspection of mineral surfaces using modern spectroscopic methods. With this background, precipitation and dissolution reactions of importance in soil systems are reviewed, with the introduction of the general form of a rate law for heterogeneous mineral surface reactions that can account for the influence of pH, temperature, ionic strength, and ligands on mineral stability in soils. No attempt is made to provide a comprehensive review of published observations dealing with precipitation and dissolution of primary and secondary minerals found in soil systems. Rather the intent is to summarize current theories dealing with precipitation and dissolution reactions as they may apply to soils with sufficient depth and historical perspective to serve the immediate needs of the reader, and to provide a stimulus for further investigation.

II. BACKGROUND

A. Solubility Product

Most primary and secondary minerals found in soil systems can be considered sparingly soluble in the presence of water: that is, the amount of mass transferred from the bulk phase to hydrated ions in solution is negligible compared to the total mass of the solid phase present. Standard state assumptions concerning the activity of the bulk solid phase are thus applicable, and the solubility of a solid phase can be referenced to an equilibrium constant. However, these same assumptions concerning the activity of the solid phase ignore the surface reactivity of the solid phase in contact with aqueous solutions, and the fact that atoms at the surface of a solid reside in a gradient of different environments as opposed to atoms in the bulk structure.[19] Failure to account for differences in surface reactivity may lead to false conclusions when mineral solubility is reviewed only from an equilibrium (thermodynamic) perspective.

1. Relative Saturation

When a solid phase is brought into contact with the soil solution, one of two possible conditions can exist:

1. The solid phase is in equilibrium with the ionic species in solution and neither dissolves nor grows in volume
2. The solid phase is not in equilibrium with the ionic species in solution and either undergoes dissolution, or experiences growth due to precipitation at its surface

In either case, the reaction can be expressed mathematically in relation to the Gibbs energy change as:

$$\Delta G_r^\circ = -RT \ln K \tag{1}$$

where ΔG_r° is the standard Gibbs energy change,[20] R is the molar gas constant, T is absolute temperature, and K is the equilibrium constant when all reactants and products are in their standard states. For sparingly soluble compounds, K is written as K_{so} and defined as the solubility product when the interaction of the solid with the soil solution is expressed as a dissolution reaction. For example, consider the dissolution of strengite, an iron phosphate, that may be present in soils and sediments:[21]

$$FePO_4 \cdot 2H_2O_{(s)} \rightleftharpoons Fe^{3+}_{(aq)} + H_2PO^-_{4(aq)} + 2OH^-_{(aq)} \tag{2}$$

$$K_{so} = \frac{\left(Fe^{3+}\right)\left(H_2PO^-_4\right)\left(OH^-\right)^2}{\left(FePO_4 \cdot 2H_2O\right)_{(s)}} = 10^{-34.9} \tag{3}$$

where Equation 3 is written in terms of activities for the aqueous and solid phases. Under standard state conditions, the solid phase activity is defined as being equal to one, and Equation 3 reduces to the ion activity product (IAP) of the solid-phase constituent ions in solution. At equilibrium, K_{so} is defined as being equal to the IAP. For nonequilibrium conditions the state of the soil solution with respect to a solid phase can then be defined as:

$$IAP > K_{so} \quad \text{(supersaturated)}$$

or

$$IAP < K_{so} \quad \text{(undersaturated)}$$

Solutions which are over supersaturated with respect to a solid phase favor precipitation, while those that are undersaturated promote dissolution. When $IAP = K_{so}$ the solution is said to be saturated.

Often it is not possible to determine conclusively whether a given solid phase is present in a soil. Comparing the IAP calculated from the analysis of the soil solution to the K_{so} of a given solid phase provides an indirect measure of its possible presence. For Equation 2, an acid soil (pH = 4.7) with solution activities of 10^{-14} for $Fe^{3+}_{(aq)}$ (Lindsay[22]), and $10^{-5.8}$ for $H_2PO^-_{4(aq)}$ has an IAP for strengite of $10^{-38.4}$ which is $\ll K_{so}$. As noted by Sposito,[23] failure to observe $IAP = K_{so}$ leads to one or more of the following conclusions concerning the possible presence of a solid phase (in this case strengite):

1. The dissolution reaction has not yet attained equilibrium
2. Strengite is not present in the soil
3. The composition of the soil solution is in equilibrium with an iron phosphate solid phase, but it has either a different crystalline structure (different K_{so}), or the strengite solid phase is not in the standard state assumed in computing K_{so}

If the presence of strengite could be identified by a separate analytical technique (e.g., X-ray powder diffraction[24]), then failure to observe $IAP = K_{so}$ would be due to conditions one, three, or both. Often in such controlled equilibrations, the soil is equilibrated with solutions which are both supersaturated and undersaturated with respect to the solid phase of interest. The ability of the IAP to approach the same constant value from both a supersaturated and an undersaturated state is often considered further evidence of the presence of a solid phase controlling the composition of the soil solution at equilibrium. While this is a reasonable assumption for relatively simple solids like calcium carbonate or iron phosphates (Equation 2), extrapolation to more complicated structures like layered phyllosilicates is open to question.[11]

2. Macro vs. Micro Scale

The dissolution of strengite depicted in Equation 2 can be expressed as a rate equation, as follows:[25]

$$J = k\left[Fe^{3+}_{(aq)}\right]^a\left[H_2PO^-_{4(aq)}\right]^b \theta S \tag{4}$$

where J is the rate of dissolution in moles s^{-1}, k is a rate constant, $[Fe^{3+}_{(aq)}]$ and $[H_2PO^-_{4(aq)}]$ are the concentrations of the solution constituent ions of strengite that contribute to the rate of dissolution (at constant pH), and a and b are the orders of the reaction with respect to the constituent ions. The θ and S terms are added to reflect the fact that the interaction of any solid with its surroundings must take place starting at its surface. The S term refers to the surface area (m^2), and the θ term is a measure of surface reactivity as defined by Helgeson et al.[26] The θ term is added to acknowledge the fact that mineral surfaces are only rarely flat (uniform) over large areas. Mineral surfaces are made up of atomic- and molecular-scale steps, undulations, microcracks, trapped impurities, point defects, and dislocations that vary in their change in free energy of interaction with the surrounding aqueous solution. These differences in the change in the free energy of interaction with water molecules (and other dissolved solutes) will give rise to different rates of reaction along a mineral surface. The θ term, therefore, is a correction factor to reflect the fact that the rate of dissolution is not a simple function of the total surface area.

Inclusion of the θ and S terms in Equation 4 also serves as a demarcation between macro- and micro-scale processes when considering precipitation and dissolution reactions in soils. Exclusion of these two terms from Equation 4 means an acceptance of the assumption that the solid phase meets the standard state criteria (activity of the bulk solid phase equal to one) and that consideration of the actual surface composition of the solid in contact with the aqueous phase is not necessary to describe the rate of reaction on a macro scale. Inclusion of the terms implies an acceptance that differences in dissolution mechanisms across a mineral surface (micro scale) control the observed rate of dissolution, and either limit, or totally invalidate attempts to understand precipitation and dissolution reactions via the approach illustrated in Equation 3.

a. Surface Microtopography

The surface reactivity of a mineral is dependent upon three surface characteristics:[19] (1) chemical composition, (2) atomic structure, and (3) surface microtopography. Items 1 and 2 are essentially mineral dependent, but characterization of surface microtopography can be extended across most solid surfaces found in soil systems. Following the convention used by a number of authors, the features that are thought to constitute the surface microtopography of mineral surfaces are shown in Figures 2 to 4. The individual cubes represent ions which are constituents of the solid phase. Maximum stability (lowest free energy) for an individual ion should occur when it is part of a three-dimensional lattice, or, using the cube analogy, when all six sides of the cube are in contact with other cubes which comprise the solid matrix.[93] Stability should decrease as the sides not in contact increase progressively from 1 (a "terrace"), 2 (a "step"), 3 (a "kink"), 4 (a "notch"), and 5 (a "hole"). A hole in a terrace is also called a vacancy. A hole that occurs several layers deep within a terrace is an etch pit, which may extend in size to form a pore within the solid matrix. Addition of single atoms or molecules to a terrace, or which extend out of atomically rough surfaces are called adatoms or admolecules. Growth at the surface by continuous additions of adatoms or admolecules (precipitation) may deviate from the original pattern of the bulk solid phase. Under these circumstances the growth will appear to spiral about a central axis out away from the original surface. Physical evidence for the presence of these surface microtopographical features has been provided by a number of researchers using modern analytical surface techniques.

The cube analogy of ions at a mineral surface depicted in Figures 2-4 provides a visual impression that these ions exist at a different energy level from ions in the bulk solid matrix. This difference in energy is available to promote interaction with the surrounding water molecules. Dislocations within the bulk mineral lattice also provide additional energy that can promote dissolution. A dislocation is a line defect in the solid matrix. Distortion of the crystal lattice in the vicinity of a dislocation strains the crystal bonds, resulting in an area of excess strain energy within

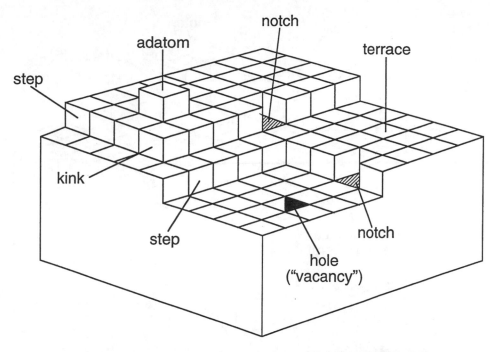

Figure 2 Pictorial representation of the different types of sites on a mineral surface. Stability of constituent ions (represented as single cubes) decreases in the order: terrace > step > kink > notch > hole ("vacancy"). Strength of bonding for adsorbed ions decreases in the reverse order.

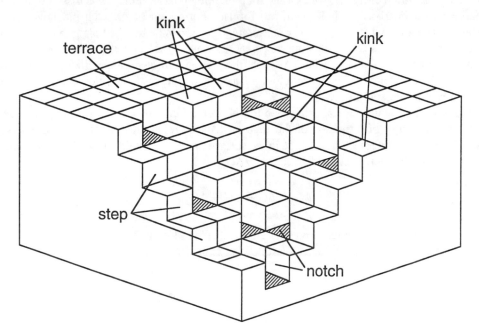

Figure 3 Pictorial representation of dissolution and subsequent formation of an etch pit on a mineral surface. Note the increase in surface area adding to the degree of "roughness" of the mineral surface.

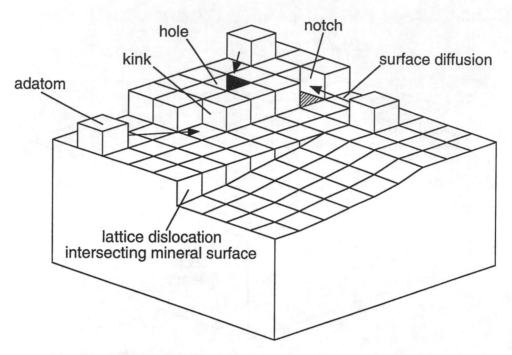

Figure 4   Pictorial representation of precipitation with surface diffusion of adsorbed ions (adatoms) to preferred sites. Representation of a line defect in the solid matrix (dislocation) intersecting the mineral surface. Dislocation can serve as a site for precipitation or dissolution.

the immediate vicinity of the dislocation.[27] When the dislocation intersects the mineral surface, the additional excess strain energy provides a site for preferential dissolution. Upon dissolution, energy is released and a microscopic hole (etch pit) forms.[28] Formation of the etch pit, however, also increases surface area, which is an energy consuming process. Whether the hole grows will depend on the energetics of these two competing processes, as well as the degree of saturation of the bulk solution in contact with the solid phase. As such, the presence of etch pits in silicates can serve as an indication of the degree of undersaturation of the surrounding soil solution and the degree of weathering within the soil profile.

Formulation of a mathematical expression for the formation of an etch pit starts by considering that the overall energy of formation of the edge pit (ΔG_p) is related to the dissolution of a given volume of the bulk solid (ΔG_v), the change in the surface area of the solid caused by the dissolution (ΔG_s), and the strain energy of the dislocation[29] (ΔG_{disl}):

$$\Delta G_p = \Delta G_v + \Delta G_s + \Delta G_{disl} \tag{5}$$

Expressions for each of the ΔG terms on the right-hand side of Equation 5 have been derived[27,28,30,31] and substitution into Equation 5 yields the following expression for ΔG_p:

$$\Delta G_p \frac{\pi r^2 hRT \ln\left(\dfrac{c}{c_o}\right)}{\overline{V}} + 2\pi rh\delta - \frac{hK'B^2 \ln\left(\dfrac{r}{r_o}\right)}{4\pi} \tag{6}$$

where r is the radius of the pit, h is the depth of the pit, c and c_o are the concentration of constituent ions in solution at nonequilibrium and equilibrium, respectively, $\overline{V}$ is the molal volume, δ is the surface free energy, K' is an energy factor, and B is the Burgers dislocation core. The Burgers vector is used to define the magnitude and type of dislocations within a mineral solid.[32] Both the δ and B terms influence the magnitude of ΔG_p. For a given mineral, the faces of the mineral will have different δ values and will behave differently in etch pit formation. For any given mineral face, the number of etch pits will be directly related to the types of dislocations present.[28]

The ΔG_{disl} term has the effect of removing the barrier to the nucleation of etch pit growth so that etch pits easily nucleate and grow. The exact form of the expression for the ΔG_{disl} term depends on the type of dislocation being considered[28] (screw vs. edge); however, in either case, the maxima for Equation 6 that are possible due to various combinations of the listed variables represent energy barriers to outward expansion of the etch pits.[27] It should not be assumed that, once formed, etch pits will continue to enlarge as long as $c < c_0$. Using Monte Carlo simulations, a number of investigators have demonstrated that Equation 6 can yield a number of maxima and minima for input to the listed variables. Multiple maxima and minima predict that edge pit development will be restricted once a certain r value is obtained. Not surprisingly, limitations on etch pit growth are directly related to the free energy of the bulk solution (c/c_0), and the existence of a critical concentration (c_{crit}) can be defined, above which there should be little or no pit nucleation, and slower growth of pits already present on the surface even if $c < c_0$. From the perspective of reactive surface area (θ term in Equation 4), c_{crit} defines when portions of the mineral surface are not undergoing dissolution, or when these same areas may control the rate of dissolution observed.

Measurements of c_{crit} for quartz were determined by Brantley et al.,[30] and were found to be $0.75c_0 \pm 0.15$ for samples hydrothermally etched at 300°C. Their observations suggest that for $c > c_{crit}$, dissolution of quartz precedes predominantly at edges and kinks on the surface, while for $c < c_{crit}$, the formation and growth of the etch pits contribute to the overall dissolution rate. Brantley et al.[30] acknowledged, however, that elevated temperatures enhance etch pit formation in quartz. To test their hypothesis at low temperatures, they analyzed the surfaces of quartz grain samples from a soil profile developed *in situ* over a granite. A shift in surface morphology for the quartz particles was observed between the 60 and 80 cm depth from angularly pitted surfaces to rounded surfaces. Such a change in microtopology is consistent with the weathering environment of the soil which favors $c < c_{crit}$ in the upper portion (<68 cm) of the soil profile (mineral dissolution by formation of etch pits), vs. deeper within the soil profile where $c > c_{crit}$ (lack of etch pit formation). Solution composition at the deeper depths was still such that $c < c_0$ for quartz, which meant that dissolution of quartz was proceeding via dissolution of fines, edges, cracks, etc., but not via the formation of etch pits.

The observed dissolution morphologies for minerals obtained from field sites, as reported by Brantley et al.,[30] and others, suggests that dissolution rates of many such minerals may be substantially influenced by the formation of etch pits. This in turn implies that dislocation density within the mineral matrix will be directly related to dissolution rate. Laboratory studies, however, have failed to demonstrate a strong correlation between dislocation density and dissolution kinetics.[33] Using highly undersaturated solutions, Blum et al.[33] were unable to discern differences in dissolution rate for quartz either in distilled water (80°C), or in 0.2 *M* HF at 22°C. The dislocation densities of the quartz samples ranged from $<10^5$ cm^{-2} to $\sim5 \times 10^{10}$ cm^{-2}, which spans the range of dislocation densities found in most natural quartz. Blum et al. concluded that, even though increasing dislocation density does promote etch pit formation, especially in 0.2 *M* HF, the accelerated dissolution rate promoted by the dislocations is a small contribution to the overall dissolution rate of quartz. The presence of dissolution pits on natural mineral surfaces, therefore, may in fact provide a qualitative indication of the weathering environments in which the mineral occurs, but the dislocation density of the mineral is probably not an important factor in controlling dissolution, even though severe etch pitting is most commonly observed.[33]

Anbeek et al.[34] have proposed that dissolution of naturally weathered feldspars and quartz is in fact controlled by the density of micropores and/or mesopores which form at the surface of mineral particles. Micropores and mesopores (diameters ≤50 nm) may form at crystal dislocations intersecting the surface. The density of these pores at the surface, and their diameters, are a function of crystalline structure, mineral composition, temperature, pH, and time.[35] In the mineral assemblage studied (100 to 1000 μm in particle size), Anbeek et al. observed that, for similar particle diameters, the micropore/mesopore density increased in the order quartz < microcline < albite < oligo-close/andesine. They noted that this sequence is also the same as the well-known sequence of relative weatherability of these minerals.

b. Particle Size

The ΔG_s term in Equation 5 was included to reflect the change in energy due to the increase in surface area through formation of an etch pit. This change in energy due to dissolution or precipitation at the mineral surface can be extended to the individual mineral particles, and it can be shown that changes in particle size can influence the overall free energy of the dissolution or precipitation reaction.

Following the approach outlined by Stumm[1] for a single particle that has its surface (S) defined by $S = kd^2$ and its volume (V) equal to kd^3, the change in surface with change in volume can be expressed as:

$$\frac{dS}{dV} = \frac{2S}{3V} \tag{7}$$

where d is the particle diameter, and k is a constant. Defining the molar surface as $\overline{S} = NS$, where N is Avogadro's number, and $\overline{\delta}$ as the mean free surface energy (interfacial energy), the ΔG_s term in Equation 6 can be rewritten as:

$$\Delta G_s = \frac{2}{3} \overline{\delta}\, \overline{S} \tag{8}$$

The ΔG_{rp}^o for an individual particle at equilibrium therefore becomes[36]

$$\Delta G_{rp}^o = \Delta G_r^o + \Delta G_s \tag{9}$$

or

$$\log K_{so(\bar{s})} = \log K_{so(\bar{s} \to o)} + \frac{\tfrac{2}{3}\,\overline{\delta}\,\overline{S}}{2.3\ RT} \tag{10}$$

The dependence of $\overline{S}$ on particle size is most easily seen through Equation 11:

$$\overline{S} = \frac{M\alpha}{pd} \tag{11}$$

which relates $\overline{S}$ to the formula weight of the particle (M), the density of the particle (ρ), geometry of the particle (α), and its size (d). The inverse relationship between $\overline{S}$ and d essentially characterizes the transition between the micro vs. macro scale when describing dissolution or precipitation reactions using Equation 1. The relative stability of different sized particles of the same composition has long been of interest in quantitative chemical analysis,[37] but there have been relatively few studies detailing the importance of $\overline{S}$ in Equation 10 when describing mineral dissolution or precipitation in soil or aquatic systems. As summarized by Stumm and Morgan,[36] these studies indicate that particle size effects become significant for $\overline{S}$ values greater than 10,000 to 20,000 m^2, and particle sizes less than 0.1 μm.

An alternative expression relating particle size to mineral stability can be derived between c and c_o and r, the radius of curvature of the particle surface:[31]

$$\ln\left(\frac{c}{c_o}\right) = \frac{2}{3}\frac{10^{-7}\,\overline{\delta}\,V\,B^1}{r\,R\,T} \tag{12}$$

where B^1 is a geometric factor related to shape (e.g., 16.8 for spheres), and V is volume. Equation 12 is a discontinuous function at r = O. For quartz and amorphous silica[31] ($\overline{\delta}$ = 335 – 385 mJ m^{-2}; Steefel and Van Cappellen[38]), positive values of r (convex surface) result in a rapid increase in solubility when r is less than 0.1 μm. For extremely small particles, solubility as defined by ln c/c_o approaches infinity. Thus, Equation 12 lays the foundation for the experimental observation that quartz is seldom found in the fine clay fraction of soils. Conversely, for concave surfaces (surfaces with a negative radius of curvature) such as in a crack within the mineral solid, values of r less than –0.1 μm favor precipitation along the crack, resulting in its eventual closer. Over time, Equation 12 predicts the growth of large quartz particles with smooth faces, with at least a tendency to be spherical in shape.[31]

The relationship between rate of dissolution and particle size can be defined by rewriting Equation 4 in the general form:

$$J = k[A]^a[B]^b\,\theta\,S \tag{13}$$

and assuming the rate constant k (and the formula weight of the mineral) does not vary with particle size. Equation 13 predicts that a decrease in particle size, and its corresponding increase in S, will result in an increase in the rate of dissolution, supporting the common assumption that the rate of dissolution (weathering rate) of oxides and silicates is proportional to the exposed surface area. While this is generally true for simple ionic salts, is not necessarily true that the rate of dissolution of minerals in soils fits a surface area controlled model.[39] This is because of the dependence of J on the surface reactivity term θ as well as on surface area. The situation is further confounded since it is unclear exactly what measure of surface area is most appropriate to use in Equation 13, even if a means of estimating θ were indeed available.[26] The density of surface features such as edges, dislocations (edge pits), cleavage planes, twin boundaries, micropores, and mesopores, which are represented by the θ term, cannot be expected to remain constant with changes in particle size, nor under varying conditions of sample preparation. In addition, mineral dissolution, like crystal growth, is anisotropic and dissolution rates will differ at different crystallographic faces.[25] It is not surprising, therefore, that there is often lack of agreement between published investigations concerning mineral dissolution rate and particle size.

Zhang et al.[40] determined the relationship between bulk dissolution rates (mol g^{-1} s^{-1}) and particle size, and measured specific surface area (m^2 g^{-1}) for four particle-size fractions (0.045 to 0.075, 0.075 to 0.11, 0.11 to 0.25, and 0.50 to 1.00 mm sieve sizes) of hornblende. For Al, Fe, and Mg, the relationship between dissolution rate and surface area was linear. Holdren and Speyer,[39] on the other hand, noted that bulk dissolution rates (mol g^{-1} h^{-1}) for five fractions (>600 microns to <38 microns) of an alkali feldspar varied by less than a factor of two. In a subsequent study using eight specimens of feldspars,[41] these same investigators concluded that the measured reaction rates are not related to surface area in any single fashion, a result they thought consistent with the inclusion of the θ term in Equation 13. Holdren and Speyer[39] developed a model that divided the bulk dissolution rate-surface area relationship into three regions. For coarse particle size, the mean particle dimensions are significantly greater than the distance between adjacent surface defects (θ

term in Equation 13 was in fact constant), and bulk dissolution rate will increase linearly with surface area (region 1). As particle size decreases, the mean particle dimensions are smaller than, or of the same order as, the distance between adjacent defects, such that increases in surface area do not produce corresponding increases in the numbers of defects intersecting the surface (region 2). The θ term is no longer constant in Equation 13 and becomes a function of particle size, and dissolution rate no longer varies solely with changes in surface area. With continued decrease in particle size, the distance between adjacent surface defects is greater than the mean particle dimensions, and interfacial energy becomes increasingly important (Equation 9). Bulk dissolution rate will increase with decrease in particle size, but probably not in a linear fashion (region 3). The observations of Zhang et al.[40] are consistent with region 1 with the model proposed by Holdren and Speyer.[41]

B. Congruent vs. Incongruent Dissolution

Precipitation and dissolution reactions at mineral surfaces found in soils must involve two steps: (1) the transport of ions to or from the mineral surface through the aqueous phase, and (2) surface reactions that actually incorporate or remove constituent ions from the crystal lattice.[25,43] The physical representation of the different reactive sites on a mineral surface, as depicted in Figures 2 to 4, ignores the inherent chemical structure of the solid phase and assumes that the dissolution reaction is congruent as demonstrated by Equation 2: there is a stoichiometric release of ions to solution.

Congruent dissolution requires that the rate of release of each chemical species represented by the formula weight leaves the mineral surface and enters solution at the same time. Studies measuring the rate of dissolution of minerals, therefore, provide direct information as to the validity of using Equation 1 to describe precipitation/dissolution reactions in soils. Most such studies report that, at least for the initial dissolution phase, the rates of dissolution for various ions in the mineral structure vary and are incongruent: the ratio of ions released to solution is not the same as that found in the bulk phase of the mineral.[25]

A number of chemical reactions are involved during dissolution of a mineral surface[43] (hydration, hydrolysis, ion exchange, and condensation reactions), all of which are dependent on the chemical structure (bonding environment) of the bulk solid phase itself. If dissolution is viewed as a ligand exchange reaction,[43] it should become apparent that incongruent dissolution should be the norm rather than the exception for many solid phases in soils.

Since mineral dissolution in soils occurs in an aqueous medium, hydration is the first step in the dissolution process. The effects of hydration range from the simple penetration of the water molecule into the mineral solid phase to disruption of chemical bonds (structure) between the mineral constituent ions.[35] Factors affecting this process include (1) structural pores, (2) the degree of covalent bonding between constituent ions, and (3) the hydration energies for the constituent cations. Subsequent hydrolysis reactions with the constituent cations following penetration of water molecules into the bulk mineral structure can be represented as:[35]

$$-M-O-M-+H_2O \rightleftharpoons -M-OH+HO-M- \qquad (14)$$

where M refers to a structural cation. A series of these hydrolysis reactions results in eventual migration of the hydrated metal ion from the mineral surface to the bulk solution. Equation 14 is essentially a ligand exchange reaction. Since the rate coefficients for water exchange around hydrated metal ions are well known,[43] they can provide an indication of which metal cations will likely undergo congruent, vs. incongruent, dissolution. When arranged by their first-order rate coefficients of water exchange (slowest to fastest), metal ions commonly found in mineral structures have the following order:

$$Cr(III) < Al^{3+} < Fe(III) < Ti^{3+} < Mg^{2+} < Fe^{2+}(II) < Mn(II)$$

$$< Zn^{2+} < Ca^{2+} < K^+ < Na^+ Cs^+$$

Nonstoichiometric release of metal ions like Na^+, Ca^{2+}, Mg^{2+}, and Fe^{2+} are often observed from primary silicate minerals (rate coefficients of water exchange typically <1 μsec). Whereas metals ions such as Al^{3+} and Cr(III) with rate coefficients of water exchange ranging from seconds to hours may in fact accumulate at mineral surfaces.

Enhancements in mineral dissolution rates with decrease in pH and increase in temperature are also consistent with characterization of metal release from mineral surfaces as a ligand exchange reaction.[35] An increase in the hydronium ion (H_3O^+) concentration in solution results in a more rapid diffusion into the mineral structure, and promotes hydrolysis through a simple S_n1 (proton-promoted) mechanism:[43]

$$\overset{H^+}{-M-O-M- +H_3O^+} \rightleftharpoons -M-O-M- +H_2O \qquad (15)$$

$$\overset{H^+}{-M-O-M-} \rightarrow -M-OH + \oplus M- \qquad (16)$$

$$-M \oplus H_2O \rightarrow -M-OH + H^+ \qquad (17)$$

The rate-limiting step, Equation 16, is enhanced by an increase in temperature.

Dissolution by the S_N1 mechanism appears to be dominant when pH <4, and accelerates markedly in extremely acidic solutions (pH <2).[44] When pH >4, which is more common for most mineral soil systems, hydrolysis of the metal ions in the mineral structure proceeds via an S_N2 mechanism, as depicted in Equation 18:

$$\overset{H_2O}{-M-O-M- +H_2O} \rightarrow -M-O- \quad M \rightarrow -M-OH + HO-M- \qquad (18)$$

As opposed to the S_N1 mechanism (Equations 15 to 17), which relies on the dissociation of the M–O bond to form a coordinatively undersaturated metal site, the S_N2 mechanism in Equation 18 proceeds by the association of a ligand (H_2O), thereby increasing the coordination of the metal site as the intermediate step. Surface depletion of metal cations from mineral surfaces is much less when pH >5, suggesting dissolution via the S_N2 reaction is slower and/or diffusion of water into the mineral structure is less. The S_n2 reaction depicted in Equation 18 can involve any species (ligand) having sufficient electron density to associate with a metal ion. Since the hydroxyl ion is more nucleophilic than the water molecule, the rate of mineral dissolution should increase via the S_N2 mechanism with increase in pH, especially above pH 7.0. Plots of the rate of mineral dissolution vs. pH are typically parabolic in shape with minima near neutral pH.[45-47]

Viewing dissolution as a ligand exchange reaction involving water molecules provides one explanation for the nonstoichiometric release of ions from mineral surfaces. The restrictions on the movement of water molecules (both vibrational and translational) within the confined spaces of the crystalline structure at the mineral surface may also act to depress mineral dissolution rates[48] through reduction of water exchange around metal cations. Alternative explanations for reduced mineral dissolution rates, as well as continued incongruent dissolution, however, have been proposed. These, together with the assumption that mineral dissolution is essentially a ligand exchange reaction, can be summarized as follows:[49,50]

1. The surface reaction hypothesis — rate of dissolution is controlled by reactions (hydrolysis) along mineral surfaces at distinct surface features (dislocations, etch pits, micropores)
2. The armoring precipitate hypothesis — incongruent dissolution leads to reprecipitation of secondary minerals at the original surface, which limits diffusion to and from solution
3. The leached layer hypothesis — incongruent dissolution leads to the development of a depleted rind of repolymerized silica, which in turn limits diffusion to and from the underlying bulk mineral to solution.

A variety of modern analytical techniques[48] have made possible the evaluation of these three hypotheses, especially the armoring precipitate and the leached layer hypotheses. Spectroscopic techniques commonly used for such investigations have included X-ray photoelectron spectroscopy,[50-53] secondary ion mass spectrometry[52,54] (SIMS), scanning Auger microscopy,[55] Raman spectroscopy,[56] resonant nuclear reaction[49] (RNR), and high-resolution transmission electron microscopy[19,57,58] (HRTEM), as well as scanning electron microscopy and transmission electron microscopy. Application of these techniques to a variety of primary minerals has confirmed the presence of metal-depleted silica-rich layers occurring on most mineral surfaces when compared to the bulk solid. The depth of the leached silica surface rinds can be extensive depending on treatment. In highly acid media (pH <4), the altered layer around feldspar particles can extend from hundreds to thousands of Ångstroms (Å) into the bulk solid phase.[52,59] The repolymerization reactions which result in the formation of these relatively thick silica-rich surface layers, however, result in a structure which is highly porous and unlikely to limit diffusion to and from the surrounding aqueous solution.[43] Under less acid conditions (pH >5), the formation of the silica-enriched layer is reduced to one or two unit cells[51,59] (~30 to 80Å). These thin layers will also present little impediment to diffusion to and from the surrounding aqueous solution.

Studies using a variety of mineral types have also failed to consistently find evidence for formation of secondary solid phases which act as protective layers on weathering minerals.[25,51,60] Clay-sized material found adhering to mineral surfaces in soils (Figure 1) are easily removed by ultrasonic treatment, suggesting no strong surface interaction with the bulk mineral solid.[51] Velbel[60] has observed the presence of protective layers on weathered garnets (almandine and spusartene) as distinguished by laterally continuous, nonporous surface layers underlain by smooth, rounded reactant-mineral surfaces. In this case, protective layers do appear to be controlling, or at least strongly influence, the garnet mineral dissolution rate. In general, however, it appears that for most primary minerals found in soils (feldspars, pyroxenes, amphiboles, and olivines), weathering carried out under well-leached, oxidizing conditions will not result in the formation of protective surface layers.[43,57]

Microscopic characterization of weathered mineral surfaces appears to support the conclusion that the rate of dissolution is controlled by reactions along mineral surfaces at distinctive surface features: the surface reaction hypothesis. Certainly hydrolysis will figure prominently in the reactions at the surface and subsequent release of metal cations to solution. This in turn favors incongruent dissolution as being the norm for most primary minerals in soils. Even if the original dissolution reaction for a given mineral is congruent, the total weathering reaction within soils is often highly incongruent, especially when secondary phases form in close proximity to the primary mineral. In such cases, only the constituents from the primary mineral not contributing to the nearby secondary phase will migrate into the bulk solution contained in soil pores.[48]

C. Activity Diagrams

The fact that most primary and secondary minerals in soils can be considered sparingly soluble in the presence of water was used as a starting point for defining the degree of saturation of the soil solution (Equation 1). A common question confronting soil chemists about soil systems that contain several solid phases which share the same constituent ion(s) is which solid phase controls the activity of the cation or anion of interest in solution. From a thermodynamic point of view, the answer is the solid phase that results in the smallest value of the aqueous solution activity of the ion in solution[23] at saturation (IAP = K_{so}). A simple comparison between K_{so} (ΔG_r^0) values for different soil minerals is often not sufficient to determine the controlling solid phase because of the dependence of the activity of the other constituents which comprise the solid on such factors as pH or the partial pressure of carbon dioxide in the soil atmosphere.

The degree of saturation of the soil solution with respect to a given solid phase can be evaluated quickly for a range of factors such as pH or partial pressure of carbon dioxide using a graphical construct termed an activity-ratio diagram. An activity-ratio diagram is a graph of the negative logarithm of the solution activity of the solid-phase constituent ion vs. a critical soil solution parameter such as pH. Plotted in this fashion, the degree of saturation for a given constituent ion if controlled by a given solid phase usually appears as a straight line. Data points falling above the line represent soil solutions supersaturated with respect to the solid phase. Data points below the line represent undersaturation with respect to the solid phase. Solid phases containing the same constituent ion but with differing composition will result in straight lines with different slopes. These lines typically intersect, demarking conditions where one solid phase becomes thermodynamically the most stable.

Activity-ratio diagrams, and other similar graphical constructs, have long been used by geochemists[61] and soil chemists[22] to predict which solid phases are likely to control the activity of an ion in solution at equilibrium. Acceptance that most mineral dissolution reactions are controlled by surface-controlled dissolution steps (i.e., preceded most likely by incongruent dissolution), however, would argue that application of activity-ratio diagrams to describe soil systems is inappropriate because the assumptions inherent in Equation 1 are not valid. Indeed, as noted by Rai and Kittrick,[11] solubility methods have in general been downgraded in importance because of the inability to control parameters such as solid/solution separation, measurement of aqueous concentrations of specific constituent ions across a wide range of pH, and a slow if not incomplete approach to thermodynamic equilibrium. Since monitoring the lack of change in soil solution concentration over relatively long periods of time does not necessarily guarantee that a given system has reached equilibrium, the use of activity-ratio diagrams to predict or identify the solid phase controlling the activity of an ion in solution must be approached with due caution and an awareness of alternative explanations for the observed activity of an ion in solution.

Soil chemists have dealt with the inherent limitations in the use of activity-ratio diagrams by noting that very slow rates of dissolution (and precipitation) inhibit equilibrium between sand- and silt-sized minerals and the soil solution in many soils.[62,63] Clay-sized particles, on the other hand, because of the influence of surface area on mineral dissolution rate (Equation 13) have been assumed to be often at near equilibrium (or partial equilibrium) with the soil solution, making the use of activity-ratio diagrams acceptable. More recently, Schnoor[64] has argued that a congruent surface-controlled dissolution will in fact follow after an initial incongruent dissolution period for sparingly soluble solids. Schnoor's argument centers primarily on the assumption that the rate of mass diffusion of a constituent ion that readily hydrolyses becomes slower as it is forced to migrate through an increasing depleted layer at the surface. A pseudo-steady state is attained when the rate of mass diffusion through the depleted surface layer equals the rate of surface-controlled dissolution of the remaining constituent ions. The initial incongruent dissolution period may last over a time scale of hours to days.

As noted in the previous section, formation of a leached silica-rich layer around a solid phase is pH dependent, and there is little evidence to support the assumption that this layer of repolymerized silica restricts diffusion of ions to and from solution. Schnoor's argument differs in that it focuses more on the rate of detachment of the activated complex formed during the surface-controlled dissolution reaction. The depleted layer thickness need only extend 2 to 17 Å in order to have the necessary effect on the rate of mass diffusion of a constituent ion.

Schnoor's hypothesis concerning the formation of a pseudo-steady state lends credence to the long-held belief by soil chemists that activity-ratio diagrams can be used to successfully describe soil systems. Nevertheless, their use and interpretation must still be approached with caution. Experience suggests that activity-ratio diagrams are suitable for predicting and identifying controlling solid phases which are sulfates, carbonates, sulfides, and phosphates in neutral to alkaline soils, metal oxides and hydrousoxides, and relatively simple phyllosilicate minerals (e.g., kaolinite). Extension to more complicated mineral structures, especially for phyllosilicate minerals, has proven less successful, primarily because of additional reactions which follow the initial dissolution reactions, and the need for additional information on constituent ion activities which increases the uncertainty in the lines plotted on the activity-ratio diagrams.

1. Fundamental Concepts

Activity-ratio diagrams are constructed through the mathematical manipulation of dissolution reactions such as illustrated in Equation 3. The four-step approach detailed by Sposito[20,65] will be presented here to derive an activity-ratio diagram for the gibbsite, kaolinite, and silica mineral assemblage.

Step 1. Identify the minerals that contain the constituent ion of interest and any other mineral phases that may indirectly have an effect on the activity of the constituent ion in solution. Write dissolution reactions for each mineral phase that may be controlling the solubility of the constituent ion of interest. The stoichiometric coefficient of the activity of the constituent ion of interest in solution must be equal to 1.0:

Gibbsite; log K_{dis} = 8.11

$$Al(OH)_{3(s)} + 3H^+_{(aq)} \rightleftharpoons Al^{3+}_{(aq)} + 3H_2O_{(1)}$$ (19)

Kaolinite; log K_{dis} = 7.43

$$Al_2Si_2O_5(OH)_{4(s)} + 6H^+_{(aq)} \rightleftharpoons 2Al^{3+}_{(aq)} + 2H_4SiO^0_{4(aq)} + H_2O_{(1)}$$ (20)

Silica (amorphous); log K_{so} = -2.74

$$SiO_{2(amorp)} + 2H_2O_{(1)} \rightleftharpoons H_4SiO^0_{4(aq)}$$ (21)

Silica (quartz); log K_{so} = -4.00

$$SiO_{2(quartz)} + 2H_2O_{(1)} \rightleftharpoons H_4SiO^0_{4(aq)}$$ (22)

Here the subscript (s) refers to the solid phase, (aq) to the hydrated ion in solution, and (l) to the bulk aqueous phase. The dissociation constant, K_{dis}, reflects the addition of the hydrolysis reaction of water to the overall dissolution reaction as defined by K_{so}. This is a mathematical convenience to allow pH to be easily incorporated as a master variable in the dissolution reactions. The dissolution of silica (Equations 21 and 22) is independent of pH.

Selection of the solid phases that contain the constituent ion of interest and may be controlling its solubility is not a trivial task. Often it is necessary to consult or conduct a detailed study to identify possible solid phases that may be present (e.g., see Graham et al.[66]) in the system of interest, using techniques such as X-ray diffraction or scanning and transmission electron microscopy. Not only the presence, but the degree of crystallinity of the solid phases is also in question. Equations 21 and 22 reflect the fact that even in the simple system used in this example, the nature of the final activity-ratio diagram will change substantially depending which solid phase of silica is assumed present.

Step 2. Compile values of K_{dis} for each solid phase. Rewrite the dissolution reactions as log (activity) for each reactant and product. Rearrange the equation to have log [(solid phase)/(constituent ion of interest)] equal to the log (activity) of the remaining variables in the equation. The log [(solid phase)/(constituent ion of interest)] is termed the activity ratio:

Gibbsite

$$\log\left[\left(Al(OH)_{3(s)}\right)\middle/\left(Al^{3+}_{(aq)}\right)\right] = -8.11 + 3pH + 3\log\left(H_2O_{(1)}\right) \tag{23}$$

Kaolinite

$$\log\left[\left(Al_2Si_2O_5(OH)_{4(s)}\right)\middle/\left(Al^{3+}_{(aq)}\right)\right] = -3.72 + 3pH + \log\left(H_4SiO^0_{4(aq)}\right) + \tfrac{1}{2}\log\left(H_2O_{(1)}\right) \tag{24}$$

Silica (amorphous)

$$\log\left[\left(SiO_{2(amorp)}\right)\right] = 2.74 + \log\left(H_4SiO^0_{4(aq)}\right) - 2\log\left(H_2O_{(1)}\right) \tag{25}$$

Silica (quartz)

$$\log\left[\left(SiO_{2(quartz)}\right)\right] = 4.00 + \log\left(H_4SiO^0_{4(aq)}\right) - 2\log\left(H_2O_{(1)}\right) \tag{26}$$

In this example, the aluminum ion (Al^{3+}) is the constituent ion of interest. The K_{dis} values for gibbsite and kaolinite are those used by Sposito,[20] while the K_{so} values for amorphous silica and quartz are from Lindsay.[22] These K_{dis} and K_{so} values reflect average values that are commonly accepted for these solid phases as determined by a number of investigators. Locating reliable estimates of K_{dis} values for solid phases that may be controlling the activity of the constituent ion of interest in solution is necessary to ensure the accuracy of the activity-ratio diagram. A compilation of K_{so} values for a variety of minerals can be found in Lindsay.[22] It is incumbent upon the investigator constructing the activity-ratio diagram to verify the K_{dis} values for the solid phases they have selected. Often this will entail careful review of the literature with attention to experimental design and analytical methodology used in determining K_{dis}: in particular, verification of the composition of the solid phase, pretreatment of the solid phase, verification of equilibrium conditions, analytical methodology used to determine the concentration of the constituent ions in solution, and suitable documentation related to the calculation of the activities of the constituent ions in solution at equilibrium. Suitable documentation should include verification of the hydrolysis constants, complexation constants, etc., used in making the activity calculations.

Step 3. Select the independent variable, expressed as log (activity) for the x-axis of the activity-ratio diagram. In this example the independent variable is log ($H_4SiO^0_{4(aq)}$). The resulting activity-ratio diagram, therefore, will provide an estimate of which solid phase would be expected to control the activity of aluminum in solution ($Al^{3+}_{(aq)}$) as a function of the leaching environment (activity of $H_4SiO^0_{4(aq)}$ maintained in soil solution).

Step 4. Assign values for the log (activity) of the other constituent ions in the dissolution reactions. Use these values to derive the final form of the linear equation relating the activity ratio to the log (activity) of the independent variable. Plot all of these equations on the same graph.

The only other constituent ion of concern in this example is H^+. Gibbsite and kaolinite are common in the subsoils throughout the southeast U.S., and the pH of these subsoils typically ranges between 4 and 5. Assuming that the activity of the solid phase and water ($H_2O_{(l)}$) is one, the final linear equations that will be used to construct the activity-ratio diagram are as follows:

Gibbsite; pH 4.0

$$\log\left[\left(Al(OH)_{3(s)}\right)\Big/\left(Al^{3+}_{(aq)}\right)\right] = 3.89 \tag{27}$$

Gibbsite; pH 5.0

$$\log\left[\left(Al(OH)_{3(s)}\right)\Big/\left(Al^{3+}_{(aq)}\right)\right] = 6.89 \tag{28}$$

Kaolinite; pH 4.0

$$\log\left[\left(Al_2Si_2O_5(OH)_{4(s)}\right)\Big/\left(Al^{3+}_{(aq)}\right)\right] = 8.28 + \log\left(H_4SiO^0_{4(aq)}\right) \tag{29}$$

Kaolinite; pH 5.0

$$\log\left[\left(Al_2Si_2O_5(OH)_{4(s)}\right)\Big/\left(Al^{3+}_{(aq)}\right)\right] = 11.28 + \log\left(H_4SiO^0_{4(aq)}\right) \tag{30}$$

Silica (amorphous)

$$\log\left(H_4SiO^0_{4(aq)}\right) = -2.74 \tag{31}$$

Silica (quartz)

$$\log\left(H_4SiO^0_{4(aq)}\right) = -4.00 \tag{32}$$

Since the solid-phase activity is defined as being equal to one, the activity ratio can be rearranged as follows:

$$\log\left[(solid)\Big/\left(Al^{3+}_{(aq)}\right)\right] = -\log\left(Al^{3+}_{(aq)}\right) \tag{33}$$

Substitution of $-\log(Al^{3+}_{(aq)})$ into Equations 27 to 30, and transposing of the negative sign allows the y-axis of the activity-ratio diagram to be plotted with activity increasing along both axes. The resulting activity-ratio diagram is shown in Figure 5.

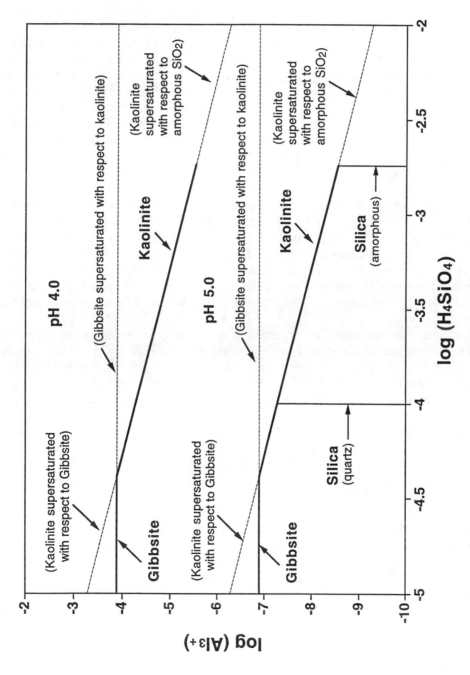

Figure 5 Activity-ratio diagram for the gibbsite, kaolinite, silica mineral assemblage for pH 4.0 and pH 5.0. Solid lines represent saturation with respect to indicated solid phase. Dashed lines represent supersaturation. Dashed line representing supersaturation of kaolinite with respect to quartz is not shown.

Recalling that the controlling solid phase will be the one that results in the smallest value for the constituent ion of interest in solution,[23] Figure 5 predicts that kaolinite will be the controlling solid phase for aluminum ($Al^{3+}_{(aq)}$) in this simple mineral assemblage until the activity of $H_4SiO_4^0$ is less than 4.07×10^{-5} (log ($H_4SiO^0_{4(aq)}$) = -4.39). Furthermore, only at a value of log ($H_4SiO^0_{4(aq)}$) = -4.39 can both kaolinite and gibbsite thermodynamically coexist. When the activity of $H_4SiO_4^0$ is greater than 4.07×10^{-5}, gibbsite is thermodynamically unstable with respect to kaolinite. This relationship between gibbsite and kaolinite is unaltered by a change in pH.

Inclusion of the dissolution of silica in Figure 5 relates to the conditions stipulated in steps 1 and 4. Silica does not contain the constituent ion of interest ($Al^{3+}_{(aq)}$) but is ubiquitous in soil systems, and its dissolution will influence log ($H_4SiO^0_{4(aq)}$) in soil solution, which will in turn influence the solubility of kaolinite. Under true thermodynamic equilibrium, the vertical lines define the limits for log ($H_4SiO^0_{4(aq)}$) in soil solution as a function of crystallinity of the silica solid phase that is present. Therefore, most activity diagrams dealing with the solubility of silicates include silica, even though the mineral does not contain the constituent ion of interest. Other such minerals often included, either directly or indirectly, in activity-ratio diagrams are gibbsite (α-Al(OH)$_3$) and hematite (α-Fe$_2$O$_3$) or goethite (α-FeOOH) in acid soils, and calcite (CaCO$_3$) and fluorite (CaF$_2$) in neutral to alkaline soils. Fluorite is commonly assumed to fix the activity of fluoride ($F^-_{(aq)}$) in soil solution when one of the solid phases under consideration is fluorapatite (here written as Ca$_{10}$(PO$_4$)$_6$F$_2$). When selecting values for the log (activity) of the other constituent ions in the dissolution reactions (step 4) it is common to either assume that these activities are indeed controlled by another solid phase, or to assign a constant value for the log (activity) based on other measurements or knowledge gained about the system under study. The solid phases selected may in turn also respond to changes in the independent variable selected in step 3 (e.g., selection of CaCO$_3$ to control the log ($Ca^{2+}_{(aq)}$)) in solution, adding to the complexity in generating the final activity ratio diagram.

The final issue to be addressed in this section deals with the assignment of unit activity for water and the solid phases in Equations 27 to 32. Since natural minerals always contain substitutional impurities,[76] the assignment of unit activity may be in error and the activity of the solid phase should be accounted for in the dissolution reaction. This issue is addressed in the next section dealing with solid solutions. The remainder of this section focuses on the assumption of unit activity of water ($H_2O_{(l)}$).

The activity of water in bulk solution is given by the ratio of the vapor pressure of the aqueous solution of interest to the vapor pressure of pure water.[36] This is essentially a measure of relative humidity such that the activity of water in soil solutions is equal to the relative humidity of the soil atmosphere, when expressed as a decimal fraction.[65] Thus, it is a reasonable assumption that in most soil systems the activity of water is near or equal to one as the relative humidity of the soil atmosphere is >0.99.[68] Relative humidities as low as 0.5 can occur in surface soils, and even lower in many salt-affected arid soils.[68] Under these conditions, mineral solubility is governed not only by the solute activities in the ambient soil solution, but also the activity of the aqueous phase. The influence of an activity of water <1 for the gibbsite, kaolinite, silica mineral assemblage is shown in Figure 6 for ($H_2O_{(l)}$) = 0.5. Figure 6 illustrates another common form of the activity-ratio diagram in which pH is included on one or both of the axes. The equations used to generate Figure 6 are as follows (note that the activity of the soil phases has been set equal to 1):

Gibbsite; ($H_2O_{(l)}$) = 1

$$\log\!\left(Al^{3+}_{(aq)}\right) + 3pH = 8.11 \tag{34}$$

Gibbsite; ($H_2O_{(l)}$) = 0.5

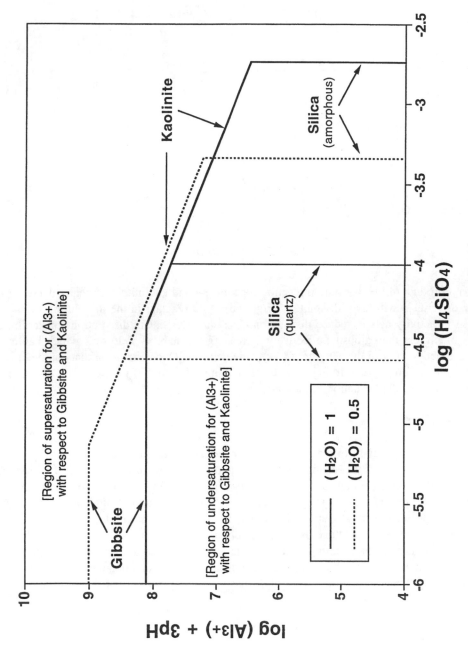

Figure 6 Influence of the activity of water on the solubility of gibbsite, kaolinite, and silica. Solid lines represent saturation when the activity of water is set equal to one. Dashed lines represent saturation when the activity of water is set equal to 0.5.

$$\log\left(Al_{(aq)}^{3+}\right) + 3pH = 9.01 \tag{35}$$

Kaolinite; $(H_2O_{(l)}) = 1$

$$\log\left(Al_{(aq)}^{3+}\right) + 3pH = 3.72 - \log\left(H_4SiO_{4(aq)}^0\right) \tag{36}$$

Kaolinite; $(H_2O_{(l)}) = 0.5$

$$\log\left(Al_{(aq)}^{3+}\right) + 3pH = 3.87 - \log\left(H_4SiO_{4(aq)}^0\right) \tag{37}$$

Silica (amorphous); $(H_2O_{(l)}) = 0.5$

$$\log\left(H_4SiO_{4(aq)}^0\right) = -3.34 \tag{38}$$

Silica (quartz); $(H_2O_{(l)}) = 0.5$

$$\log\left(H_4SiO_{4(aq)}^0\right) = -4.60 \tag{39}$$

Water is a product of the dissolution reaction for gibbsite and kaolinite such that a decrease in the activity of $(H_2O_{(l)})$ will favor dissolution (Figure 6). As a reactant in the dissolution of silica, the decrease in activity acts to reduce dissolution. The enhancement in dissolution for kaolinite, however, is much less than gibbsite, effectively increasing the stability field over which kaolinite would be assumed to control the activity of $(Al_{(aq)}^{3+})$ in solution. One conclusion that can be drawn from Figure 6, therefore, is that drying of soil (relative humidity <0.99) promotes the dissolution of gibbsite in favor of the formation of kaolinite.[68]

2. Solid Solutions

Assignment of unit activity to the solid phases in Equations 27 to 32 is in keeping with the thermodynamic definition of the standard state for solids. For pure solids the mole fraction (x) approaches 1 and the contribution to the chemical potential of the solid from the free energy of mixing and the potential free energy is equal to zero.[36] However, deviation from a pure solid phase (x = 1) is to be expected in natural systems due largely to the isomorphous replacement of a foreign constituent ion in the crystalline lattice. If the isomorphous replacement results in no concentration gradient within the lattice, the resulting solid phase is a homogeneous solid solution. A homogeneous solid solution is a stable solid mixture of two or more solids (components).[69]

Assuming congruent dissolution, each solid phase which is part of the solid solution will maintain its own IAP in solution, which can be expressed as:

$$IAP_i = g_i x_i K_{SO_i} \tag{40}$$

where g_i is the activity coefficient of solid i in the solid solution, x_i is the mole fraction of solid i, and K_{SO_i} is the solubility product of the pure solid. If the constituent ion of interest is part of the major component of the solid phase, the IAP that denotes saturation of the solid solution will only be slightly different from that defined by the K_{so} for the pure mineral phase. If the constituent ion of interest is part of the minor component of the solid phase, the resulting IAP at saturation can be substantially less than predicted from the K_{so} for the pure phase.[69] In regards to the activity-ratio diagram, plotted data would appear undersaturated with respect to the pure phase of the minor component of the solid solution.[36]

It is also possible to define an IAP for the entire solid phase that is a mixture of two or more components. For a binary solid solution $(B_{1-x}C_xA)$, the overall dissolution reaction can be written as:[70]

$$B_{1-x}C_xA \rightarrow (1-x)B^+ + xC^+ + A^-$$
(41)

where, in this example, two different monovalent cation species (B and C) share a common monovalent anion (A) to comprise the two components of the solid solution. The IAP_{ss} for this reaction can be defined as:

$$IAP_{ss} = \left(C^+\right)^x \left(B^+\right)^{1-x} \left(A^-\right)$$
(42)

Equation 41 is not written as an equilibrium reaction in acknowledgment of the fact that most solid solutions are thermodynamically unstable at room temperatures:[61] the chemical potential of each of the constituents are not equal everywhere within the solid matrix and not equal to the corresponding chemical potential of the constituent ions in the aqueous solution. In other words, the dissolution of solid solutions cannot be treated the same as a pure mineral phase, even though the corresponding mass balance equations appear similar in form. This presents a problem in predicting the solubility of a solid solution, and current theories dealing with this issue require adoption of one of two possible hypotheses.[70] The first assumes the solid solution can in fact be considered as a pure phase that is relatively insoluble but does undergo congruent dissolution in a sufficiently short period of time. The requirement that the solid solution be relatively insoluble means that no significant recrystallization of the initial solid or precipitation of a secondary solid-phase occurs during the dissolution reaction. Under these conditions, Equation 42 is said to represent stoichiometric saturation:[71] a limiting equilibrium state characterized by the assumption that a solid solution may behave as a single-component solid of invariant composition. The condition of stoichiometric saturation satisfies the saturation state of the soil solution with respect to a solid phase that is a homogeneous solid solution. It does not necessarily satisfy the distribution function for constituent ions of the solid solution that can be generated from the K_{so} of the pure components,[72] as is often derived in many textbooks.[1,36]

Stoichiometric saturation for the dissolution reaction in Equation 41 can be described as follows:

$$K_{ss} = \frac{\left(C^+\right)^x \left(B^+\right)^{1-x} \left(A^-\right)}{\left(B_{1-x}C_xA\right)}$$
(43)

where K_{ss} is the dissolution constant for the solid phase at the given mole fractions of the components of the solid solution. The assumption that the binary solid acts as a single component means that its composition represents the expected composition at standard temperature and pressure: i.e., standard state. The binary solid phase $(B_{1-x}C_xA)$ can therefore be assumed to represent unit mole fraction of the mixture and be assigned unit activity.[20] This allows K_{ss} to be expressed as:

$$K_{ss} = \left(C^+\right)^x \left(B^+\right)^{1-x} \left(A^-\right) = IAP_{ss}$$
(44)

If a value for K_{ss} can be determined, steps 1 to 4 could be followed in the derivation of an activity-ratio diagram that would describe soil solution saturation with respect to Equation 44. The derivation of Equation 44 illustrates that the assumption of unit activity for the solid phases in Equations 27 to 32 is in fact generally valid, even if the selected mineral phases are not pure solids, provided the assumption of congruent dissolution is correct, the solid phases are relatively insoluble, and the solution phase is at stoichiometric saturation.

The range of values the K_{ss} can assume for a given binary solid solution can be determined by defining a solubility constant (K_{SO}^{ss}) for the solid phase as:[20]

$$K_{SO}^{ss} = \left[K_{SO(BA)}g_{BA}(1-x)\right]^{1-x}\left[K_{SO(CA)}g_{CA}x\right]^{x} \tag{45}$$

After assuming an ideal solid solution ($g_i \rightarrow 1$) and taking log values, Equation 45 further simplifies to:

$$\log K_{SO}^{ss} = (1-x)\log\left[K_{SO(BA)}(1-x)\right] + x\log\left[K_{SO(CA)}x\right] \tag{46}$$

For an ideal solid solution, Equation 46 defines that the limiting values for K_{SO}^{ss} will be the respective K_{so} values, and that (K_{SO}^{ss}) can assume a value anywhere between these two limits. What this means for interpretation of the controlling phase from an activity-ratio diagram is that data points which fall between two lines for pure phases which could form a homogeneous solid solution, may in fact represent saturation with such a solid phase. For example, fluorapatite (FA: $Ca_{10}(PO_4)_6F_2$) and hydroxyapatite (HA: $Ca_{10}(PO_4)_6OH_2$) could form a binary solid solution with an IAP for each component (assuming $g_i \rightarrow 1$) expressed as:

$$IAP_{HA} = \left(Ca^{2+}\right)^{10}\left(PO_4^{3-}\right)^6\left(OH^-\right)^2 = K_{SO(HA)}x_{HA} \tag{47}$$

$$IAP_{FA} = \left(Ca^{2+}\right)^{10}\left(PO_4^{3-}\right)^6\left(F^-\right)^2 = K_{SO(FA)}x_{FA} \tag{48}$$

The following equation was used to generate the activity-ratio diagram in Figure 7, which shows the degree of saturation with respect to calcium phosphate solid phases of soil solutions extracted from eight different soils to which no phosphate fertilizer had been added for at least 5 years prior to sampling:[73]

$$\left[pH + pH_2PO_4\right] = 3\left[pH - \tfrac{1}{2}pCa\right] + \frac{1}{6}\left[pK^1 - px_{FA}\right] \tag{49}$$

The p in Equation 49 stands for –log(activity) and K^1 is a combined constant which includes the respective K_{so} values for the pure solid phase and the dissociation constants for phosphoric acid.

Three lines are drawn for fluorapatite in Figure 7 to reflect three conditions selected for the value for the log (activity) of fluoride (step 4). Line 4 assumes that the log (activity) of fluoride is controlled by fluorite (CaF_2), which, at this activity of fluoride in solution results in $x_{FA} \rightarrow 1$. Therefore, line 1, which represents hydroxyapatite, and line 4 represent the end-members of the homogeneous solid solution for hydroxyapatite-fluorapatite. One interpretation of the position of the plotted data points in Figure 7 is that the activity of phosphate in the soil solution for these soils is controlled by a homogeneous solid solution which has fluorapatite as its primary component. The validity of this interpretation, however, is limited by the assumption of an ideal solid solution. Values of g_i seldom approach a value of unity for most solid solutions.

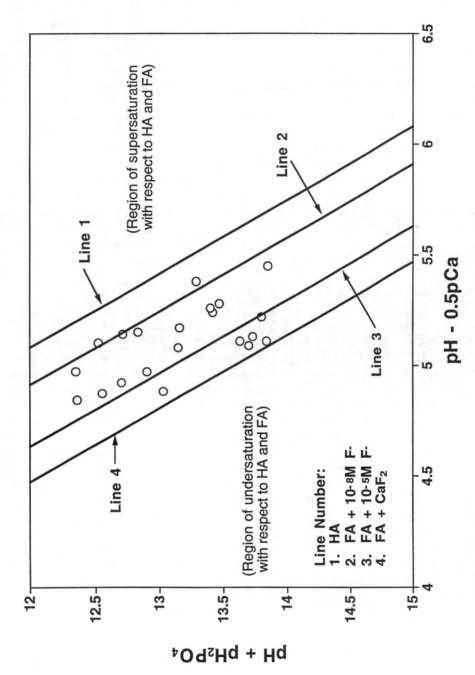

Figure 7 Solubility of phosphate in eight soils that had not received phosphate fertilizer additions for at least 5 years. HA, hydroxyapatite; FA, fluorapatite; open circles are experimental data points. (Data obtained from Murrmann and Peech.[73])

The second hypothesis that can be adopted when predicting the solubility of solid solutions is that the solid phase should indeed be treated as a multicomponent solid solution which, if given a sufficient equilibration period will adjust its bulk composition in response to the soil solution composition.[70] This hypothesis requires that the solid phase be relatively soluble and sufficient time for reaction with the soil solution be allowed to reach equilibrium. If the equilibration period is short, it still may be possible to have the outer layer of the solid undergo rearrangement or recrystallization, leading to a state of primary saturation for the soil solution:[71] the condition where the soil solution is at saturation with respect to a secondary solid solution which has formed as the first step in the dissolution of a solid phase that is a homogeneous solid solution. Since most solid phases in soil systems can be considered relatively insoluble, the applicability of the second hypothesis to soils may be limited when compared to the conditions for formation of stoichiometric saturation. Glynn[70] has argued, however, that there are currently insufficient data to determine when which of these two hypotheses should be applied, especially under field conditions. It probably should be assumed that in many instances, neither hypothesis will be correct for predicting equilibrium end-points in the solid-solution aqueous-solution. This may be especially true for heterogeneous solid solutions, which do have a concentration gradient in the solid phase from the center to the periphery, reflecting the changing environments under which the solid phase was formed.[36] There is also continued uncertainty as to the formulations of the equilibrium expressions that should be applied when dealing with the dissolution of solid phases and predicting equilibrium conditions. The reaction of solid solutions in aqueous media, therefore, continues to be a subject area of active research[70-72,74-76] and discussion[77] among geochemists and soil chemists alike.

3. Covariance Between Axes

Figures 5-7 provide a sampling of the types of activity-ratio diagrams that can be generated by mathematical manipulation of the assumed dissolution reactions. The construct demonstrated in Figure 7 ([pH + pH$_2$PO$_4$] vs. [pH – ½pCa]) is often used to allow plotting of soil solution data across a wide range of conditions, without the need for generating a series of lines as was done in Figure 5. There is an inherent limitation in using constructs such as shown in Figure 7, however, because of the obvious autocorrelation between the y and x axes (pH appears in both terms). In fact, upon closer inspection even the basic form of the activity-ratio diagram suffers from this problem if it is necessary to use the independent variable selected for the x-axis to calculate the activity of the constituent ion on the y-axis. One example of this is the often used activity-ratio diagrams of log ($Al^{3+}_{(aq)}$) vs. pH to estimate the degree of saturation with respect to gibbsite in soil solutions and in stream waters.[78-80] Because it is not possible to measure directly the activity of ($Al^{3+}_{(aq)}$) in solution, it must be calculated using appropriate hydrolysis constants, sample pH, and the analytically determined total dissolved concentration of monomeric aluminum species in solution. The plotted data therefore are autocorrelated with the x-axis.

Neal and coworkers[81-84] have argued that the inherent autocorrelation in typical activity-ratio diagrams for such species as Al^{3+} negates their use in characterizing the controlling solid phases in field systems. They further argue that conclusions concerning the presence of various solid phases that appear to be controlling the activity of ions in solution because of the plotted data being parallel to or overlaying the linear lines in these diagrams are in error. These investigators have promoted the use of plots with independent axes to evaluate the possible presence of solid phases that control the activity of ions in soil solutions and stream waters. For the dissolution of gibbsite (assuming the absence of fluoride and dissolved organic matter in solution), an equation with the appropriate axes can be derived as follows:[81]

$$\log K_{so(gibbsite)} = \log[Al_T] - \left[\log f(H_1T) + 3\log\left(H^+_{(aq)}\right)\right] \qquad (50)$$

where $[Al_T]$ is the concentration of total dissolved aluminum in solution, and $f(H_1T)$ represents the summation of terms comprised of the hydrolysis constants, activity coefficients, and activity of $H^+_{(aq)}$ in solution for the monomeric aluminum species assumed present. Plotting field data on plots generated using Equation 50, Neal and coworkers have concluded that the field systems they monitored were inconsistent with a simple $Al(OH)_3$ solubility control, as suggested when using activity-ratio diagrams constructed with Equation 23.

Xu and Harsh[85] have also evaluated the inherent autocorrelation in activity-ratio diagrams involving $\log (Al^{3+}_{(aq)})$ and concluded that the constructs are useful in describing unknown systems, at least within certain pH ranges. Using an approach based on Monte Carlo simulation, these investigators concluded that for a simulated $\log [Al_{(T)}]$ distribution, the inherent autocorrelation between $\log (Al^{3+}_{(aq)})$ and pH does not interfere with interpretation for gibbsite dissolution if pH <5.5. For pH >5.5, autocorrelation will interfere with interpretation, primarily because $[Al_{(T)}]$ remains essentially constant, and the hydrolysis species of $Al^{3+}_{(aq)}$, which are a direct function of pH, are the dominant monomeric aluminum species in solution. These investigators have concluded that activity-ratio diagrams are a useful tool to study solid-liquid interactions in soils if used with appropriate caution: uncertainties in measured pH, total dissolved aluminum concentrations, and constants for hydrolysis and complexation reactions, together with variations in temperature,[81] will always ensure scatter in the data about a given line representing a given solid phase.

III. PRECIPITATION

Solid phases in soil systems represent a selective accumulation of at least two or more constituent ions into an organized solid matrix which is often crystalline in nature. The process by which this selective accumulation occurs to form a distinct solid phase is termed precipitation. A precipitate can be considered a particulate phase which separates from a continuous medium.[86] The fact that solid phases form in soil systems means that the overall free energy of formation is negative for the combined physical processes operating during the period of the formation. The actual steps leading to the formation of a separate solid phase, however, must occur at the microscale level: the joining together of the constituent ions or molecules that will eventually be recognized as a distinct separate phase. Under classical nucleation theory, three steps are generally considered necessary for these microscale processes to result in the formation of crystals that will persist and survive over relatively long periods of time.[1] A simplified scheme for the sequence of these three steps, as they apply to most solid systems, is shown in Figure 8.

Step 1 refers to the collisions that occur between constituent ions of the solid phase to be formed due to their thermal motion in solution. A certain percentage of these collisions will result in the formation of critical clusters which in turn form a population of nuclei. A nucleus in which the number of ions reaches a critical number is known as a critical nucleus.[87] The remaining nuclei which do not contain a critical number of ions are termed subnuclei or embryos and redissolve, while the population of critical nuclei continues to grow to form crystallites (step 2). For a slightly soluble ionic precipitate such as barium sulfate, the critical nuclei may contain as little as six or seven ions each.[88] Through ripening the population of crystallites changes in favor of a smaller population of stable crystals (step 3) which can be defined by an IAP and continue to grow in solution until ideally IAP = K_{so} of the new solid phase. In a supersaturated solution, the processes that trigger the nucleation sequence that leads to the eventual formation of a precipitate can proceed spontaneously (homogeneous precipitation), or be triggered by the presence of foreign particles which provide adsorption sites for constituent ions (heterogeneous precipitation).[89] Both processes are of concern in soil systems.

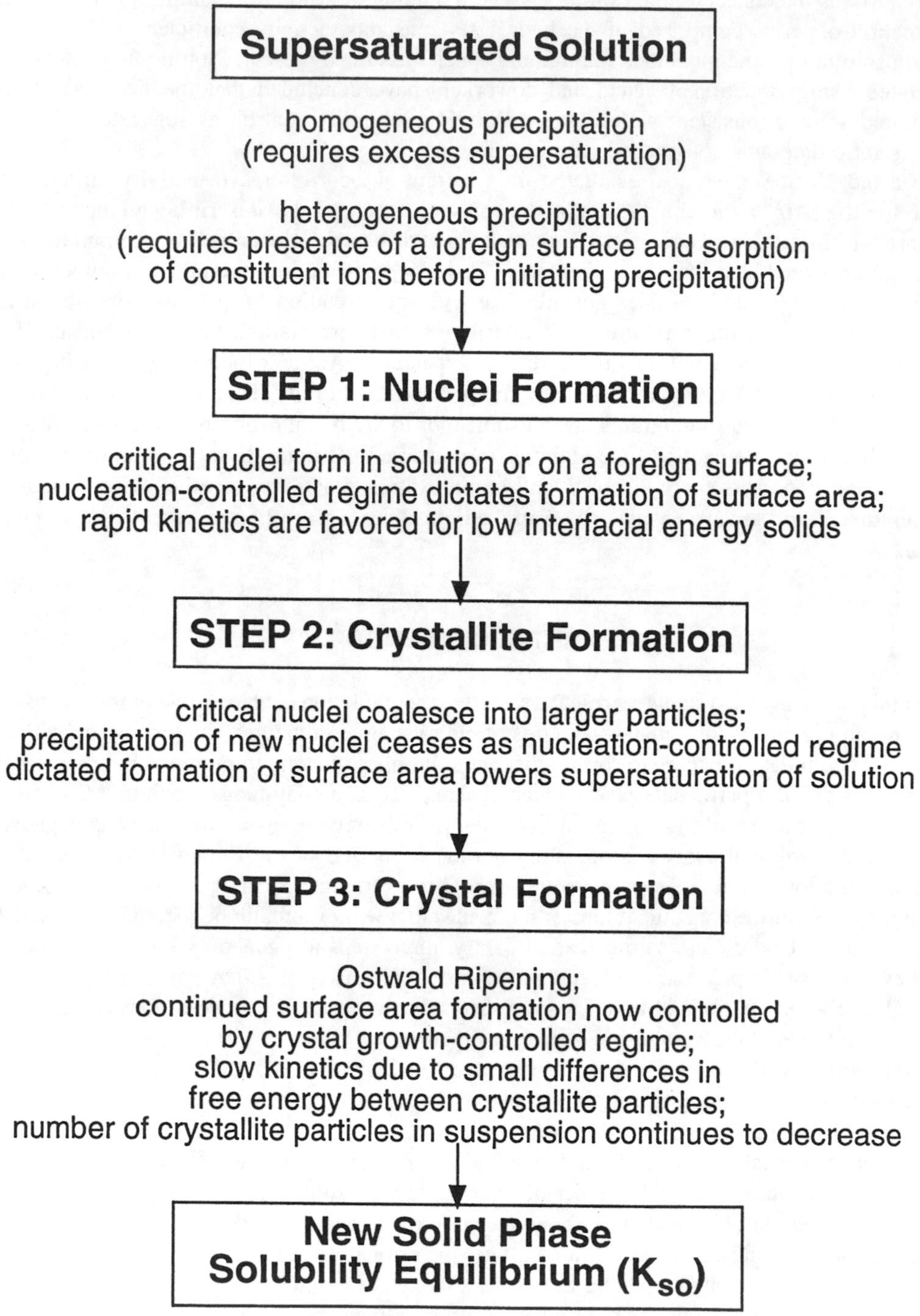

Supersaturated Solution

homogeneous precipitation
(requires excess supersaturation)
or
heterogeneous precipitation
(requires presence of a foreign surface and sorption
of constituent ions before initiating precipitation)

STEP 1: Nuclei Formation

critical nuclei form in solution or on a foreign surface;
nucleation-controlled regime dictates formation of surface area;
rapid kinetics are favored for low interfacial energy solids

STEP 2: Crystallite Formation

critical nuclei coalesce into larger particles;
precipitation of new nuclei ceases as nucleation-controlled regime
dictated formation of surface area lowers supersaturation of solution

STEP 3: Crystal Formation

Ostwald Ripening;
continued surface area formation now controlled
by crystal growth-controlled regime;
slow kinetics due to small differences in
free energy between crystallite particles;
number of crystallite particles in suspension continues to decrease

New Solid Phase
Solubility Equilibrium (K_{so})

Figure 8   Step diagram for processes involved with homogeneous or heterogeneous precipitation from a supersaturated solution. (*Modified from Stumm.*[1])

A. Homogeneous Precipitation

Spontaneous generation of nuclei can occur in supersaturated solutions because of local variations in enthalpy in solution. The number of nuclei formed per unit volume of solution per unit time (nucleation rate, J) for spherical particles can be expressed as:[89]

$$J = A\exp\left[-16\pi\overline{\delta}^3\overline{V_s}^2 \big/ 3k^3T^3(\ln\Omega)^2\right] \tag{51}$$

where A is a constant (magnitude 10^{23} to $10^{32.5}$), $\overline{\delta}$ is the interfacial energy, $\overline{V}_s$ is the molar volume of the solid, k is the Boltzmann constant, and Ω is the supersaturation ratio (IAP/K_{so}) for the solid phase of interest. At a constant temperature, the nucleation rate is directly dependent on both $\overline{\delta}$ and Ω. Below a critical value of Ω (Ω_m), the value of J is near zero, meaning that the solution can remain stable for long periods of time without the eventual formation of a precipitate. For values of $\Omega > \Omega_m$, there is a rapid, almost linear increase in J. Thus, once nucleation has initiated, the eventual formation of a precipitate is directly dependent on the degree of excess saturation above Ω_m.

Acting against this effect of Ω on J is the influence of forming a new interface in contact with the aqueous phase. Relatively small changes in $\overline{\delta}$ have a substantial effect on J for a constant value of Ω. The term within the exponential of Equation 51, therefore, represents the overall free energy of formation of a nucleus in a supersaturated solution ($\Delta G^0_{r(nucleus)}$), and is equal to the net sum of the energy released by chemical bond formation within the nuclei (ΔG_{bulk}) and the energy required to form a new surface in contact with the solution (ΔG_s):

$$\Delta G^0_{r(nucleus)} = \Delta G_{bulk} + \Delta G_s \tag{52}$$

For any nucleus, ΔG_{bulk} can be shown to be equal to:

$$\Delta G_{bulk} = -jkT\ln\Omega \tag{53}$$

where j is the number of molecular units in the nucleus. For values of $\Omega > 1$, the ΔG_{bulk} term will always be negative, which is consistent with the definition of supersaturation (IAP > K_{so}).

The ΔG_s term has already been defined (Equation 8) in relation to particle size and solubility. Here it is defined in more general terms:

$$\Delta G_s = \overline{\delta}dS \tag{54}$$

to emphasize the importance of interfacial tension in the formation and growth of nuclei in a supersaturated solution. Interfacial tension is a measure of how many ergs are required to increase the interfacial area (dS) by 1 cm². The work required to form the interface between a solid and solution as represented by Equation 54, will always be positive in value, and thus will act to inhibit nucleation. Substituting Equations 53 and 54 into Equation 52, an expression for $\Delta G^0_{r(nucleus)}$ for a spherical particle with radius r can be written as:[1]

$$\Delta G^0_{r(nucleus)} = -\frac{4}{3}\frac{\pi r^3}{V}kT\ln\Omega + 4\pi r^2\overline{\delta} \tag{55}$$

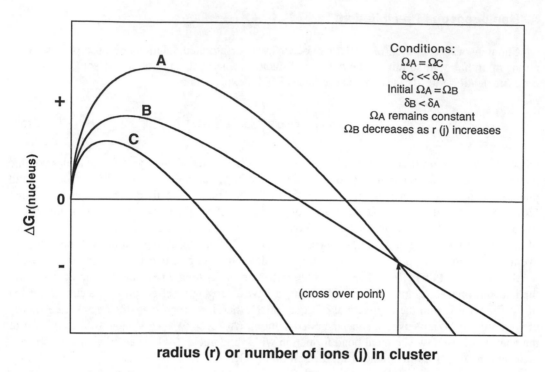

Figure 9   Change in free energy of nuclei formation as a function of size of the nuclei. Curves A and C illustrate the influence of interfacial energy (δ) on precipitation for the same degree of supersaturation (Ω). Curves B and A illustrate eventual shift to a more highly crystalline solid phase (crossover point).

The relationship between $\Delta G^0_{r(nucleus)}$ and r as expressed in Equation 55 is shown in Figure 9. For two solids that can form from the same constituent ions, the difference in $\Delta G^0_{r(nucleus)}$ (at constant Ω) will be directly related to the difference in $\bar{\delta}$ between the two solid phases. The interfacial tension of solids, therefore, directly influences the probability of their formation from supersaturated solutions, both on the free energy of formation and on the rate of nuclei formation in solution. The relative value of Ω (degree of supersaturation) can act to lessen or enhance this effect, but it is still the interfacial tension of a solid phase that exerts a controlling influence on its potential formation from saturated solutions.

The dependence of $\Delta G^0_{r(nucleus)}$ on $\bar{\delta}$ illustrated in Equation 55 offers a quantitative explanation for the common experimental observation that the solid phases initially formed from supersaturated solutions are not those predicted solely by thermodynamic equilibrium.[90] This has come to be known as the Ostwald step rule or the rule of steps,[1,20,36] which can be stated as:

> If a reaction can result in several products, it is not the stablest state with the least amount of free energy that is initially obtained, but the least stable one, lying nearest to the original state in free energy.[90]

Those solids which tend to be most thermodynamically stable, also have higher values of interfacial energy. Thus, while an amorphous phase of a solid may maintain a higher activity of a constituent ion in solution at equilibrium, the value of Ω required to initiate nucleation from a supersaturated solution is actually lower. This means that the nucleation of an amorphous phase will be favored over that of a crystalline phase. Once nucleation has initiated, the rapid expansion in solid surface area that is nucleation-controlled will act to lower Ω further, reducing the probability of nucleation of the more crystalline phase.

Equations 51 and 55 are derived from classical nucleation theory, and while sufficient for characterization of the onset of nucleation in a supersaturated solution, do not predict when nucleation will end in solution.[89] The presence of the radius term in Equation 55 means that a population of nuclei can be envisioned to form from a supersaturated solution, as long as the combination of r and $\bar{\delta}$ terms results in a negative value for $\Delta G^0_{r(nucleus)}$.[87] The formation of this relatively stable population of nuclei will begin to compete with the nucleation process by effectively lowering the value of Ω. Below a critical value of Ω, nucleation will effectively cease, but crystal growth will continue because of the relative differences in stability within the population of nuclei generated. The small nuclei and subsequent crystallites formed in solution should be considered imperfect and will contain defects. As these imperfect particles grow in size, their value of $\Delta G^0_{r(nucleus)}$ will change with respect to those nuclei and crystallites with a more ordered lattice, and will become unstable relative to the more ordered structures. The net result will be a shift in population of nuclei and crystallites in solution, with a reduction in the total number of such particles in solution as their average size (r) increases. This crossover point as a function of r is illustrated graphically in Figure 9. The overall shift in the population of nuclei and crystallites formed from a supersaturated solution proceeds by what is formally termed Ostwald ripening:[91] an increase in crystal size with a corresponding decrease in the number of crystals and crystallites due to the net transfer of material from fine to coarse crystals via dissolution and reprecipitation. The change in surface area generation controlled by crystal growth as opposed to formation of nuclei (nucleation-controlled growth) effectively defines the Ω value at which nucleation ceases in supersaturated solutions.[38] Surface area generation as controlled by crystal growth is described by Equations 10 and 11, and is most important for particles <1 μm in size.[91]

B. Heterogeneous Nucleation

Heterogeneous nucleation has been defined as a first-order phase transition in which constituent ions nucleate onto surfaces.[92] In soil systems, with an abundance of various reactive surfaces, heterogeneous nucleation can be expected to be the norm, with homogeneous precipitation being the exception.[93] Evaluated from the perspective of Equation 52, heterogeneous nucleation results in a decrease in ΔG_s because of the interfacial tension that arises from the creation of an interface between the nucleating constituent ions and the substrate upon which they are sorbed.[38] A decrease in ΔG_s arises when the interaction energies among the constituent ions and between the constituent ions and the substrate are comparable, and the lattice distances of the substrate and nucleating phase are reasonably similar.[94] Since it is reasonable to assume that the constituent ions will select and sorb onto soil surfaces that result in the formation of the strongest chemical bond (interaction energies are similar[38]), the primary effect upon nucleation by sorption of constituent ions onto a foreign surface is through a change in the interfacial area of the forming nuclei in contact with the aqueous solution. Letting $\Delta G_{interface} = \Delta G_s$ for heterogeneous nucleation, the expression for $\Delta G_{interface}$ can be written as:[1,38]

$$\Delta G_{interface} = \bar{\delta}_{NW} S_{NW} + \left(\bar{\delta}_{NS} - \bar{\delta}_{SW}\right) S_{NS} \qquad (56)$$

where NW, NS, and SW refer to the nuclei-water, nuclei-substrate, and substrate-water interactions, respectively. Equation 56 accounts for the free energy that arises because of the development of an interface between the nucleus and water, and between the nucleus and the substrate.[38]

If the lattice distances for the nuclei and substrate are comparable, than $\bar{\delta}_{NS} < \bar{\delta}_{NW}$, and $\bar{\delta}_{NW} \approx \bar{\delta}_{SW}$. This allows Equation 56 to be simplified to:

$$\Delta G_{interface} = \bar{\delta}_{NW}\left(S_{NW} - S_{NS}\right) \qquad (57)$$

which decreases $\Delta G_{interface}$ through a decrease in S_{NW}. This type of surface interaction is defined as a surface-catalytic effect[1] and would favor precipitation of solids with similar $\overline{\delta}$ values (degree of crystallinity) as the sorbing substrate. Growth of the nuclei that would minimize $S_{NW} - S_{NS}$ would be favored (two-dimensional growth along the substrate surface) until the difference reached a limiting value. Nuclei growth should then proceed in three dimensions.[38]

It is possible, however, to have heterogeneous precipitation when the $\overline{\delta}$ values for the substrate and resulting nuclei are dissimilar ($\overline{\delta}_{NW} \ll \overline{\delta}_{SW}$). Under this situation, Equation 56 reduces to:

$$\Delta G_{interface} = \overline{\delta}_{NW} S_{NW} - \overline{\delta}_{SW} S_{SW} \tag{58}$$

assuming that $\overline{\delta}_{NS} \ll \overline{\delta}_{SW}$. Stipulating $\overline{\delta}_{NS} \ll \overline{\delta}_{SW}$ is the same as saying that nucleation would favor the formation of an amorphous coating across the more crystalline substrate:[1] the value of $\Delta G_{interface}$ approaches a zero or even negative value as $S_{NW} \rightarrow S_{SW}$.

The net effect of heterogeneous precipitation is to reduce the Ω_m required to initiate nucleation. This increases the probability for nucleation of more crystalline solid phases than would be possible under conditions of homogeneous precipitation. It is even possible that the initial phase nucleated for a given value of Ω during homogeneous precipitation could in turn serve as a precursor surface for the heterogeneous precipitation of a more crystalline phase, effectively circumventing the need for direct nucleation from solution. With time this more crystalline phase would control the activity of constituent ions in solution. Steefel and Van Cappellen[38] refer to this type of heterogeneous precipitation as "cannibalism" of the more amorphous precursor surface area by the more crystalline phase, and cite as an example fluorapatite formation following the homogeneous precipitation of a calcium phosphate precursor from artificial seawater. It follows that the ability of heterogeneous precipitation to lower $\Delta G^0_{r(nucleus)}$ will probably favor the formation of a number of metastable species that might not otherwise form under conditions of homogeneous precipitation. The relative stability of these metastable species in turn, would follow the dictates associated with Ostwald ripening.

Heterogenous precipitation differs from homogeneous precipitation by the requirement of the sorption of the constituent ions onto a foreign substrate prior to nucleation. The sorption of all the constituent ions which will form the new solid phase is necessary,[95] thus changes in parameters such as pH which act to limit sorption of one of the constituent ions will in turn have a negative impact on nucleation. Like homogeneous precipitation, heterogeneous precipitation will only be favored when $\Omega > 1$ (Equation 53) even if ΔG_s ($\Delta G_{interface}$) is zero. This in turn requires that the surface coverage of the constituent ions on the foreign substrate reach a minimum value (defined by $\Omega = 1$) before heterogeneous precipitation can be initiated. The need for $\Omega > 1$ separates heterogeneous precipitation from the phenomenon of surface precipitation, which has been defined by Stumm[1,36] as the formation of a solid solution between the surface of the foreign substrate and constituent ions. As defined by Equation 40, the formation of a surface precipitate would appear to occur when $\Omega < 1$. This raises the possibility of a new solid phase forming by using the solid solution that develops when $\Omega < 1$ as a precursor for nucleation, with the continued addition of constituent ions to the substrate surface eventually excluding substrate ions that were initially present in the solid solution to form a homogeneous three-dimensional crystallite or crystal. The extent to which this latter sequence of reactions will occur will be dependent on the nature (lattice structure) of the foreign substrate, and the degree of interaction between the constituent ions and the substrate ions (isomorphous substitution).[1,36]

C. Precipitation in Soil Systems

Weathering reactions in soils are often written as the precipitation of a secondary mineral[96] (e.g., gibbsite, kaolinite, hematite, montmorillonite) from the dissolution of a primary mineral. The source of the primary mineral is usually assumed to be the soil parent material. The general form of this reaction can be written as:[97]

$$\begin{array}{l} \text{mineral A + water} \rightleftharpoons \text{mineral B + solution species C} \\ \text{(primary} \qquad\qquad \text{(secondary} \\ \text{mineral)} \qquad\qquad \text{mineral} \end{array} \qquad (59)$$

Unlike other dissolution reactions written in this chapter, Equation 59 emphasizes the direct transition of one solid phase to another. Thermodynamically this approach is valid and consideration of the phase rule for Equation 59 allows for only one degree of freedom. Thus, if mineral A is known to be present, and the activity of C is known in solution and is assumed to be at equilibrium, then mineral B must be assumed to be present.[97] Since Equation 59 only focuses on the formation of a new solid phase, concerns about congruent dissolution, metastable states, etc., are not required, provided the assumption of the activity of C representing an equilibrium state is valid.[97]

If mineral A is a primary mineral such as a feldspar or an olivine, then the assumption that Equation 59 represents a true equilibrium situation (reversible reaction) is not possible, as the probability is essentially zero that such high-temperature minerals would form under low-temperature conditions.[11] If, however, the condition of partial equilibrium is assumed, the implication for characterizing the phase transitions (precipitation) of secondary minerals in soils implied by Equation 59 is possible. Partial equilibrium was defined qualitatively in Section II.C. Here a more formal definition is given as a state in which a system must be in equilibrium with at least one process, or reaction, in a given system.[98] As applied to Equation 59, partial equilibrium can arise when the reaction between the solution and the primary mineral (a nonreversible reaction) results in the formation of a secondary mineral that is in equilibrium with the solution. That is, the precipitation of the secondary mineral will necessarily force soluble species C to be in equilibrium with the new solid phase. It is a simple logical step to extend the possibility of one system at partial equilibrium to a series of successive partial equilibrium steps, each describing the precipitation of a new solid phase in equilibrium with the neighboring solution. Thus, an irreversible process, such as the dissolution of a primary mineral in a soil environment, may result in a local equilibrium for the local environment in the immediate vicinity of the mineral, even though the overall dissolution reaction cannot be considered a true equilibrium reaction.[98]

Extending this concept to the weathering environment around any single mineral, or even within a soil profile, one can envision a series of secondary phases being precipitated by a series of successive partial equilibrium states, each with their own unique solution composition. Each secondary mineral precipitated would in turn undergo dissolution to form the next secondary mineral in the sequence. For example, rain water infiltrating a soil and coming in contact with a primary mineral would pass through successive zones of precipitated secondary minerals (e.g., gibbsite $\rightarrow$ kaolinite $\rightarrow$ muscovite) before coming in actual contact with the primary mineral surface.[38] Within the soil profile, one would expect to see successive zones of precipitated secondary minerals as water moved from the leached surface horizons to within the C horizon. Construction of activity-ratio diagrams, such as in Figure 5, provide a predictive tool for determining which mineral phases are likely to be present in a given sequence for a given weathering environment (concentration of $H_4SiO_4^0$ maintained in solution). Typically what is assumed is that minerals far removed from the environment in which they formed will be the least stable.[11] Those minerals most likely to precipitate are assumed to be the most thermodynamically stable for the present leaching environment.[11]

The physical presence of a range of secondary minerals in soils is taken as a measure of proof that Equation 59 is an accurate approximation describing mineral precipitation reactions in soils. The assumption of partial equilibrium implicit in Equation 59, however, is only appropriate if the rate of precipitation of the secondary phase exceeds the rate of primary mineral dissolution.[38,99] Furthermore, as presented by Steefel and Van Cappellen,[38] the approach to mineral precipitation reactions in soils represented by Equation 59 provides no direction for choice of the secondary minerals which are permitted to precipitate (other than the assumption based on the most thermodynamic stable phase), nor is there a mechanism provided for the formation of these crystalline (high interfacial energy) phases. A more appropriate model of mineral precipitation in soils may be one based on kinetics in which poorly crystalline or amorphous materials nucleate preferentially, and are gradually replaced by minerals with higher crystallinities and lower solubilities.[38] Weathering and the subsequent precipitation reactions in soils would then best be considered a quasi-steady-state phenomenon where the processes of heterogeneous nucleation, nuclei and crystallite growth and dissolution, and Ostwald ripening (with the possibility of forming solid solutions) would act to control which surface area is being generated and modified. This approach does not negate the formation of stable crystalline end-members in a given mineral sequence, but does provide for the formation of metastable solid-phase species which may persist because of kinetic considerations for relatively long periods of time.

This approach to precipitation reactions in soils would mean that the formation of amorphous precursors would be the norm for most low-temperature aqueous systems. A precursor phase is usually amorphous, and may have a chemical composition different from that of a more crystalline phase.[38] The formation of the precursor is favored because of the influence of interfacial tension through the process of heterogeneous precipitation (Equation 52), and because the rate of nucleation is not limiting. This in turn means that solutions in contact with dissolving solid phases in soils will reach a degree of supersaturation above Ω_m for the precursor for only short periods of time before the nucleation reaction will be initiated. One can therefore envision a process by which the precursor is continuously generated over geologic time, as long as the initial dissolving phase is present and the Ω_m value for the precursor is exceeded in solution. A more stable solid phase may eventually nucleate via heterogeneous precipitation and grow to a viable solid phase in the system, but its presence alone will not dictate the continued existence of the metastable solid phases. This is controlled more by the kinetics of the nucleation-controlled regime of surface area generation, rather than precursor dissolution and reprecipitation as the more stable phase (crystal-growth regime of surface area generation). Using a model incorporating the processes of heterogeneous precipitation, Steefel and Van Cappellen[38] estimated that a Ω_m only 5% higher than saturation with amorphous silica was necessary for its nucleation in the presence of foreign surfaces. They concluded that it is the ability of amorphous silica to nucleate easily which explains why amorphous silica saturation is an upper limit for $H_4SiO_4^0$ in natural aqueous systems (Figure 5).

An illustration of the influence of interfacial energy and nucleation kinetics on dissolution and precipitation reactions among secondary minerals can be seen in the work of May et al.[100] When kaolinite was placed in aqueous solutions either undersaturated or supersaturated with respect to K_{SO}, the initial conditions migrated towards saturation with kaolinite (Figure 10), indicating that the kaolinite solid phase was the most appropriate for the free energy of the system. When smectite was placed in a similar range of aqueous solutions, the initial conditions migrated towards the gibbsite solubility line, suggesting formation of an amorphous aluminum hydroxide species (Figure 11). Formation of this solid phase is consistent with the Ostwald step rule in that smectite is not the most obvious stable phase appropriate to the system. The precipitation of the amorphous aluminum hydroxide solid phase proceeded most likely via heterogeneous precipitation and was favored because of the kinetics of nucleation for the amorphous phase.

1. Sorption and Precipitation

The work of May et al.[100] also serves to illustrate that the processes of heterogeneous nucleation cannot be ignored whenever solid phases are brought in contact with aqueous solutions, even when the objective of the experimental design is ostensibly for another purpose. This applies especially to the sorption of specifically sorbed anions and cations on most reactive soil surfaces including

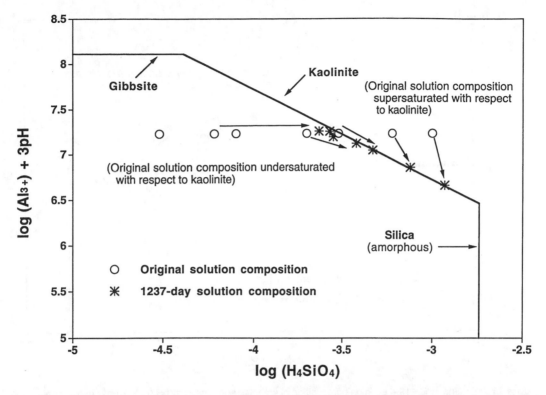

Figure 10   Approach to saturation by suspensions originally undersaturated and supersaturated with respect to kaolinite. Arrows indicate migration of original solution concentrations during 1237-d equilibration period. (Data from May et al.[100])

organic matter.[95] If the formation of nuclei on a foreign surface is the rate-limiting step, then the rate of nucleation can be expressed as:

$$J = k_{AB}\left(\theta_A^n \cdot \theta_B^m\right) \tag{60}$$

where k_{AB} is the rate constant, n and m the stoichiometric coefficients for the constituent ions A and B that comprise the nuclei, and θ is the degree of surface coverage. Equation 60 emphasizes again that the constituent ions must be sorbed to the foreign surface as a prerequisite for nucleation. Since most specifically sorbed metal ions of interest to soil chemists can form oxides and hydrousoxides, the potential for heterogeneous precipitation will always be present in aqueous systems because of the presence of the constituent OH^- ion as coordinated water molecules on the foreign surfaces. Anion sorption studies are not immune because of the possibility of potential constituent ions forming inner sphere complexes at the foreign surfaces. The source of these metal constituent ions either being from dissolution of the solid phase serving as the foreign surface, or from other ions in solution. Sorption, therefore, can be considered as the precursor of heterogeneous precipitation and the formation of surface nuclei should be anticipated in any aqueous system dealing with the sorption of specifically sorbed anions and cations on foreign surfaces. Possible indicators of heterogeneous precipitation reactions in these systems include continued sorption of constituent ions from solution over time, and/or the formation of a distinct new solid phase in suspension.

Examples of heterogeneous precipitations for specifically sorbed anions can be found for the interaction of phosphate with aluminum oxide[14,15,101] surfaces and calcium carbonate.[16,102] Maintaining the concentration of phosphate in solution below K_{so} values for known aluminum phosphates,

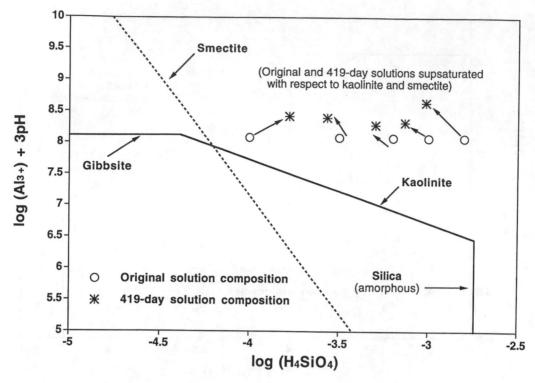

Figure 11   Original and 419-d solution composition of smectite suspensions (pH 5.0). Arrows indicate migration of original solution composition during 419-d equilibration period. Smectite saturation line from Sposito:[20] $(Fe^{3+}) = 10^{-13}$, $(Mg^{2+}) = 6 \times 10^{-3}$, $(H_2O) = 1$. (Data from May et al.[100])

van Riemsdijk et al.[14] noted the appearance of sterrittite ($[Al(OH)_2]_3HPO_4H_2PO_4$) in suspension after a 40-d equilibration period, while Veith and Sposito[101] reported the presence of X-ray amorphous analogs of variscite ($Al(OH)_2H_2PO_4$) and montebrasite ($AlOHNaPO_4$). Griffin and Jurinak[16,102] observed that approximately 1 mg P l^{-1} (0.03 mM P) in solution was adequate for the eventual formation of a crystalline calcium phosphate in a calcium carbonate suspension. The nature of the calcium phosphate formed was a function of the concentration of P in solution. The reaction did not proceed if calcium carbonate was not present, even though the calcium activity in solution was kept constant.

More recently, reports have been published concerning heterogeneous precipitation of metal ions on foreign surfaces, especially for those studies incorporating modern spectroscopy techniques (such as synchrotron-based X-ray absorption spectroscopy) in their experimental design. Reported observations include zinc precipitates on calcite,[103] cluster formations of manganese (II) on goethite (α-FeOOH) and magnesium on boehmite[104] (δ-AlOOH), cobalt precipitates on α-Al$_2$O$_3$,[105] polynuclear chromium (III) hydroxide structures on silica[106] and goethite,[17] and the formation of mixed nickel-aluminum hydroxides[107,108] on pyrophyllite. In all of these studies, the surface precipitates were found to occur above a given pH and degree of saturation of the foreign surface by the sorbed metal ion. This is consistent with Equation 60, in that the increase in pH would favor an increase in the concentration of OH$^-$ at the foreign surface (or CO$_3^{2-}$ in the case of calcium carbonate), effectively increasing the probability ($\theta_A^n \cdot \theta_B^m$) of finding neighboring sites occupied by constituent ions. The study by Scheidegger et al.[107,108] dealing with nickel sorption by pyrophyllite is especially telling in that dissolution of the pyrophyllite mineral itself contributed the aluminum necessary for formation of the nickel-aluminum hydroxide nuclei. With longer equilibration times, the appearance of a more Ni(OH)$_2$-dominated precipitate was detected by X-ray absorption fine structure obtained from X-ray absorption spectroscopy. This indicates that the surface crystallites were becoming more selective in their crystal growth, which is consistent with the anticipated sequence for heterogeneous precipitation reactions outlined in Section III.B.

IV. DISSOLUTION

Dissolution can be defined as the detachment of constituent ions from the surface of a mineral and their subsequent transport to the bulk solution. The rates at which this relatively simple reaction occurs for primary and secondary minerals have implications that extend to a global scale, leading essentially to the development of life on the earth as we know it.[109,110] It is no wonder, therefore, that the scientific literature is replete with studies on mineral dissolution as a function of various parameters (e.g., pH, temperature, solution composition), and that this plethora of research, as noted by May and Nordstrom[111] for the case of feldspar, has sometimes lead to a "... confusing array of experimental observations and explanatory hypotheses." During the past 20 years, however, a consistent theme has emerged regarding the theory of mineral dissolution reactions, due largely to (1) the acceptance of the surface reaction hypothesis for describing dissolution (Section II.B), and (2) the realization that many earlier studies were probably confounded by experimental artifacts[47,112] that resulted in erroneous conclusions.[111] There are a number of extensive reviews summarizing reported mineral dissolution reactions and calculated rates of dissolution for primary and secondary minerals,[25,113-118] and no attempt will be made in this section to repeat this effort. Here, the remainder of this section will be devoted to explanation of a proposed general rate law for mineral dissolution (and precipitation) that embodies much of the reported experimental observations.[18]

A. General Rate Law

If we describe dissolution as a ligand exchange reaction[119] (Section II.B), then we can state that during dissolution the coordination environment of a constituent metal ion changes as it is detached from the mineral surface and transported into solution.[120] The most important chemical processes which influence the coordination of metal ions in solution are hydrolysis, complexation with inorganic and organic ligands, and changes in oxidation state due to interaction with reductants and oxidants.[121] It follows, therefore, that a necessary precursor to the formation of an activated surface complex involving a constituent metal ion and ligands that will result in a change in the coordination environment of the metal ion, is the sorption of these ligands at the mineral surface. Viewed schematically, the two sequences required for dissolution via a change in the coordination environment of the constituent ions are[121]

$$\text{Surface sites + reactants} \xrightarrow{\text{fast}} \text{surface species}$$
$$\left(H^+, OH^-, \text{ligands}\right) \tag{61}$$

$$\text{Surface species} \xrightarrow[\text{migration to form activated complex}]{\text{slow}} \text{constituent ion}_{(aq)} \tag{62}$$

Equation 61 is assumed to be a fast reaction because it is basically a sorption reaction, which is known to be relatively rapid for H^+, OH^-, and most inorganic and organic ligands. The resulting formation of inner-sphere surface complexes (monodentate, bidentate, mononuclear, binuclear, bidentate mononuclear) between the sorbed ligands and constituent ions at the mineral surface occurs at a slower rate often involving migration of the sorbed ligand across the surface to a preferred site. The formation of the surface coordination bonds with the sorbed ligands acts to weaken the internal bonds between the constituent ions, leading to the eventual formation of a surface activated complex. The detachment of the activated complex from the surface results in release of a constituent ion to the bulk solution. The nature of the coordinated bonds that will eventually form the surface activated complex is a function of the stereochemistry of both the ligands and the mineral lattice structure. The differences in stereochemistry between ligands (especially for organic ligands) will be reflected in differences in rates of dissolution for the same mineral surface.[120]

The formation of the surface activated complex is dependent upon the concentration of preferred surface sites and the concentration of the sorbed ligand(s). If we assume that after detachment of the activated complex the mineral surface rapidly reprotonates to form another preferred site[120] (i.e., the concentration of preferred sites remains constant), then the rate of dissolution (J) for the solid phase becomes directly proportional to the concentration of sorbed ligand:[121]

$$J \propto [\text{surface species}] \qquad (63)$$

Equation 63 is important for two reasons. First, it reinforces the concept that it is the concentration of sorbed ligands that controls the rate of dissolution, and that complexation reactions occurring in the bulk solution will have no direct effect on the dissolution reaction.[120] Second, it emphasizes only the formation of a surface species and does not place any restrictions on what the composition of the final surface activated complex may entail. Thus, other sorbed ions may interact in the formation of the surface species and the activated complex (and thereby influence the rate of dissolution) that otherwise would not be associated with the aqueous constituent ion in the bulk solution. The possible influence of these other ions on the measured rate of dissolution should not be ignored.

As noted in Section II.A.2, mineral surfaces are not uniform planes, and the presence of edges, kinks, pits, etc., will result in several different populations of potential preferred sites for dissolution. A general schematic representation of the rate of dissolution of a mineral surface, therefore, must account for both the concentration of sorbed ligands and the surface density of preferred sites:[120]

$$J \propto [\text{sorbed ligands}] \cdot [\text{surface density of sites}]^n \qquad (64)$$

The exponent n is added in Equation 64 to reflect the number of ligands required to form the activated complex at a preferred site.

Equation 64 forms the basis for many rate laws developed by numerous researchers studying dissolution of primary and secondary minerals. Lasaga[18] has integrated these various formulations into a general form he has termed the rate law for heterogeneous mineral surface reactions:

$$J = k_0 S_{min} \exp^{(-E_a/RT)} a_{H^+}^{N_{H^+}} g(I) \Pi_i a_i^{N_i} f(\Delta G_r) \qquad (65)$$

where k_0 is the intensive rate constant (mol cm^{-2} s^{-1}), S_{min} is the reactive surface area, and E_a is the overall activation energy for the reaction. The possible effects of pH ($a_{H^+}^{N_{H^+}}$), ionic strength ($g(I)$) and other ligands, reductants, oxidants, or sorbed ions ($\Pi_i a_i^{N_i}$) are written as products to emphasize that these effects can be studied separately or in concert with one another, but that one cannot ignore the influence of all of these variables on J (i.e., their value cannot be set equal to zero). The final term, $f(\Delta G_r)$, is added in recognition that the observed J is also a function of Ω, the degree of supersaturation or undersaturation in solution. Thus, J can represent dissolution or precipitation reactions (depending on Ω), and is equal to zero when $\Delta G_r = 0$ ($\Omega = 1$).

Equation 65 does not strictly follow the general format of Equation 64 as the variables representing pH, and the presence of other ligands, reductants, and oxidants are expressed in terms of bulk solution activities which are readily measured or calculated. A more correct statement of the general rate law would be:[18]

$$J = k_0 S_{\min} \exp^{(-E_a/RT)} x_{H^+}^{N_{H^+}} g(I) \Pi_i x_i^{N_i} f(\Delta G_r) \qquad (66)$$

where x represents the mole fraction of the respective sorbed species on the mineral surfaces. With the development of models that describe the concentration of sorbed species on a surface (for example, the triple-layer model[122]), as well as the availability of the more traditional adsorption isotherms, a number of researchers have substituted expressions for the x terms into Equation 66 deriving rate expressions that contain a direct estimate of the sorbed ligands.[18,47,123-128]

Stumm and Wieland[121] have also derived a general rate law for the dissolution of minerals due to proton- or ligand-promoted dissolution:

$$J = k_o x_i P_j S' \qquad (67)$$

The k_0 term is the rate constant, x_i the mol fraction of preferred sites for dissolution, P_j is the probability of finding a preferred site with the coordinating ligand to result in the formation of the activated complex, and S' is the surface concentration of all sites (mol m^{-2}). The exponential dependence on n in Equation 64 is arrived at through the derivation of an expression of P_j for a specific activated complex using surface complexation equilibria for the proton- or ligand-promoted dissolution reactions.[129]

This general rate law differs from that of Lasaga[18] in that it focuses only on proton- and ligand-induced dissolution as separate rate expressions which are added together to arrive at the overall rate of dissolution. Reference to Equation 67 is included here for the sake of completeness, and also because of its relationship to Equation 13, Section II.A.2.a. The product of $x_i P_j$ is essentially an estimate of the reactive surface θ term in Equation 13, provided S' is directly related to S, the mineral surface area.

B. Dissolution Rates Near Equilibrium

A formal dependence between J and $f(\Delta G_r)$ (by way of the Ω term) can be derived by evaluating the rate equations for the dissolution of a solid. Following the example of Sposito,[20] a general dissolution equation can be written as:

$$C_s \underset{k_p}{\overset{k_d}{\rightleftharpoons}} A_{(aq)} + B_{(aq)} \qquad (68)$$

where A and B are the constituent ions of the solid C, k_d is the rate coefficient for dissolution, and k_p is the rate coefficient for precipitation. Defining K_{so} as the ratio of the two rate coefficients (k_d/k_p), and assuming unit activity for C_s, an expression for the rate of change of species $A_{(aq)}$ in solution can be written as:

$$\frac{d[A]}{dt} = k_d(1 - \Omega) \qquad (69)$$

Assuming congruent dissolution, d[A]/dt is a direct estimate of J such that:

$$J \cong k_d(1 - \Omega) \qquad (70)$$

When the surrounding solution is undersaturated (IAP $\ll$ K_{so}):

$$J \cong k_d \tag{71}$$

and the rate of dissolution of the solid phase becomes constant. This is frequently observed in mineral dissolution studies and is referred to as the dissolution plateau, although the exact value for Ω to arrive at this point varies from solid to solid.[43] At values of IAP $\approx$ K_{so} (near equilibrium), the validity of Equation 70 becomes less certain, especially at values of Ω very near to one. A more valid empirical expression for J near equilibrium is:

$$J \cong k_d(1-\Omega)^n \tag{72}$$

where n is not related to reaction stoichiometry, but is an empirical parameter applied to the overall reaction in recognition of the role surface microtopography plays in controlling J as $\Omega \to 1$. Equation 72 still goes to zero at $\Omega = 1$, but allows for experimental observations that J is not a smooth function near equilibrium.

The reason for the asymmetry in J as $\Omega \to 1$ is due to ΔG_{dis} as discussed in Section II.A.2.b. Like precipitation (Equation 53), the value of Ω represents the energy available in solution to drive the dissolution reaction. A similar definition (c/c_0) was used to relate etch pit formation to a c_{crit} necessary to overcome the interfacial energy barrier as an etch pit attempted to grow in size. The only energy available to drive the dissolution reaction when $\Omega \to 1$ ($c_{crit} < c < c_0$) is from the lattice defects that intersect the mineral surface. Although generally assumed to be sufficient in number to promote dissolution, the experimentally observed asymmetry in J as $\Omega \to 1$ suggests different reaction pathways are at work than when $\Omega \ll 1$.[20] This effectively invalidates the derivation of Equation 70 for processes near equilibrium, and serves to illustrate that rate laws describing dissolution of heterogeneous surfaces in nature are by definition empirical in design.[20]

C. Field vs. Laboratory Rate Estimates

Dissolution rates at near equilibrium have been suggested as one reason why there is a large discrepancy between rates of mineral dissolution reported for laboratory studies vs. those obtained from whole catchment studies.[18] Typically laboratory rate estimates are 100 to 1000 times greater than rate estimates for the same mineral assemblage in the field.[110] The suggestion that near-equilibrium dissolution may account for this discrepancy is based on the assumption that solution exchange near weathering surfaces under field conditions is much less than a stirred laboratory experiment, and that the degree of undersaturation in solution would be less than under laboratory conditions. Velbel[130] has argued effectively against this assumption, and the measurements of solution composition in the field support his position that near-equilibrium conditions are not the norm during weathering of minerals.

Another explanation that has been proposed to account for the observed discrepancy is that related to the mineral surface area (Equation 65) for laboratory vs. field studies.[131] It may be argued that artificially treated mineral surfaces are more reactive than their field counterparts, and hence the difference in values for J. This, combined with the difficulty in estimating reactive mineral surface area in a watershed, could account for the reported differences. A corollary to this hypothesis is the possibility of surface inhibition reactions in natural systems that would be absent from laboratory studies. Sorbed inorganic anions have been observed to inhibit iron oxide dissolution by organic ligands when the system pH favors bidentate coordination of the inorganic anions with the mineral surface.[132-134] A lower average temperature for natural systems (<25°C) would also favor slower dissolution rates because of the inverse relationship between temperature and the activating energy for the dissolution reaction (Equation 65). It is also possible that microbial weathering processes may play an important role in dictating rates of dissolution observed for whole catchment studies.[135]

Brady and Zachara[110] have considered all of the above arguments and still have found them wanting as suitable explanations for the reported differences between laboratory and field mineral rates of dissolution. They cite work by Drever and Swoboda-Colberg,[137] who have taken minerals from natural soils and demonstrated that they dissolve at the same rate as those prepared for laboratory dissolution experiments, as evidence that differences in mineral surface area is not a credible hypothesis for explaining the reported discrepancies. They argue further than an underestimation of the surface area for minerals in the field would act to offset the effect of lower temperature. Brady and Zachara[110] and others[64,136] have concluded that the only remaining explanation to account for the difference between laboratory- and field-derived rates of dissolution is one based on hydrology, specifically, the diminished degree of fluid exposure to soil minerals during unsaturated flow. This is not to be confused with the basic assumption of the surface reaction hypothesis of mineral dissolution: i.e., mineral dissolution is controlled by the rate of formation and detachment of the activated complex from the surface vs. transport control as the detached constituent ion diffuses away from the mineral surface.[1] The hydrologic factor being set forth to account for the differences in rates of dissolution does not require limiting diffusion of detached constituent ions away from mineral surfaces. Rather, estimates of field rates of mineral dissolution obtained from whole catchment studies differ from laboratory measured values because there is insufficient flow (unsaturated flow) in large portions of the watershed to wet all the available mineral surfaces and carry away the dissolved solutes.[64] While there is uncertainty in estimating the surface area of the reactive minerals in the field, there is even more uncertainty concerning hydrologic flow with a watershed, and this alone can account for the magnitude of the difference between field-derived and laboratory-measured rates of dissolution.

REFERENCES

1. Stumm, W., *Chemistry of the Solid-Water Interface. Processes at the Mineral-Water and Particle-Water Interface in Natural Systems,* John Wiley & Sons, New York, 1992.
2. Sparks, D.L., Dynamics of soil potassium, *Adv. Soil Sci.,* 6, 1, 1987.
3. Bredemeier, M., Quantification of ecosystem-internal proton production from the ion balance of the soil, *Plant Soil,* 101, 273, 1987.
4. Homann, P.S., Van Miegroet, H., Cole, D.W., and Wolfe, G.V., Cation distribution, cycling, and removal from mineral soil in Douglas-fir and red alder forests, *Biogeochemistry,* 16, 121, 1992.
5. Johnson, D.W. and Lindberg, S.E., Acidic deposition on Walker Branch Watershed, in *Acidic Precipitation, Vol. 1, Case Studies,* Adriano, D.C. and Salomons, W., Eds., Springer-Verlag, Berlin, 1989, 1.
6. Johnson, D.W., Kelly, J.M., Swank, W.T., Cole, D.W., Van Miegroet, H., Hornbeck, J.W., Pierce, R.S., and Van Lear, D., The effects of leaching and whole-tree harvesting on cation budgets of several forests, *J. Environ. Qual.,* 17, 418, 1988.
7. Van Breeman, N., Mulder, J., and Driscoll, C.T., Acidification and alkalinization of soils, *Plant Soil,* 75, 283, 1983.
8. Schnoor, J.L. and Stumm, W., Acidification of aquatic and terrestrial systems, in *Chemical Processes in Lakes,* Stumm, W., Ed., John Wiley & Sons, New York, 1985, 311.
9. Ulrich, B., A concept of forest ecosytem stability and of acid deposition as driving force for destabilization, in *Effects of Accumulation of Air Pollutants in Forest Ecosystems,* Vol. 29, Ulrich, B., and Pankroth, J., Eds., D. Reidel, Dordrecht, 1983, 1.
10. De Vries, W., Van Grinsven, J.J.M., Van Breeman, N., Leeters, E.E.J.M., and Jansen, P.C., Impacts of acid deposition on concentrations and fluxes of solutes in acidy sandy forest soils in the Netherlands, *Geoderma,* 67, 17, 1995.
11. Rai, D. and Kittrick, J.A., Mineral equilibria and the soil system, in *Minerals in Soil Environments,* 2nd ed., Dixon, J.B. and Weed, S.B., Eds., Soil Science Society of America, Madison, WI, 1989, 161.

12. Allen, B.L. and Hajek, B.F., Mineral occurrence in soil environments, in *Minerals in Soil Environments,* 2nd ed., Dixon, J.B. and Weed, S.B., Eds., Soil Science Society of America, Madison, WI, 1989, 199.

13. Talibudeen, O., Precipitation, in *The Chemistry of Soil Processes,* Greenland, D.J. and Hayes, M.H.B., Eds., John Wiley & Sons, 1981, 81.

14. Van Riemsdijk, W.J., Weststrate, F.A., and Bolt, G.H., Evidence for a new aluminum phosphate phase from reaction rate of phosphate with aluminum hydroxide *Nature,* 257, 473, 1975.

15. Van Riemsdijk, W.H., Weststrate, F.A., and Beck, J., Phosphates in soils treated with sewage water. III. Kinetic studies on the reaction of phosphate with aluminum compounds, *J. Environ. Qual.,* 6, 26, 1877.

16. Griffin, R.A. and Jurinak, J.J., The interaction of phosphate with calcite, *Soil Sci. Soc. Am. Proc.,* 37, 847, 1973.

17. Fendorf, S.E., Surface reactions of chromium in soils and waters, *Geoderma,* 67, 55, 1995.

18. Lasaga, A.C., Fundamental approaches in describing mineral dissolution and precipitation rates, *Rev. Mineral.,* 31, Chapter 2, 1995.

19. Hochella, M.F., Jr., Atomic structure, microtopography, composition, and reactivity of mineral surfaces, *Rev. Mineral.,* 23, 87, 1990.

20. Sposito, G., *Chemical Equilibria and Kinetics in Soils,* Oxford University Press, New York, 1994.

21. Nriagu, J.O., Solubility equilibrium constant of strengite, *Am. J. Sci.,* 272, 476, 1872.

22. Lindsay, W.L., *Chemical Equilibria in Soils,* John Wiley & Sons, New York, 1979.

23. Sposito, G., *The Thermodynamics of Soil Solutions,* Oxford University Press, New York, 1981.

24. Whittig, L.D. and Allardice, W.R., X-ray diffraction techniques, in *Methods of Soil Analysis,* Pt. 1, *Physical and Mineralogical Methods,* 2nd ed., Klute, A., Ed., Soil Science Society of America, Madison, WI, 1986, 331.

25. Bloom, P.R. and Nater, E.A., Kinetics of dissolution of oxide and primary silicate minerals, in *Rates of Soil Chemical Processes,* Sparks, D.L. and Suarez, D.L., Eds., Soil Science Society of America, Madison, WI, 1991, 151.

26. Helgeson, H.C., Murphy, W.M., and Aagaard, A., Thermodynamic and kinetic constraints on reaction rates among minerals and aqueous solutions. II. Rate constants, effective surface area, and the hydrolysis of feldspar, *Geochim. Cosmochim. Acta,* 48, 2405, 1984.

27. Blum, A.E. and Lasaga, A.C., Monte Carlo simulations of surface reaction rate laws, in *Aquatic Surface Chemistry. Chemical Processes at the Particle-Water Interface,* Stumm, W., Ed., John Wiley & Sons, 1987, 255.

28. Lasaga, A.C. and Blum, A.E., Surface chemistry, etch pits and mineral-water reactions, *Geochim. Cosmochim. Acta,* 50, 2363, 1986.

29. Lin, C.C. and Shen, P., Incubation time of etch pits at dislocation outcrops, *Geochim. Cosmochim. Acta,* 59, 2955, 1995.

30. Brantley, S.L., Crane, S.R., Crerar, D.A., Hellmann, R., and Stallard, R., Dissolution at dislocation etch pits in quartz, *Geochim. Cosmochim. Acta,* 50, 2349, 1986.

31. Dove, P.M. and Rimstidt, J.D., Silica-water interactions, *Rev. Mineral.,* 29, 259, 1993.

32. Adamson, A.W., *Physical Chemistry of Surfaces,* 2nd ed., John Wiley & Sons, New York, 1967.

33. Blum, A.E., Yund, R.A., and Lasaga, A.C., The effect of dislocation density on the dissolution rate of quartz, *Geochim. Cosmochim. Acta,* 54, 283, 1990.

34. Anbeek, C., Van Breeman, N., Meijer, E.L., and Van Der Plas, L., The dissolution of naturally weathered feldspar and quartz, *Geochim. Cosmochim. Acta,* 58, 4601, 1994.

35. Casey, W.H. and Bunker, B., Leaching of mineral and glass surfaces during dissolution, *Rev. Mineral.,* 23, 397, 1990.

36. Stumm, W. and Morgan, J.J., *Aquatic Chemistry. Chemical Equilibria and Rates in Natural Waters,* 3rd ed., John Wiley & Sons, New York, 1996.

37. Kolthoff, I.M., Sandell, E.B., Meehan, E.J., and Bruckenstein, S., *Quantitative Chemical Analysis,* 4th ed., Macmillan, London, 1969.

38. Steefel, C.I. and Van Cappellen, P., A new kinetic approach to modelling water rock interactions: the role of nucleation, precursors, and Ostwald ripening, *Geochim. Cosmochim. Acta,* 54, 2657, 1990.

39. Holdren, G.R., Jr. and Speyer, P.M., Reaction rate-surface area relationships during the early stages of weathering. I. Initial observations, *Geochim. Cosmochim. Acta,* 49, 675, 1985.

40. Zhang, H., Bloom, P.R., and Nater, E.A., Change in surface area and dissolution rates during hornblende dissolution at pH 4.0, *Geochim. Cosmochim. Acta,* 57, 1681, 1993.

41. Holdren, G.R., Jr. and Speyer, P.M., Reaction rate-surface area relationships during the early stages of weathering. II. Data on eight additional feldspars, *Geochim. Cosmochim. Acta,* 51, 2311, 1987.

42. Bloom, P.R., Nater, E.A., and Zhang, H., Reply to the comment by C. Anbeek on "Change in surface area and dissolution rates during hornblende dissolution at pH 4.0", *Geochim. Cosmochim. Acta,* 58, 1851, 1994.

43. Casey, W.H., Surface chemistry during the dissolution of oxide and silicate materials, in *Mineral Surfaces,* Vaughan, D.J. and Pattrick, R.A.D., Eds., Chapman and Hall, London, 1995, 185.

44. Casey, W.H., Westrich, H.R., Arnold, G.W., and Banfield, J.F., The surface chemistry of dissolving labradorite feldspar, *Geochim. Cosmochim. Acta,* 53, 821, 1980.

45. Chou, L. and Wollast, R., Steady state kinetics and dissolution mechanisms of albite, *Am. J. Sci.,* 285, 963, 1985.

46. Walther, J.V., Relation between rates of aluminosilicate mineral dissolution, pH, temperature, and surface charge, *Am. J. Sci.,* 296, 693, 1996.

47. Amrhein, C. and Suarez, D.L., Some factors affecting the dissolution kinetics of anorthite at 25°C, *Geochim. Cosmochim. Acta,* 56, 1815, 1992.

48. Hochella, M.F., Jr. and Banfield, J.F., Chemical weathering of silicates in nature: a microscopic perspective with theoretical considerations, *Rev. Mineral.,* 31, 353, 1995.

49. Schott, J. and Petit, J.C., New evidence for the dissolution of silicate minerals, in *Aquatic Surface Chemistry: Chemical Processes at the Particle-water Interface,* Stumm, W., Ed., John Wiley & Sons, 1987, 293.

50. Inskeep, W.P., Nater, E.A., Bloom, P.R., Vandervoort, D.S., and Erick, M.S., Characterization of laboratory weathered labradorite surfaces using X-ray photoelectron spectroscopy and transmission electron microscopy, *Geochim. Cosmochim. Acta,* 55, 787, 1991.

51. Berner, R.A. and Schott, J., Mechanism of pyroxene and amphibole weathering. II. Observations of soil grains, *Am. J. Sci.,* 282, 1214, 1982.

52. Muir, I.J., Bancroft, G.M., Shotyk, W., and Nesbitt, H.W., A SIMS and XPS study of dissolving plagioclase, *Geochim. Cosmochim. Acta,* 54, 2247, 1990.

53. Hellmann, R., Eggleston, C.M., Hochella, M.F., Jr., and Crerar, D.A., The formation of leached layers on albite surfaces during dissolution under hydrothermal conditions, *Geochim. Cosmochim. Acta,* 54, 1267, 1990.

54. Nesbitt, H.W. and Muir, I.J., SIMS depth profiles of weathered plagioclase and processes affecting dissolved Al and Si in some acidic soil solutions *Nature,* 334, 336, 1988.

55. Hochella, M.F., Jr., Ponader, H.B., Turner, A.M., and Harris, D.W., The complexity of mineral dissolution as viewed by high resolution scanning Auger microscopy: labradorite under hydrothermal conditions, *Geochim. Cosmochim. Acta,* 52, 285, 1988.

56. Casey, W.H., Westrich, H.R., Banfield, J.F., Ferruzzi, G., and Arnold, G.W., Leaching and reconstruction at the surfaces of dissolving chain-silicate minerals, *Nature,* 366, 253, 1993.

57. Veblen, D.R., Banfield, J.F., Guthrie, G.D., Jr., Heaney, P.J., Ilton, E.S., Livi, K.J.T., and Smelik, E.A., High-resolution and analytical transmission electron microscopy of mineral disorder and reactions, *Science,* 260, 1465, 1993.

58. Banfield, J.F., Ferruzzi, G.G., Casey, W.H., and Westrich, H.R., HRTEM study comparing naturally and experimentally weathered pyroxenoids, *Geochim. Cosmochim. Acta,* 59, 19, 1995.

59. Casey, W.H., Westrich, H.R., Massis, T., Banfield, J.F., and Arnold, G.W., The surface of labradorite feldspar after acid hydrolysis, *Chem. Geology,* 78, 205, 1989.

60. Velbel, M.A., Formation of protective surface layers during silicate-mineral weathering under well-leached, oxidizing conditions, *Am. Mineralogist,* 78, 405, 1993.

61. Garrels, R.M. and Christ, C.L., Solutions, Minerals and Equilibria, *Harper and Row,* New York, 1965.

62. Van Breemen, N. and Brinkman, R., Chemical equilibria and soil formation, in *Soil Chemistry. A. Basic Elements,* Bolt, G.H. and Bruggenwert, M.G.M., Eds., Elsevier, Amsterdam, 1976, 141.

63. Brinkman, R., Clay transformations: aspects of equilibrium and kinetics, in *Soil Chemistry. B. Physicochemical Models,* Bolt, G.H., Ed., Elsevier, Amsterdam, 1979, 433.

64. Schnoor, J.L., Kinetics of chemical weathering: a comparison of laboratory and field weathering rates, in *Aquatic Chemical Kinetics. Reaction Rates of Processes in Natural Waters,* Stumm, W., Ed., John Wiley & Sons, New York, 1990, 475.

65. Sposito, G., *The Chemistry of Soils,* Oxford University Press, New York, 1989.

66. Graham, R.C., Weed, S.B., Bowen, L.H., and Buol, S.W., Weathering of iron-bearing minerals in soils and saprolite on the North Carolina Blue Ridge Front. I. Sand-size primary minerals, *Clays Clay Miner.,* 37, 19, 1989.

67. Glynn, P.D., Reardon, E.J., Plummer, L.N., and Busenberg, E., Reaction paths and equilibrium end-points in solid-solution aqueous-solution systems, *Geochim. Cosmochim. Acta,* 54, 267, 1990.

68. Mattigod, S.V. and Kittrick, J.A., Temperature and water activity as variables in soil mineral activity diagrams, *Soil Sci. Soc. Am. J.,* 44, 149, 1980.

69. Van Riemsdijk, W.J. and Van Der Zee, S.E.A.T.M., Comparison of models for adsorption, solid solution and surface precipitation, in *Interactions at the Soil Colloid-Soil Solution Interface,* Bolt, G.H., Ed., Kluwer Academic, Dordrecht, 1991, 241.

70. Glynn, P.D., Modeling solid-solution reactions in low- temperature aqueous systems, in *Chemical Modeling of Aqueous Systems,* Vol. 2, Melchoir, D.C. and Bassett, R.L., Eds., American Chemical Society, Washington, D.C., 1990, 74.

71. Glynn, P.D. and Reardon, E.J., Solid-solution aqueous-solution equilibria: thermodynamic theory and representation, *Am. J. Sci.,* 290, 164, 1990.

72. Gresens, R.L., The aqueous solubility product of solid solutions. I. Stoichiometric saturation; partial and total solubility product, *Chem. Geol.,* 32, 59, 1981.

73. Murrmann, R.P. and Peech, M., Reaction products of applied phosphate in limed soils, *Soil Sci. Soc. Am. Proc.,* 32, 493, 1968.

74. Konigsberger, E., Hausner, R., and Gamsjager, H., Solid-solute phase equilibria in aqueous solution. V. The system $CdCO_3$-$CaCO_3$-CO_2-H_2O, *Geochim. Cosmochim. Acta,* 55, 3505, 1991.

75. Bohn, H.L., Chemical activity and aqueous solubility of soil solid solutions, *Soil Sci.,* 154, 357, 1992.

76. Bohn, H.L. and Bohn, R.K., Solid activity coefficients of soil components, *Geoderma,* 38, 3, 1986.

77. Konigsberger, E., Gamsjager, H., Glynn, P.D., and Reardon, E.J., Comment and Reply. Solid-solution aqueous-solution equilibria: thermodynamic theory and representation, *Am. J. Sci.,* 292, 199, 1992.

78. Driscoll, C.T. and Schecher, W.D., The chemistry of aluminum in the environment, *Environ. Geochem. Health,* 12, 28, 1990.

79. Johnson, J.M., Driscoll, C.T., Eaton, J.S., Likens, G.E., and McDowell, W.H., 'Acid rain', dissolved aluminum and chemical weathering at the Hubbard Brook Experimental Forest, New Hampshire, *Geochim. Cosmochim. Acta,* 45, 1421, 1981.

80. May, H.M., Helmke, P.A., and Jackson, M.L., Gibbsite solubility and thermodynamic properties of hydroxy-aluminum ions in aqueous solution at 25°C, *Geochim. Cosmochim. Acta,* 43, 861, 1979.

81. Neal, C., Aluminum solubility relationships in acid waters — a practical example of the need for a radical reappraisal, *J. Hydrol.,* 104, 141, 1988.

82. Neal, C. and Williams, R., Towards establishing hydroxy aluminum silicate solubility controls for natural waters, *J. Hydrol.,* 97, 347, 1987.

83. Neal, C., Skeffington, R.A., Williams, R., and Roberts, D.J., Aluminum solubility controls in acid waters: the need for a reappraisal, *Earth Planet Sci. Lett.,* 86, 113, 1987.

84. Neal, C., Smith, C.J., Walls, J., and Dunn, C.S., Major, minor and trace element mobility in the acidic upland forested catchment of the upper River Severn, Mid Wales, *J. Geol. Soc. London,* 193, 635, 1986.

85. Xu, S. and Harsh, J.B., Influence of autocorrelation and measurement errors on interpretation of solubility diagrams, *Soil Sci. Soc. Am. J.,* 59, 1549, 1995.

86. Buckle, E.R., General introduction, *Discussions Faraday Soc.,* 61, 7, 1976.

87. Lieser, K.H., Steps in precipitation reactions, *Angew. Chem. Int. Ed.,* 8, 188, 1969.

88. Gunn, D.J., Mechanism for the formation and growth of ionic precipitates from aqueous solution, *Discussions Faraday Soc.,* 61, 33, 1976.

89. Leubner, I.H., Crystal formation (nucleation) under kinetically controlled and diffusion-controlled growth conditions, *J. Phys. Chem.,* 91, 6069, 1987.

90. Morse, J.W. and Casey, W.H., Ostwald processes and mineral paragenesis in sediments, *Am. J. Sci.,* 288, 537, 1988.

91. Stoffregen, R., Numerical simulation of mineral-water isotope exchange via Ostwald ripening, *Am. J. Sci.,* 296, 908, 1996.

92. Lazaridis, M. and Ford, I.J., A statistical mechanical approach to heterogeneous nucleation, *J. Chem. Phys.,* 99, 5426, 1993.

93. Corey, R.B., Adsorption vs. precipitation, in *Adsorption of Inorganics at Solid-Liquid Interfaces,* Anderson, M.A. and Rubin, A.J., Eds., Ann Arbor Science, Ann Arbor, MI, 1981, 161.

94. Furedi-Milhofer, H., Spontaneous precipitation from electrolyte solutions, *Pure Appl. Chem.,* 53, 2041, 1981.

95. Stumm. W., Coordinative interactions between soil solids and water — an aquatic chemist's point of view, *Geoderma,* 38, 19, 1986.

96. Kittrick, J.A., Mineral equilibria and the soil system, in *Minerals in Soil Environments,* Dixon, J.B. and Weed, S.B., Eds., Soil Science Society of America, Madison, WI, 1977, 1.

97. Marshall, C.E., *The Physical Chemistry and Mineralogy of Soils, Vol. 2, Soils in Place,* John Wiley & Sons, New York, 1977, 12.

98. Helgeson, H.C., Evaluation of irreversible reactions in geochemical processes involving minerals and aqueous solutions. I. Thermodynamic relations, *Geochem. Cosmochim. Acta,* 32, 853, 1968.

99. Helgeson, H.C., Garrels, R.M., and MacKenzie, F.T., Evaluation of irreversible reactions in geochemical processes involving minerals and aqueous solutions. II. Applications, *Geochim. Cosmochim. Acta,* 33, 455, 1969.

100. May, H.M., Kinniburgh, D.G., Helmke, P.A., and Jackson, M.L., Aqueous dissolution, solubilities and thermodynamic stabilities of common aluminosilicate clay minerals, *Geochim. Cosmochim. Acta,* 50, 1667, 1986.

101. Veith, J.A. and Sposito, G., Reactions of aluminisilicates, aluminum hydrous oxides, and aluminum oxide with phosphate; the formation of X-ray amorphous analog of variscite and montebrasite, *Soil Sci. Soc. Am. J.,* 41, 870, 1977.

102. Griffin, R.A. and Jurinak, J.J., Kinetics of the phosphate interaction with calcite, *Soil Sci. Soc. Am. Proc.,* 38, 75, 1974.

103. Zachara, J.M., Kittrick, J.A., Dake, L.S., and Harsh, J.B., Solubility and surface spectroscopy of zinc precipitates on calcite, *Geochim. Cosmochim. Acta,* 53, 9, 1989.

104. Bleam, W.F. and McBride, M.B., Cluster formation vs. isolated-site adsorption. A study of Mn(II) and Mg(II) adsorption on boehmite and goethite, *J. Colloid Interface Sci.,* 103, 124, 1985.

105. Katz, L.E. and Hayes, K.F., Surface complexation modeling. II. Strategy for modeling polymcr and precipitation reactions at high surface coverage, *J. Colloid Interface Sci.,* 170, 491, 1995.

106. Fendorf, S.E., Lamble, G.M., Stapleton, M.G., Kelley, M.J., and Sparks, D.L., Mechanisms of chromium(III) sorption on silica. I. Cr(III) surface structure derived by extended X-ray absorption fine structure spectroscopy, *Environ. Sci. Technol.,* 28, 284, 1994.

107. Scheidegger, A.M., Fendorf, M., and Sparks, D.L., Mechanisms of nickel sorption on pyrophyllite: macroscopic and microscopic approaches, *Soil Sci. Soc. Am. J.,* 60, 1763, 1996.

108. Scheidegger, A.M., Lamble, G.M., and Sparks, D.L., Investigation of Ni sorption on pyrophyllite: an XAFS study, *Environ. Sci. Technol.,* 30, 548, 1996.

109. White, A.F. and Brantley, S.L., Chemical weathering rates of silicate minerals: an overview, *Rev. Mineral.,* 31, Chapter 1, 1995.

110. Brady, P.V. and Zachara, J.M., Geochemical applications of mineral surface science, in *Physics and Chemistry of Mineral Surfaces,* Brady, P.V., Ed., CRC Press, Boca Raton, FL, 1996, 307.

111. May, H.M. and Nordstrom, D.K., Assessing the solubilities and reaction kinetics of aluminous minerals in soils, in *Soil-Acidity,* Ulrich, B. and Summer, M.E., Eds., Springer-Verlag, Berlin, 1991, 125.

112. Eggleston, C.M., Hochella, M.F., Jr., and Parks, G.A., Sample preparation and aging effects on the dissolution rate and surface composition of diopside, *Geochim. Cosmochim. Acta,* 53, 797, 1989.

113. Brady, P.V and House, W.A., Surface-controlled dissolution and growth of minerals, in *Physics and Chemistry of Mineral Surfaces,* Brady, P.V., Ed., CRC Press, Boca Raton, FL, 1996, 225.

114. Drever, J.I., Ed., *The Chemistry of Weathering,* D. Reidel, Dordrecht, 1985.

115. Nagy, K.L., Dissolution and precipitation kinetics of sheet silicates, *Rev. Mineral.,* 31, Chapter 5, 1995.

116. White, A.F., Chemical weathering rates of silicate minerals in soils, *Rev. Mineral.,* 31, Chapter 9, 1995.

117. Dove, P.M., Kinetic and thermodynamic controls on silica reactivity in weathering environments, *Rev. Mineral.,* 31, Chapter 6, 1995.

118. Blum, A.E. and Stillings, L.L. Feldspar dissolution kinetics, *Rev. Mineral.,* 31, Chapter 7, 1995.

119. Casey, W.H. and Ludwig, C., Silicate mineral dissolution as a ligand-exchange reaction, *Rev. Mineral.,* 31, Chapter 3, 1995.

120. Stumm, W. and Furrer, G., The dissolution of oxides and aluminum silicates; examples of surface-coordination-controlled kinetics, in *Aquatic Surface Chemistry. Chemical Processes at the Particle-Water Interface,* Stumm, W., Ed., John Wiley & Sons, New York, 1987, 197.

121. Stumm, W. and Wieland, E., Dissolution of oxide and silicate minerals: rates depend on surface speciation, in *Aquatic Chemical Kinetics. Reaction Rates of Processes in Natural Waters,* Stumm, W., Ed., John Wiley & Sons, New York, 1990, 367.

122. Goldberg, S., Use of surface complexation models in soil chemical systems,*Adv. Agron.,* 47, 233, 1992.

123. Brantley, S.L. and Stillings, L., Feldspar dissolution at 25°C and low pH, *Am. J. Sci.,* 296, 101, 1996.

124. Casey, W.H. and Sposito, G., On the temperature dependence of mineral dissolution rates, *Geochim. Cosmochim. Acta,* 56, 3825, 1992.

125. Stillings, L.L. and Brantley, S.L., Feldspar dissolution at 25°C and pH 3: reaction stoichiometry and the effect of cations, *Geochim. Cosmochim. Acta,* 59, 1483, 1995.

126. Wieland, E. and Stumm, W., Dissolution kinetics of kaolinite in acidic aqueous solutions at 25°C, *Geochim. Cosmochim. Acta,* 56, 3339, 1992.

127. Chorover, J. and Sposito, G., Dissolution behavior of kaolinitic tropical soils, *Geochim. Cosmochim. Acta,* 59, 3109, 1995.

128. Dove, P.M. and Elston, S.F., Dissolution kinetics of quartz in sodium chloride solutions: analysis of existing data and a rate model for 25°C, *Geochim. Cosmochim. Acta,* 56, 4147, 1992.

129. Furrer, G. and Stumm, W., The coordination chemistry of weathering. I. Dissolution kinetics of δ-Al_2O_3 and BeO, *Geochim. Cosmochim. Acta,* 50, 1847, 1986.

130. Velbel, M.A., Effect of chemical affinity on feldspar hydrolysis rates in two natural weathering systems, *Chem. Geol.,* 78, 245, 1989.

131. Velbel, M.A., Geochemical mass balances and weathering rates in forested watersheds of the southern Blue Ridge, *Am. J. Sci.,* 285, 904, 1985.

132. Stumm, W., From surface acidity to surface reactivity; inhibition of oxide dissolution,*Aquatic Sci.,* 55, 1, 1993.

133. Biber, M.V., Dos Santos Afonso, M., and Stumm, W., The coordination chemistry of weathering. IV. Inhibition of the dissolution of oxide minerals, *Geochim. Cosmochim. Acta,* 58, 1999, 1994.

134. Bondietti, G., Sinniger, J., and Stumm, W., The reactivity of Fe(III) (hydr) oxides: effects of ligands in inhibiting the dissolution, *Colloids Surfaces A. Physicochem. Eng. Aspects,* 79, 157, 1993.

135. Berthelin, J., Microbial weathering processes in natural environments, in *Physical and Chemical Weathering in Geochemical Cycles,* Lerman, A. and Meybeck, M., Eds., Kluwer Academic, Dordrecht, 1988, 33.

136. Van Grinsven, J.J.M. and Van Riemsdijk, W.H., Evaluation of batch and column techniques to measure weathering rates in soils, *Geoderma,* 52, 41, 1992.

137. Drever, J.I. and Swoboda-Colberg, N.G., in *Water-Rock Interaction WRI-7,* Kharaka, Y. and Maest, A., Eds., Balkema, Rotterdam, 1992, 211.

CHAPTER 6

The Chemistry of Soil Organic Matter

Nicola Senesi and Elisabetta Loffredo

CONTENTS

0-87371-883-6/99/$0.00+$.50
© 1999 by CRC Press LLC

I. INTRODUCTION

Soil organic matter (SOM) is one of the most important natural resources and the basis of soil fertility. The term SOM generally refers to the total organic carbon (C)-containing substances in soil. The total mass of organic C in soils has been estimated at 22×10^{14} kg, similar to the total of all other C reservoirs on the earth's surface (21×10^{14} kg).[1] SOM contents range from less than 1% in sandy and desert soils to 1%-5% (w/w) in the surface horizons (top 15 cm) of typical mineral and agricultural soils, to almost 100% in organic soils.[2] Even at the lowest levels, the role of SOM in all processes occurring in soil is ascertained to be highly relevant.

The ultimate source of SOM is the C fixed through photosynthetic reactions, which may reach the soil primarily from plant remains and root exudates. Relatively small inputs are also contributed from animal remains, autotrophic soil bacteria and burning fossil fuels. Anthropogenically produced C that can be incorporated into SOM is generally insignificant.

Essentially any compound synthesized by living organisms can be detected in SOM. Mineralization processes of SOM recirculate C to the atmosphere as CO_2, whereas some of the C is assimilated into microbial tissues, i.e., the soil biomass, and part is converted through the humification process into stable forms, i.e., humic substances, which can be mineralized concurrently. In this way, total SOM is maintained at some steady-state level typical of the ecosystem considered.[3,4]

In its broadest sense the term SOM encompasses all organic materials contained in soil and is made up of live organisms, their undecomposed and partly decomposed remains, and microbially and/or chemically resynthesized products resistant to further biological attack.[5,6] More specifically, the term SOM refers to the nonliving organic components, which are largely composed of products resulting from microbial and chemical transformations of organic debris. Stevenson [3,4] defined SOM as the total of organic components in soil, excluding undecayed plant and animal tissues, their "partial decomposition" products (the organic residues), the soil biomass (living microbial tissues), and macrofauna and macroflora. The terms SOM and "humus" are thus generally interchangeable, as suggested by Waksman.[7]

To simplify this very complex system, SOM is generally divided into two groups designated as nonhumic substances and humic substances.[3-5,8-10] The nonhumic substance group comprises organic compounds that belong to chemically recognizable classes and are not unique to the soil. These include polysaccharides and simple carbohydrates, amino sugars, proteins and amino acids, fats and waxes, lignin, resins, pigments, nucleic acids, hormones, a variety of organic acids, and so on. Most of these substances are relatively easily degradable and can be utilized as substrates by soil microorganisms, and as such have a transient existence in the soil. In contrast, humic substances comprise a heterogeneous mixture of chemically unidentifiable macromolecules that are distinctive to, and synthesized in the soil, and are relatively resistant to chemical degradation and microbial attack. Common definitions and terminology for SOM constituents are given in Table 1.[3]

The complexity of SOM, in conjuction with the fact that a significant portion of it is bound to the soil mineral fraction, has made SOM separation from the inorganic soil components an essential prerequisite to its selective chemical characterization. Isolation is generally followed by fractionation that lessens SOM heterogeneity so that the chemical, molecular, and structural properties of the obtained fractions can be studied. In principle, the ideal isolation and fractionation method should be able either to extract the entire organic component from the soil or to isolate selectively a well-defined SOM fraction, in both cases in an unaltered state. Because of the wide range of chemical compounds that constitute SOM it appears doubtful that an extractant that responds to the above requirements will be ever available. Recently, however, advanced analytical methods such as pyrolysis mass spectrometry and solid-state ^{13}C nuclear magnetic resonance (NMR) spectroscopy, which can eliminate the need for separating SOM from the mineral soil components, are being more and more frequently used to study the chemical properties of SOM directly in the soil mineral matrix.[11-18]

Table 1   Terms and Definitions of Soil Organic Matter (SOM)

Term	Definition
Organic residues	Undecayed plant and animal tissues and their partial decomposition products
Soil biomass	Organic matter present as live microbial tissue
Humus	Total of the organic compounds in soil exclusive of undecayed plant and animal tissues, their "partial decomposition" products, and the soil biomass
Soil organic matter	Same as humus
Humic substances	A series of relatively high-molecular-weight, brown- to black-colored substances formed by secondary synthesis reactions; the term is used as a generic name to describe the colored material or its fractions obtained on the basis of solubility characteristics; these materials are distinctive to the soil (or sediment) environment in that they are dissimilar to the biopolymers of microorganisms and higher plants (including lignin)
Nonhumic substances	Compounds belonging to known classes of biochemistry, such as amino acids, carbohydrates, fats, waxes, resins, and organic acids; humus probably contains most, if not all, of the biochemical compounds synthesized by living organisms
Humin	The alkali insoluble fraction of soil organic matter or humus
Humic acid	The dark-colored organic material which can be extracted from soil by various reagents and which is insoluble in dilute acid
Fulvic acid	The colored material which remains in solution after removal of humic acid by acidification
Hymatomelanic acid	Alcohol-soluble portion of humic acid

From Stevenson, F., *Humus Chemistry. Genesis, Composition, Reactions,* John Wiley & Sons, New York, 1982. With permission.

The objective of this chapter is to present an account of our current knowledge of the composition, genesis, isolation methodology, structural chemistry, and chemical properties of SOM. In the first part of the chapter the major nonhumic components of SOM, which include carbohydrates, nitrogenous (N) compounds, organic phosphorus (P) and sulfur (S) compounds, and lipids are discussed. In the second part of the chapter special attention is given to humic substances, the major SOM components. After a brief review of the main theories regarding their synthesis, humic substances are discussed in detail in terms of methods of extraction and fractionation, analytical characteristics, molecular structure, and physical and chemical properties. For space reasons, this chapter will not deal with interactions of humic substances with metal ions, clay minerals, and organic xenobiotics. Also the study of SOM without any separation from the mineral soil component will not be treated in this chapter.

II. NONHUMIC SUBSTANCES

Recent estimates of the average composition of SOM are the following:[2] carbohydrates (including uronic acids), 10%; N components (including proteins and amino acids, amino sugars, purines and pyrimidines, and unidentified compounds), 10%; lipids (including alcanes, fatty acids, waxes, and resins), 15%; humic substances, 65%. However, different soils may contain widely different amounts of nonhumic and humic substances. The amount of carbohydrates can range from 5% to 25%; proteins may vary from 15% to 45%; lipids from 2% in forest SOM to 20% in acid peat soils; and humic substances from 33% to 75% of the total SOM.[3,4]

A. Carbohydrates

1. Content and Origin

The carbohydrates in soil are estimated to account for about 5% to 25% of SOM, with an average of about 10% of the total soil organic C content, thus forming one of the largest fractions of SOM.[2,4] In general, soil carbohydrate levels parallel SOM contents, being highest in organic-rich surface horizons and lowest in subsoil horizons low in organic matter.[19] The proportion of carbohydrate is higher in soils rich in undecomposed or partially decomposed plant debris, such as peat soils and humus horizons of forest soils, whereas SOM contains relatively less carbohydrates in podzol B horizons. In general, with increasing humification, the carbohydrate content of SOM progressively decreases.[19]

Carbohydrates are contributed to the soil system from a variety of sources, including live plants, microorganisms and animals, and their remains. Carbohydrates comprise 50% to 70% of the dry weight of most plant tissues, and hence are the most abundant materials added to soil in the form of plant residues. Further, carbohydrates are important structural constituents and exocellular and intracellular components of soil microorganisms. Exudates of plant roots and animal residues also serve as a minor source of carbohydrates for soil.

Factors controlling carbohydrate levels in soil are the same that influence inputs of organic matter and its rate of decomposition. Thus, the carbohydrate content in virgin soils will be determined by the natural SOM controlling factors at any particular location. In cultivated soils, carbohydrate content can be greatly dependent on the agronomic practices used. However, the proportion of carbohydrates in SOM remains relatively constant.[19]

More detailed information regarding soil carbohydrates can be found in several reviews and books.[3,4,19-23]

2. Isolation and Composition

Total carbohydrates in soil are generally extracted by mineral acid (generally H_2SO_4) hydrolysis of whole soil. A variety of methods have been used for the determination of monosaccharides in soil acid hydrolysates, including (1) the classical Fehling's reagent and phenol-concentrated H_2SO_4 for the determination of total reducing sugars, often recorded as total carbohydrates;[24,25] (2) colorimetric determination of specific types of monosaccharides;[3,4] and (3) chromatographic techniques, including paper chromatography, gas-liquid chromatography (GLC), and high-performance anion-exchange chromatography for the determination of single monosaccharides.

Individual monosaccharides identified in whole soil hydrolysates are essentially the same for all soils studied. Further, only minor variations are measured in the proportions of monomers identified, whereas considerable variation is observed in the monomer distribution between individual soil fractions.[19] Selected results on neutral sugar distribution in whole soil hydrolysates are presented in Table 2.[19] Hexose sugars range from 4% to 12% and pentose sugars are lower than 5% of total SOM. Glucose is consistently the most abundant monosaccharide and galactose, mannose, arabinose, xylose, and rhamnose are always present.[22] Smaller amounts of fucose and ribose are also commonly found, as well as traces of a number of minor constituents, some of which have been identified, such as methylated sugars. In addition, amino sugars (hexosamines, mostly glucosamine and galactosamine) and uronic acids appear to be of universal occurrence in soils, ranging, respectively, from 2 to 5% and 1 to 5% of SOM.[22]

Most of the carbohydrates in soil occur in the form of polysaccharides but small amounts of monosaccharides (free sugars) and oligosaccharides are also present.[23] The total carbohydrates in a soil can be partitioned into free, water-soluble monosaccharides and oligosaccharides, and complex, water-insoluble polysaccharides that can be extracted and separated from other soil constituents by proper procedures, and further characterized for their monomer composition following hydrolysis.[19] However, the polysaccharide fractions that are strongly attached to clay and/or humic materials cannot be easily isolated and purified.[3,4]

3. Free Sugars

Free sugars and oligosaccharides are generally present in the soil solution in very small quantities. This is not surprising, because they are most readily available to microorganisms as

Table 2 Neutral Sugar Distribution in Whole Soil Hydrolysates (Expressed in Percent of Total Neutral Sugars Recovered)

Soil		Glucose	Galactose	Mannose	Arabinose	Xylose	Rhamnose	Fucose	Ribose	Ref.
Longterm cultivated		41.3	14.4	10.8	12.7	12.7	5.1	1.1		26
Longterm pasture		39.2	14.3	10.4	11.5	16.9	5.8	2.0		26
Marshall Ap — developed under prairie		27.6	17.5	17.5	15.0	12.9	5.4	2.3	1.8	27
Grundy Ap — developed under prairie		29.3	16.7	16.8	15.5	10.7	6.4	2.6	2.0	27
Mexico A_1 — developed under prairie		33.1	15.6	19.5	11.2	9.5	6.4	3.4	1.3	27
Menfro Ap — developed under forest		30.7	18.7	16.1	12.4	11.4	6.4	3.2	1.1	27
Weldon Ap — developed under forest		30.1	17.4	13.5	15.1	12.7	6.9	3.3	1.0	27
Marion Ap — developed under forest		28.5	20.0	16.8	11.2	11.3	6.5	4.1	1.6	27
Urrbrae pasture (0–6 cm)		39.3	12.2	19.4	11.7	7.5	4.4	3.6	1.9	28
Urrbrae fallow-wheat (0–6 cm)		43.1	13.0	18.5	9.8	8.7	3.5	2.6	0.8	28
Orthic grey luvisol	LH	44	15	12	12	9	5	4[a]		29
(grey wooded)	AH	48	14	13	10	8	5	2[a]		29
	Bt	33	19	14	13	8	9	4[a]		29
Orthic podzol	LH	54	15	15	5	4	7	5[a]		29
	B	35	16	16	9	9	9	6[a]		29
Dark brown solodized solonetz	Ah	28	16	18	17	9	8	4[a]		29
	Bnt	24	14	18	18	10	8	8[a]		29
Orthic black	Ah	36	14	16	15	8	8	3[a]		29
	Bm	34	15	14	12	8	12	5[a]		29

[a] Includes ribose.

From Lowe, L. E., Carbohydrates in soil, in Soil Organic Matter, Schnitzer, M. and Khan, S. U., Eds., Elsevier, New York, 1978, 65. With permission.

primary sources of food and energy, and are thus easily subject to decomposition by microbial attack.

Free sugars are generally extracted from soil with water or, better, 80% ethanol that eliminates soil dispersion and can easily be removed by evaporation at low temperature under reduced pressure. The sugar residue is then redissolved in water, centrifuged, clarified by various means, and analyzed by chromatographic procedures.[3,4] Glucose is the predominant component identified for a variety of mineral and organic soils.[20] Galactose, mannose, fructose, arabinose, xylose, fucose, rhamnose, and ribose are also reported to be present. Free sugars appear to be somewhat more abundant in cool climatic conditions where decomposition rates are low.

4. Polysaccharides

Carbohydrates in SOM occur mainly as polysaccharides,[23] which may be different from the original plant analogs. Most polysaccharides in soil are believed to be formed by recombination of monomeric units derived from plant and microbial polysaccharides, which produces more complex structures that are less susceptible to further biodegradation.[30] However, in soils rich in undecayed or partially decomposed plant remains, a portion of the polysaccharide fraction will still exist as cellulose, which is the main polysaccharide present in higher plants.

Evidence to support that most soil polysaccharides are products of microbial metabolism is provided by the relatively low contents of xylose and glucose — a particularly abundant sugar in plant polysaccharides — and proportionally higher contents of other monosaccharides identified in hydrolysates of crude soil polysaccharides, which are not found in higher plants but are common constituents of exocellular and endocellular polysaccharides of microorganisms. These include chitin, which is the main structural component of fungal cell walls.[3,4]

In contrast to simple sugars, soil polysaccharides exhibit greater resistance to enzymatic attack due to a combination of several factors. These include size and structural complexity, formation of insoluble salts or chelate complexes with polyvalent cations, adsorption on clay minerals and oxide surfaces, and incorporation into humic substances as integral molecular components bound to the structural core through ester linkages and possibly other covalent bonds.[3,4] In general, the greater the different types of binding and the greater the branching of the polysaccharide structure, the greater is the resistance to enzymatic degradation in soil. This resistance may account for accumulation of polysaccharides in soils relative to other carbohydrates. However, Cheshire et al.[31] and Cheshire[32] suggest that the persistence of polysaccharides in soil may be related to inaccessibility caused by chemical combinations, complexing, and insolubilization, and not to a biologically stable molecular structure. For example, adsorption of polysaccharides by expanding clays in intermicellar and interlayer spaces renders them inaccessible to microbial attack.[33] Evidence has also been obtained that complex reactions with metal cations such as Cu, Fe, and Zn may inhibit enzymatic decomposition of soil polysaccharides.

In order to characterize soil polysaccharides, they must first be extracted from the soil by an extractant that should ideally be: (1) equally effective for all soils, (2) selective for polysaccharides, even if extracting easily removable impurities, (3) nondestructive, and (4) providing a sufficiently complete recovery of the extracted material to be representative of the total.[22] Such an extractant has so far not been found.

In practice, polysaccharides are extracted from soil by: (1) hot water (70 to 100°C) using a Soxhlet procedure;[34] (2) 98% methanoic acid under reflux conditions;[35] (3) dilute mineral acids such as H_2SO_4[36] or HCl;[37,38] (4) dipolar aprotic solvents;[39] (5) organic solvents such as formic acid;[35] (6) aqueous buffers at pH 7;[40] (7) complexing agents such as EDTA, Amberlite resin, urea, or dimethylformamide;[36] and (8) 0.5 M NaOH.[38,41-43] The highest yields with a single extractant have been obtained using hot anhydrous formic acid containing 0.2 N LiBr.[35] A sequential extraction procedure including a double treatment with 1 N HCl and 0.5 N NaOH followed by a third treatment using acetic anhydride in 2.5% H_2SO_4 at 60°C for 2 h, and extracting in chloroform, increased the yield to 80%.[21] Several reviews dealing with the extraction of soil polysaccharides have been provided.[3,4,6,21-23,44]

Subsequent to extraction from the soil, polysaccharides can be recovered by various methods, including dialysis, filtration through activated charcoal, precipitation by addition of acetone, ether, or alcohol, and deproteinization by solvent-solvent extraction.[3,4]

Crude polysaccharide preparations recovered from soil extracts require further fractionation treatment to obtain products suitable for characterizations. Fractionation methods used include (1) fractional precipitation with ethanol and quaternary ammonium compounds[40] or with cetavalon;[35] (2) filtration on Sephadex gels;[36,38,45,46] (3) combined chromatography on charged cellulose and Sephadex;[47] and (4) anion exchange processes, employing resins or anion exchange Sephadex, and anion exchange cellulose preparations.[36,39,46] A procedure based on whole soil methylation for the extraction of soil polysaccharides, and successive characterization by infrared spectroscopy (IR) and gas chromatography-mass spectrometry (GC-MS), has been described by Cheshire et al.[48]

Complete recovery of polysaccharides from soil cannot, however, be achieved even using rather elaborate procedures, thus suggesting that part of this material is covalently bound to soil humic substances and/or irreversibly adsorbed to mineral matter.

The monomer composition of soil polysaccharides, as determined by chromatographic and colorimetric analyses, appears to be similar for all soils examined. The most common sugars identified are the hexoses glucose, galactose, and mannose, the pentoses arabinose, xylose, and ribose, and the deoxyhexoses rhamnose and fucose.[23,35,37,41] Several other sugars, including fructose, traces of methylated sugars, the amino sugars glucosamine, galactosamine, and muramic acid, the sugar alcohols mannitol and inositol, and the uronic acids glucuronic and galacturonic acids have also been identified.[2]

Marked differences in monomer composition of polysaccharides are reported, however, between the light fraction and heavy fraction of soils.[26] The light fraction is generally richer in glucose and xylose than the heavy fraction, possibly due to the predominance of plant debris in the former fraction. The composition of crude polysaccharides isolated from various soils is reported in Table 3.[4]

Table 3   Composition of Crude Polysaccharides Isolated from Various Soils

Soil type	Agricultural soil[a]	Sandy loam[b]	Meadow soil[b]	Spodosol [b]
Extractant	0.5N NaOH	Anhydrous formic acid containing LiBr		
Purification	charcoal filtration	Cetavalon precipitation		
Ash (%)	n.d.[c]	11.7	6.0	5.7
Carbohydrate (%)	80	26.3	37.8	24.7
Nitrogen (%)	0.34	3.16	2.65	3.02
Uronic anhydride (%)	15.8	15.8	17.8	16.3
Amino sugars (%)	0	5.2	5.0	3.1
Monosaccharides [d]				
Glucose	20.8	33.6	37.7	33.6
Galactose	20.0	19.1	20.1	23.1
Mannose	21.9	18.0	18.2	17.9
Arabinose	11.7	7.9	7.6	6.7
Xylose	23.6	7.9	8.2	7.5
Ribose	1.5	—	—	—
Rhamnose	0	13.5	8.2	11.2

[a] Data from Reference 41.
[b] Data from Reference 35.
[c] n.d.; not determined.
[d] Individual monosaccharides are expressed as a percentage of total neutral sugars in the preparation.

Studies with ultracentrifuge, gel filtration, and electrophoretic methods show that polysaccharides, as recovered from soil, are extremely polydisperse. Molecular weights (MW) ranging from less than 4000 Da up to 450,000 Da are estimated using sedimentation, viscosimetric and gel filtration methods.[3,4,19] A continuum of molecular sizes ranging from oligosaccharide sizes to components with MW approaching 200,000 Da are indicated by Sephadex-gel filtration studies.[3,4]

Soil polysaccharides exhibit marked viscosity in aqueous systems, which is related to their size, shape, and molecular structure.[19] Viscosity increases with MW, thus it is used to estimate the MW of soil polysaccharides. Viscosity is also shown to be inversely related to the precipitability of polysaccharides from aqueous media with organic solvents.[49]

Another property of soil polysaccharides in aqueous medium is optical rotation, which has been used as an index of heterogeneity in fractionation studies.[40] Information on structural features and origin of polysaccharides can be obtained from specific optical rotation. Strong dextrorotation is associated with the predominance of α-glycosidic linkages, whereas levorotation indicates predominance of β-glycosidic linkages.[50] Soil polysaccharides are shown to be levorotatory,[23] which suggests a predominance of β-linkages between the sugar units and, thus, the existence of linear helical structures. On the contrary, the polysaccharides would assume more globular or random coil configuration where α-glycosidic linkages predominate. The β-glycosidic linkage configurations would allow more intimate contact than those having α-configurations, with flattened sorbent surfaces, such as those of clays.[51]

Infrared spectra of soil polysaccharides indicate the presence of: (1) hydroxyl groups and C–C and C–O bonds typical of sugar rings and glycosidic linkages; (2) carboxyl groups possibly attributed to uronic acids; and (3) N–H bonds possibly due to amino sugars.[19,52]

Results of titration with bases and electrophoretic mobility studies have revealed the presence of carboxyl groups and the existence of a charge that increases with increasing pH, and arises from the ionization of carboxyl groups possibly related to uronic acids present in soil polysaccharides.[52]

5. Properties and Behavior

Carbohydrates influence soil physical conditions, cation exchange and metal complexing reactions, anion retention, synthesis of humic substances, carbon metabolism, and biological activity in general.

The chemical behavior of carbohydrates, both simple sugars and polysaccharides, in soil is largely dependent on their reactive groups, especially ring hydroxyls, carboxyls of uronic acids, and amino groups of hexosamines. The presence of these groups provides ample possibility for interactions with metal ions and mineral colloids. In the case of metals, both salt formation and complex formation are possible. Carbohydrates may also react with lignin and amino acids in soil, thus contributing to the formation of complex macromolecules such as humic substances.

All such interactions may deeply modify the chemistry and behavior in soil of both the carbohydrate fraction and total SOM, and the mineral fraction. The contribution of polysaccharides to soil cation exchange capacity (CEC) is probably related to the content of uronic acids and is pH-dependent. The solubility and rate of decomposition of polysaccharides in the soil can be dependent on their interactions with metal ions.

Adsorption of mono- and polysaccharides on clay minerals is markedly influenced by the type of clay, pH, and exchangeable cations.[53] This property will enhance the aggregation and structural stability of soil particles, and hence influence a number of related soil physical properties. By interaction with soil clays, polysaccharides may change the adsorption properties for water of the clay surfaces. Adsorption onto clay, especially montmorillonite, and binding to humic substances is shown to retard the normally relatively rapid rates of carbohydrate decomposition in soil.[54]

Polysaccharides are generally strongly hydrophilic, and as such will enhance water holding capacity of soils. However, swelling of soil polysaccharides under wet conditions may contribute to pore-clogging and reduce water infiltration and water permeability in soil.

Carbohydrates are known to prevent the precipitation of phosphate by Fe and Al, depending on pH levels.[55] Further, phosphate fixation by clays is expected to be decreased by coating by polysaccharides. The presence of readily metabolizable carbohydrates in soil may contribute to increased microbial immobilization of nitrogen and, possibly, stimulate denitrification.[19] Alternatively, such materials may lead to accelerated breakdown of stable SOM fractions, thus increasing N mineralization. Free sugars and polysaccharides are known to form stable complexes with borate ions, thus affecting soil boron availability to plants.[19]

Carbohydrates with metal chelating properties can accelerate mineral weathering, with consequent release of nutrients.[56] Further, carbohydrates may have a role in the downward translocation of Fe, Al, and clays.[19] Finally, degradation of organic pesticides and petroleum products spilled on soil can be accelerated by the presence of carbohydrate-rich materials.

B. Nitrogenous Components

1. Content and Origin

Nitrogen occurring in organic forms is estimated to account for over 90% of the total N in the surface layer of most soils, with most of the remainder being accounted for as fixed ammonium in the clay fraction.[3,4,57,58] Nitrogen found in biomass is estimated to be 10% to 15% of total soil N.[59]

A great variety of types and quantities of N-containing organic compounds may reach the soil in the form of the major N constituents of plant, microbial, and animal tissues after death. These include (1) amino acids, proteins, nucleic acids, and phospholipids, which are common to all types of organisms; (2) amino sugars and teichoic acids, which are the main structural components of bacterial cell walls; (3) chitin, found in the cell walls of fungi and green algae, in the exoskeleton of insects and the crustaceans, and in the cuticle of arthropods; (4) vitamins; and (5) several other compounds of minor quantitative importance.[57] Besides tissue constituents of dead organisms, living organisms are known to release to soil various N compounds in plant root exudates, microbial metabolites and wastes, and animal excretions. These include (1) amino acids secreted by higher plant roots and some microorganisms; (2) antibiotics produced in trace amounts by fungi and actinomycetes; (3) amino sugars; and (4) a variety of other substances.

The amount of N in SOM varies with a number of factors which include soil properties, vegetation, climate, and management practices. Organic soils contain much higher concentrations of N than mineral soils, and the N content in subsurface horizons is generally considerably lower than in topsoil layers.[60] The N content of most soils is found to decline but not to change composition when the soil is subjected to intense cultivation.[4]

Among the major nutrient elements, N is unique in that soil reserves are almost entirely in the organic forms. Therefore, the quantity and nature of SOM have a marked influence on the dynamics, chemistry, and biochemistry of this element in the soil, and on its availability to plants and microorganisms.

Background information on nitrogenous components in soil may be found in several reviews and books.[3,4,57,60,61]

2. Isolation and Composition

The distribution of organic N forms in soils has been studied mostly in extracts obtained by refluxing the whole soil with hot 6 M HCl for 12 to 24 h, after which the N forms are separated into various fractions and analyzed by methods described by Bremner[62] and Stevenson.[63] The hydrolysis procedure is not standardized and many variations in conditions have been used. Besides 6 M HCl, H_2SO_4 has been employed, and various acid concentrations have been used. Mixtures of acids,[35] partial hydrolysis,[64] alkaline hydrolysis,[65] and superheated water[66] have also been employed. The most frequently used temperatures of hydrolysis are 100°C or reflux temperatures for times

Table 4   Fractionation of Soil N by Acid Hydrolysis Procedures

Form	Definition/method	% of soil N
Amino acid N	α-amino acid-N, usually determined by ninhydrin-CO_2 or ninhydrin-NH_3 method	30–40
Amino sugar N	Steam distillation with phosphate-borate buffer at pH 11.2	5–10
NH_3 N	Ammonia recovered from hydrolysate by steam distillation with MgO	20–35
Hydrolyzable unknown N	(HUN) Hydrolyzable N not accounted for as amino acids, amino sugars, or NH_3	10–20
Acid insoluble unidentified N	N remaining in soil residue following acid hydrolysis (boiling 6 *N* HCl for 12 h)	20–35

From Kelley, K.R. and Stevenson, F.J., in *Humic Substances in Terrestrial Ecosystems*, Piccolo, A., Ed., Elsevier, Amsterdam, 1996, 407.

usually between 12 and 24 h.[57,62,67,68] Autoclaving soils with 6 *M* HCl at 120°C and 15 lb/in.[2] for 6 h is another alternative.[69] Ratios of acid to soil from 3:1 to 21:1 have been used.[70] Pretreatment with HF has been found to decrease the interference of mineral soil matrix with the hydrolysis.[71,72]

Once the soil extract is obtained, several methods have been applied to determine various groups of N compounds quantitatively and identify individual constituents within each group. These include paper, thin layer, ion exchange, gas-liquid and liquid chromatographies, and electrophoresis.[60]

The main N fractions generally obtained by using these procedures, and their percentages relative to total soil N, are listed in Table 4.[58] Acid insoluble N usually refers to N forms not solubilized by acid hydrolysis. The percentage of the N in acid-insoluble forms has been found to increase with an increase in degree of humification[73] and in degree of decomposition.[74] Part of this N is now believed to occur as a structural component of humic substances.[4] Pretreatment of the soil with HF prior to hydrolysis has been shown to reduce the amount of N recovered as acid-insoluble forms, since HF is able to release, at least partially, amino acids bound to clay minerals and humic substances.[75] Extraction with dilute base and subsequent solubilization by acid hydrolysis also can reduce the acid-insoluble fraction of N to about 15% of the total soil N.[76]

The NH_3-N is the fraction recovered by distillation with MgO. Some of the NH_3 is believed to derive from indigenous clay-fixed NH_4^+, part to be contributed by amino sugars and the amino acid amides, asparagine and glutamine, and part can arise from the hydrolytic breakdown of certain amino acids such as tryptophan, serine, and threonine.[4]

The hydrolyzable unknown (HUN) fraction refers to acid-soluble N not accounted for as NH_3 or amino acid N and amino sugar N. From one fourth to one half of the hydrolyzable unknown N has been suggested to occur as non-α-amino acid N, such as arginine, tryptophan, lysine, hystidine, and proline,[77] whereas a small proportion (1% to 3%) of it is suggested to occur in purines and pyrimidines.[2]

The distribution of the different N forms determined by acid hydrolysis procedures in soils from widely differing climatic conditions is shown in Table 5[2] and Figure 1.[58] Although the total N contents of the soils investigated varies relatively widely, the proportions of the various N forms measured are very similar. Soils from the warmer climates contain percentages of amino acid N and amino sugar N greater than soils from colder regions, whereas the reverse is true for NH_3-N. Unidentified hydrolyzable N and nonhydrolyzable N show similar percentages for all soils, whereas the content of unknown N decreases with increase in temperature.

In conclusion, the main N components identified in SOM are amino acids, peptides, proteins, amino sugars, ammonia-N, purine and pyrimidine bases in nucleic acids and derivatives, N-containing heterocyclic ring structures in chlorophyll and chlorophyll-degradative products, amines, vitamins, and various nitrogenated residues and degradation products of soil applied pesticides.[2-4,57,60,78] Most of these compounds are present in the soil both as free forms and as forms associated to a various extent, and by bonds of various types and strength, to organic and mineral soil constituents, including humic substances, lignin, clays, tannins, polysaccharides, and polyphenols.

Table 5 Nitrogen Distribution in Soils from Widely Differing Climatic Zones (Means and Standard Deviations)

Climatic zone	Total soil N (range)	Total hydrolyzable N	Amino acid N	Percentage of total soil N				
				Hydrolyzable amino sugar N	Ammonia N	Unidentified N	Nonhydrolyzable N	"Unknown" N
Arctic	0.02–0.16	86.1 ± 6.6	33.1 ± 9.3	4.5 ± 1.7	32.0 ± 8.0	16.5	13.9 ± 6.6	46.4
Cool temperate	0.02–1.06	86.5 ± 6.4	35.9 ± 11.5	5.3 ± 2.1	27.5 ± 12.9	17.8	13.5 ± 6.4	45.1
Subtropical	0.03–0.30	84.2 ± 4.9	41.7 ± 6.8	7.4 ± 2.1	18.0 ± 4.0	17.1	15.8 ± 4.9	41.9
Tropical	0.24–1.61	88.9 ± 3.8	40.7 ± 8.0	6.7 ± 1.2	24.0 ± 4.5	17.6	11.1 ± 3.8	40.7

From Schnitzer, M., *Soil Sci.*, 151, 41, 1991. With permission.

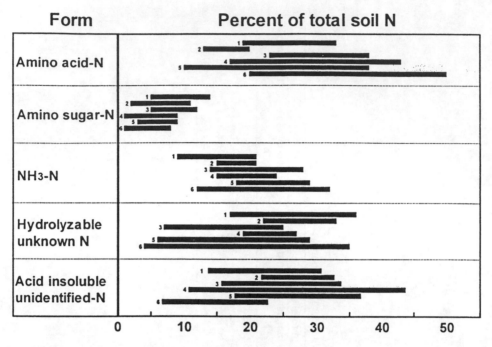

Figure 1 Reported ranges of various N forms as percentages of total soil N recovered in acid hydrolysates of soils from: **1**, Africa (n = 12 soils); **2**, Argentina (n = 4); **3**, Canada (n = 34); **4**, Japan (n = 6); **5**, U.S. (n = 28); **6**, West Indies (n = 7). (From Kelley, K.R. and Stevenson, F.J. in *Humic Substances in Terrestrial Ecosystems,* Piccolo, A., Ed., Elsevier, Amsterdam, 1996, 407. With permission.)

However, approximately one half of soil organic N cannot be adequately characterized and accounted for in known compounds.[2,4] Also, little is still known as to the structural components of soil N complexes and the chemical bonds that link N compounds to soil organic and mineral colloids.

3. Amino Acids

Amino acid N contents in soils range mostly between 30% and 45% of the total N (Table 4), with extremes of 8.9% and 58.5%.[79]

Amino acids mostly occur in soils as single compounds, peptides and proteins bound to clay minerals on external and internal surfaces, and to humic colloids by hydrogen and covalent bonds. Amino acids also exist in soil as mucoproteins, combined with glucosamines, uronic acids and other sugars, as muramic acid-containing mucopeptide, and as teichoic acids, combined with polyols or glycerophosphates.[4]

Bound amino acids are extracted from soil by using hydrolysis procedures described above. Methods for estimating amino acid N in soil hydrolysates are based on the classical ninhydrin-CO_2 or ninhydrin-NH_3 methods,[62,67] and colorimetric analysis.[80,81] A less used method applies nitrous acid as reactant, but this method is less efficient, being subject to several interferences by many compounds.[4]

Small amounts of free amino acids occur in the soil liquid phase. They may originate during the conversion of protein N to NH_3 by heterotrophic organisms, as excretory products of invertebrates, and secreted by microorganisms and plant roots. Free amino acids have only a transient existence in soil, being readily decomposed by microorganisms. Thus, the amount present in soil at a certain time represents a balance between formation and use by microorganisms, with highest levels found where microbial activity is intense.[4] Further, some free amino acids may be held in small voids or micropores inaccessible to microorganisms and difficult to extract.

Free amino acids are extracted from soils using distilled water, 80% ethanol, dilute aqueous solutions of Ba(OH)$_2$ or ammonium acetate, ether and water sequentially, and water in the presence of carbon tetrachloride.[82-84] Water and ethanol have been shown to be rather inefficient in extracting amino acids, whereas the other solvents cited above can extract larger quantities and a greater variety of amino acids. In particular, Ba(OH)$_2$ and ammonium acetate appear to be the most efficient in removing large amounts of amino acids adsorbed on clay particles.[3,4]

Free amino acid level and distribution appear to be fairly dynamic in the soil system and strongly dependent on several factors, including climate, soil moisture, soil temperature and freezing, type and growth stage of plants, addition of organic residues, application of pesticides, treatment with fungicides, cultivation conditions, soil horizon, and land use.[4,60] Free amino acid content of rhizosphere soil is normally many times higher than that for nonrhizosphere soil.[85] Plant roots excrete amino acids into the soil, whose kinds and amounts vary with plant type and maturity, and environmental conditions affecting growth, such as temperature, light intensity, moisture, and nutrient availability.[85]

A variety of chromatographic techniques have been used to identify soil extracted amino acids, including paper partition chromatography[86] and ion-exchange chromatography.[87,88] High-pressure liquid chromatography (HPLC) and GC methods are now available for rapid amino acid analysis.[75,89] Up to 57 amino compounds have been isolated in acid hydrolysates of soils,[90] and up to 30 have been identified.[68] Data on the precise distribution patterns of amino acids isolated from soil are difficult to evaluate because of incomplete extraction, losses, possible improper identification, and other causes of analytical errors. An extreme variability is reported for the amino acid composition of soils and divergent results are recorded for individual amino acids.[4]

Analytical results show that many of the amino acids found in soils are not normal constituents of proteins, and that amino acids present in the cell walls of microorganisms are predominant.[91,92] Several of these compounds may represent waste products of microbial metabolism and products synthesized by microorganisms. Glycine, alanine, aspartic acid, glutamic acid, ornithine, lysine, and diaminopimelic acid are often the dominant amino acids in bacterial cells. Ornithine and β-alanine are constituents of certain antibiotics, and the latter is also a component of the vitamin pantothenic acid. In general, the amino acid composition of soils shows a great similarity to that of bacteria, thus indicating that much of the amino acids present in soil is of microbial origin.[2,4]

4. Amino Sugars

Nitrogen-containing carbohydrates, or amino sugars, can account for from 5% to 10% of N in the surface layer of most soils.[4] Amino sugars are commonly extracted from soil by acid hydrolysis methods described above, and then analyzed by a standard colorimetric method or by alkaline distillation.[62,67] Both methods give similar results.[63] Individual amino sugars have been isolated from soil hydrolysates by paper partition chromatography,[86] ion-exchange chromatography,[93,94] and GLC.[95,96] These studies indicate that the predominant soil amino sugars are D-glucosamine and D-galactosamine, with the former occurring in greatest amounts. Muramic acid, D-manosamine and, possibly, D-fucosamine have also been identified in soil extracts.[94,97]

Amino sugars in soil appear not to occur in the "free" state, but as structural components of mucopolysaccharides, in combination with mucopeptides and mucoproteins, in antibiotics and in chitin, which is an important polymeric component of cell walls, structural membranes, and skeletons of fungal mycelia.[4] On these bases, and because they are rarely found in tissues of higher plants, it is generally assumed that the amino sugars in soil are of microbial origin.

Amino sugar content of soil appears to depend primarily on climate, cultivation, and depth. Subtropical and tropical soils generally contain levels of amino sugars higher than those found in soil of temperate and cold climates.[74] An increase of the content of amino sugar N is also observed when soils are subjected to intensive cultivation,[4] and, in some cases, with increasing depth in the soil profile, reaching a maximum in the B horizon.[98-100]

5. Nucleic Acids and Nitrogen Bases

Nucleic acids occur in the cells of all living organisms and contain N in the form of purine and pyrimidine bases, thus it is not surprising to find these constituents in soils. The N in such bases is generally considered to account for less than 1% of the total soil N.[78]

Studies using chromatographic and spectroscopic techniques on hydrolyzed soil extracts have allowed the identification of all the nucleic acid bases, which include adenine, guanine, cytosine, uracil, thymine, and 5-methylcytosine. Apparently, these N-bases mostly occur as polynucleotides derived primarily from bacterial DNA, with only a minor contribution of higher plant DNA.[101,102]

6. Chlorophyll and Chlorophyll Derivatives

Significant amounts of chlorophyll and its derivatives may reach the soil as plant remains and animal feces.[103-105] Chlorophyll and its derivatives are estimated by measuring the intensity of absorption peaks near 665 nm in 90% aqueous acetone extracts of soil.[4]

The persistence of these materials in soil depends on a variety of soil conditions, including moisture content and pH.[4] Poorly drained soils have been shown to contain larger amounts of chlorophyll-type compounds than well-drained soils, probably due to a deficiency of oxygen, which inhibits chlorophyll degradation.[4] The chlorophyll content of acidic soils has been found to be higher than that in neutral soils.[106,107]

7. Phospholipids, Vitamins, Amines, Antibiotics, and Other Compounds

Phospholipids contain ester phosphate, and may also include N when choline, ethanolamine, and serine are integral components. Small amounts of N occurring in the form of glycerophosphatides, which accounts for only a very small proportion of the total organic N, have been extracted from soil, and identified as phosphatidyl choline and phosphatidyl ethanolamine as the major components.[108-110] The microbial and/or plant origin of these components cannot be determined from the types of phospholipids extracted, although phosphatidyl ethanolamine is known to represent over one third of the lipid material in some bacteria.[4]

Vitamins are important constituents of organisms and thus may easily reach the soil. Several water-soluble, N-containing B vitamins have been detected in soils, including biotin, thiamine, nicotinic acid, pantothenic acid, and cobamide coenzyme.[4] Levels of B vitamins in soil appear directly related to factors influencing microbial activity. These constituents may act as growth factors for numerous organisms, thus the physiological effects they may exert are more important to soil systems than their N content.

A wide variety of amines and other organic N compounds have been detected in trace amounts in soil extracts. These include choline, ethanolamine, trimethylamine, urea, histamine, creatinine, allantoin, cyanuric acid, and α-picoline-γ-carboxylic acid.[4] The formation and preservation of amines in soil appear to be favored by anaerobic or water-logged conditions.[111] Secondary amines can react with nitrous acid to form N-nitrosamines that are carcinogenic and mutagenic at low concentrations. Thus, these compounds may represent a potential health hazard for man and animals when they are formed in soil and leached into water supplies, or taken up by man and animal food plants. Although trace quantities of nitrosamines have been detected in simulated soil experiments, evidence is lacking that the synthesis of these compounds in common soils may represent an actual hazard.[4]

Many N-containing pesticide residues, such as S-triazines, substituted ureas, amides, phenylcarbamates and others, and their partial degradation products are present in cultivated soils, and may persist for years in forms stably associated with humic substances and clay minerals.[112]

Table 6   Some Values for Organic P in Soil

Location	Organic P (μg/g)	Percentage of total P
Australia	40–900	—
Canada	80–710	9–54
Denmark	354	61
England	200–920	22–74
New Zealand	120–1360	30–77
Nigeria	160–1160	—
Scotland	200–920	22–74
United States	4–100	3–52
Tanzania	5–1200	27–90

Adapted from Halstead, R. L. and McKercher, R. B., in *Soil Biochemistry*, Vol. 4, Paul, E. A. and McLaren, A. D., Eds., Marcel Dekker, New York, 1975, 31. With permission.

8. Properties, Reactivity, and Stability

Several N-containing compounds, including amino acids, peptides, and proteins, are able to form biologically resistant complexes in soil by reacting chemically with other organic constituents such as humic substances, lignins, tannins, quinones, poliphenols, and reducing sugars. The products thus formed have been shown to possess a high stability in soil and be highly resistant to microbial attack and mineralization.

Further, adsorption and entrapping of organic N compounds by clay minerals has been shown to enhance their stability and to protect them from microbial decomposition in soil. These phenomena explain the higher N content of fine-textured soils, particularly those rich in montmorillonitic clays, with respect to coarse-textured soils.

Several organic N compounds can also form biologically stable complexes with polyvalent cations, such as Fe and Al. Further, some of the organic N may occur in small pores of soils and, as such, becomes physically inaccessible to microorganisms.

C. Phosphorus Compounds

1. Content and Origin

Organic P forms are estimated to account for from 15% to 80% of the total P in soils. A great variety of types and quantities of P compounds contained in plants, animals, and microorganisms may reach the soil. These include phosphoproteins, sugar phosphates, nucleic acids, phospholipids, vitamins, teichoic acids, and others.[60] Most of these compounds are subject to rapid decomposition in soil and new products are synthesized by microorganisms.

The specific amount of organic P in soils depends on a number of factors, including the type and nature of the soil, vegetation, climate, and management practices.[3,4,61] In most mineral surface soils between one half and two thirds of the total P is organic, whereas P in organic soils is almost entirely organic.[113] The organic P content of soil is shown to concentrate in the silt and, especially, clay fractions, and to decrease with depth in almost the same way as organic C. Some values for total organic P and the percentage of the total P in organic forms are presented in Table 6.[114] The higher percentages of organic P are found in tropical soils, Histosols, and uncultivated forest soils.

The P content of SOM varies from 1.0% to over 3.0%. This variation accounts for the variable organic C:P ratio commonly observed for soils of different type and nature, which has been found to range up to from 46 to 648 for 38 Canadian soils.[115] Factors which influence the proportion of P in SOM include parent material, climate, drainage, cultivation, pH, texture, and soil depth.[4] The

P content of SOM appears to be higher in fine-textured soils than in coarse-textured ones. The C:P ratio is found to narrow with decreasing soil aggregate size.[116] The C:P ratio is lower in the low-MW than in the high-MW components of SOM.[117] The wide C:P ratios measured relative to C:N and C:S ratios may be ascribed to the relatively lower P amount occurring as structural components of humic substances.[4]

Estimates of total soil organic P can be made by direct or indirect methods. The direct method of Anderson and Black[118] has not been widely used, whereas indirect measures are much more common. Two main types of indirect methods are applied for the determination of total organic P in soils, the extraction and the ignition methods.

In extraction methods, total P, namely organic and mineral forms of P, is extracted from the soil with a base, generally diluted NaOH. Pretreatment of the soil with a mineral acid, generally HCl, is necessary for removing polyvalent cations, such as Ca, which render P compounds insoluble. Since this treatment may cause some hydrolysis of organic P compounds, other extractants, such as 8-hydroxyquinoline and EDTA have been used to attempt to eliminate this problem. Organic P in the alkaline extract is then converted to inorganic P by oxidation, and the content of organic P is calculated by subtracting total P from inorganic P determined separately in a dilute acid extract of the original sample.

Ignition methods involve conversion of organic P to inorganic P by ignition of the soil, and calculation of organic P as the difference between inorganic P in acid extracts of ignited and nonignited soil samples.

Several problems and shortcomings are associated with both methods, including: (1) incomplete extraction and/or hydrolysis of organic P during extraction, for the extraction methods; and (2) incomplete conversion of organic to inorganic P, loss of P by volatilization, changes in the solubility of native inorganic P by ignition, incomplete extraction of mineralized organic P, and hydrolysis of labile organic P during extraction of inorganic P, for the ignition methods.[60] Because of these analytical problems, and because organic P is measured by the difference between total and inorganic P, these methods can be regarded as approximate at best, and accuracy is poor especially when the soil is low in organic P.

A more complete discussion on the determination of organic P in soils is available elsewhere.[114,119-121]

2. Composition and Properties

Despite the progress made in determining the nature of organic P compounds in soils, much of the organic P has yet to be identified in known compounds. Less than 50% of the organic P in most soils can be accounted for. Principal known forms of organic P and correspondent approximate recoveries are inositol phosphates, 2% to 50%; phospholipids, 1% to 5%; nucleic acids, 0.2% to 2.5%; phosphoproteins, traces; metabolic phosphate, traces.[4] Part of the unaccounted P has been suggested to be in the form of monophosphorylated carboxylic acids, as teichoic acids and sugar phosphates.[122,123] From 2% to 5% of the organic P in cultivated soils occurs in the biomass.[4] The distribution of the forms of organic P in soils from various geographical areas is listed in Table 7.[122,124]

Difficulty in accounting for the total organic P forms in soils may be due to the occurrence of: (1) chemical association and incorporation of P as structural components of humic substances; and (2) polymeric P-containing compounds such as teichoic acids from bacterial cell walls and phosphorylated polysaccharides.[4]

The distribution of organic P in soil extracts appears to be quite different from that found in organisms, with inositol phosphates being by far the most abundant form. The discrepancy in composition of soil organic P, with respect to organisms, can be explained by the different stabilization achieved in the soil by the various classes of P compounds. As organic P is released into the soil by living and dead organisms, it may be rapidly recycled by the biomass, or stabilized in

Table 7 Distribution of the Forms of Organic P in Soils

	Percentage of organic P		
Source	Inositol phosphates	Nucleic acids	Phospholipids
Australia	0.4–38	—	—
Bangladesh[a]	9–83	0.2–2.3	0.5–7.0
Britain	24–58	0.6–2.4	0.6–0.9
Canada	11–23	—	0.9–2.2
New Zealand	5–43	—	0.7–3.1
Nigeria	23–30	—	—
United States	3–52	0.2–1.8	—

[a] Data from Islam and Ahmed.[124]

From Dalal, R.C., *Adv. Agron.*, 29, 83, 1977. With permission.

the soil matrix. In particular, the high abundance of inositol phosphates in soils, which is different from organisms, may be due to their tendency to strongly adsorb to clay minerals and oxyhydroxides, and to form insoluble complexes with polyvalent cations, such as Fe and Al in acid soils, and Ca in neutral and alkaline soils.[4] Other organic P compounds that are held less strongly by the soil matrix are more available to enter the biological cycle. For example, phosphodiesters are less easily adsorbed, making them less resistant to enzymatic attack in the soil. Such different stabilization determines the organic P composition of most soils.[125]

Detailed reviews are available on the composition and forms of organic P found in soils.[3,4,60,61,114,122,123,125-129]

3. Inositol Phosphates

Inositol phosphates are quantitatively the most significant organic P forms in soils. Absolute amounts of inositol phosphate P in soils can range from a trace amount up to 460 mg kg^{-1}.[114] A range of inositol phosphates from mono- to hexaphosphates is chemically possible. They can also occur as stereoisomers with nine possible forms.[114]

The inositol phosphates are generally extracted from soils with alkali, generally 0.5 *N* NaOH, after pretreatment of the soil with 5% HCl. After oxidation of the extracted organic matter with hypobromide, the inositol phosphates are precipitated as insoluble Fe salts from acid medium, or Ba salts from alkaline medium. Paper partition chromatography, paper electrophoresis, and anion-exchange chromatography are then applied to separate, identify, and quantify the inositol phosphates previously converted from Fe or Ca salts to the corresponding Na salts or free acids.[130,131] Although the hexaphosphate ester is generally the most abundant, lower phosphate esters are also present in percentages that vary widely in soils.

Chromatographic analysis has allowed the isolation and identification of a number of unusual isomers of inositol phosphates occurring in soil. These include *myo-*, *scyllo-*, *neo-*, and D- and L-*chiro*-inositol hexaphosphates, and pentaphosphates of *myo-*, *chiro-* and *scyllo*-inositol.[132-134] By paper partition chromatography, the presence in soils of inositol tri- and tetraphosphate has also been shown.[135]

For many years, the inositol phosphates in soil were thought to be derived from the phytin of higher plants, but results of the above-cited studies and others[131,136] have shown that the isomer composition of inositol phosphates in soil is different from that of either plants or microbes, thus most of them are thought to be synthesized *in situ* by microorganisms.[137,138] These compounds, different from other organic P forms in soil, are mostly stabilized as insoluble complexes with metal ions, such as Ca, and oxides, and as components bound to humic substances, proteins and carbohydrates. Thus, inositol phosphates gradually accumulate in soil to form the largest fraction of soil organic P. However, a significant portion of inositol polyphosphates is suggested to exist in soil as "free" forms, that is, as moieties separate from other SOM constituents.[139]

4. Phospholipids

Relatively small amounts of phospholipids, or phosphatides, are found in soil, usually less than 5 mg kg^{-1}, which account for from 0.5% to 7.0% of the total organic P, with a mean value of 1% (Table 7).

Removal of phospholipids from soil by organic solvents is not easy because these compounds are stabilized by adsorption on clay minerals and incorporation in soil microbial biomass.[60,108] Sequential extraction with ethanol-benzene and methanol-chloroform has been shown to improve recovery of phospholipids from soil.[140]

The presence of glycerophosphate, choline, and ethanolamine in hydrolysates of lipid-solvent (alcohol or ether) extracts of soil has provided evidence for the existence of phospholipids in soil.[110] Phosphatidyl choline has been isolated and identified as the predominant soil phospholipid, followed by phosphatidyl ethanolamine.[109] The phospholipids in soil are believed to be of microbial origin.[4]

The combined hydrophobic-hydrophilic properties of phospholipids can play a significant role in some chemical and biochemical processes in soil, even if these components occur in minute amounts.

5. Nucleic Acids and Derivatives

Although nucleic acids are essential components of all organisms and can be synthesized by soil microorganisms during the decomposition of plant and animal residues, chromatographic and spectrophotometric studies of soil hydrolysates have shown that only a small percentage, up to 3%, of the total soil organic P occurs in the form of nucleic acids and their derivatives.[124,140] Nucleoside diphosphates[141] and bacterial DNA[142,143] have been isolated from soil, and large part of nucleic acids in soil is thought to be of microbial origin.[101]

Nucleic acids are known to be adsorbed on clay minerals and to associate with humic substances, which contributes to difficulties in extraction and uncertainties in measurements.

6. Other Organic P Compounds

Traces of sugar phosphates, phosphoproteins, and glycerophosphates,[144] and small amounts of phosphonates[145] and phosphorylated polysaccharides[146] have been found in soil extracts. Bacterial lipopolysaccharides or bacterial enzyme-induced reaction products have also been suggested as possibly present in soil.[147] A number of phosphate esters, including several monophosphorylated carboxylic acids and two esters containing glycerol have been identified in alkaline extracts of soil.[147] Further, two organic phosphates, one of which is possibly glucose-1-phosphate, have been detected in citric acid extracts of soil.[148]

The chemical nature of the remaining soil organic P, which amounts to over one half of the total, is still unidentified. Probably most of the unidentified organic P occurs in forms tightly associated with clay minerals, humic substances, and other high-MW organic colloids.[137,138,140,149-152]

However, some high-MW forms of organic P are thought not to be connected with other constituents of SOM.[60] Attempts to characterize these forms of organic P in soil extracts have been made by the use of several separation techniques, including anion-exchange chromatography,[131] gel filtration,[152,153] and HPLC.[154]

7. ^{31}P Nuclear Magnetic Resonance

Although present in low amounts in SOM, a great potential exists for the characterization of soil organic P by ^{31}P NMR spectroscopy, based on the high sensitivity and 100% natural abundance of the ^{31}P nucleus.[155] Initial studies by ^{31}P NMR spectroscopy have provided both qualitative[145] and quantitative[156] estimates of the various forms of P in soil alkaline extracts.

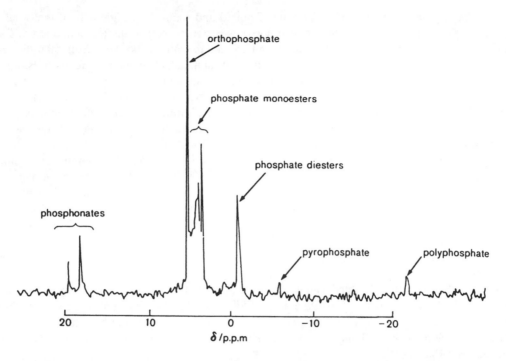

Figure 2   ^{31}P NMR spectrum of an alkali extract of McKerrow soil showing the different forms of P. (From Newman, R. H. and Tate, K. R., *Commun. Soil Sci. Plant Anal.*, 11, 835, 1980. With permission.)

Separate ^{31}P signals from orthophosphate inorganic P, orthophosphate monoester and diester organic P, phosphonate, polyphosphate, and pyrophosphate have been distinguished in alkaline extracts from a range of native grassland soils (Figure 2).[145] Further studies by ^{31}P NMR have confirmed that most of the soil extractable organic P is present as phosphate monoesters, with small amounts of diesters, phosphonates — in which P is bound to C rather than O — and teichoic acids.[157-161]

The ^{31}P NMR spectroscopy has also been used to study the relationships between the chemical forms and dynamics of P in soils and various factors, such as soil type, water regime, climate, vegetation, cultivation and long-term fertilization.[129,162] For example, the predominance of diester forms of organic P under cold, wet climatic conditions,[156] their marked decrease as a result of long-term cultivation,[157,159] and the accumulation of monoester P forms in pasture soils as a consequence of long-term fertilization,[157,158] have confirmed that diester forms of organic P are mineralized in the soil more readily than P in monoester forms.

Direct application of high-resolution, cross-polarization solid-state ^{31}P NMR to whole soil has also been used.[163,164] However, the use of ^{31}P NMR on whole soils is somehow restricted by the low P concentrations of soils and by the presence of paramagnetic metals such as Fe and Mn, which may cause marked signal broadening.[165,166]

Future developments of solid-state ^{31}P NMR spectroscopy is expected to have a considerable potential for application in the analysis, identification, and transformations of organic P forms in soil, because of the typical noninvasive characteristics of this technique, which allows for almost undisturbed observations.

D. Sulfur Compounds

1. Content and Origin

Total S contents in soils vary extremely widely, ranging from 20 to 35000 mg kg^{-1}.[167] However, in mineral soils S contents generally vary in a more restricted range, between 20 and 2000 mg kg[-1].[168] The normal range of total S for soils of humid and semihumid regions is of the order of 100 and 500 mg kg[-1],[61] whereas the organic horizons of forest soils of the temperate regions are estimated to contain between 1000 and 2000 mg kg[-1], and the mineral horizons from 50 to 500 mg kg[-1].[169] Organic soils, tidal marsh and swamp soils, and saline, alkaline and calcareous soils of arid and semiarid regions are known to contain the largest amounts of S.[170]

Both inorganic and organic forms of S occur in soils, with the organic S accounting for more than 90% of the total S in most surface soils of temperate, humid, and semihumid regions of the earth.[168,171-177] Saline and gypsiferous soils represent exceptions in that inorganic sulfate or sulfide can account for a larger proportion of the total S.[170]

The ultimate origin of organic S forms in soil can be ascribed to many S-containing compounds ubiquitous in living organisms which can be released to soil both while they are alive and after death. These compounds include the *S*-amino acids cisteyne and methionine and their derivatives, coenzymes and vitamins, such as coenzyme A, biotin, thiamine and lipoic acid, iron-sulfur proteins, thioredoxins, sulfolipids, and others.[178] Several different biological and chemical processes are involved in the conversion of the S originated from living organisms into soil organic S, with soil microorganisms playing an important role.[170] Microbial biomass S has been estimated to account for from 1% to 3% of the organic S in soil.[179-182]

Organic S is usually calculated as the difference between total soil S measured by dry- or wet-ashing methods, and inorganic S (sulfate and sulfide) extracted from the untreated soil by dilute HCl or $NaHCO_3$.[183] The S content in soil digests or extracts is commonly determined by the methylene blue method.[184] Methods for the determination of the various forms of soil S have been reviewed by Blanchar[183].

The absolute amount of organic S in soils generally follows closely that for organic C, and is related to the same factors affecting SOM content.[4] The mean C:S ratios recorded for the soils from different regions of the world are remarkably similar and close to 100:1, commonly ranging between 60 and 120.[61,168] More detailed investigations have shown that differences in the C:S ratio between soil types can be attributed to several factors that also influence the contents of the total S and organic S fraction in soils. These include physical and chemical properties of soil, such as pH, moisture, parent material, SOM content and mineralogical composition, depth, climate, vegetation, and agricultural management.[4,170]

Trace amounts of volatile organic S compounds may be released from soils in anaerobic conditions, such as poorly drained soils, which favor the microbial decomposition of soil organic S compounds.[4] Volatile S compounds synthesized by soil microorganisms include H_2S, mercaptans, alkyl sulfides, and others.[185,186]

2. Isolation and Composition

Several S-containing compounds are expected to be present in soil in a heterogeneous mixture of relatively free forms and forms combined in complex organic colloids and associated to mineral components. The complex nature of soil organic S is far from being completely elucidated, mainly because of a lack of suitable methods for extracting and separating the major S organic compounds in a quantitative and representative way, thus avoiding alteration of their chemical structure and composition.

Several approaches have been used in extracting and fractionating organic S compounds from whole soil and SOM extracts, to be successively identified and characterized chemically. These include hydrolysis, chemical extraction and fractionation, mild extraction and fractionation, and physical fractionation of organic S from whole soil, and chemical fractionation of S in extracted SOM.

a. Whole Soil Hydrolysis

The amino acids cystine and methionine and their derivatives, such as methionine sulfoxide, methionine sulfone, and cysteic acid, have been detected in whole soil hydrolysates.[88] Amino acid

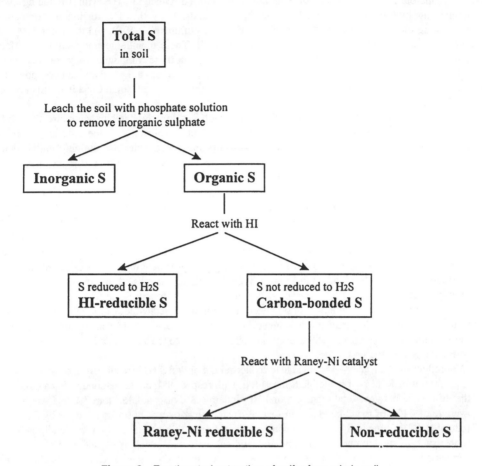

Figure 3   Fractionated extraction of sulfur from whole soil.

S content in such hydrolysates account for 21% to 30% and 11% to 15% of total organic S, respectively, in two Australian soils[187] and in some Scottish soils.[188]

Small and variable amounts of sulfolipids have been found in soils, equivalent to 0.3% to 0.5% and 0.5% to 3.5% of the total organic S, respectively, in ten Iowa soils[189] and in 27 British Columbia soils.[190] Recovered sulfolipids have been further separated into three main fractions by using silicic-acid based column chromatography: polar lipids (7% to 36% of total lipid S), glycolipids (35% to 50%), and less polar lipids (13% to 53%).[191] Sulfolipids in soil are thought to originate from higher plants.[189]

Sulfated polysaccharides have also been identified in very small concentrations in the surface horizons of several soils in Canada.[192] Little is known, however, about the actual structures, and distribution and origin of these S compounds in soil.[60]

b. Chemical Extraction and Fractionation

A procedure commonly used for the fractionated extraction of organic S forms from whole soil is schematized in Figure 3.[4,170] The soil is first leached with a phosphate solution to remove inorganic sulfates, which is followed by a treatment of the leached soil with HI. The S is thus separated in a HI-reducible fraction and a HI-nonreducible fraction. The latter is further treated with Raney-Ni and fractionated into Raney-Ni reducible S and nonreducible S, or residual S.

The chemical nature of the three S fractions obtained can be described as follows:

1. "HI-reducible S" is the fraction of S in soil which can be reduced to H_2S with a reducing agent containing hydriodic, formic, and hypophosphoric acids. Only the S occurring in compounds containing the organic ester and ether sulfate (C-O-S), sulfamate (C-N-S), and the second S* in S-sulfocysteine (-C-S-S*-) linkages can be reduced by HI. These include phenolic sulfates, sulfated polysaccharides, choline sulfate, and sulfated lipids,[175,193] but these compounds may not account for all of this fraction.[60] Evidence has been provided that ester sulfates are the dominant component of the HI-reducible S in soils.[61] Sulfated polysaccharides extracted from some Canadian soils account for less than 2% of the total S.[192]

2. "HI-nonreducible S" comprises all compounds in which S is not HI reducible, being directly bonded to C. This fraction is also termed "carbon-bonded S", and is calculated as the difference between total organic S and HI-reducible S. Included in this fraction are the S-containing amino acids, disulfides, mercaptans, sulfones, and sulfonic acids.[172] This fraction can be further separated into Raney-Ni reducible and nonreducible S.

3. "Raney-Ni reducible S".[194] Results of detailed investigations have shown that Raney-Ni reacts with both reduced C-bonded S forms (R-C-SH) and oxidized S in the form of sulfoxides (R-C-SO-CH$_3$), sulfinic acids (R-C-SO-OH), and sulfonic acids attached to aromatic nuclei (R-C$_6$H$_4$-SO$_2$-OH).[195] Raney-Ni, however, does not reduce the C-bonded S of some organic compounds, including the S combined with humic substances. Further, metals such as Fe and Mn can interfere with the Raney-Ni reduction process and lead to low results.[195] The good agreement obtained between the amounts of Raney-Ni reducible S and amino acid S in some soils has suggested that this fraction consists mainly of amino acid S.[188,196] However, exceptions have been found. For example, the amino acid S have been found to account for only 39% of the Raney-Ni reducible S in humic acid of several Canadian soils.[197] This may be due, however, to the recovery in this fraction of elemental S, thiosulfate, and some mineral sulfides and sulfites, as well as to the above-cited possible interference of Fe and Mn in the procedure.

4. "Nonreducible S". Raney-Ni is not able to reduce all C-bonded S. This results from the considerably less amount of Raney-Ni reducible S measured, with respect to the total C-bonded S, calculated as the difference between total organic S and HI-reducible S. For example, from 7% to 55%, with an average of 44%, of the C-bonded S in several Australian soils was not accounted for by Raney-Ni reducible S.[195] Organic S-containing compounds not reducible by Raney-Ni include aliphatic sulfones, such as methionine sulfone, and aliphatic sulfonic acids, such as cysteic acid.[195] It may be concluded that a chemically unreactive or inert fraction of C-bonded S compounds do exist in soils.[195,198,199]

The average distribution of different fractions of organic S have been determined in several soils from different regions of the world (Table 8). In general, the percentage of HI-reducible S in soils is comparable to that of HI-nonreducible S forms. Soils of humid and temperate regions generally contain between 30% and 70% of total S in the HI-reducible form, although the percentage can range from 18% to 93%.[172] The S fraction in HI-reducible forms appears to be higher in grassland soils than in forest soils where it represents a smaller proportion (10% to 30%) of total S than C-bonded S (40% to 80%).[203-208] In general, HI-reducible S increases with depth in the soil profile, whereas the percentage of C-bonded S decreases concomitantly.[169,175,201,208-210] The SOM in the fine clay fraction shows a greater proportion of HI-reducible S than SOM associated with sand and silt fractions.[211] The proportion of total S in the HI-reducible fraction in some Canadian soils has been found to decrease along a gradient of decreasing temperature and increasing rainfall.[212] Raney-Ni reducible S accounted for 7% to 30% of the total organic S in mineral soils, whereas a much greater percentage (46% to 50%) has been reported for organic soils.[213]

c. Mild Extraction and Fractionation

Extraction of unaltered soil with water or other mild solvents, such as neutral ammonium acetate, allows for the recovery of only trace amounts of S-containing amino acids and their derivatives, including cysteine, cystine, methionine, methionine sulfoxide, taurine, and cysteic acid.[4]

Table 8 Distribution (Range and Mean) of Different Chemical Fractions of Organic S in Soils (Expressed in % of Total S)

Location (no. of soils)	HI-reducible S Range	HI-reducible S Mean	Raney-Ni-reducible S Range	Raney-Ni-reducible S Mean	Nonreducible S Range	Nonreducible S Mean	Ref.
Quebec, Canada (3)	40–65	55	12–32	24	0–43	14	200
Quebec, Canada							201
Chernozemic (8)	47–67	54	12–22	17	7–30	20	
Grey wooded (7)	16–33	24	20–32	26	31–45	41	
Australia (15)	32–63[a]	47[a]	22–54	30	3–31	23	195
Iowa (37)	31–63	50	3–20	11	21–53	37	175
Saskatchewan, Canada (54)	28–54	52	—	—	—	—	173
Iowa (6)	43–60	50	7–18	11	30–39	34	202
Brazil (6)	20–65	40	5–12	7	24–59	42	202
Scotland[b]							198
Acid soils (40)	—	64	—	19	—	17	
Calcareous soils (10)	—	23	—	26	—	51	

[a] Including inorganic sulfate S.
[b] Percentages expressed on the basis of total organic S.

The concentration of relatively free S-containing amino acids may be higher in the rhizosphere soil than nonrhizosphere soil, since plant roots are known to excrete these compounds.[4]

The low and variable amounts of uncombined organic S compounds found in soil depend on their high susceptibility to microbial attack and ease of decomposition. Because of this, organic S compounds cannot persist in the free form when they are produced in the soil, and the low amounts found in soil at any time represent a balance between production and destruction by microorganisms.[4]

Another approach consists of using 0.2 M aqueous acetylacetone at pH 8, in combination with an ultrasonic treatment.[214,215] The organic S extracted by this procedure amounted to 80%–100% of the total organic S in soil, and can be separated by gel permeation chromatography into two fractions of MW lower and higher than 200,000 Da.[215] Results showed that the amount of HI-reducible S increases with increasing MW, reaching a value of more than 75% of the total S in the fraction with a nominal MW >200,000 Da.

d. Physical Fractionation

Physical fractionation of soil organic S has been performed by ultrasonic dispersion of the soil, followed by particle size separation.[211] The distribution and composition of S in the various size fractions obtained show marked differences. Most of the S (about 70%) is recovered in the coarse and fine clay fractions, where most of the organic S (about 70%) occurred in HI-reducible forms. Sand and coarse silt fractions contained more than 75% of S in C-bonded forms, whereas in the fine silt fraction HI-reducible and nonreducible forms were present in comparable amounts.

e. Chemical Fractionation of S in Extracted Humic Substances

Humic substances are believed to contain part of the C-bonded S, especially the fraction not reduced with Raney-Ni, as a structural component mostly resulting from the reaction of thiol groups with quinones and reducing sugars.[4] The existence of such C-S bonds in humic substances may be, at least partly, responsible for soil organic S resistance to microbial attack.

Fractionation of organic S has been attempted in extracted SOM but results from such procedures should be treated with caution, because of the possibility of uncertainty resulting from different extractants and conditions used for extracting SOM and side reactions such as oxidation, bond rupture, and hydrolysis, which are likely to occur during extraction procedures.

Less than half of the total S can be extracted from soils by treatment with $Na_4P_2O_7$, $NaHCO_3$-Na_2CO_3 and Chelex 100 (Na form of the resin).[216] The percentage of HI-reducible S in any of these extracts has been found to differ from that of the original soils, probably because the organic fractions extracted are not fully representative of the whole SOM. Much of the HI-reducible S occurred in high MW compounds (humic acids).

Classical extraction of SOM with 0.5 M NaOH at pH 13.5 under N_2 appeared to cause hydrolysis of HI-reducible S probably in both humic and fulvic acid fractions.[217] A combination of extraction with 0.1 M NaOH–0.1 M $Na_4P_2O_7$ at pH 13 under N_2 and ultrasonic dispersion, sedimentation, and centrifugation produced six organic fractions from some Canadian soils, which contained between 63% and 72% of total soil S.[212,218] In any case, the lighter fulvic acid fractions contained greater proportions of HI-reducible S.

3. Properties and Reactivity

HI-reducible S can be easily hydrolyzed by acids and bases, whereas C-bonded S can resist several reactants.[171,201,216,219] Forms of S not reducible by either HI or Raney-Ni are very stable chemically, resistant to treatment with strong chemical reactants.[200,201] HI-reducible S forms are therefore considered to be the most labile fraction of soil organic S. This fraction is thought to be largely associated with active side chain components of fulvic and humic acids.[176]

A high proportion (75%) of S mineralized in soils subject to long-term cultivation consists of C-bonded forms.[220] These and similar results obtained from other researchers[221-224] suggest that HI-reducible S has a more transitory nature in soil, whereas C-bonded S converts to HI-reducible forms prior to being mineralized.

Long-term soil fertilization with S has been shown to cause accumulation of organic S occurring largely in C-bonded forms,[210,221,224] thus indicating that C-bonded S acts as a reservoir for applied S.

Variations in stable S isotope ratio ($^{34}S/^{32}S$) have provided valuable information on S transformations in soil.[225] It is well known that in many chemical and biological reactions the lighter isotope (^{32}S) reacts preferentially, thus enriching the residue in ^{34}S. Because of this, organic S forms which are readily degraded and have a high turnover rate in soil should result and be enriched in the ^{34}S isotope. The ester sulfate fraction of several surface Canadian soils has been found to be enriched in ^{34}S relative to the C-bonded S fraction.[226] These results thus support the hypothesis that HI-reducible organic S is more labile and readily mineralizable than C-bonded S.

E. Lipids

1. Content and Origin

Lipids represent a convenient analytical group of organic compounds that are chemically different but have the common property of being soluble in various organic solvents, the so-called "lipid solvents," such as benzene, ether, chloroform, methanol, ethanol, acetone, etc., and mixtures of these solvents.

Structurally lipid materials are very diverse, ranging from relatively simple compounds, such as fatty acids and glycerol, to very complex substances, such as fats, waxes, resins, sterols, terpenes, chlorophyll, and polynuclear hydrocarbons.

Lipids appear to be the least studied of the SOM components, probably because they represent only a small percentage of total SOM in mineral soils and their extraction from soil may be complicated by a number of factors.[227-229] Most of the soil lipids have been shown to occur as fats, waxes, and resins, which usually range from 1% to 6% of SOM in most mineral soils.[4,230] However, the lipid content of organic soils may represent 10% to 20% of SOM, and some mineral soils, such as Spodosols, may contain up to 16% of SOM in the lipid fraction. Many lipid constituents, such as low-MW organic acids and sterols are present in soils in extremely low amounts.[4]

The bulk of lipid material in soils is believed to originate from plant and animal inputs, although microorganisms certainly contribute to the total. A variety of lipids, both saponifiable, such as triglycerides and other fatty acid esters, phospholipids and glycolipids, and unsaponifiable, such as long-chained and polynuclear hydrocarbons, terpenoids and sterols, are present in plant residues.[231] Triterpenoid acids, the so-called resin acids, are also common plant constituents and are frequently associated with polysaccharide gums in gum resins. The cuticle waxes of plants may represent a major source of soil waxes and long-chain acids, alcohols and alkanes, which are particularly resistant to microbial decomposition and can survive essentially unchanged over long periods.[4] Steroids and terpenoids occurring in soil are also believed to be of plant origin, whereas glycerides and phosphatides, which can be readily subject to microbial decomposition in soil, probably originate mainly from microorganisms.[4]

Important factors that influence the lipid content of SOM are pH, aerobic or anaerobic conditions, vegetation, and the chemical nature of the lipid material.[4,230] The high proportion of SOM occurring in lipid forms in Spodosols may be ascribed to their typical acidic nature, whereas the low content of fats, waxes, and resins found in prairie soils possibly depends on their higher pH values.[4] A similar relationship between pH and lipid content appears to hold for organic soils. Highly acidic (pH 3.4 to 4.3) Scottish peat soils contain the highest amounts (12% to 20%) of lipids, whereas the lowest amounts (<1% to 2%) occur in Florida peat soils with pH ranging from 7.1 to 7.5.[228] The high lipid content found in SOM of highly acidic soils may result from the larger amounts of lipids synthesized by microorganisms, especially fungi that predominate in these soils, and/or from the inability of microorganisms to decompose completely the lipids of plant residues.[4] Fats and waxes have been shown to accumulate noticeably in waterlogged soils and under anaerobic conditions. The decomposition of sterols has been shown to be inhibited in conditions of extreme acidity, poor aeration, and excess water, which were considered to be responsible for high sterol contents in peat soils.[232] Plant residues differ greatly in lipid content and in their susceptibility to decomposition, which depends on the species and age of the residue. These factors influence the content and distribution of lipid materials in soil.

2. Isolation and Composition

Extraction of lipids from soil is not an easy task because some of these materials may be associated with inorganic soil constituents and other organic components, such as proteins and carbohydrates, thus rendering them insoluble in common lipid solvents.

Lipids have been recovered from soil by using a sequential extraction procedure with ether (fats and waxes) and alcohol (resins). Pretreatment of the soil with an acid mixture of HF and HCl has been used to increase extraction efficiency, after which the lipids have been extracted with a solvent mixture of chloroform-methanol (2:1, v/v).[233] Extraction may also be enhanced by using ultrasonic vibration techniques.

The separation of components from complex lipid mixtures of soil has been obtained by both column and layer adsorption chromatographies with the use of silica gel or other materials, such as neutral alumina, as an adsorbent.[4] Crude lipid extracts have been fractionated on silica gel, eluting successively with n-heptane, carbon tetrachloride, benzene, and methanol.[234] The first three solvents eluted about 20% of the lipid material, which consisted mainly of saturated hydrocarbons in heptane and CCl_4, and of waxes in benzene. About 70% of polar lipids were recovered in methanol eluates, and about 10% were not recovered.

A modified procedure for the separation of components from a complex lipid mixture, and their identification by use of thin-layer chromatography (TLC) and GLC has been proposed by Stevenson,[4] based on the original procedure outlined by Eglinton and Murphy.[235] Briefly, the chloroform extract of soil is shaken with an aqueous NaOH solution that removes the organic acids as sodium salts. These are reconverted into the acid form by acidification, extracted with ether, methylated with diazomethane, and separated by TLC into saturated and unsaturated fatty acids, which are finally identified by GLC. The chloroform layer is fractionated by TLC into three components corresponding to alkanes, aromatic hydrocarbons, and alcohols, which are successively identified by GLC. Alternatively, HPLC may be used, and gas chromatographic separation combined with high resolution mass spectrometry is recommended when appropriate.[4]

Humic substances are known to contain lipid constituents of various classes.[2] Recently, humic acids, fulvic acids, and fine-clay fractions from several agricultural soils have been extracted first with n-hexane, then with chloroform, and finally with supercritical n-pentane, and the extracts obtained have been analyzed by ^{13}C NMR and pyrolysis mass spectrometry.[236,237] The ^{13}C NMR spectra of all extracts were dominated by strong signals due to long-chain and other aliphatic groups. Pyrolysis-field ionization mass spectra showed the presence in the extracts of n-alkanes (C_{17} to C_{101}), n-fatty acids (C_{15}–C_{34}), n-alcohols (C_{18}–C_{35}), diols (C_{16}, C_{24}, C_{31}, C_{32}), sterols (C_{27}–C_{29}), and neutral alkyl monoesters (C_{40}–C_{68}). In addition to the above-listed compounds, pyrolysis-field desorption mass spectra showed the presence of n-alkyl diesters (C_{56}–C_{79}) and n-alkyl triesters (C_{75}–C_{94}). Most of the compounds identified were considered to be indicative of the presence of waxes in SOM. The paraffins identified possibly originated from fungi, bacteria, algae, earthworms, insects, and other small soil animals.[238]

3. Waxes

A significant portion of soil lipids are suggested to be waxes,[4] namely, esters of long-chain fatty acids and higher aliphatic alcohols, which are usually unbranched and saturated, both having a preponderance of even C-numbered atoms, although cyclic alcohols may also be present.

The waxes recovered by elution with benzene represented 5% of the extracted material and were examined in detail by mass spectrometry.[234] The waxes ranged from C_{36} to C_{52}, and even C-numbered (C-even) waxes were predominant (90%) with respect to odd C-numbered (C-odd) homologs (10%).

Gas chromatographic analysis of the acids and alcohols produced by saponification of a neutral wax obtained from an Australian soil allowed the identification of acids ranging from C_{12} to C_{30}, with the C-even predominating.[239] The acids present in greatest amounts were C_{22} (13%), C_{24} (22%), and C_{26} (21%).

Being hydrophobic compounds, waxes are believed to resist biodegradation and have long residence times in soil.

4. Organic Acids

Several organic acids have been isolated from soils and most of them have been identified as high-MW fatty acids.[228] These include α-hydroxystearic, cerotic, triacosanoic, dihydroxystearic, lignoceric, pentacosanoic, crotonic, and other acids.

A purified lipid preparation from a garden soil contained free fatty acids of n-C_{20} to n-C_{34}, with 80% of even-numbered C atoms.[240] Fatty acids were also found to be among the major components of the ether extracts of some forest soils.[241]

Fatty acids, as well as alkanes, have been selectively recovered by supercritical gas extraction of soil with n-pentane.[242,243] The major acids identified were n-fatty acids with n-C_7 to n-C_{29} and lower amounts of branched fatty acids (C_{12}–C_{19}), unsaturated fatty acids (n-C_{16}, n-C_{18}, and n-C_{20}), hydroxy fatty acids (C_{12}–C_{16}), and α-ω-diacids (C_{15}–C_{25}). These were considered to be primarily of microbial origin.

Water extracts from different horizons of a series of forest soils (Ultisols, Entisols, Spodosols) were found to contain several low-MW aliphatic organic acids, including oxalic and formic acids in highest concentrations, and citric, acetic, maleic, aconitic, and succinic acids in trace amounts.[244]

Organic acids are believed to have a transitory existence in aerobic soils, whereas anaerobic conditions would favor their preservation and accumulation, as observed for butyric acid in rice fields.[228]

5. Hydrocarbons

The unsaponifiable fraction of ether, alcohol, and benzene-alcohol extracts of soil are shown to contain several long-chain normal alkanes. A preponderance of alkanes with C-odd to C-even atoms, with main alkanes identified being n-C_{29} (21%), n-C_{31} (31%), and n-C_{33} (15%), has been found in some soils.[240]

In contrast, C-even to C-odd atoms were dominant, with C_{24} and C_{26} being the most abundant, in n-alkanes recovered from the fine-clay fractions of several soils.[237]

The difference observed for the relative abundance of C-even and C-odd alkanes may reflect their different origin in soils and SOM. A predominance of C-odd to C-even atoms is reported for n-alkanes of higher plants, whereas an almost similar amount of C-odd and C-even atoms is found in bacteria and fungi.[245]

A number of simple aromatic hydrocarbons have also been identified in soils, including benzene, toluene, ethylbenzene, o-, p-, and m-xylene, and naphthalene.[246]

6. Polycyclic Hydrocarbons

Small quantities of polycyclic aromatic hydrocarbons have been detected by GC-MS of methanol and methanol-benzene extracts of soils. These include phenanthrene, fluoranthrene, pyrene, chrysene, perylene, benzanthracene, 1,2- and 3,4-benzopyrene, and others.[4]

Although the major sources of polycyclic aromatic hydrocarbons in soils are suggested to be anthropogenic, mostly resulting from combustion, some of them may be of natural origin.[247-249] A significant amount of dehydroabietin, possibly derived from a conifer resin, has been detected in an Alaskan forest soil.[250]

7. Steroids and Terpenoids

Steroids are a group of compounds having molecules composed of three fused cyclohexane rings in the phenanthrene-type arrangement fused to a terminal cyclopentane ring.

These compounds are common constituents of plants and animals, thus they are expected to occur in soils. Two sterols, β-sitosterol, a compound commonly found in higher plants, and β-sitostanol have been found in soil (Figure 4).[229,230]

Sterols and several triterpenoids, including taraxerol, friedelin, α-amyrin, and taraxerone, have been detected in the benzene extracts of several mineral soils (Figure 4).[234]

8. Carotenoids and Porphyrines

Carotenoids represent the orange pigment of many plants and are associated with chlorophyll. About 20 different carotenoids have been detected in wet sediments and they are undoubtedly present in soils. However, their occurrence in agricultural soils has not yet been demonstrated experimentally, probably because these compounds are readily oxidized and decomposed in aerobic conditions.[4]

Porphyrines are generally considered as being derived from chlorophyll. The occurrence of pheophytins and other chlorophyll derivatives in soil and their distribution patterns as a function of soil type have been reported.[251]

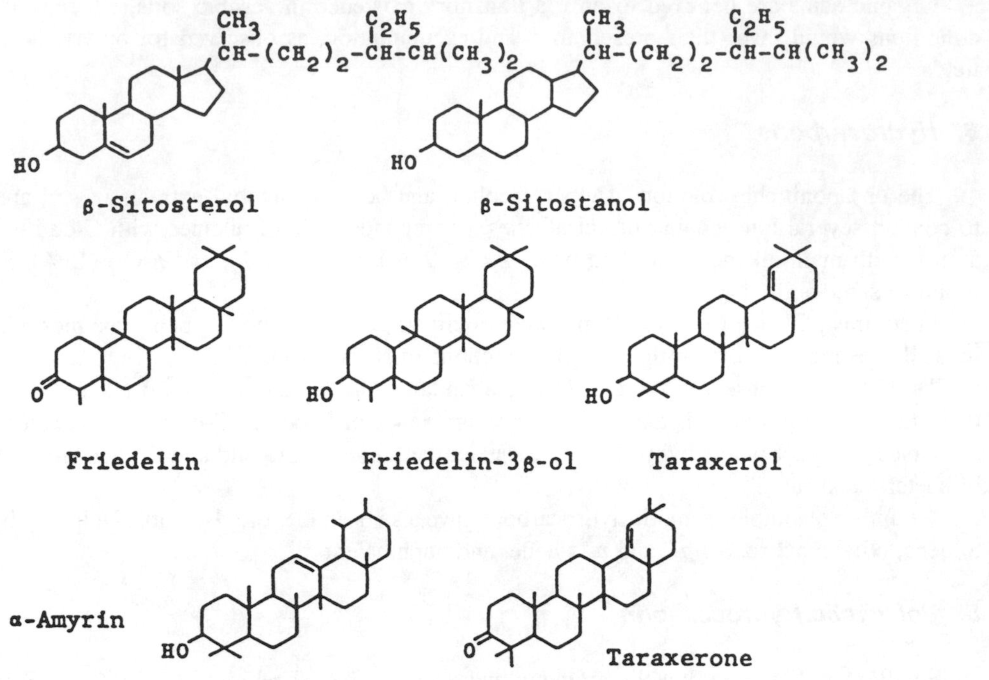

Figure 4 Sterols and terpenoids detected in soils.

The chlorophyll content of woodland soils was found to be an order of magnitude lower than that of freshwater sediments, possibly because oxygen deficiency inhibits chlorophyll decomposition.[107] It is thus expected that poorly drained soils might contain amounts of chlorophyll relatively higher than those of well-drained, aerated soils.

9. Other Compounds

The occurrence and nature of phospholipids in soils have been discussed previously in this chapter. Glycerol, the key component of all fats, has been detected in saponified soil extracts.[4]

A number of long-chain methyl ketones were isolated from soil and analyzed by GLC, showing that the components ranged from n-C_{19} to n-C_{35}, with 81% odd-C numbers.[230]

Many other compounds of lipid nature may be present in soils in very small amounts, including aliphatic alcohols,[252] phenolic acids,[253] polyphenols, and complex quinones.[254]

The occurrence of natural phthalates in soils has been reported,[255,256] although the possibility of anthropogenic sources (plastic materials) cannot be excluded.

10. Properties and Functions

Many compounds belonging to the soil lipid classes are physiologically active, showing either a depressing effect or a promoting effect on plant growth.

Ethylene, methane, and other volatile organic compounds may be produced in soils by organic residue decomposition under anaerobic conditions, such as those existing in the organic horizons of forest soils.[257-260] Although evolved in very low amounts, these compounds may reach toxic concentrations and adversely affect plant growth.

The concentration of phenolic acids in some English soils was shown to be sufficiently high to adversely affect plant growth.[261] Many agricultural soils of Taiwan were found to contain phytotoxic substances of lipid nature.[233,252,253]

Allelophatic effects in higher plants have been shown to be caused by allelophatic materials of lipid nature released into the soil by root excretion, leaf washing, and microbial decomposition of plant residues. A wide variety of phytotoxic compounds have been isolated from soil, including several aldheydes, acids, ketones, flavonoids, terpenoids, steroids, alkaloids, coumarins, glycosides, and organic cyanides.[262-264]

Although many lipids are phytotoxic, in most cultivated soils they usually undergo rapid microbial decomposition, thus losing their toxicity to higher plants. However, accumulations are possible under conditions of slow and/or incomplete decomposition, such as in poorly aerated and heavy-clay soils.[4]

Several growth-promoting substances in soil are of lipid nature as well. Vitamins and other growth factors, such as those derived from decaying lignin,[265] are common constituents of soils, being released from organic residues, plant roots, and microorganisms.

The effectiveness of physiologically active compounds in soil appears to depend on several factors, which include temperature, light, water, aeration, and nutrient availability.[265]

Most soil lipids, especially waxes, are hydrophobic and water repellent substances. Thus, their presence may alter some soil physical properties, such as wetting capability and aggregate stability. Some paraffins have been suggested to exert positive effects on the structural stability of soils by forming continous films over the surfaces of soil aggregates.[266]

III. HUMIC SUBSTANCES

Humic substances are the most widespread and ubiquitous natural nonliving organic materials in all terrestrial and aquatic environments, and represent a significant proportion of total organic C in the global C cycle. They constitute the major fraction of SOM (up to 80%) and the largest fraction of natural organic matter (NOM) in aquatic systems (up to 60% of dissolved organic C).[2-4,267]

Soil HS comprise a physically and chemically heterogeneous mixture of naturally occurring, biogenic, relatively high-MW, yellow-to-black colored, amorphous, colloidal, polydispersed organic polyelectrolytes of mixed aliphatic and aromatic nature, formed by secondary synthesis reactions (humification) during the decay process and transformation of biomolecules that originate from dead organisms and microbial activity (Table 1). These materials are distinctive of the soil system and exclusive of undecayed plant and animal tissues, their partial decomposition products, and the soil biomass.[3]

Similar to nonhumic substances, HS in soil are present predominantly in forms associated with mineral colloids, charge-neutralizing cations, and other organic compounds. Thus, an extraction step that allows separation of HS from soil components, followed by a fractionation step that decreases heterogeneity of bulk HS, must be achieved before detailed structural and chemical studies can be performed.

Based on solubility in acids and alkalis, HS can be divided into several fractions: (1) humic acid (HA), the portion that is soluble in dilute alkaline solution and is precipitated upon acidification to pH 2; (2) fulvic acid (FA), the portion that is soluble at any pH value, even below 2; (3) humin, the portion insoluble in both alkalis and acids; and (4) hymatomelanic acid, the alcohol-soluble portion of HA (Table 1). Further separations in more defined analytical fractions are reported,[5] but these are not of general value and/or application. In any case these terms should be regarded as operational terms useful for distinguishing between different analytically separated fractions of HS, and not as referring to definite, chemically identifiable compounds.

Humic substances consist of a chemically heterogeneous mixture of compounds, thus they cannot be regarded as single chemical entities described by unique, chemically defined molecular structures. Although it is virtually impossible to describe uniquely the molecular formula of HA, FA, and other HS fractions, it is possible to depict the general structure of a "typical" molecule of HA and FA on the basis of available compositional, structural, functional, and behavioral data. The model structures thus constructed contain the basic structural moieties and types of functional groups that are common to all the single, indefinitely variable and unknown molecules of HS. In Figure 5, two typical model structures proposed for soil HA and FA are presented.[3,268] The macromolecular structure consists of aromatic, phenolic, quinonic, and heterocyclic "building blocks" that are randomly condensed or linked by aliphatic, oxygen, nitrogen, and sulfur bridges. The macromolecule bears aliphatic, glucidic, amino acidic, and lipidic surface chains as well as chemically active functional groups of various nature (mainly carboxylic, but also phenolic and alcoholic hydroxyls, carbonyls, etc.), which render the humic polymer acidic. The models feature both hydrophilic and hydrophobic sites, a highly polyelectrolytic character, and several sites and functional groups potentially able to bind with metal ions, mineral surfaces, and other organic compounds. The structure and composition of HA are more complex than those of FA (more details are provided in the following sections). Although these "type" structures, and several others proposed[269] may explain acceptably several chemical and functional properties of HA and FA, none of them can be considered complementary since each emphasizes certain particular properties but neglects others.

Recently, a state-of-the-art two-dimensional (2D) structural model for HS has been proposed based on an integrated approach using previously published, comprehensive results of geochemical, wet-chemical, biochemical, spectroscopic, agricultural, and ecological studies, combined with data obtained from extensive investigations by analytical pyrolysis methods, ^{13}C NMR, chemical, oxidative, and reductive degradation, and electron microscopy on soil HS (Figure 6).[270,271] The proposed structure has been assumed to possibly trap within its voids approximately 10% carbohydrates and 10% proteinaceous materials, thus increasing the contents of O and N relative to C.[15,272] Successively, the use of a dedicated software (Hyperchem®) has allowed the production of a computer-assisted design of a geometry-optimized and energy-minimized three-dimensional (3D) structure of the 2D model of HA (Figure 7).[15]

The ultimate objective in the study of soil HS is to relate structural and chemical information obtained to the biogenesis, roles, and functions of HS in the soil environment. Humic substances, in fact, contribute substantially in improving the global soil fertility status by exerting, besides several general fertility functions that they possess in common to other SOM and soil components, a number of functions that are specific and typical of humified SOM. These include, among others, slow release of nutrients such as N, P, and S, high CEC, pH buffer capacity, specific physiological effects on plant growth, and an extended capacity of interactions with micronutritive and/or microtoxic metal ions and xenobiotic organic molecules such as pesticides.

A. Genesis

Several pathways have been proposed for the formation of soil HS, the major ones being schematized in Figure 8.[3,4] The classical theory of Waksman,[273] the so-called "lignin-protein theory," considers plant lignin as the main source of soil HS, with the involvement of amino compounds produced by microbial synthesis (pathway 4 in Figure 8). In pathway 1 (Figure 8), reducing sugars and amino acids formed as by-products of microbial metabolism are assumed to be the only precursors of HS. The current concepts of HS genesis mostly favor a mechanism, the so-called "polyphenol theory," which involves polyphenols and quinones, either derived from lignin (pathway 3 in Figure 8), or synthesized by microorganisms (pathway 2 in Figure 8).

In practice, all four pathways may operate for the synthesis of HS in all soils, but not to the same extent, one pathway usually being prominent. For example, lignin may represent the predominant precursor of HS formed in poorly drained and wet soils. Polyphenols synthesized by microorganisms may be predominant in HS formation in forest soils. Sugars and amines may be important precursors for HS synthesis in soils where frequent and sharp fluctuations in continental climate factors occur.[3,4]

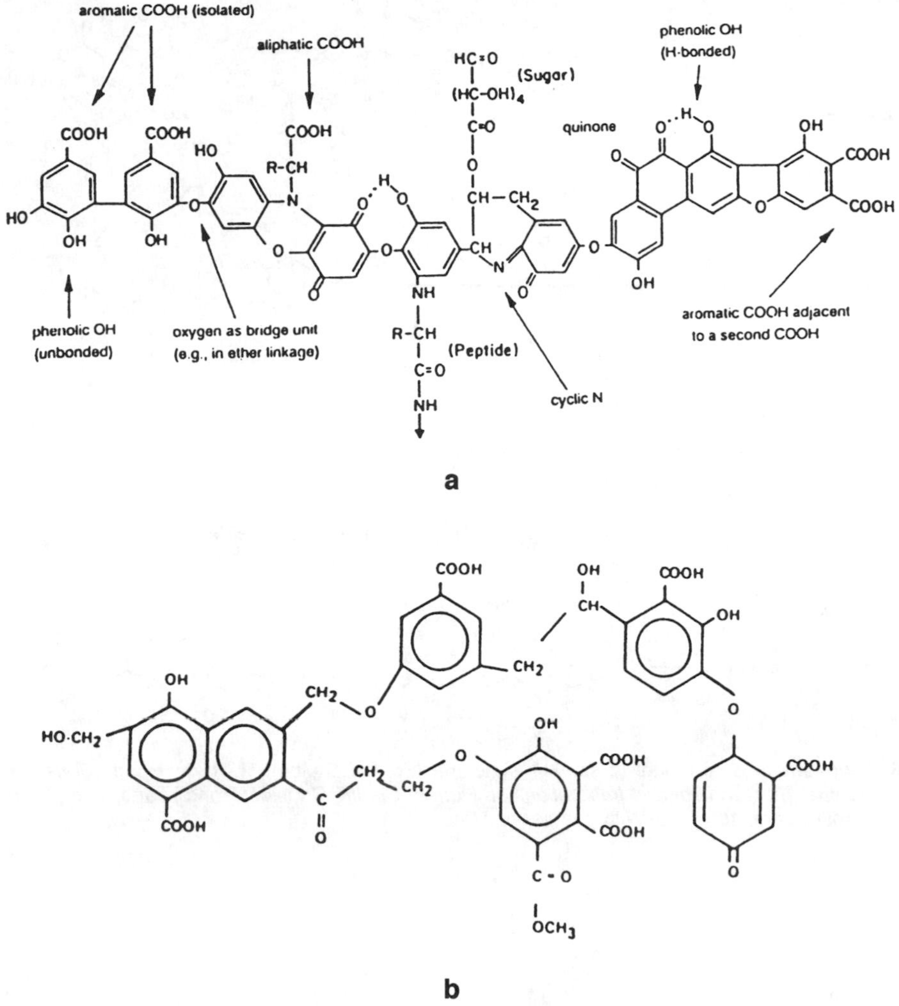

Figure 5 Model structures proposed for soil HA (a) and soil FA (b). (Figure 5(a) from Stevenson, F. J., *Humus Chemistry. Genesis, Composition, Reactions,* John Wiley & Sons, New York, 1982. With permission.) (Figure 5(b) from Langford, C. H. et al., in *Aquatic and Terrestrial Humic Materials,* Christman, R. F. and Gjessing, E. T., Eds., Ann Arbor Science, Ann Arbor, MI, 1998, 219. With permission.)

Further, in any given soil, not all HS components may form by the same mechanism. For example, in some soils HA may originate from plant-lignin and microbially synthesized polyphenols, whereas FA may arise preferentially from the sugar-amine condensation mechanism.[3,4]

However, a completely satisfactory mechanism for explaining the chemistry and biochemistry of HS formation in soil has not yet been developed, and represents one of the most challenging aspects of HS studies. Detailed information on the research conducted and information available on the subject can be found in several excellent reviews.[9,274-278]

1. The Lignin-Protein Theory

Lignins are constituted of phenylpropane units (C_6-C_3) including coniferyl alcohol, *p*-hydroxycinnamyl alcohol, and sinapyl alcohol, the proportion of each being dependent on the type of plant. For example, the sinapyl alcohol unit is contained in relatively high amount in the lignin of conifers, whereas in the lignin of herbaceous plants the three-alcohol units are all well represented.

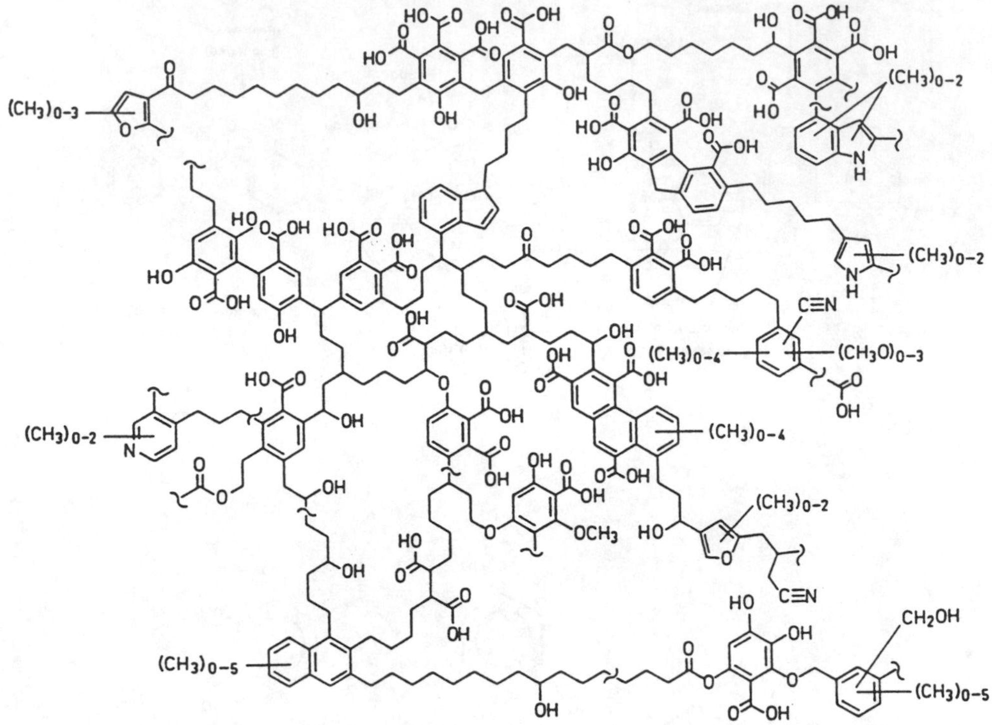

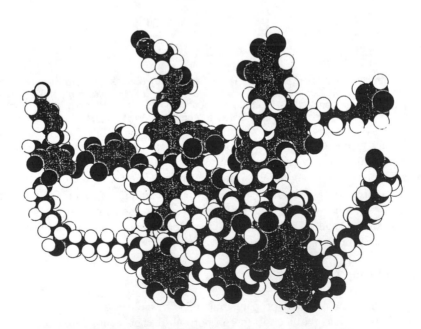

Figure 6 Two-dimensional model of soil HA structure. (From Schulten, H.-R., in *Humic Substances in the Global Environment and Implications on Human Health,* Senesi, N. and Miano, T. M., Eds., Elsevier, Amsterdam, 1994, 43. With permission.)

Figure 7 Geometry-optimized and energy-minimized three-dimensional structure of soil HA. (From Schulten, H.-R., The three-dimensional structure of humic substances and soil organic matter studied by computational analytical chemistry, *Fresenius J. Anal. Chem.,* 351, 62, 1995. With permission.)

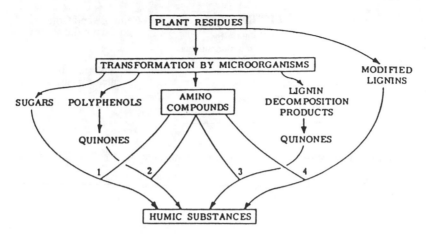

Figure 8   Major pathways proposed for the formation of soil humic substances. (From Stevenson, F. J., *Humus Chemistry. Genesis, Composition, Reactions,* 2nd ed., John Wiley & Sons, New York, 1994. With permission.)

For many years it has been thought that lignin was the main source of HS in soil.[273] According to this theory, lignin is incompletely utilized by soil microorganisms, and can undergo a preliminary series of modifications, including loss of methoxyl (OCH_3) groups, generation of *o*-hydroxyphenols, and oxidation of terminal aliphatic side chains to form carboxylic (COOH) groups.[232,276,279] Successively, the *o*-dihydroxybenzene units resulting from demethylation of lignin would further oxidize to quinones capable of undergoing condensation reactions with amino compounds and NH_3 would be produced by microorganisms during the decay of N-containing organic substances. This process would yield first humin, then HA, and finally FA.

Apparently, lignin can be demethylated without further degradation by a large number of bacteria.[275] Demethylated and oxidized lignins contain lower C and higher O contents than the original lignins, and may be further enriched in COOH groups arising from aromatic ring cleavage.

A modified Waksman's lignin-protein theory has been proposed, which is extended to include refractory macromolecules other than lignin.[277] In this model, besides lignin, additional resistant biopolymers originated from plants, such as cutin and suberin, and microorganisms, such as melanins and paraffinic macromolecules, are considered as possible precursors of HS. The refractory biopolymers would combine to form the humin fraction of HS. Increasing degradation would lead to macromolecules enriched in O-containing functional groups that would increase their solubility in aqueous alkali and evolve, first, to HA and then to FA.

Although lignin is more resistant to microbial attack than other plant constituents, in common aerobic soils lignin may be completely decomposed into low-MW compounds susceptible to HS synthesis. Since oxygen is required for the complete microbial breakdown of lignin, it is feasible that in poorly drained soils modified lignins will contribute substantially to HS formation.[3,4] Recently, the importance of lignin as a precursor of HS has been emphasized.[278]

2. The Sugar-Amine Theory

Nonenzymatic, purely chemical condensation of reducing sugars and amino compounds formed as by-products of microbial metabolism, and further polymerization reactions, which are known to form typical brown nitrogenous polymers, have been postulated to play an important role in the formation of HS in soil.[280-282]

The initial reaction of this pathway involves the condensation of an amino group, e.g., of amino acids, with the aldehyde group of a sugar to form a Schiff-base and n-substituted glycosylamine. Successive rearrangements and reactions may produce fragmentation and loss of water molecules

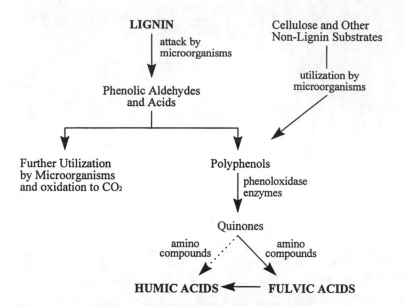

Figure 9   Schematic representation of polyphenol theory of humic substance formation. (From Stevenson, F. J., *Humus Chemistry. Genesis, Composition, Reactions*, 2nd ed., John Wiley & Sons, New York, 1994. With permission.)

to form 3-C chain aldehydes and ketones, such as acetol, glyceraldehyde and dihydroxyacetone, reductones, and hydroxymethylfurfural.[283] In addition, aromatic and furanoid molecules may also form from carbohydrates.[284] All these compounds are highly reactive and can readily polymerize in the presence of amino compounds to form brown-colored products similar to HS.

According to this theory, HS may be formed by purely chemical reactions in which microorganisms do not play a direct role, except to produce sugars from carbohydrates and amino acids from proteins. Although these reactants are produced abundantly by soil microorganisms, the condensation reaction apparently proceeds rather slowly at common soil temperatures, and strong competition exists from microorganisms for the use of these compounds. However, in soils where lignins are not prevalent, and/or drastic and frequent changes, such as wetting and drying, freezing and thawing, occur in the soil environment, this pathway, possibly facilitated by the catalytic action of soil minerals, may be realistic for HS synthesis.[4]

3. The Polyphenol Theory

Nowadays, the most accepted view is that the major building blocks of HS originate from polyphenols of lignin origin or synthesized by microorganisms (Figure 9).[4] Thus, HS are believed to form in soil prevalently through a pathway in which the first step consists of the breakdown of all plant biopolymers, including lignin, into their monomeric structural units. Possible sources of phenols utilized in the HS formation include plant lignin, glycosides and tannins, and microbial synthesis.

According to the hypothesis of Flaig et al.,[279] after lignin is freed from its linkage with cellulose during decomposition of plant residues, the side chains of its building units are oxidized and demethylated, yielding polyphenols that are converted to quinones by polyphenoloxidase enzymes. Quinones that originate from lignin, and possibly from other sources, then react with N-containing compounds and polymerize to produce humic macromolecules of increasing complexity. Thus, the order of formation of HS would be: FA → HA → humin.

Fungi are considered to be the most important group of microorganisms responsible for cleavage of lignin.[274] The initial step in lignin degradation by fungi would involve release of the dilignol components guaiacylglycerol-β-coniferyl ether, pinoresinol, and dehydrodiconyferyl alcohol, and the formation of primary phenylpropane (C_6–C_3) units. Among phenolic aldehydes and acids released by further decomposition of dilignols are, for example, guaiacylglycerol, coniferyl alcohol, coniferaldehyde, and ferulic acid. The C_6–C_3 units also undergo oxidation in the side chains to yield a variety of low-MW aromatic acids and aldehydes, including vanillin, vanillic acid, syringaldehyde, syringic acid, p-hydroxybenzaldehyde, p-hydroxybenzoic acid, protocatechnic acid, and gallic acid. Additional OH groups may be introduced and decarboxylation may also occur during this stage, whereas the fungi involved appear to have a limited capacity for aromatic ring cleavage of lignin.[274]

The role of microorganisms as sources of polyphenols of nonlignin origin, e.g., from cellulose and other carbohydrates, has been emphasized in early work by Kononova.[9] The observation that HS can accumulate in soils without lignin being present provides evidence for the formation of HS without the participation of lignin. For example, in cold, wet regions HS are apparently formed from lower plants that do not contain lignin, e.g., the mosses. Actinomycetes and fungi are known to be able to degrade cellulose and other organic plant constituents, besides lignin, and to form numerous phenolic and hydroxy aromatic acids, i.e., melanins.[274,285]

Phenolic components typical of both lignins, such as syringyl and guaiacyl derivatives, and microorganisms, such as flavonoids, have been identified in products of oxidative and reductive cleavage of HA, which provides indirect evidence that both systems contribute to HS formation.[4] Soil properties and conditions may determine the relative importance of lignins and microorganisms as sources of phenolic materials for HS synthesis.

Phenolic products are not stable in soil, and may be subject to oxidative conversion to quinones, which may occur either chemically, e.g., in alkaline media, or more likely operated by polyphenoloxidase enzymes. The consequent, prevalently enzymatic, oxidative polymerization of mono-, di-, and trihydroxyphenols, quinones, and aromatic acids may occur in the absence or in the presence of amino acids, peptides, proteins, and amino sugars, which may undergo condensation reactions with quinones, and thus be covalently incorporated into humic macromolecules. Oxidative polymerization also involves several other side reactions, including demethylation, oxidation of aldehyde components to carboxylic acids, decarboxylation, hydroxylation, and coupling of various intermediates.

In conclusion, HS in soil may be formed by all the mechanisms mentioned above, although most researchers favor the polyphenol theory in which the starting material consists of low-MW organic compounds from which humic macromolecules originate mainly through oxidation, condensation, and polymerization reactions. The number of possible single precursor molecules and the types of chemical reactions involved is very large, and the number of possible combinations is extremely high, thereby accounting, on one side, for the infinite chemical variety of single humic macromolecules that can be obtained, and, on the other side, for the fundamental similarity in their structural building blocks.

B. Extraction, Fractionation, and Purification Procedures

In the soil environment, HS are part of a complex system in which they interact with metal ions, mineral colloids, and nonhumified organic materials. Thus, the structural, chemical and functional properties of HS can be studied in detail only when they are isolated from the soil in the free state. As a consequence, the first objective is to isolate HS from the inorganic matrix, and separate HS from nonhumic compounds. Subsequently, a fractionation step is necessary in order to decrease the physical and chemical heterogeneity of the bulk HS obtained. Finally, the crude fractions should be subjected to a purification procedure to eliminate impurities and remove coextracted materials.

The choice of the extractant and the fractionation procedure for HS is limited by the complex heterogeneity of the material and the rigorous demands placed upon the extractant. Further, this choice is complicated by the existence of no clear-cut distinctions between the physicochemical properties of the molecules within the HS gross material. Although HS have similar origin, composition, and structure, they are a family of macromolecular compounds exhibiting a wide range of values for any given molecular property. It is thus pointless to try to design solvent systems that are able to isolate relatively homogeneous HS from soils. The general approach is then to use a solvent able to isolate the maximum amount of the less altered and contaminated HS, and then to rely on fractionation procedures to decrease the heterogeneity of the system as much as possible by subdividing the HS according to some property related to their molecular composition.

An ideal extractant should ensure universal applicability for the complete isolation of an unaltered and uncontaminated material. Further, in order that the structural and chemical information obtained for HS truly reflects the processes in natural systems, it is of great importance that the compounds fractionated are representative of those in the environment. Unfortunately, the isolation process itself involves unavoidable alterations in the chemistry of HS, such as disruption of interactions with metal ions, and coextraction of tightly bound inorganic and nonhumified organic materials. On the other hand, studies of metal-HS complexes require the extraction of the complexes intact, but removal of the metals is generally a declared objective of the isolation procedure.

Presently, there is no universally applicable and accepted method for the extraction, fractionation, and purification of HS. Tremendous efforts are being made by the International Humic Substances Society (IHSS) to encourage the adoption of a standardized, unique, comprehensive procedure for the isolation of HS from soils, sediments, and waters.

In the following sections the most reliable and widely used methods for the extraction, fractionation, and purification of soil HS will be briefly discussed, giving special emphasis to the limitations and unresolved problems encountered in these procedures. These include production of artifacts, coextraction and removal of contaminants, decrease in sample heterogeneity, effectiveness, representativeness, and reproducibility. A number of useful detailed reviews on this subject have been published.[3-5,286-288]

1. Extraction Methods

Any attempt to design an effective procedure for the isolation of HS from soil should take due account of the properties of the extractants to be used, of the materials to be extracted, and of the types of associations that exist between these materials and other soil constituents.[286]

Four general criteria for solvents for HS have been proposed:[289] (1) a high polarity and high dielectric constant to assist in the dispersion of the charged molecules; (2) a small molecular size to penetrate into the humic structures; (3) the ability to disrupt the existing hydrogen bonds and to provide alternative groups to form humic-solvent hydrogen bonds; and (4) the ability to immobilize metallic cations.

According to Stevenson,[3,4] the ideal extraction method would meet the following objectives: (1) the method leads to the isolation of unaltered material; (2) the extracted HS are free of inorganic contaminants, such as clay and polyvalent cations. (3) extraction is complete, thereby ensuring representation of fractions from the entire molecular weight range; and (4) the method is universally applicable to all soils.

The objectives listed above have yet to be realized. In addition there is the troublesome problem of removing coadsorbed organic impurities from "true" HS.

In Table 9, a number of solvents and reagents used for the extraction of HS from soil are listed, together with the total yield obtained.[3,4,290]

a. Alkaline Solvents

Alkaline solvents were the earliest reagents used for extracting HS, and remain the most efficient and widely used, although with some modifications. The data in Table 9 show that the alkaline

Table 9   Solvents and Reagents Used for Extraction of Soil Humic Substances and Yields of Extraction

Extractant	Yield
Strong bases	
NaOH	to 80%
Na_2CO_3	to 30%
Neutral salts	
$Na_4P_2O_7$, NaF	to 30%
Organic acid salts	to 30%
Organic chelates	
Acetylacetone	to 30%
Cupferron	
8-quinolinol (8-hydroxyquinoline)	
Na-EDTA 1 mol l^{-1}	16%
EDA 2.5 mol l^{-1} (pH 12.6)	63%
EDA (anhydrous)	5%
Organic solvents	
Pyridine	36%
DMF	18%
Sulfolane	22%
DMSO	23%
Formic acid	to 55%
Acetone-H_2O-HCl solvent	to 20%

Data from Stevenson[3] and Hayes et al.[290]

solvents, 0.5 or 0.1 mol l^{-1} NaOH, and 2.5 mol l^{-1} EDA in water at pH 12.6, are able to extract more HS than do the other solvents. A soil to extractant ratio from 1:2 to 1:5 (g ml^{-1}) is commonly used and repeated extraction is required to obtain maximum recovery. Pretreatment of the soil sample with dilute HCl or HF-HCl, which remove Ca and other polyvalent cations, increases the efficiency of alkaline extraction. However, a certain amount, generally less than 5%, of low-MW humic macromolecules are removed in the pretreatment process.

Alkaline extraction of HS presents some undesirable features, including contamination by silica dissolved from the mineral components, coextraction of nonhumic organic constituents, uptake of oxygen (autooxidation), and other chemical modifications, including some breakdown of humic macromolecules and amino-carbonyl condensation (browning reactions), leading to production of artifacts.

The more alkaline the solution and the longer the extraction period the greater will be the extraction efficiency but also the chemical changes produced in extracted HS.[291] For example, O_2 uptake during extraction sensibly increases with pH of extractant.[291] However, damage through oxidation can be minimized when alkaline extraction is carried out in the presence of an inert gas, usually N_2.

Comparison of elemental analysis data for the two principal HS fractions, HA and FA extracted with 0.5 mol l^{-1} NaOH and 2.5 mol l^{-1} EDA show that the C and N contents of the EDA-extracted HS are significantly higher than for the NaOH extracts, while the reverse is true for the O content (Table 10).[290] The enrichment in N of HS extracted in EDA suggests the occurrence of a condensation reaction between the amino groups of the solvent and the carbonyl group of humic macromolecules to form Schiff base structures. This reaction is likely to occur also for any other primary amine solvent (Table 10), thus suggesting that these solvents are not appropriate for extraction of HS.

b. Aqueous Salt Solutions

Sodium and potassium salts of inorganic and organic acids and complexing agents have been used as milder and more selective extractants for HS. Although these extractants may produce less

Table 10   Carbon, Oxygen, Hydrogen, and Sulfur Contents, on a Dry Ash-Free Basis, of Humic Acids (HA) and Fulvic Acids (FA)

Solvent	Carbon		Oxygen		Hydrogen		Nitrogen		Sulfur	
	HA	FA	HA	FA	HA	FA	HA	FA	HA	FA
EDA 2.5 mol l^{-1}	56.8	51.2	29.2	30.3	5.9	5.7	6.4	11.1	—	—
Pyridine	55.9	47.1	32.9	39.9	5.1	5.3	4.4	6.0	—	—
DMSO	55.0	55.0	35.5	37.1	4.2	4.4	3.3	2.2	2.0	1.3
Sulfolane	54.4	53.2	35.2	37.4	4.8	4.4	3.2	3.3	2.4	1.7
DMF	54.3	52.3	36.8	38.8	4.6	4.1	2.6	3.2	1.7	1.6
NaOH 0.5 mol l^{-1}	53.1	45.0	36.3	43.0	6.0	6.0	2.9	4.3	—	—
Na-EDTA 1 mol l^{-1}	52.1	48.4	—	—	4.1	4.2	—	—	—	—
$Na_4P_2O_7$ 0.1 mol l^{-1} (pH 7)	50.9	37.3	41.1	50.9	3.3	5.1	3.0	5.0	—	—

From Hayes M. H. B. et al., *Geoderma*, 13, 231, 1975. With permission.

alteration, they are much less effective than alkaline solvents in the isolation of HS. However, the yield can be increased by HCl or HCl-HF pretreatment of the soil sample.

A 0.1 mol l^{-1} solution of $Na_4P_2O_7$ is the most widely used and most efficient, with amounts of HS recovered up to 30%. To minimize chemical alteration, extraction should be carried out at pH 7. The HS isolated with this solvent usually contain high amounts of Fe and Al.

Other inorganic salts used for the extraction of HS include, in order of decreasing efficiency, NaF, $(NaPO_3)_6$, Na_3PO_4, $Na_2B_4O_7$, NaCl, NaBr, NaI. The order of decreasing extraction efficiency for the sodium salts of organic acids is oxalate, citrate, tartrate, malate, salicylate, benzoate, succinate, and ethanoate.[286]

The efficiency of these solvents depends on their capacity to replace with sodium the polyvalent cations that neutralize the charges on the HS and bind HS to mineral colloids, and to form complexes with these cations. Salt concentration in the extractant solution is a crucial factor, since excess electrolyte concentrations curtail hydration of HS and their dissolution may not take place.

c. Organic Solvents

Data in Table 9 show that, in the absence of water, EDA is a very poor solvent for HS, and that pyridine extracts higher amounts of HS than do DMF, DMSO, or sulfolane. The enhanced solubilization by pyridine can be attributed partially to a pH effect. On the other hand, the amounts of HS extracted by the dipolar aprotic solvents DMF, DMSO, and sulfolane depend entirely on the extent to which these solvents can solvate the HS macromolecules. DMSO is the best of the organic solvents tested for HS, and its effectiveness is improved by the presence of small amounts of HCl.

The elemental analysis data (Table 10) for the HAs isolated by organic solvents and by 0.5 mol l^{-1} NaOH are comparable, while FAs from the alkaline solution have lower C and higher O contents. These results suggest that some uptake of O_2 occurs under the alkaline conditions.

Up to 55% of organic matter in mineral soils can be extracted with formic acid containing LiF, LiBr, or HBF_4, which are able to disrupt hydrogen bonds or complex metal ions.[289] Advantages of anhydrous formic acid are that it is a polar compound which exhibits neither oxidizing nor hydrolytic effects. A disadvantage is the large amount of Ca, Fe, Al, and other inorganic components dissolved by formic acid along with organic matter, and these are hard to remove.

d. Chelating Agents

EDTA, acetylacetone, cupferron, 8-quinolinol (8-hydroxyquinoline), and other organic compounds that are able to form chelate complexes with polyvalent metal ions have been used for extracting HS from some types of soils, but they are rather ineffective when used for other soil types. The HS isolated, however, still contain residual metal cations.[289]

e. Other Extractants and Extraction Procedures

Aqueous HCl mixtures of such solvents as acetone, dioxane, cyclopentanone, dimethylformamide, and others have been used for the extraction of organic matter from soils, but the yields obtained are generally low.[292]

Up to 23% of the total C can be extracted from soils with an acetone-H_2O-HCl solvent mixture, but the extracted organic matter contains appreciable amounts of contaminating mineral components.[3] Various combinations of DMSO, HCl, and water have been used with some success in the extraction of HS from some soils.[286]

Other approaches that have been employed for the extraction of organic matter from soils involve treatment with chelating resins, ultrasonic dispersion, and supercritical gases.

2. Fractionation Procedures

Humic substances are comprised of a broad spectrum of macromolecules exhibiting a wide range of values for any given molecular property. It is therefore necessary to subdivide humic fractions according to some property related to their molecular composition and obtain distinct fractions that exhibit a much narrower variation in the property being measured. Fractionation procedures are commonly used as a preliminary step to subsequent physical, chemical, and physicochemical measurements and analyses of humic fractions so obtained.

A wide range of methods has been used for fractionation, but generally they exploit physicochemical differences in solubility, reactions with metal ions, molecular size, charge or charge density, and adsorption characteristics.

a. Fractionation Based on Solubility and Precipitation

pH Adjustment — The adjustment of pH is the most used technique for crude fractionation of HS extracted from soil by alkaline solvents (Figure 10). The main fractions obtained include (Table 1): HA, soluble in alkali, insoluble in acid (precipitated on adjustment to pH $\approx$1); FA, soluble in alkali, soluble in acid; humin, insoluble in alkali and in acid. Although these definitions were initially based on the alkaline crude extract of soil, the same terminology is used when the extractant is neutral, acidic, or organic. Humin, the portion of organic matter not extracted with alkali, is believed to exist as high-MW polymers or as humates strongly associated with oxyhydroxides and clays.

It should be clear that HA and FA are not only extremely heterogeneous, but each fraction contains intermediate forms which have characteristics overlapping each other. HA and FA are therefore not to be regarded as independent groups of HS, but as part of a unique sequence linked by interlocking types.[3]

Several factors influence the separation based on pH manipulation. The HA fraction obtained by freeze-drying the wet precipitate at pH 1 is a soft, friable powder which dissolves readily in neutral or slightly alkaline aqueous solvents, whereas air-dried HA preparations are brittle and difficult to redissolve. The HA/FA ratio is affected by metals contained in the crude HS extracts and by the NaOH (or $Na_4P_2O_7$) concentration used for extraction.[293] Prolonged standing at high pH can induce a partial conversion of FA to HA, whereas storage in acidic solutions may induce with time polymerization of FA to yield HA.

Organic Solvents, Salts and Metal Ions — Organic solvents, salts, and metal ions have been used to further separate HA and FA in subfractions. For instance, HA may be extracted with ethanol to obtain its ethanol-soluble fraction, named hymatomelanic acid (Figure 10).[294] Acetone, methanol, methylisobutylketone, and diethyl ether can also be used successfully to fractionate HS.

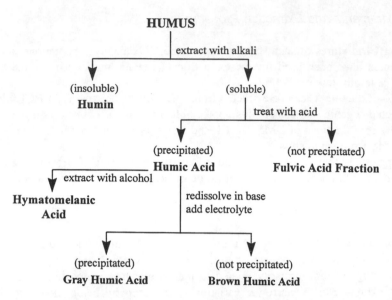

Figure 10   Classical scheme for the fractionation of HS.

The presence and concentration of background electrolytes (salts) affect the behavior of the charged humic macromolecules (polyelectrolytes). If the salt concentration is increased, the extension of the double layer is confined much closer to the surfaces of the macromolecules, which thus approach one another more closely so that intermolecular attractive forces predominate and coagulation and precipitation may occur. Suppression of charge interaction by background electrolyte to the extent that precipitation occurs is referred to as "salting out." Base-redissolved HA may be separated into a precipitated fraction, the "gray" HA, and a nonprecipitated fraction, the "brown" HA, by the addition of a salt, usually KCl (Figure 10).[295] Ammonium sulfate at pH = 7 has been used successfully to fractionate HA.[296]

Many divalent and polyvalent ions are able to form insoluble salts or complexes with HS, thus producing the precipitation of a certain amount of HS from solution. Copper ions have been used to fractionate FA into "apocrenic" and "crenic" acids.[297] The fractional precipitation of HA or FA by using either gradually increasing concentrations of the metal ion causing precipitation or changing the type of the precipitation ion has also been attempted, with some success.[287]

The use of organic solvents, salting-out and polyvalent metal ions is, however, unlikely to produce fine fractionations, and likely to produce rather crude or gross fractions. In fact, most researchers prefer to use only the three main fractions, HA, FA, and, sometimes, humin, in their analytical studies of HS.

Sequential Extraction/Fractionation Procedures — Humic substances may also be fractionated by gradual extraction of the soil sample with increasingly powerful extractants obtained by changing the pH and the nature of the extractant anion,[298] or by using a range of organic solvents in sequence.[290,299-302] Scheffer et al.[299] showed that the sequential extraction of soil with organic solvents in a Soxhlet apparatus under a nitrogen atmosphere and under reduced pressure did not significantly alter the isolated organic materials, and the solvents were easily removed by evaporation under vacuum. Sequential extraction thus presents a further refinement of procedures for obtaining less heterogeneous fractions of soil organic matter and nonpolar solvents followed by dipolar aprotic solvents and then by alkaline solutions would provide a good solvent series for the simultaneous isolation and fractionation of soil organic matter.[6]

A series of mild organic solvents of different polarity (Table 11) have been used for the fractional extraction of soil HS.[301,302] The single organic fraction isolated from the three soils were substantially

Table 11   Properties of the Solvents Used for Sequential Extraction Procedure of SOM

Solvent	Chemical formula	Dielectric constant (25°C)	E_T (Kcal mol^{-1}) (25°C)	Boiling point (p = 760 mmHg)
Ethyl ether	$(C_2H_5)_2O$	4.22	34.6	34.6
Benzene	C_6H_6	2.27	34.5	80.10
Acetone	$(CH_3)_2CO$	20.7	42.2	56.24
Dioxane	$(CH_2)_4O_2$	2.21	36.0	101.32
THF	$(CH_2)_4O$	7.39	37.4	65.0
Ethanol	C_2H_5OH	24.3	51.9	78.32
DMF	$HCON(CH_3)_2$	37.6	—	153.0
Pyridine	C_6H_5N	12.3	40.2	115.58
DMSO	$(CH_3)_2SO$	46.7	45.0	189
Formamide	$HCONH_2$	109.5	56.6	210.5

From Senesi, N. et al., *J. Soil Sci.,* 34, 801, 1983. With permission.

similar to each other in chemical character but showed different chemical and structural properties which varied according to the progression of the solvent series, with some similarities between two successive extracts. Along each series of extracts aliphaticity appeared to decrease, while aromaticity, polar oxygenated and nitrogenated functional groups, transition metal ion content, and organic free radical concentrations increased. Although the yields of extracted organic material were not high, the fractions could be considered representative of SOM composition and furnish new chemical information on its nature and properties.

b. Fractionation Based on Molecular Size

Gel Chromatography — Gel chromatography, also known as gel permeation chromatography or gel filtration, is a separation technique in which solute molecules are fractionated on the basis of their molecular dimensions. The separation is obtained in columns by elution of the sample through a bed of porous beads or granules which are composed of cross-linked polymer gels such as dextran, polyacrylamide, agarose, and porous glass. Small particles have complete access to solvent both within and outside the gel beads, medium-sized molecules have a limited access to the inside spaces, and large molecules cannot enter the pores of the gel and are therefore forced to remain in the solvent outside the gel beads, that is, in the outer, or excluded volume. Therefore, the excluded, large molecules move through the column faster than the medium-sized molecules, and small molecules are the most retarded and last eluted from the column. In conclusion, when a sample is applied to the top of the column, its components are eluted in order of decreasing molecular size. The general principles of gel chromatography are discussed in Yau et al.[303]

Gel chromatography has been extensively applied to the separation of HAs and FAs into less heterogeneous fractions and to the determination of molecular sizes of HAs and FAs, as discussed later in this chapter. Most studies have used Sephadex gels, which are modified cross-linked dextran polysaccharides available in several different pore-size grades depending on the degree of cross-linking. The wide range of MW within a HS sample can present difficulties in choosing a suitable gel. The successive use of gels of various exclusion limits and reapplication of the excluded or included portion from a given gel to another with a higher or lower exclusion limit, can solve this problem. For example, six different fractions have been obtained for a soil FA using five different pore-size Sephadex gels in sequence,[304] whereas only two polydispersed fractions have been obtained on single Sephadex columns for HAs.[43,305-307] A review on the principles and limitations of gel chromatography and its applications to HS has been provided by De Nobili et al.[308]

Although the technique is rapid, cheap, and very versatile, a number of problems are encountered which, if not overcome, can invalidate the results. The only interaction between the solute and gel must be a size-dependent interaction, whereas all other physical and chemical interactions, such as adsorption and electrostatic interactions must be absent or very weak.[303] However, HAs have been shown to interact both electrostatically (charge effects) and by adsorption with the most commonly used gels such as Sephadex, especially at low pH and low ionic strength.[43] To suppress these effects and to obtain accurate molecular size distributions, fractionation was recommended using basic buffer of relatively high ionic strength.[43] For HS the type and degree of adsorption appear to depend largely on the sample origin, degree of gel cross-linking, and composition of the eluent.[308]

The formation of molecular aggregates of HAs in solution, as a function of both pH and concentration, is another factor of uncertainty in the behavior of HA in gel chromatography.[309] Swift and Posner[43] found that the elution patterns of HAs were not independent of sample concentrations, as required by theory. With increasing dilution, a greater percentage of the sample appeared in the excluded or near-excluded regions. A decrease in the ionic strength caused by a decrease in the concentration of the sample was suggested to enhance the repulsion of negatively charged HA molecules and increase their exclusion from the gel.[43] Further, gel chromatography of HAs applied in the acid forms (charges suppressed) showed an increased percentage in the excluded fraction.[310] In addition, the elution patterns were strongly dependent on the pH of the applied sample. These results show that, besides concentration, pH, and ionic strength, the chemical form of the HS sample also has a profound influence on its behavior in gel chromatography experiments. Despite these handicaps, gel chromatography has proved to be a particularly useful technique for the fractionation of HS on the basis of molecular size.

Ultrafiltration — Ultrafiltration is a useful and simple technique which allows the rapid fractionation of HS by the use of a suitable series of membranes.[287,311] Membranes with pore sizes ranging over a few nanometers can be used to filter molecules in solution on the basis of molecular size. The fractionation can be obtained either in order of ascending or descending molecular weight.

One problem is that the pore size distribution within these membranes is not completely uniform so that the molecular weight cut-off is not as sharp as expected. In addition, the actual MW value at which the cut-off operates for a given substance will depend on the charge and molecular configuration of the substance. This results in charge-charge interactions between humic macromolecules and the membrane that can interfere with the filtration process so that it is no longer based solely on molecular size.

Ultracentrifugation — Fractionation of HS can be carried out successfully by ultracentrifugation by using density gradient or zonal centrifugation techniques.[312] The procedure is somewhat laborious when compared with gel chromatography and ultrafiltration. Furthermore, the suppression of intermolecular charge repulsion by the addition of electrolyte is essential in any attempt to fractionate HS by ultracentrifugation techniques,[6] as discussed later in this chapter.

c. Fractionation Based on Charge Characteristics

Electrophoresis and Related Techniques — Electrophoresis describes the movement of charged solute molecules in an electric field. In the traditional electrophoretic experiment, the HS sample is dissolved in an alkaline buffer and placed on a support medium such as cellulose in flat beds or on gel beads in column systems. Migration occurs as an inverse function of the molecular size and direct function of charge density. As expected, discrete fractions are not obtained, but rather there is a gradation of fractions with one fraction merging into another. In general, the fractionation efficiency of the traditional type of electrophoresis is inferior to that obtained on the basis of MW outlined previously. A complete discussion of electrophoretic procedures can be found in books by Deyl et al.[313] and Everaerts et al.[314]

More recent developments in electrophoretic techniques, including polyacrylamide gel electrophoresis (PAGE), isoelectric focusing (IEF), and isotachophoresis (ITP), have been usefully applied for fractionation of soil HS, often in association with gel permeation chromatography.[287]

In PAGE a polyacrylamide gel is used as the support medium and the sample is subjected simultaneously to fractionation by electrophoresis, on the basis of charge, and by gel permeation chromatography, on the basis of molecular size. PAGE is a very versatile technique, since experimental conditions can be modified by altering a wide number of variables to optimize the fractionation obtained.

In IEF, a pH gradient is set up within the gel support by incorporation of a range of amphoteric substances called ampholines. Then, when a range of macromolecules are subjected to electrophoresis in the system, they will migrate to the pH of the isoeletric point and then cease to move. The separations obtained should be interpreted with extreme caution, however, because of several difficulties encountered with pH gradient electrophoresis of soil HS. One problem is the interaction of the humic molecules with the organic ampholines used to form the pH gradient. Another important factor of uncertainty is intermolecular aggregation of the humic molecules themselves, which would probably also be pH dependent. Further, the electromigration of soil HS components in pH gradients appears to be controlled by the acidic functional groups rather than any amphoteric characteristics which they may possess. Thus, the separation of HS components appears to be largely the result of electrophoresis in a pH gradient rather than true isoelectric focusing.[315]

Isotachophoresis is similar to IEF but includes the use of additional ampholines or multiphasic buffer systems to improve the separation and resolution. Each of these techniques can be run using columns, tubes, thin layers, or slabs and is very promising for further application in the fractionation of HS.

Ion-Exchange — Anion-exchange resins consisting of solid, nonporous polystyrene beads have been used to fractionate soil HS.[316] Some of the humic material is readily retained and a fractionation can be obtained by elution with a salt gradient and/or an alkaline reagent (usually NaCl and NaOH, respectively). In practice, the fractionation obtained is not satisfactory, probably because of the limited surface area of the resin beads, which limits their ability to interact with all charged sites of the humic macromolecules. As a consequence, charge differences between the molecules will be less well defined, and the fractionation will lose resolution.

Better fractionations are obtained when porous anion-exchange gels and anion-exchange cellulose are used.[317] After adsorption of the humic molecules onto the gel, a fractionation can be obtained by eluting with buffer solutions and salt gradients, and then, if necessary, with an alkaline reagent.

These materials offer a relatively simple but sensitive means of fractionation based primarily on electric charge properties and deserve more development in their application to HS.

d. Fractionation Based on Adsorption

A number of adsorbent media, such as charcoal, alumina, various gels, and, more recently, the macroporous methylmethacrylate resin XAD-8, have been used for the fractionation of HS, especially fulvic acids.[287] Desorption is achieved by a number of organic solvents and acidic and basic reagents. In some cases, it is difficult to distinguish between fractionation and purification (see next section) when using these procedures, which are probably best used for the separation of HS from nonhumic impurities, such as polysaccharides, rather than for the fractionation itself.

One problem is that adsorption of HS is often so strong that they can be desorbed only with the use of rather strong or potentially damaging reagents. The XAD-8 resin is an exception, being a relatively weak adsorbent, and is expected to be suitable for fractionation rather than purification of HS.

3. Purification Procedures

Both inorganic impurities (salts, clay, oxyhydroxides) and organic impurities (carbohydrates, proteins, lipids, etc.) are initially present and must be removed from crude HS fractions, as normally recovered from soils, by procedures that do not cause alteration of the HS.

Considerable removal of clay impurities from crude HA can be achieved by repeated alkaline dissolution, followed by high-speed centrifugation, and acid reprecipitation. Reduction in ash content can be obtained by repeated precipitation with mineral acids and by passage through ion-exchange resins.[318] Another method to reduce ash content is treatment of crude HA with dilute solutions of HCl/HF followed by dialysis against distilled water in the presence of the H^+ form of an ion-exchange resin contained in a second dialysis bag.[319]

Alternate dialysis against 0.3 N HF and 0.02 N $Na_4P_2O_7$ reduces the ash content of HA preparations to less than 1%. The effect of HF can be ascribed to its ability to dissolve hydrated clay minerals and to form complexes with polyvalent cations. However, chemical modification and significant weight losses of HA may result from HF treatments.

As for mineral impurities, attempts to separate coprecipitated or coadsorbed nonhumic organic impurities is not entirely successful. Lipids (fats, waxes, resins, etc.) can be rather easily removed with ether or alcohol-benzene. Boiling HA (or FA) with water can eliminate polysaccharides and polypeptides.[320] Acid hydrolysis with 6 mol l⁻¹ HCl is also able to remove coadsorbed nonhumic materials, but weight losses of over 40% and chemical changes can occur. The treatment has little effect on HA, whereas drastic changes are observed for FA. The C content increases, while total acidity decreases, apparently because of decarboxylation. Hydrolysis procedures appear to be satisfactory for the purification of HA, but not FA.

In general, purification (and recovery) of the FA fraction from the acid solution following precipitation of HA presents major problems. The FA solution contains appreciable amounts of mineral matter dissolved during extraction, several soluble nonhumic compounds, and large amounts of salts formed by neutralization of the base with the acid. Inorganic cations can be eliminated by selective adsorption, by passage through a cation exchange resin, or by repeated dialysis.

4. Comprehensive Isolation Procedure Proposed by the IHSS

Tremendous efforts have been made in the last few years by the International Humic Substances Society (IHSS) in order to standardize the isolation and fractionation methods of HS from soil, sediment, and water by introduction of a procedure that should be adopted universally.

On this basis, Thurman[321] has recently proposed a comprehensive conceptual view of the isolation procedure and terminology for HA and FA of any source (Figure 11). The right side of Figure 11 refers to solid materials (soil, sediments, etc.) and the left side to aquatic samples. During pretreatment and extraction (steps I and II), samples are treated in different ways. Pretreatment involves preparation of the sample for the extraction of HA and FA. For soil materials, typical pretreatments are demineralization by HCl and/or HF, flotation to separate plant fragments, etc. The typical extraction for soil involves the removal of the alkaline soluble fraction, commonly by NaOH and pyrophosphate, or dipolar aprotic solvents, such as dimethylsulfoxide. In step III (acidification) the procedures join, thus permitting the use of a final, common terminology, i.e., the use of the terms HA and FA. This step consists in the acidification of the alkali-soluble extract to pH ≤ 2 with HCl, and precipitation. Centrifugation is needed to separate the sample into two fractions, a FA fraction and a HA fraction. These two fractions are not solely HA and FA but may contain specific compound classes such as polysaccharides and amino acids, as well as inorganic ions, which are removed in the purification step. Step IV describes the "purification" of HA and FA fractions to yield the final products, HA and FA. Since the structure of HS is not known precisely, the term "purification" refers essentially to the removal by physical methods of substances that are "physically" but not "chemically" bound in humic extracts, e.g., salts, metals, lipids, polysaccharides, fatty acids, etc. Thurman[321] concludes his scheme by proposing a final "verification" step (step V) for the isolated HA and FA. This is done by comparing the identity of the products isolated

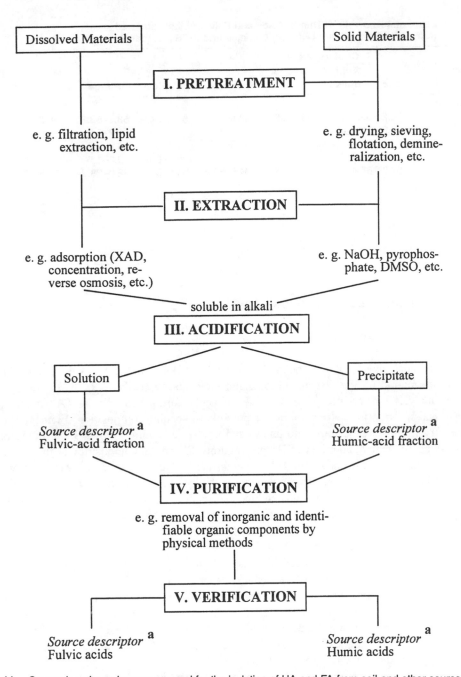

Figure 11   Comprehensive scheme proposed for the isolation of HA and FA from soil and other sources.[a] The source from which the sample is isolated should be included in the specific nomenclature. (From Thurman, E. M., in *Humic Substances and their Role in the Environment*, Frimmel, F. H. and Christman, R. F., Eds., Wiley-Interscience, Chichester, 1988, 31. With permission.)

on the basis of the simple elemental and acid functional group composition, or on a more complex basis, e.g., NMR, MW determinations, etc. The efforts of IHSS to publish values and ranges for average soil and aquatic HA and FA of its standard and reference collection will be greatly useful for such comparative verification.

Table 12   Elemental Analysis of Humic Acids and Fulvic Acids Extracted from Soils from Widely Differing Climates and IHSS Standards

Element (%)	Arctic	Cool, temperate Acid soils	Neutral soils	Subtropical	Tropical	Range	Average	IHSS standards
Humic acids								
C	56.2	53.8–58.7	55.7–56.7	53.6–55.0	54.4–54.9	53.6–58.7	56.2	58.1
H	6.2	3.2–5.8	4.4–5.5	4.4–5.0	4.8–5.6	3.2–6.2	4.7	3.7
N	4.3	0.8–2.4	4.5–5.0	3.3–4.6	4.1–5.5	0.8–4.3	3.2	4.1
S	0.5	0.1–0.5	0.6–0.9	0.8–1.5	0.6–0.8	0.1–1.5	0.8	0.4
O	32.8	35.4–38.3	32.7–34.7	34.8–36.3	34.1–35.2	32.8–38.3	35.5	34.1
Fulvic acids								
C	47.7	47.6–49.9	40.7–42.5	42.2–44.3	42.8–50.6	40.7–50.6	45.7	50.6
H	5.4	4.1–4.7	5.9–6.3	5.9–7.0	3.8–5.3	3.8–7.0	5.4	3.8
N	1.1	0.9–1.3	2.3–2.8	3.1–3.2	2.0–3.3	0.9–3.3	2.1	2.7
S	1.6	0.1–0.5	0.8–1.7	2.5	1.3–3.6	0.1–3.6	1.9	0.6
O	44.2	43.6–47.0	47.1–49.8	43.1–46.2	39.7–47.8	39.7–49.8	44.8	43.7

Data from Reference 323 and the IHSS data collection.

C. Elemental and Functional Group Analyses

1. Elemental Composition

Elemental analysis is probably the most commonly used tool in the characterization of HS and provides information on the distribution of major elements, typically C, H, N, O, and S, in HS. Elemental analysis does not provide an absolute molecular formula for HA and FA but only general compositional information setting limits for possible molecular composition. Elemental analysis data are very useful in describing and understanding the geochemistry of HS and in establishing the processing efficiency and purity of HS preparations. Elemental composition is also often useful in distinguishing different classes of HS, although care should be used in evaluating results of elemental analyses, the most serious problem being the lack of reproducibility of analytical results.

An excellent review of the methods and techniques commonly used in the elemental analyses of HS has been recently provided by Huffman and Stuber.[322] An interlaboratory analysis of three carefully prepared, freeze-dried HS preparations has been conducted by these authors[322] with the purpose of evaluating both the applicability of the usual methods of elemental analysis to HS and the overall reliability of results obtained. The results of the interlaboratory study ranged from rather good agreement for C and H to poor agreement for other elements and water content in the analyzed HS samples.[322] The authors concluded that the moisture status of the samples should be carefully determined and that it is desirable to perform analysis on air equilibrated samples and then correct the results to a dry basis using the determined moisture value. The establishment of standard and reference HA and FA samples by the IHSS was considered to be of great help in validating the applicability of methods for the elemental analysis of HS.

The elemental composition (on ash- and moisture-free basis) of HAs and FAs isolated from soils from widely differing climatic zones and the mean elemental composition of a "model" HA and FA are shown in Table 12.[323] Table 12 also shows the elemental composition of a Mollisol HA and FA belonging to the IHSS collection of standard and reference HAs and FAs. Neutral soils appear to have narrow ranges of C, H, and O, while acid soils show broader ranges. Compared to HAs, FAs contain more O and S but less C, H, and N.

Based on these data, and disregarding S, average minimal formulas of $C_{10}H_{12}O_5N$ for HA and $C_{12}H_{12}O_9N$ for FA were calculated.[324] Although these chemical formulas are of no absolute value, they are useful to set boundaries for probable chemical structures to be confirmed by other chemical and physicochemical properties.

Table 13 Functional Group Analysis of Humic Acids and Fulvic Acids Extracted from Widely Differing Climates

Functional group (cmol kg $^{-1}$)	Arctic	Cool, temperate		Subtropical	Tropical	Range	Average
		Acid soils	Neutral soils				
Humic Acids							
Total acidity	560	570–890	620–660	630–770	620–750	560–890	670
COOH	320	150–570	390–450	420–520	380–450	150–570	360
Phenolic OH	240	320–570	210–250	210–250	220–300	210–570	390
Alcoholic OH	490	270–350	240–320	290	20–160	20–490	260
Quinonoid C=O	230	10–180	450–560	80–150	140–260	10–560	290
Ketonic C=O	170				30–140		
OCH$_3$	40	40	30	30–50	60–80	30–80	60
Fulvic Acids							
Total acidity	1100	890–1420	ND[a]	640–1230	820–1030	640–1420	1030
COOH	880	610–850	ND	520–960	720–1120	520–1120	820
Phenolic OH	220	280–570	ND	120–270	30–250	30–570	300
Alcoholic OH	380	340–460	ND	690–950	260–520	260–950	610
Quinonoid C=O	200	170–310	ND	120–260	30–150	120–420	270
Ketonic C=O	200		ND		160–270		
OCH$_3$	60	30–40	ND	80–90	90–120	30–120	80

[a] Not determined.

From Schnitzer, M., in *Soil Organic Matter,* Schnitzer, M. and Khan, S. U., Eds., Elsevier, Amsterdam, 1978, chap. 1. With permission.

The elemental composition of HS is expected to be affected by such factors as pH, parent material, vegetation, and age of the soil.[4] Methods used for extraction and fractionation of HS, sample preparation and handling, and moisture determination may also affect the analytical results for elemental determination of HS samples, and may cause lack of reproducibility. Acid hydrolysis has been often used to remove ash, coprecipitated materials, and peripheral groups, such as peptide and polysaccharide chains, from HS. Although this method should leave intact the "core" of HA or FA, structural changes such as depolymerization and acid-catalyzed condensation may occur especially for FA, causing loss of H and N and a related C content increase.[3]

2. Atomic Ratios

Atomic ratios of H/C, O/C, and N/C (Table 12) can be useful in: (1) identifying types of HS; (2) monitoring structural changes of HS; and (3) devising structural formulas for HS.[324]

The O/C and H/C ratios are considered the best indicators of HS types. Typically, O/C and H/C ratios of soil HAs are about 0.5 and 1.0, respectively, while those of soil FAs are 0.7 and 1.4, respectively.[324] The difference in O/C ratios may reflect higher amounts of oxygenated functional groups, such as COOH groups, and carbohydrates in FA. Low H/C ratios indicate a high contribution of aliphatic components in HS.

3. Functional Groups

The major oxygen-containing functional groups in HS are carboxyls, phenolic and alcoholic hydroxyls, carbonyls, and methoxyls. Analytical data for these groups in HAs and FAs extracted from soils from widely differing climatic zones and their ranges and means are shown in Table 13.[323]

Because of the existing difficulties in estimating functional groups in HS and lack of specificity, absolute values reported for the various groups must be regarded with due reservation. A discussion of constraints on assignment of COOH, phenolic OH, and C=O groups has been given by Perdue.[325] However, sufficient information has now accumulated which allows one to draw tentative conclusions on the nature and trends of oxygen-containing functional groups in HS of various types.

The main acidic groups in HAs and FAs are carboxyl and phenolic OH groups. Alcoholic OH and carbonyl (quinonoid and ketonic C=O) groups are also well represented, whereas methoxyl (OCH_3) groups are found in smaller amounts. The total acidity, and especially the COOH content, and alcoholic OH group content of FAs are appreciably higher than those of HAs. The COOH content of HS appears to be inversely related to MW. Both HAs and FAs contain approximately the same concentration of phenolic OH, total C=O, and OCH_3 groups. However, the C=O content varies relatively widely, especially in the case of HAs. The quinone content of HAs is generally higher than for FAs and a higher percentage of the total C=O in HAs occurs in quinones than in FAs. Essentially all of the oxygen in most soil FAs can be accounted for in O-containing functional groups, whereas a smaller fraction, somewhat less than 75%, has been recovered in these groups in most HAs.

During humification, COOH and C=O groups have been found to increase, while phenolic and alcoholic OH, and OCH_3 decreased.[326,327] During diagenesis, HS are subjected to a partial carbonification with an increase in C content and a decrease in O and N contents, and losses of COOH and OH groups. In the coalification process, COOH groups disappear first, followed by OCH_3 and C=O groups.[4]

a. Total Acidity

The most common method for determining total acidity of HS is based on the reaction of a known amount of HS sample with excess 0.1 M $Ba(OH)_2$, followed by potentiometric titration of the unused reagent with standard 0.5 M HCl to pH 8.4.[5] The main advantage of the method is its simplicity, whereas its main weakness is the assumption made that all of the HS sample is precipitated and removed as the insoluble Ba-salt during filtration, which is not always the case.

Another approach to measure total acidity of HS consists in the estimate by the Zeisel method of OCH_3 formed after methylation with diazomethane (CH_2N_2).[328] Diazomethane reacts with a wide variety of acidic groups, including COOH, phenolic and enolic OH, and acidic N-H groups. However, H-bonded OH groups may not react and side reactions may occur. To overcome the difficulty of obtaining complete methylation, a three-step methylation procedure has been proposed.[329]

Total acidity of HS has also been determined by estimating the H_2 produced by reaction of HS with diborane (B_2H_6).[330,331] The method is believed to be independent of pK values and to allow for the determination of sterically hindered protons.

b. Carboxyl Groups by Wet Chemical Methods

Carboxyl groups in HS are generally determined by the Ca-acetate method.[5] The acetic acid liberated during the reaction is titrated potentiometrically with a standardized 0.1 M NaOH solution of pH 9.8. Strictly quantitative values may not be achieved due to the release of protons by Ca^{2+} from sites other than COOH groups.[332] Another pitfall of the method may be the incomplete removal of HS during filtration. Modified procedures involve recovery of acetic acid by steam distillation[333] or ultrafiltration.[332] In these procedures weaker COOH groups would not be included in the COOH content determined, thus values for the COOH content would be lower than those obtained using the original method.

The estimate of COOH groups in HS has also been performed by methylation of HS and subsequent saponification of the methyl esters obtained.[5] Several methylation procedures have been used, including CH_2N_2 and CH_3OH in dry HCl, and saponification procedures, including distillation of the liberated CH_3OH and determination of the increase in OCH_3 content. However, the latter methods have not been widely accepted because quantitative recovery of material following saponification is difficult.

An iodometric method based on ion exchange has been used to determine COOH groups in HS, but this method gave higher values than the Ca-acetate method.[316]

Another method to determine COOH groups in HS, which gives results comparable to the Ca-acetate method, involves decarboxylation of HS when heated with quinoline in the presence of suitable catalysts.[316]

c. Total OH Groups

The total OH content of HS has been determined by methylation with dimethyl sulfate or acetylation with acetic anhydride. In the former method, the HS sample is methylated in an alkaline solution, after which the resulting precipitate is analyzed for OCH_3 by the Zeisel method.[328] Only phenolic and alcoholic OH, but not COOH, are assumed to be methylated. The validity of the method is questionable, however, because of possible side reactions in strongly alkaline solution. Less drastic procedures have been attempted in which the HS sample is refluxed with dimethyl sulfate over anhydrous K_2CO_3 in acetone[334] or methanol is used as the solvent.[335]

Acetylation with excess acetic anhydride to form acetate esters has been used more extensively for the determination of total OH content in HS.[5] The excess anhydride is hydrolyzed to acetic acid, which is then titrated with standard 0.1 M NaOH. Since COOH groups are also acetylated with acetic anhydride, a correction has to be made for COOH, which is determined by an independent method.

Alternatively, a procedure has been proposed in which COOH groups are methylated prior to acetylation.[335] Other functional groups in HS which can also react with acetic anhydride include primary and secondary amines and sulfydryl groups, but interference from these compounds is not considered to be serious.[4]

d. Phenolic OH Groups

The amount of phenolic OH in HS is commonly calculated as the difference between the total acidity and COOH content.[5] However, values obtained in such a way should not be regarded as accurate for phenolic OH content in HS.

A modified Ubaldini procedure, has been used as the only direct method for the estimation of phenolic OH in HS.[5] The method involves formation of K salts of COOH and phenolic OH groups by refluxing the HS sample with alcoholic KOH. After removal of the excess alkali by filtration, the residue is washed and suspended in 85% alcohol, and saturated with CO_2. The K^+ released as K_2CO_3 is then estimated by titration with standard acid. The K^+ is assumed to originate from phenolic OH groups, whereas no K^+ should be released from K^+-salts of COOH groups. The nonspecificity of the reaction and the risk of hydrolysis during the refluxing with alcoholic KOH render the method seriously questionable.

e. Alcoholic OH Groups

No direct method is available for estimating alcoholic OH groups in HS. Estimates of these groups obtained indirectly from the difference between total OH and phenolic OH must be accepted, for obvious reason, with reservation.

f. Total Carbonyl Groups

The carbonyl groups present in HS have been determined by methods mostly based on the formation of derivatives such as oximes and phenylhydrazones obtained by reaction with reagents such as hydroxylamine, phenylhydrazine, and 2,4-dinitrophenylhydrazine.[5] The formation of derivatives is usually ascertained by measurement of the increase in N content of the derivative, or by analyses of the excess reagent. The most common procedure is based on the reaction of the HS sample with an excess hydroxylamine followed by potentiometric back-titration of the unused hydroxylamine with standard perchloric acid solution.[5]

A polarographic method for determining C=O groups in HS has been developed based on refluxing of the HS sample with an excess of 2,4-dinitrophenylhydrazine in acidified ethanol followed by the polarographic determination of the excess reagent.[336] The procedures based on the formation of derivatives are, however, criticized because the reagents employed may react with groups other than C=O groups.

Another method for estimating C=O groups in HS is based on reduction to alcoholic groups with excess sodium borohydride ($NaBH_4$) in alkaline solution, followed by manometrical estimation of the H_2 liberated from the unused $NaBH_4$ after decomposition with HCl.[5] The method is reported to be highly specific for C=O groups, and values obtained for some HS compare favorably with those found by the oxime method.[331]

g. Quinone Groups

Specific estimates of quinone groups in HS are based on selective reduction by $SnCl_2$ in acid or alkaline solution, and by Fe^{2+} in alkaline triethanolamine solution. The first method involves heating of the HS sample for 4 h in a sealed tube with a dilute HCl solution of excess $SnCl_2$ followed by filtration of the residue and titration of the excess Sn^{2+} in the filtrate with standard I_2 solution, using starch as an indicator.[337]

This method has been criticized since (1) side reactions may take place under the drastic experimental conditions used; and (2) oxidation, adsorption, and/or complexing of Sn^{2+} may occur, thus yielding unrealistically high quinone values for HS.[338-340] To overcome these difficulties, a modified method has been proposed based on reduction with $SnCl_2$ of HS dissolved in 0.1 M NaOH under N_2, followed by potentiometric back-titration of excess Sn^{2+} with standard $K_2Cr_2O_7$ solution.[338]

Another method proposed for the specific determination of quinone groups in HS is based on reduction by Fe^{2+} in alkaline triethanolamine, followed by amperometric back-titration of the excess Fe^{2+} with standard dichromate solution.[341] Excellent agreement was obtained between results of reduction with $SnCl_2$ and Fe^{2+}-triethanolamine in alkaline solution.[339]

h. Other Functional Groups

Methoxyl content in HS is generally determined by the classical Zeisel method. Free amino (NH_2) groups in HS have been measured by the nitrous acid method or by reaction with fluorodinitrobenzene or phenylisocyanate.[4] The nitrous acid method can account for a significant fraction of N in HS (up to 30%). However, because of interferences due to phenolic and lignin components in HS, the results are considered questionable. The other two methods mentioned also do not give satisfactory results.

D. Electrochemistry and Charge Characteristics

1. Titrations

Humic and fulvic acids behave like weak-acid polyelectrolytes and as such are amenable to be studied by methods based on the ionization of acidic functional groups. The acidic nature of HS is usually attributed to ionization of COOH and phenolic OH groups, although other groups, such as enol and imide groups, may be involved. Base titration is a rapid and easy technique for determining the acidic properties of HAs and FAs.

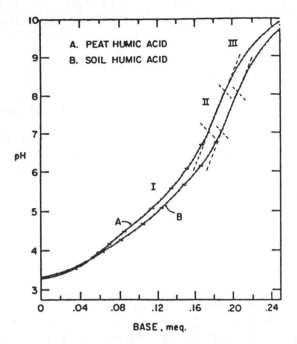

Figure 12   Titration curves of a soil and peat humic acid. The small wavy lines on the curves indicate end
points for ionization of weak-acid groups having different, but overlapping ionization constants.
(From Stevenson, F. J., *Humus Chemistry. Genesis, Composition, Reactions,* 2nd ed., John Wiley
& Sons, New York, 1994. With permission.)

Essential requirements for a proper description of the acid-base properties of HS are believed
to be (1) the identification and quantification of acidic functional groups by rigorous and repro-
ducible means; and (2) the description of pK_a values of HS by a suitable and rigorous model.[325]
Because of the almost certain existence in HS of a "continuous" and complex spectrum of "non-
identical" acidic functional groups with pK_a values that span a very wide range, neither of these
objectives has been satisfactorily met up to date. Consequently, it is not generally possible to compare
results obtained by different researchers if different methods and/or HS samples have been used.[325]

Besides electrometric (potentiometric and conductometric) titrations, high-frequency and thermometric titrations have been employed for the determination of acidic functional groups in HS.
These methods are amply discussed in several reviews and books.[3-5,279,325]

a. Potentiometric Titrations

Several procedures have been employed for potentiometric acid-base titrations of HS, including
continuous and discontinuous titrations and titrations in aqueous and nonaqueous solvents.[4,325]

Typical titration curves of HAs are shown in Figure 12.[4] The gradual rise in pH with addition
of base illustrates the large buffering capacity of HS, which is apparent over a wide pH range, and
is consistent with the concept that HS behave as weak-acid polyelectrolytes. The titration curves
in Figure 12 can be divided into three zones (I, II, and III). Zone I, the most acid region, represents
dissociation of COOH groups; zone III represents dissociation of phenolic OH and other very weak
acid groups; and zone II is an intermediate region where ionization of weak COOH and very weak
acid groups, such as phenolic OH, overlap.[4]

Factors that can affect the dissociation-association of protonated functional groups in HS include
(1) electrostatic attraction or repulsion due to charges located on the molecule; (2) inter- and intramolecular H-bonding, and (3) steric hindrance of acidic groups in hydrophobic regions of the
macromolecule.[342]

Further, several difficulties are encountered in the preparation and interpretation of titration curves of HS, including:

1. The method of sample preparation, e.g., dissolution of the sample prior to titration and partial flocculation of the sample especially at high ionic strength[4]
2. The occurrence of secondary chemical reactions at both low and high pHs, e.g., nucleophilic and electrophilic attack on unsaturated carbonyl groups and acid- and base-catalyzed keto-enol transformations
3. Counterion condensation at low pH, e.g., retention of the cation used for titration[343]
4. Electrostatic charge accumulation on the macromolecule as neutralization proceeds, which causes the remaining acidic groups to become weaker in acidity
5. The effect of dilution in raising the pH and reducing the dissociation of COOH groups
6. The consumption of alkali at high pH (>8) values, where phenolic OH groups dissociate, through autooxidation of HS in the presence of even traces of O_2, with a corresponding drop in pH upon standing[344,345]
7. The difficulty of making accurate pH measurements at high pH, due in part to electrode interferences arising from the very high concentrations of alkali metal ions, and the consequent difficulty in determinig the amount of base that has reacted at a particular high pH[325]

Another problem in the interpretation of titration curves of HS consists in the selection of the appropriate end point for ionization of COOH groups, which is rather arbitrary. Posner[344] used the point at which the rate of change of pH with added alkali became maximum (initial portion of region II in Figure 12), which occurred at pHs between 7.0 and 7.6, depending on the ionic strength of the solution. An arbitrarily selected pH, often a value of 8.0, has been used by other researchers. Graphical methods, such as those introduced by Gran,[346,347] have also been applied which allow for the assessment of end points for COOH groups by use of first- or second-derivative plots.

An additional difficulty in the base titration of HS is that the respective end points of the several acidic groups present on the macromolecule may not be evident because the titration of one type begins before that of the other ends, that is, the dissociation of protons from the two groups may partially overlap. This may be the reason why several workers,[5,348] have reported a single inflection point, whereas the curves in Figure 12 show several inflections.

Titration curves have also been utilized to obtain the ionization constants, K_a, of HS.[4] The ionization constant of an acid reveals the proportion of the sample which occurs in ionized and unionized forms at any chosen pH. At half-neutralization, when the concentrations of the two forms are equal, the pK_a of the acid, or acidic group, is equal to pH. Applying the simple mass-action law (the Henderson-Hasselbach equation) to weak acid polyelectrolytes such as HS, variable dissociation constants are obtained which change markedly with concentration of the polyelectrolyte and addition of neutral salt. Along the titration curve the apparent ionization constant, K_{app}, decreases with an increase in the degree of dissociation, α, due to a greater stretching of the macromolecule as repulsive forces are enhanced through the generation of COO^- from COOH as base is added. Due to this expanding effect, the remaining acidic functional groups ionize with increasing difficulty, thus causing a continuous decrease in K_{app}.[4] To minimize this effect a great amount of a neutral electrolyte, usually KCl, is added to keep the macromolecule in a contracted form.

Although there is disagreement about the pK_a values of HS that are recorded in the literature, the pK_a provides a convenient means of comparing the strength of COOH groups, both between HS from various sources and for any given HA or FA as affected by neutral salts. In general, the lower the pK_a the stronger is the acidic group. The pK_a values may also give an indication of the expected degree of ionization at higher and lower pH values.[4]

A model based on the assumption that COOH groups in HS can be described in terms of a mixture of low-MW organic acids has been proposed by Perdue et al.[349] In this model the dissociation constants are normally distributed so that they can be described by a symmetrical Gaussian distribution curve. A similar description can be obtained for OH groups. Other mathematical models applied to describe proton binding by HS include continuous distribution models[350] and affinity spectrum models.[351] A detailed discussion of these models is beyond the scope of this chapter and the reader is referred to the review by Perdue[325] for additional information.

b. Conductometric Titrations

Conductometric titrations have been conducted with variable results to evaluate the end points for acidic groups in HS. From one to four breaks have been shown to occur in conductometric titration curves of HS.[279] Two end points have been obtained for a FA sample, the first believed to be due to COOH groups ortho to phenolic OH groups, and the second to COOH not adjacent to phenolic OH groups.[350]

c. Titration in Nonaqueous Solvents

Titration in nonaqueous, aprotic solvents such as pyridine, dimethylformamide, and ethylene-diamine, by use of a very strong base such as the ethoxide ion has been attempted to determine the end points for acidic functional groups in HS. A single inflection point that corresponds to the total COOH and phenolic OH content was obtained by titrating some HAs and FAs with sodium aminoethoxide in the above-mentioned solvents.[316]

Sodium isopropylate has been used as titrant in dimethylformamide as a solvent for conducto-metric and high-frequency titration aiming to determine COOH groups in some HAs.[352] Several distinct breaks were evident on titration curves of HAs obtained using potassium methoxide as titrant and pyridine as solvent, although no attempt was made to assign the inflections to particular acidic groups on the HA macromolecules.[353]

Unfortunately, any comparison of "nonaqueous total acidity" and "aqueous total acidity" and any efforts to subdivide nonaqueous total acidity into functional group classes by analogy to pK_a values of acidic groups in aqueous solution are theoretically unfounded.[325]

d. High-Frequency Titrations

High-frequency titrations in which the electrodes are placed outside the titration vessel have been attempted for both aqueous and nonaqueous HS.[352] The titration curves obtained agree well with those obtained by conductometric titration but positions of end points were more pronounced.

e. Thermometric Titrations

The neutralization of an acidic functional group with OH⁻ is usually accompanied by the evolution of heat. The reaction is quantitative and the heat evolved is directly proportional to the amount of base added for acidic functional groups with $pK_a < 9$. The slope of heat vs. moles of reacted base is the enthalpy of neutralization of the acidic functional group. However, the reaction with added OH⁻ is incomplete and the amount of heat is not proportional to the amount of added base if the acidic functional group is rather weak ($pK_a > 9$).[325]

The technique of thermometric titration consists in monitoring the temperature change of the solution as a function of the amount of titrant added, or as a function of time when the titrant is added at a constant rate.[4] Titration calorimetry has been applied to detect the acidic functional groups of HS.[354-358] The enthalpy of neutralization of the acidic functional groups of HS was found to be constant over most of the titration and equal to the expected value for neutralization of COOH groups.[357,358] At higher levels of base added the heat evolved was no longer proportional to the amount of base added because weakly acidic groups, possibly including weaker COOH and phenolic and enolic OH groups, were subsequently neutralized. Thus, these groups could not be quantified by this method.

Table 14   Contribution of Organic Matter and Clay Fractions to
Soil Cation Exchange Capacity as Influenced by pH

Buffer pH	Clay fraction (cmol kg⁻¹ clay)	Organic fraction (cmol kg⁻¹ SOM)	% of CEC due to SOM
2.5	38	36	19
3.5	45	73	28
5.0	54	127	37
6.0	56	131	36
7.0	60	163	40
8.0	64	215	45

From data in Reference 362.

2. Cation Exchange Capacity (CEC)

Soil organic matter in general, and HS especially, are a major contributor to the CEC of most soils, especially organic soils. For many soils a direct relationship is held between CEC and SOM content.[359] Even though the amount of HS present in mineral soils may be quite low relative to clay, from 25% to 90% of the total CEC of the top layer of these soils is estimated to be due to HS,[4] whereas from 66.4% to 96.5% of the CEC of sandy soils is attributed to SOM.[360]

The occurrence of charged sites, e.g., COO^- and O^-, accounts for the ability of HS to retain cations in nonleachable forms prevalently by coulombic and electrostatic forces, although the bonds may be partly covalent for some polyvalent cations. Humic substances are a variable-charge soil component with a low point-of-zero charge, about 3. Thus, HS are negatively charged at pH >3. Since COOH and phenolic OH groups can deprotonate at pHs common in many soils, these groups are the major contributors to the negative charge of soils. Up to 55% of the CEC from SOM is estimated to be due to COOH groups ($pK_a < 5$),[361] while about 30% of the CEC of SOM up to pH 7 is due to phenolic and enolic groups.

The capacity of HS for binding exchangeable cations is not fixed but varies widely according to their chemical nature and properties. For organic soils, the greater the degree of humification of SOM the higher is the CEC.[4] In general, the CEC of HS increases dramatically with increasing pH, due to increased negative charge deriving from increased deprotonation and ionization of acidic functional groups at higher pH (Table 14).[362]

In the natural soil, CEC values for SOM and HS are expected to be somewhat less than for the isolated components, for the reason that some exchangeable sites are lost through association with clay minerals and polyvalent cations. Two types of CEC for SOM should thus be considered:[363] (1) "measured" CEC as determined experimentally by exchange with NH_4^+ or any other appropriate cation; and (2) "potential" CEC, the sum of the above and CEC due to "blocked" sites, which are exposed when SOM is extracted from soil.

E. Molecular Weight, Size and Shape

The molecular weight (MW), size, and shape are basic properties of HS which are important for several reasons: (1) to establish approximate molecular formulas in conjunction with elemental and functional group composition and other chemical data; (2) to allow the conversion of weight to molar concentrations for stoichiometric use; (3) to compare HS of different nature and origin; and (4) to better evaluate the interactions of HS with other soil constituents, such as clays, metal ions, organic and inorganic xenobiotics, and organisms.

Several problems are encountered when dealing with MW, size, and shape of HS because of the colloidal, polyelectrolytic, polydisperse, and physically and chemically heterogeneous nature of HS. In addition, these properties are greatly dependent on the physical state of the HS preparation, their concentration, and the pH and ionic strength of the medium. Thus, it is not too surprising if considerable disagreement exists in the literature between measurement methods and discrepancies are apparent between published results on MW, size, and shape of HS obtained by different researchers, even if the same types of HS samples have been used in different laboratories.

The literature on this subject has been covered in several reviews [279,308,311,315,323,364-369] and books.[3-5]

1. Molecular Weight

Unlike "monodisperse" macromolecules that exhibit a single value for MW, HS are "polydisperse", that is, they exhibit a range of MW that may vary from as low as a few hundred daltons for FAs to as much as several hundred thousand daltons for HAs. Because of the polydispersed nature of HS, methods should be applied which could provide distribution patterns for MW of HS. However, this has rarely been done and most methods applied to HS have recorded values that represent an "average" MW. Further, factors affecting MW measurements, such as HS concentration, pH, ionic strength, and type of polyvalent cations present in the medium, have not always been taken into account in determining MW of HS.

The "average" MW of polydispersed systems, including HS, can be expressed in several ways and each of these represents a different way of obtaining a mean value of a particular MW distribution, depending on the physical method of determination. The three most commonly used MW averages for HS are the following:

1. The number-average MW ($\overline{M}_n$), which is derived by summing the number of ith molecules (n_i) each multiplied by its MW (M_i) and dividing by the total number of molecules, is defined as

$$\overline{M}_n = \frac{\Sigma n_i M_i}{\Sigma n_i} \qquad (1)$$

The methods used to determine $\overline{M}_n$ involve estimates of the total number of molecules in solution, regardless of size, by the measurement of a colligative property such as osmotic pressure, vapour pressure, and freezing point depression.

2. The weight-average MW ($\overline{M}_w$), which is derived by summing the weight of ith molecules ($w_i = n_i M_i$) each multiplied by its MW (M_i), and dividing by the total weight of the molecules, is defined as

$$\overline{M}_w = \frac{\Sigma w_i M_i}{\Sigma w_i} = \frac{\Sigma n_i M_i^2}{\Sigma n_i M_i} \qquad (2)$$

The $\overline{M}_w$ is generated by methods, such as sedimentation velocity by ultracentrifugation and light scattering, which depend on the masses of sample in different fractions.

3. The z-average MW ($\overline{M}_z$), which is derived from the third moment of the MW (M_i), is defined as

$$\overline{M}_z = \frac{\Sigma n_i M_i^3}{\Sigma n_i M_i^2} \qquad (3)$$

The $\overline{M}_z$ may be calculated from sedimentation data, particularly those derived from equilibrium ultracentrifugation.

Although $\overline{M}_n$, $\overline{M}_w$, and M_z are the MW averages most commonly used, a viscosity-average MW ($\overline{M}_v$) has been also defined and measured relatively simply.[366]

Table 15   Molecular Weights Recorded for Humic Substances

Method	Type of Material		
	Humic acid	Fulvic acid	Unspecified
Ultracentrifugation			
Sedimentation-diffusion	53,000–100,000[a]		
Sedimentation-viscosity	22,000–28,000		
Equilibrium sedimentation	24,000–230,000		
Viscosimetric properties	36,000		
Freezing-point depression	25,000	640–1,000	670–1,680
Osmotic pressure measurements		951[b]	47,000–53,800
Light scattering	65,000–66,000		
Small-angle X-ray scattering	200,000–1,000,000		

[a] Range of 2,000–1,360,000 for highly fractionated samples.
[b] Range of 275–2,110 following fractionation by gel filtration.

Data from Stevenson.[4]

In general, for homogeneous, monodisperse systems, $\overline{M}_n = \overline{M}_w = \overline{M}_z$, whereas for heterogeneous, polydisperse systems like HS, $\overline{M}_n > \overline{M}_w > \overline{M}_z$, and $\overline{M}_v$ will lie somewhere between $\overline{M}_n$ and $\overline{M}_w$. There is a substantial difference between $\overline{M}_n$ and $\overline{M}_w$. In a polydisperse system, $\overline{M}_n$ tends to be strongly influenced by lower MW components, whereas $\overline{M}_w$ tends to emphasize the contribution of the heavier molecules in the mixture, and the $\overline{M}_z$ does this to an even greater extent, thus resulting in higher MW values than M_n. As a result, M_w is generally considered a most representative average MW value for use with polymeric, macromolecular materials and to correlate better with molecular properties of HS. In any case, these averages are not directly comparable and it is important to know the methodology used and the type of average MW reported to evaluate MW values of macromolecules such as HS.

An overview of results obtained for MW of HS is presented in Table 15.[4] Data presented in Table 15 show that there is little agreement between MW determined by the various methods. As expected, MW based on changes in colligative properties (numbers of particles) are lower than those obtained by most other methods. Comparisons of MW from one laboratory to another are further complicated by different conditions used in the measurements of MW, including HS concentration, pH, and ionic strength, and the possibility that molecular associations may have occurred in some instances.

a. MW Distribution and Polydispersity

Since HS are polydisperse materials and exhibit a range of MW, any particular MW average is strongly affected by the mixture of molecules in the system and by their individual MW distribution and/or the related "degree of polydispersity," which indicate the range or extent of the MW exhibited by a particular sample. Preparations of HS of different origin and nature tend to possess different degrees of polydispersity. Figure 13 shows a "notional", but probably not unrealistic, MW distribution for a soil HA.[367] The MW distribution is shown as a continuum ranging from about 1,000 to 500,000 Da, with a maximum concentration of species around 50,000 Da. The calculated average MW values for this particular distribution are $\overline{M}_n = 154 \times 10^3$, $\overline{M}_w = 244 \times 10^3$, and $\overline{M}_z = 284 \times 10^3$ Da.[367]

The ratio $\overline{M}_w/\overline{M}_n$ can provide an indication of the degree of polydispersity of the system. For homogeneous, monodisperse systems, $\overline{M}_w/\overline{M}_n = 1$, whereas values of $\overline{M}_w/\overline{M}_n$ between 1 and 2 are indicative of substantially polydispersed systems, and values greater than 2 would indicate extremely polydisperse systems with very wide ranges of MW and large amounts of macromolecules spread throughout the MW distribution. For the polydispersed system presented in Figure 13, the ratios are $\overline{M}_w/\overline{M}_n = 1.58$ and $\overline{M}_z/\overline{M}_w = 1.16$.[367]

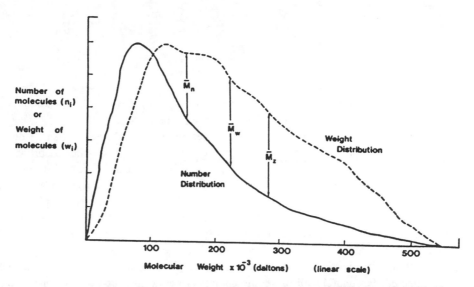

Figure 13   A notional, but plausible, molecular weight distribution for a soil humic acid extract, showing the distribution of both the number and weight of molecules and the calculated values of $\overline{M}_n$, $\overline{M}_w$, and $\overline{M}_z$. (From Swift, R. S., in *Humic Substances, Vol. 2, In Search of Structures,* Hayes, M.H.B. et al., Eds., John Wiley & Sons, New York, 1989, chap. 15. With permission.)

On the basis of the above considerations, the use of more than one technique which can provide both $\overline{M}_n$ and $\overline{M}_w$ and/or $\overline{M}_z$ is suggested, so that the degree of polydispersity can be estimated by the calculation of the ratios $\overline{M}_w/\overline{M}_n$ and, possibly, $\overline{M}_z/\overline{M}_w$. An alternative approach is to use a single average MW value in conjunction with techniques that allow the assessment of MW distributions, such as gel chromatography or gel electrophoresis (see later in this chapter). However, the best way to obtain more accurate information on the MW of HS is to measure the MW of HS fractions with a greatly decreased degree of polydispersity obtained by extensive fractionation of HS on the basis of MW.[367]

b. Ultracentrifugation

Analytical ultracentrifugation can be generally used in two distinct experimental modes to determine MW, the sedimentation velocity mode and the equilibration mode.

The molecular weight, *M,* of the sedimentary particles in a sedimentation-velocity experiment can be calculated from the Svedberg equation:[368]

$$M = \frac{RTS_0}{D_0\left(1 - \bar{v}\rho_s\right)} \tag{4}$$

where S_0 and D_0 are, respectively, the sedimentation coefficient and the diffusion coefficient at infinite dilution, ρ_s is the density of the solution, v is the partial specific volume of the solute determined separately by density measurements, R is the gas constant, and T is the absolute temperature.

In equilibrium centrifugation, the equations used are[368]

$$M = \frac{RT}{\left(1 - \bar{v}\rho_s\right)\omega^2 rc} \frac{\partial c}{\partial r} \tag{5}$$

or

$$M = \frac{2RT}{(1-\bar{v}\rho_s)\omega^2} \frac{d\ln c}{dr^2} \tag{6}$$

where ω is the angular velocity, r is the distance from the axis of rotation, c is the solute concentration, and other symbols have been identified above for Equation 4.

Equilibrium ultracentrifugation is an equilibrium measurement, and as such can be described by Equations 5 and 6, which are thermodynamically derived and have a sounder theoretical basis than Equation 4, which describes a transport phenomenon that is much less understood than the equilibrium experiment.[370,371] However, the derivation of these equations is not considered rigorous and involves a number of assumptions and approximations. The derivation strictly applies to a two-component system, in which S and D have been measured in the same solution at the same temperature, and have been extrapolated to infinite dilution, i.e., zero-concentration.[372]

The behavior and effect of various interactions that occur in the ultracentrifuge for both uncharged and charged sedimenting macromolecules, the counterions and background electrolytes,[368] are neglected. Fortunately, all these interactions are either reduced to zero or made negligible by extrapolation to zero concentration. In addition, charge effects at finite concentration are reduced greatly by making measurements in a large excess of background electrolyte (0.1 M KCl is commonly used).

The sedimentation-velocity technique is not good for handling highly polydisperse materials such as HS, because this leads to rapid spreading of the sedimenting boundary and consequent difficulties in making suitable measurements. However, a small amount of polydispersity can be tolerated. On the other hand, major advantages of the equilibrium method are that: (1) MW values can be determined without the need for measuring the diffusion coefficient; (2) a limited amount of polydispersity is tolerated; and (3) it is possible to determine $\overline{M}_n$, $\overline{M}_w$, and $\overline{M}_z$ on the same sample, thus also allowing the estimation of the degree of polydispersity in the sample. A major disadvantage of this method is that a long time is required to attain equilibrium, of the order of 16 h or more. However, mathematical modifications of the technique are available which can shorten substantially the experimental time required.

Relatively few studies have been performed on the application of ultracentrifugation to the determination of MW of HS. These studies have been reviewed by Flaig et al.,[279] Hayes and Swift,[6] Stevenson,[3,4] Wershaw and Aiken,[311] and Swift.[368]

The main problems encountered in the application of ultracentrifugation to HS are charge effects, polydispersity, and choice of optical detection conditions.[368] In particular, the combined result of charge effect is that the observed sedimentation coefficient, S, is anomalously low and, conversely, the diffusion coefficient, D, is anomalously high, thus unrealistically low MW values may be obtained. These charge effects can be readily overcome, however, or at least reduced, by adding a large excess of background electrolyte, usually 0.1 to 0.2 M.

The earliest $\overline{M}_w$ values measured by sedimentation-velocity ultracentrifugation of whole extracts of soil HAs were 53,000,[373] 25,000,[374] and 77,000 Da.[375] The $\overline{M}_w$ of a high-MW FA fraction measured by a combination of methods involving the ultracentrifuge and electrophoresis was found to be 5893, whereas the $\overline{M}_n$ of the same sample was 3570.[304] Values of $\overline{M}_w$ from 19,000 to 71,000 have been found for HAs by Orlov et al.[376]

A more detailed ultracentrifugation study was conducted on a number of fractions of very low polydispersity obtained by extensive fractionation of HAs extracted sequentially from a single soil.[377] The $\overline{M}_w$ values ranged from as low as 2,400 to as high as 1,360,000 Da. The most abundant portion of the MW distribution was found at a MW of about 100,000 Da, and the MW distribution was probably unimodal with an extended high MW tail.

Equilibrium ultracentrifugation has been used to determine $\overline{M}_n$, $\overline{M}_w$, and $\overline{M}_z$ values by a single run of a number of soil HA extracts, of which one was partially fractionated in three fractions using ultrafiltration membranes.[378] The $\overline{M}_w$ increased from 24,000 Da through 74,000 Da to >230,000 Da in the fractions, with the most reliable data believed to be those obtained for the low-MW fractions. The ratios of $\overline{M}_n/\overline{M}_w$ and $\overline{M}_w/\overline{M}_z$ were much greater than unity, therefore indicating a high level of polydispersity.

A density gradient technique applied to a peat soil HA has given a $\overline{M}_w$ value of 216,000 Da, with approximately 40% of the sample estimated to have MW between 20,000 and 160,000 Da.[379] The large differences between $\overline{M}_n$, $\overline{M}_w$, and $\overline{M}_z$ values provide, once again, evidence for the pronounced polydispersity of soil HA. In a successive study,[380] a peat soil HA partially fractionated by ultrafiltration followed by preparative ultracentrifugation gave a $\overline{M}_z$ of 23,900 Da for the low-MW fraction and 146,200 Da for the high-MW fraction.

In conclusion, all of the soil HAs studied by ultracentrifugation methods without previous fractionation were highly polydisperse and had MW ranging from 2,000 to 1,000,000 Da, with the most abundant component found in the MW range 20,000 to 150,000 Da. However, average MW of unfractionated polydisperse HAs are of limited value if not accompanied by a MW distribution measurement. Better results are obtained on HA fractions with low dispersity.

c. Viscosimetry

Viscosimetry is a measure of the frictional resistance that a flowing liquid offers to an applied shearing force. The coefficient of viscosity, η, is the proportionality constant between this shearing force and the velocity gradient in the liquid. Advantages of the technique are that measurements are easy, not very time-consuming, and made on homogeneous solutions with a simple apparatus that provides a direct numerical value.

Viscosity measurements on dissolved soil HAs and FAs have been made by capillary viscosimeters and can provide useful information on their molecular weight, size, shape, and polyelectrolytic behavior. However, most of the reported viscosity values on HAs and FAs are incompletely determined or calculated, and in most cases are not comparable because of differences in extraction, fractionation, and purification techniques.[6]

Molecular weights, M, can be calculated from viscosity measurements using the Staudinger equation:

$$[\eta] = KM^\alpha \qquad (7)$$

where $[\eta]$, the intrinsic viscosity, is related to M by two adjustable parameters, K and α, whose values depend on the nature of the polymer and the solvent, and also on the temperature. These parameters are generally determined for a homologous polymer series with standards of known MW.[371] Values of K and α in Equation 7 can also be calculated on the basis of MW that have previously been determined by a direct method such as ultracentrifugation. The value of α for a rod-shaped molecule of high MW is approximately 1.8, for wide thin discs $\alpha \approx 1.0$, whereas for an unsolvated sphere $\alpha = 0.5$.[366] For flexible, water-soluble macromolecules such as HS, reported values of α in Equation 7 vary from 0.5 to 0.8,[381] whereas for the particular case of flexible polymers which behave hydrodynamically as a sphere a value for α of 0.5 has been deduced theoretically.[382] More detailed information on the derivation of Equation 7 can be found in Clapp et al.[366] and Stevenson,[4] whereas a complete theoretical derivation can be found in Tanford.[370]

The interpretation of values of intrinsic viscosity, $[\eta]$, recorded in the literature for HAs and FAs presents several difficulties, not only because of the wide range of MW of HAs and FAs of various nature and origin, but also because, for charged macromolecules such as HS, $[\eta]$ is inevitably affected by the pH of the solution and by the nature and concentration of background salts.[383,384]

Values of the so-called viscosity-average MW, $\overline{M}_v$, calculated by use of Equation 7 with a value of $\alpha = 0.5$ for soil HAs are 36,000 Da,[385] 29,000 and 110,000 Da [374,386] 2700 Da at pH 7.0,[384] and 3700 Da at pH 6.8.[387] For FAs, $\overline{M}_v$ values of 1200 Da at pH 6.0[384] and 2300 Da at pH 6.5[387] are reported. The $\overline{M}_v$ calculated by Chen and Schnitzer[384] were found to increase from 2625 to 3145 Da for HAs, and from 1742 to 6366 Da for FAs, with a decrease in pH from 10 to 1.

To date the most comprehensive study on $\overline{M}_v$ of soil HA has been provided by Visser[385] on six MW fractions of a soil HA obtained by an ultrafiltration technique. Values of $\overline{M}_v$ ranging from 6200 to 68,000 Da have been measured at pH 7 in 0.1 M NaCl.

d. Colligative Properties

Number-average MW ($\overline{M}_n$) of HS can be generated by colligative property measurements. By definition, a colligative property is a thermodynamic property that depends only on the number of particles in solution, and not on the nature of these particles. The colligative properties of a solution and the effects utilized in MW measurements are vapor pressure lowering (vapor pressure osmometry), freezing point lowering (cryoscopy), boiling point elevation (ebulliometry), and osmotic pressure (membrane osmometry). Each of these properties is shown to be related to the lowering of the vapor pressure of the solvent by the addition of solute. In the limit of infinite dilution, each of these properties is proportional to the number of molecules of solute present, and the $\overline{M}_n$ value of the solute can be determined.

Vapor pressure lowering and freezing point depression are useful methods for determining $\overline{M}_n$ values for solutions of macromolecules up to about 50,000 Da in MW,[388] and as such have been used extensively for the determination of $\overline{M}_n$ of HS. Both methods have the advantage of simplicity of equipment required, and are useful for solutes of relatively low MW, such as FAs. The basic theory and methodology of these methods are presented in the review by Aiken and Gillam,[364] whereas more details on measurement methods and analysis and treatment of data can be found in Bonnar et al.[389] and Glover.[388]

Two significant problems are involved in the measurement of $\overline{M}_n$ of HS by colligative property methods. These are (1) deviations from ideal behavior due to the polydisperse, polyelectrolytic acidic nature of HS that do not form ideal solutions; and (2) dissociation of HS in water, which yields a $\overline{M}_n$ value lower than the actual, because the dissociated protons are "counted" as solute molecules, thus increasing the value of the number of solute species present.[390] As a consequence, suitable correction procedures and appropriate correction factors must be applied to eliminate, or at least limit, these problems.

Several procedures for correction of the dissociation of acidic functional groups of HS in aqueous solution have been described in the literature,[390-393] and reviewed by Aiken and Gillam.[364] In general, it is difficult to evaluate which method is the best because of the lack of an appropriate standard/reference HS sample with a $\overline{M}_n$ value verified by other methods. However, the most suitable correction procedures are considered those that make no assumptions about the chemistry of dissociation of HS but only require a knowledge of solution pH and instrumental constants, such as the procedure proposed by Gillam and Riley.[392]

Another approach proposed to circumvent the problem of the dissociation of HS in water is to suppress dissociation by carrying out the measurements in a suitable organic solvent that has a dielectric constant lower than 15. However, HS and especially HAs, are insoluble or only partially soluble in most organic solvents, and the potential of dimerization exists.[394] Solvents that have been used successfully for MW determinations by cryoscopy include sulfolane, dimethyl sulfoxide, and 1,4-dioxane. However, only sulfolane has been employed for HS.[395,396] Another suggested possibility[364] would be to derivatize the HS sample, e.g., by methylation, and determine $\overline{M}_n$ values in a suitable solvent such as dimethylformamide for vapor pressure osmometry.

Almost all the available data on $\overline{M}_n$ values measured by colligative properties have been obtained on FAs and very few $\overline{M}_n$ values are reported for HAs. The $\overline{M}_n$ values of soil FAs have been usually determined by cryoscopy or vapor pressure osmometry in water, and range from 640 to 999 Da.[390,391,397] Some work on HAs has been carried out using cryoscopy with sulfolane as solvent.[395,396] Values of $\overline{M}_n$ ranging from 47,000 to 53,800 Da were obtained for nondializable SOM from a Spodosol.[398] Values of $\overline{M}_n$ of 1684 and 669 Da are reported for unfractionated SOM (HA + FA) extracted from the A_0 and B_h horizons of a Spodosol, respectively.[396] Values of $\overline{M}_n$ between 300 and 900 Da have been obtained for FAs and from 9,000–10,000 to 68,000–74,000 Da for some HAs from Russian soils.[376]

In conclusion, $\overline{M}_n$ data generated by colligative property measurements for highly polydisperse systems such as HS are of limited value. At best, they can represent only an estimate of the MW of HS, and their validity can only be evaluated by determining the extent of polydispersity of the system by other suitable methods of analysis, such as ultracentrifugation and small-angle X-ray scattering.[364] Finally, another approach could be to perform experiments on well-fractionated materials.

e. Light Scattering

The determination of MW of polymeric substances such as HS can also be performed by the technique of light scattering. The measurement is based on the direct proportionality existing between the intensity of scattered light, R_0, at any given concentration, c, of the sample and the molecular weight, M, according to:

$$\frac{kC}{R_0} = \frac{1}{M} + \frac{2Bc}{RT} \tag{8}$$

where B and C are the so-called virial coefficients of osmotic pressure, and k is the solvent-dependent constant of unpolarized primary light. The parameter k depends on the refractive index of the solvent and its change with concentration of the solution, and on the wavelength of primary light.[4]

The value of M can be obtained from Equation 8 as the intercept value 1/M of the plot of k C/R_0 vs. c in a line of slope 2 B/R T. Application of this method has given MW of 65,300 and 66,200 Da for two HAs.[376]

f. Electrophoresis

As discussed earlier in this chapter, electrophoretic procedures have been mostly used for separation or fractionation of charged components of HS, whereas little effort has been made to use them for the quantitative determination of MW and molecular size of HS.[315]

The retardation of molecules in an electrophoretic experiment conducted in gel media, such as polyacrylamide gel, is dependent on the relative diameters of the gel pores and the molecules being separated. The retardation coefficient, K_R, of a molecule is related to electrophoretic mobility and gel concentration. The K_R value can also be related to molecular radius and MW of the molecule. Assuming that the molecule is spherical and nonhydrated, a simple relation of proportionality exists between K_R and the molecular weight, M, that is[315]

$$M = a + b K_R \tag{9}$$

where a and b are constants. For the determination of MW of proteins, a plot of K_R vs. M is prepared using protein standards of known MW and assuming that the shape, degree of hydration, and density of all standards and the unknown molecule are the same. Principles and reliability of MW determination by gel electrophoresis have been discussed in Rodbard.[399]

The best application of electrophoresis to the determination of molecular parameters of HS has been provided by Kasparov et al.[400] By use of isotachophoresis in Sephadex G-200, followed by zone electrophoresis in polyacrylamide gel, five discrete bands were obtained for two soil HA samples. Molecular parameters for the various fractions were calculated assuming that the HA molecules were spherical and that procedures and standards commonly used for proteins were directly applicable to HA. Molecular diameters calculated on this basis for the HA fractions ranged from 1.6 to 15.9 nm, and their MW expressed as $\overline{M}_w$ ranged from 2.2×10^3 to 2.0×10^6 Da.[400] The mean $\overline{M}_w$ of almost 6×10^4 Da measured from both soil HAs examined was well comparable with data obtained by other methods of determining MW, such as ultracentrifugation and light scattering. However, because of the assumption used, the molecular parameters obtained by electrophoresis should be considered only as approximations. Further, both complexing of HS with metals and the presence of urea, which are shown to affect the electrophoretic fractionation patterns of HS, are suggested to alter MW and molecular size measurements of HS by electrophoretic techniques.[315]

2. Molecular Size and Shape

The most common methods that have been used to measure the molecular size and shape of soil HS include ultracentrifugation, small-angle X-ray scattering, light scattering, gel chromatography, and electron microscopy. Some of these methods are often used in conjuction with separation and fractionation of HS. In most of these methods, model compounds of known MW and composition are used to estimate the molecular size of HS. Problems may arise if model compounds not sufficiently similar to HS are used. However, the choice of appropriate model compounds is made difficult by lack of detailed structural information on HS themselves.

a. Ultracentrifugation

The analytical ultracentrifugation method can be used to generate data useful for the estimation of particle shape and sizes of macromolecules. Being a transport phenomenon, sedimentation velocity is related to the frictional coefficient, f, of the sedimenting macromolecule, which is given by:

$$f = \frac{kT}{D} \tag{10}$$

where k is the Boltzmann constant, T is the absolute temperature, and D is the diffusion coefficient discussed above.

The molecular conformation of a given macromolecule can be deduced from the value of the so-called frictional ratio, f/f_{min}, where f is the observed frictional coefficient of the macromolecule and f_{min} is the hypothetical frictional coefficient of a condensed, solid sphere having the same volume as the molecule studied. The ratio $f/f_{min} = 1$ if the studied molecule has the conformation of a rigid, condensed sphere, whereas distortions from a spherical shape and solvation yield a value of f/f_{min} above unity, which increases with these effects increasing.

The equations that allow the calculation of the frictional ratio of a macromolecule based on ultracentrifugation data have been derived by Tanford[370] and summarized by Swift.[368] The knowledge of the relationship between the frictional ratio and MW can provide an approximation of the probable type of conformation adopted by a macromolecule.[368] Thus, sedimentation velocity measurements made on a number of fractions may provide useful information on both the MW and the molecular size and shape of a given macromolecule.

Some information on the molecular size can also be derived from the radius of gyration, R_G, of a molecule, which is given by:[370]

$$R_G = \frac{kT}{0.665 \times 6\pi\eta D}$$
(11)

where η is the viscosity and other symbols have been explained above.

Frictional ratios determined on unfractionated soil HA extracts are generally much greater than unity and range from 1.5 to 2.3.[373-375] Frictional ratios of 1.38 and 1.44 were obtained at pH 7 for low- and high-MW fractions of a soil HA.[380] In their study of an extensively fractionated HA, Cameron et al.[377] determined frictional ratios that increased almost linearly with increasing MW of the fractions, and ranged from 1.14 for the lowest MW fraction up to 2.41 for the highest MW fraction.

These values of f/f_{min} are indicative of a molecular conformation of either a condensed oblate ellipsoid or a highly solvated, random coil.[368] However, of the two models, the flexible, expanding, random-coil model appears to fit better with the known chemical and physical properties of soil HS. The conformation of a randomly coiled HA in solution can be visualized as bounded with a roughly spherical shape perfused throughout with solvent which can exchange with the bulk solvent outside the molecule. Hence, the shape of the molecule in solution would be essentially spheroidal, but it is not condensed nor rigid.[368] In their study on the effects of pH and metal ions on soil HA, Ritchie and Posner[380] showed that changes in molecular size could occur, but the shape was largely unaffected.

The radius of gyration, R_G, has been calculated from ultracentrifugation measurements for soil HA and HA fractions.[375,377,380] The value of R_G was shown to increase with MW of HA fractions, and to vary from about 1.5 nm for the lowest MW fractions to greater than 25 nm for higher MW fractions, with the most abundant molecules having radii ranging from about 4 to 10 nm.[377]

b. Scattering Techniques

Light scattering and small-angle X-ray scattering (SAXS) have been used for determining size and shapes of HS. In particular, the size parameter that has been obtained for HS from scattering techniques is the radius of gyration of scattering particles. When electromagnetic radiation impinges on a molecule, part of it is absorbed, part is subjected to elastic scattering, and part to inelastic scattering. Only scattered radiation having the same wavelength of the impinging radiation, i.e., elastic scattering, is considered here. It can be shown that the angular distribution of the scattered intensity is a function of the size and shape of the scattering particles. The basic principles of scattering of electromagnetic radiation have been discussed in some detail by Wershaw;[369] a more complete discussion can be found in Dover.[401]

The basic equation for particle size measurement by SAXS was developed by Guinier and Fournet[402] who showed that for an ensemble of randomly oriented, identical particles, the radius of gyration, R_G, of the particle can be calculated from the slope of a plot of ln I vs. h^2, the so-called Guinier plot, where I is the scattered intensity and h the scattering angle. In monodisperse systems where all scattering particles are of equal size, the Guinier plot is a straight line, and a single radius of gyration is obtained. In polydisperse systems the Guinier plot is concave upward. If only a few widely different-sized particles are present in the system, it may be possible to discern discrete straight-line segments of the curve, which may be used to calculate radii of gyration of the different-sized particles. In the case of highly polydispersed systems of particles a Guinier plot can yield only the range of particle sizes present.[369] The change in particle size distribution obtained by altering a property of the system, such as pH, ionic strength, or temperature, is reflected in a change in the shape of the Guinier plot. If the particle size distribution becomes more heterogeneous, the plot becomes more concave, whereas the plot flattens approaching a straight line if heterogeneity is reduced.

Measurements by SAXS on solutions of unfractionated soil sodium humate showed that this material is polydisperse and composed of particles having radii of gyration ranging from 3.6 to 13.7 nm.[403] Lindqvist[404] found that hydrolysis of a polydisperse soil sodium humate yields an apparently monodisperse system with all particles having a radius of gyration between 1.8 and 1.9 nm. Further studies[309,405] by SAXS on HA fractions isolated by adsorption chromatography on Sephadex indicated that molecular aggregates are formed in solution, and that the degree of aggregation was a function of both pH and concentration. This dependence reflected the interaction of several different bonding mechanisms, possibly including hydrogen and charge-transfer bondings. By comparing the Guinier plots of the various fractions of a soil HA with that of the unfractionated HA, an apparent reduction in the sizes of particles in the fractions was observed, and attributed to the breaking up of aggregates in the fractionation process.[406] Air oxidation of an apparently monodisperse fraction of a soil HA had the effect of reducing the size of the particles measured by SAXS in water and DMF solutions at pH 8.8.[407] At lower pH in water the fraction was shown to aggregate and the system became polydisperse.

In principle, other information can be obtained from SAXS measurements, including MW, molecular volumes, and molecular shapes. These results, however, are meaningful only for monodisperse systems and would limit the application of SAXS to highly fractionated HS samples.

The light scattering techniques allow for the measurement of the same parameters as SAXS, but the resolution obtained is inherently lower than SAXS because of the longer wavelengths used. Apparently, only one light scattering study on HAs has been published,[376] and has provided the MW of HAs, as discussed above in this chapter.

c. Gel Chromatography

As discussed above in this chapter, gel chromatography has been extensively applied to fractionation of HS on the basis of their molecular size. In a homologous series of compounds with similar shape and structure, a relationship exists between the size of molecules and their MW, and the elution volume is linearly related to the logarithm of the MW over a certain range. A calibration plot of the elution volumes of homologous compounds of known MW vs. their log MW can thus allow the determination of the unknown MW of a compound of the homologous series.

Therefore, to measure the absolute molecular sizes of the HS sample being fractionated, the column must be calibrated with standards of physical and chemical properties similar to the HS. In attempts made to calibrate Sephadex gels for use with HS, HAs, and FAs of known MW, as well as phenolic acids, benzene-carboxylic acids, and polyphenolic compounds have been used, but almost unsuccessfully.[377,378,408,409] Thus, the use of gel chromatography for accurate molecular size determinations is limited by the availability of HAs and FAs standards.

Values of MW estimated for six fractions of a soil FA by using Sephadex gels of five different grades and water as eluent were found to be 2 to 10 times higher than MW values measured by other analytical independent methods, such as vapor pressure osmometry, ultracentrifugation, and electrophoresis.[304] Possibly, elution performed with water may markedly affect ion exchange of the polyanionic FA molecules.

In conclusion, considerable difficulty is encountered in the interpretation and comparison of published results on molecular sizes of HAs and FAs deduced from gel chromatography, thus data obtained have to be regarded at best as tentative. More useful results can be obtained when the technique is used to study comparatively the effects of origin and of physical or chemical treatments on molecular size distribution of HS.

d. Electron Microscopy

The sizes and shapes of colloidal particles such as HS, i.e., their morphological conformation can be directly observed by use of electron microscopy. Two different types of electron microscopy have been applied for studying HS: transmission electron microscopy (TEM) and scanning electron microscopy (SEM). An excellent review and critical discussion on TEM and SEM techniques applied to HS has been provided by Chen and Schnitzer.[365]

One of the major problems in observing organic materials by TEM is to obtain sufficient contrast to distinguish the specimen. Thus, it is necessary to enhance contrast by an adequate staining technique that may involve the use of a heavy metal salt or the preparation of a carbon replica.

The SEM technique is based on a principle somewhat different from TEM in that the image is formed by the secondary electrons emitted from the sample surface when an electron beam is scanned across the surface. To prevent charging of the surface by electrons, it must be coated with a conductor such as gold.

Electron microscopic studies of HS are generally performed on samples obtained by drying solutions of these materials. Sample preparation, and especially drying procedures, were found to affect markedly the structural features of the HS sample, and thus to represent the most critical aspect of electron microscopy application to the study of HS. Maintenance of the original molecular shape of the sample in solution is obviously highly desirable, although probably unattainable. However, the application of improved drying techniques may avoid, or at least limit, the formation of artifacts.

Early studies in electron microscopy of HS have been conducted on air-dried samples. Air-drying the sample under ambient conditions or at constant relative humidity is the simplest method of water removal. However, changes in molecular conformation are likely to occur as a result of air-drying, and the structural features obtained are usually ill-defined and difficult to observe by electron microscopy.

At present, the most effective method of preventing shrinkage and distortion, and consequent dimensional and morphological changes, of the HS sample resulting from drying is rapid freezing followed by freeze-drying of the sample. Advantages of this procedure are to closely maintain the original volume and to limit structural damage of the sample.

TEM studies conducted by Orlov et al.[376] on soil HAs obtained by air-drying relatively concentrated HA solutions allowed the authors to distinguish large oval particles with MW of the order of millions on an almost continuous background of minute particles with diameters of approximately 3 nm. Spherical particles with diameters from 6 to 10 nm have been recorded for HAs by electron microscopy studies,[410,411] and MW greater than 20,000 Da are obtained from these dimensions.[412] SEM studies by Orlov et al.[376] on an air-dried soil HA revealed featureless strata or sheets of condensed material. Two FA samples obtained by air drying FA solutions originally at pH 2.5 and 4.5 examined by TEM[413] showed three types of particles at pH 2.5: small spheroids (diameter, 1.5 to 2 nm), aggregates of spheroids (diameter, 20 to 30 nm) tending to form elongated, irregularly shaped structures 2 to 3 μm long, and an amorphous material perforated by voids (diameter, 50 to 110 nm). Samples obtained from solutions at pH 4.5 showed flat sheet-like lamellae perforated by voids resembling the amorphous structure observed by Orlov et al.[376] A loose spongy structure with a large number of holes was previously observed by Khan[319] in micrographs of HAs in the H[+]-form. All these results have been possibly attributed to particle aggregation resulting from shrinkage occurring on air-drying the sample.

In the early 1950s, Flaig and his associates were probably the first to apply freeze-drying to TEM studies of HS. Results of this work, as summarized in Flaig et al.,[279] showed that the dimension and morphology of HA samples prepared by freeze-drying small droplets of HA solutions at various pHs on a supporting film with contrast increased by shadow coating by platinum, depended on the pH of the initial solution. Dispersed, single spherical particles with diameters up to 10 nm were observed for soil HA dried at pH 10, whereas the same material dried at pH 3 appeared as cluster-like aggregates in which single particles coalesced into large globular structures.

The effect of gradual changes in pH on the shape and particle arrangement of a HA and a FA were studied by SEM on samples dried by rapid freezing in Freon followed by freeze-drying at a temperature of −80 to −100°C.[384,414] A gradual change was observed for the morphology assumed by FA and HA macromolecules and aggregates with change in pH of the original solutions from

2 to 10 and from 6 to 10, respectively, for FA and HA. At acidic to neutral pH (from 2 to 7), FA and HA exhibited the shape of elongated, linear, or curved fibers that tended to become thinner with increasing pH, and of bundles of fibers that tended to become predominant at pH 6 and to give a fine network at pH 7. At higher pH (8, 9), the FA and HA assumed a sheet-like structure of increasing thickness, whereas at pH 10 a fine homogeneous grain-like structure was apparent. When FA samples were obtained by air-drying from solutions at pHs 2, 6, and 10, condensed thick layers were observed with no apparent morphological differences between samples obtained at different pHs.

Later reports have confirmed that HS obtained by drying solutions at neutral to acidic pH values mostly assume the shape of fibers or bundles of fibers.[416-420]

Electron micrographs obtained by Ghosh and Schnitzer[417] indicated an increase in fiber thickness and gradual decrease in particle orientation as the concentration of neutral salt in solutions of soil HAs and FAs increased. These effects were similar to those observed on increasing the pH from moderately acidic values to neutrality.

Stevenson and Schnitzer[418] studied by TEM the effect of varying the concentration of two soil FAs and HAs in solution at the time of freezing and drying on their size and shape. Single drops of dilute (100 mg l^{-1}) aqueous FA and HA solutions at pH values of 3.5, 7.0, and 10.0 were spread uniformly on freshly cleaned mica sheets to form replica samples. The sheets were frozen rapidly in Freon and freeze-dried. The procedure used permitted the formation of a concentration gradient. In the low concentration areas, the HA and FA particles were almost spheroidal, with diameters ranging from 9 to 27 nm, and there was a tendency to coalesce to form round-shaped aggregates or linear, chain-like structures. At intermediate concentrations, the spheroids and chain-like structures formed fiber-like shapes similar to those reported earlier.[414,415] At the highest concentrations, parallel arrays of filaments tended to coalesce to sheet-like shapes. The predominant type of particle and aggregate morphology appeared, therefore, to be determined by the initial concentration of FA or HA, whereas changes in pH had no effect.

The effect of dielectric constant of the solvent was studied by TEM on the structure of a soil FA dissolved at a concentration of 100 mg l^{-1} in water adjusted to pH 8–9 with NaOH, in water-methanol (1:1, v/v), or in water-dioxane (9:1, v/v).[365] Fibrous structures were apparently dominant in the aqueous system (dielectric constant, 78.5), whereas decreasing the dielectric constant to 56.3 (water-methanol) yielded some destruction of fibers, slight collapse and aggregation. The fibers appeared to reorganize into chain-like, coiling structures when a further decrease in dielectric constant to 35.25 (water-dioxane) was achieved. Significant structural changes appear, therefore, to result in decreasing the dielectric constant of the solvent, which reduced markedly the repulsion forces between negatively charged sites on humic macromolecules.

Wershaw and Aiken[311] reported unpublished results by Wershaw and Valder who observed by TEM HA particles with an approximate diameter of 2 nm, which was similar to the value measured by SAXS. The substantially larger diameters reported in previous studies were suggested to probably refer to aggregates of smaller particles rather than to single particles.

In conclusion, electron microscopy appears to be affected, more than most other techniques, by sample preparation procedures. The conformation of HS macromolecules is also strongly affected by the conditions of the initial sample, such as pH, sample concentration, ionic strength, and solvent characteristics. These factors strongly affect charge, solvation, hydrogen-bonding, dispersion, and aggregation of HS particles. Because of this, the size and shapes of HS observed by electron microscopy are unlikely to represent the size and shape of these materials in soils.

F. Fractal Nature

1. General Principles

Fractal geometry provides a powerful mathematical means for the description of structure and properties of random, heterogeneous, highly irregular systems encountered in nature, where disorder is considered as an intrinsic rather than a perturbative feature of the system.[421] Natural fractal systems exhibit "quasi" or "statistical" self similar structures characterized by dilation invariance, i.e., by scale-invariant properties. A fractal object can be described quantitatively by a nonintegral dimension, i.e., the fractal dimension, D, which reflects the actual space occupied by the system.[421,422]

In fractal analysis, a scaling or power law of the form

$$p \propto v^\gamma \tag{12}$$

is sought where p is the property, v is the variable, and the exponent γ can be related to a fractal dimension, D, which can be computed directly from experimental data. To obtain D, it is therefore necessary to find first any physical, chemical, or biochemical property of the system which can be described by a power law as shown in Equation 12.

Three types of fractal systems can be described:[422] (1) a "mass" fractal, that is a system wherein mass and surface scale are the same, and are characterized by a "mass" fractal dimension, D_m; (2) a "surface" fractal, i.e., a system for which only the surface is fractal, and is described by a "surface" fractal dimension, D_s; and (3) a "pore" fractal, i.e., a system wherein pore space and surface scale are the same, and is defined by a "pore" fractal dimension, D_p.

The most powerful and commonly used experimental means for studying fractal systems and for determining their fractal dimension are scattering techniques, which include scattering of visible or laser light, SAXS, and small-angle neutron scattering (SANS).[423] Another method to obtain the value of the fractal dimension of a system is based on the measurement of the wavelength dependence of the turbidity of the particle suspension of the sample by monitoring the nonscattered or transmitted light with a conventional spectrophotometer.[424] Optical microscopy and TEM and SEM techniques are also successfully applied for studying forms and shapes of fractal objects and for measuring their fractal dimension on the basis of detailed image-analysis of photomicrographs obtained.[425]

Fractal geometry is expected to represent a useful tool for the quantitative description of HS systems, their formation and transformation processes, and their physical and chemical behavior and properties. The reasons for which HS may be amenable to a fractal description are the following: (1) the apparently complex, irregular, and heterogeneous nature in their molecular organization; (2) the wide variety of sizes and shapes of various porosity and compactness in the solid and colloidal states; (3) the various degree of roughness of exposed surfaces and irregularity of reactive sites; (4) the occurrence of complex aggregation and dispersion phenomena; and (5) the variable extent of adsorption and catalytic interactions with mineral surfaces, metal ions, organic chemicals, plant roots, and microorganisms.

2. Applications

Two introductory reviews on the applications of the fractal concepts and methods in general soil science with focus on the study of HS have been provided recently.[426,427] The first experimental proofs of the fractal nature of HS have been obtained only recently.[428-435] More recently, detailed studies on the fractal behavior of HS in aqueous suspensions have been conducted.[436,437]

Analysis by SANS[428] of two soil HAs at concentrations of 1 to 4 g l^{-1} in water and D_2O solutions buffered at pH 5.0 and in the presence of 0.1 M NaCl, made after 48 h of standing at 11°C, showed that the measured scattering intensity, I(q), obeyed a power law decay as a function of the scattering vector, q, according to

$$I(q) \propto q^{-D} \tag{13}$$

Polydispersity was found not to affect the power law scattering observed, and thus the scattering depended only on the fractal dimension, D, which was calculated from the slope of the plot, log I(q)/log q. A corrected value of D = 2.3 ± 0.1 was obtained for both HAs, which were suggested to consist of monomers with a radial size of ≤2.5 nm aggregated into clusters with an average radius of 40 to 50 nm. The D value of 2.3 was related either to a diffusion limited aggregation (DLA) model or, more probably, to a reaction limited aggregation (RLA) model.[438] In the latter case, negatively charged functional groups in HAs at pH 5 would provide a barrier among monomers, excluding them from the dense, central part of the cluster, thus favoring the development of an open, ramified structure for HA.[428]

In two later studies,[429,430] SANS measurements of the same two HAs dissolved in D_2O at a concentration of 3.0 g l^{-1} at pH 5 and in 0.1 M NaCl, after storage at 4°C for 48 h, gave a value of D = 1.8 for both HAs. When the temperature was raised to 22°C, the value of D increased with time and reached a constant value of D = 2.35 after 60 h. The values obtained for D in this experiment indicated that at the low temperature, the initial aggregation of HA particles is mostly governed by a diffusion-limited cluster-cluster (CLA) process, whereas at 22°C initial clusters restructure with increasing time by a RLA process, from an open, ramified structure to form more compact aggregates.[430]

The SAXS technique has been used to investigate the fractal nature and measure the fractal dimension of a number of HAs and FAs from soil and other sources in the solid and solution states.[431-433] For the soil FA sample in the solid state, experimental log-log plots of the intensity of the scattered X-rays, I(q), as a function of the scattering vector, q, obeyed a power law of the form

$$I(q) \propto q^{-(6-D_s)} \tag{14}$$

where the exponent of q is in the range $3 < 6 - D_s \leq 4$, which is typical for surface fractals.[423] The calculated surface fractal dimension for the solid soil FA was $D_s = 2.5$.

The rapid, simple, and noninvasive turbidimetric technique, based on the measurement of the nonscattered, nonadsorbed, transmitted light by a conventional UV-Vis spectrometer has been used by Senesi and associates[434-437] to investigate the fractal behavior and determine the mass or surface fractal dimension of several soil and peat HAs in dilute aqueous suspension as a function of several system parameters, including pH, equilibration time, sample concentration, and ionic strength. Theoretical considerations[424,436] show the existence of a power law dependence for the turbidity, τ, on the wavelength, λ, with an exponent, β, that might be directly related to the fractal dimension, D, of the system under consideration, provided that certain assumptions are verified. According to the value of the exponent β obtained experimentally from the log-log plot of τ vs. λ, two forms of the power law can be used

$$\tau \propto \lambda^{D_m} \tag{15}$$

where $D_m = \beta < 3$ is indicative of the existence of a mass fractal, and

$$\tau \propto \lambda^{(6-D_s)} \tag{16}$$

where $D_s = 6 - \beta$ with β resulting in $3 < \beta < 4$, which indicates the existence of a surface fractal. When $\beta \approx 3$, the material is considered to have a nonfractal nature.

Two preliminary studies[434,435] conducted by using the turbidimetric method on two different soil HAs suspended in water at a concentration of 50 mg l^{-1} showed that both HAs exhibited a fractal nature with a fractal dimension that decreased with increasing either the pH from 3 to 7 or the equilibration time from 2 h to 24 h. Data from turbidimetric analysis, supported by SEM

Table 16   Mean Values ± Standard Error of the Mass Fractal Dimension, D_m (= β<3), of Elliot Soil Humic Acid in Aqueous Suspension at Various pH Values and After Different Equilibration Times

pH	2 h	4 h	8 h	16 h	24 h
3	Nonfractal[a]	Nonfractal	Nonfractal	2.81 ± 0.12aA[b]	2.79 ± 0.09aA
4	Nonfractal	Nonfractal	2.75 ± 0.14aA	2.49 ± 0.08bB	2.43 ± 0.17bB
5	2.81 ± 0.02aA	2.55 ± 0.05aB	2.53 ± 0.06bB	2.37 ± 0.01cC	2.12 ± 0.09cD
6	2.01 ± 0.10bA	2.13 ± 0.02bA	2.07 ± 0.08cA	1.94 ± 0.02dB	1.92 ± 0.02dB
7	1.76 ± 0.04cA	1.71 ± 0.06cA	1.64 ± 0.08dA	1.38 ± 0.05eB	1.21 ± 0.01eC

[a] Nonfractal system, $\beta \approx 3$, $\beta = D_m = D_s = d$.
[b] Means in the same column followed by the same lowercase letter and means in the same row followed by the same capital letter are not significantly different ($P \leq .05$, $n = 4$).

From Senesi, N. et al., *Soil Sci. Soc. Am. J.*, 60, 1773, 1996. With permission.

observations, were tentatively related to morphological features and aggregation processes of HA particles in suspension in various experimental conditions.

In two later, more detailed studies,[436,437] two soil HAs and a peat HA belonging to the IHSS Standard and Reference Collection of HA and FA samples, were used in suspensions at a concentration of 30 and 40 mg l^{-1}, at pH values of 3, 4, 5, 6, and 7, and at ionic strength obtained by addition of either NaCl, to yield a concentration of 1, 5, and 10 mM or $CaCl_2$ at a concentration of 0, 1, and 10 mM. Four replicates were prepared for each suspension to allow statistical analysis of data. Turbidimetric analysis and SEM observations have been performed on HA suspensions after equilibration at 25°C by magnetic stirring for 2, 4, 8, 16, and 24 h. Preliminary examination by particle-size analysis of the HA suspensions showed that the fundamental assumptions required for the application of the turbidity theory to the measurement of the fractal dimension of the HA suspensions were respected.

The analysis of the power-law dependence of the turbidity on the wavelength revealed that soil and peat HAs may exhibit a mass or surface fractal nature or a nonfractal status depending on their origin and experimental conditions.[436,437] One of two soil HAs showed a mass fractal nature over the entire ranges of pH and equilibration time examined, whereas the other soil HA assumed a nonfractal regime at low pH ($\leq$4) and after relatively short equilibration times ($\leq$8 h) and a mass fractal regime at higher pH (>4) and longer equilibration times ($\geq$8 h). At low pH ($\leq$5) the peat HA behaved either as a surface fractal at short standing times ($\leq$8 h), or as a nonfractal at longer standing times ($\geq$8 h), whereas at pH values $\geq$6 it assumed a mass fractal regime at any equilibration time.[436] In the presence of NaCl, the three HAs exhibited a mass fractal regime at any ionic strength and over the entire pH range. However, where the ionic strength was provided by $CaCl_2$, one soil HA maintained a mass fractal behavior, whereas the other soil HA and peat HA were described as surface fractals.[437]

In Table 16 an example is provided of the fractal and nonfractal regime assumed and D_m values measured for a soil HA as dependent on pH and equilibration time.[436] With an increase in pH or equilibration time, the mass fractal dimension of the three HAs decreased from about $D_m = 2.8$ to values close to $D_m = 1.0$. In general, a nonfractal response reflects the existence of HA particles having compact, space-filled structures with smooth surfaces; a surface fractal regime implies compact structures with highly corrugated surfaces; and a mass fractal with decreasing D_m dimension suggests increasingly porous, fragmented, and elongated structures possessing increasingly rough surfaces. These interpretations were supported by correspondent SEM observations.[436,437]

The value of the mass fractal dimension was also used to identify the possible underlying aggregation process occurring in HA suspensions in various conditions.[436,437] Low and intermediate values of the mass fractal dimension measured at near neutral pH would suggest an aggregation process of the RLA type, which implies the existence of short-range interactions between HA particles. Differently, the relatively high mass fractal dimensions measured at acidic pH values would indicate the occurrence of extended restructuration and/or reconformation of HA macromolecules which underlie a DLA type process.

In conclusion, fractal geometry provides a novel and nonconventional approach to the quantitative description by a numerical parameter, i.e., the fractal dimension, of the morphological features assumed by humic particles in aqueous systems and the underlying aggregation processes. The major weakness of the fractal approach is, however, in the averaging out of any detailed properties of the system studied and in providing results and interpretations that can be valid only in an average or statistical sense. A number of key problems involved in the fractal study of HS and several experimental constraints that may influence the determination of the parameters that characterize a fractal structure by scattering and turbidity techniques, as well as the need to standardize the appropriate use of fractal principles and methodology when applied to the calculation and interpretation of the fractal dimension of HS from experimental data have been addressed in a recent review.[427]

The intrinsic physical and/or chemical reasons for the different fractal nature (mass or surface fractal) and fractal dimension exhibited by HS as a function of the various conditions of the sample and the medium are still unknown. Further research is needed to provide a physical and chemical basis for the results of fractal analysis of HS and to elucidate at the molecular level the physical and chemical properties underlying the fractal nature and behavior of HS. Further studies are also necessary to find out which new insights into the physics and chemistry of HS may be provided by their fractal properties and by an extended application of fractal geometry.

G. Structural and Chemical Information by Degradative Methods

The primary objective of the degradation of HS macromolecules is to obtain simple compounds representative of the main structural units, which can provide information on the actual structural characteristics of HS. The usefulness of any degradative approach therefore depends on the isolation and identification of products whose chemical structure can be related, either directly or by deduction based on the mechanisms of the reaction occurring, to structural components of the starting macromolecules.

In order to reach these objectives, the degradative method chosen should not be too drastic, but yield high amounts of relatively simple degradative products, and the formation of undesirable byproducts and/or artifacts should be minimized. The use of a combination of degradative procedures with increasing severity may be desirable. An additional, useful approach consists in removing adsorbed or labile-linked compounds, such as polysaccharides and peptides, which are believed not to be integral parts of the "core" structure, e.g., by preliminary hydrolysis, thus decreasing the complexity of the products of degradation.

A variety of degradative methods have been applied for the study of soil HS, which include hydrolysis, reduction, oxidation, thermal degradation, and other procedures (Table 17). These have been reviewed in books[3-5] and several reviews.[6,323,439-444]

A set of advantages and limitations is typical of each procedure adopted. For example, a typical limitation of mild degradation methods is to yield low amounts of identifiable products, which are not sufficiently representative of the whole HS macromolecule to allow structural hypotheses. On the other hand, most drastic methods often produce small molecular fragments that do not provide any resemblance to the original macromolecule.

At present, more than 300 different compounds have been identified as degradation products of HS, whose number and kind depend on the specific method used for degradation.[4] Not all of them, however, can be considered as directly related to the actual structure of HS macromolecules. In other words, any constituent initially released from the macromolecule may be transformed or modified to other products by several reactions which may occur during the degradation processes.

Table 17   Major Degradative Procedures Applied to Humic Substances

Hydrolysis	Reduction	Oxidation	Thermal methods	Others
Water	Na-amalgam	Alkaline $KMnO_4$	Differential thermal analysis	Na-sulfide degradation
Base	Zn dust distillation	Cu-NaOH	Differential thermogravimetry	Depolymerization with phenol
Acid	Zn dust fusion	Alkaline nitrobenzene	Pyrolysis	
	Hydrogenation, hydrogenolysis	Nitric acid		
	Metal Na in liquid NH_3	Peracetic acid		
	Red P and HI	Sodium hypochlorite		
		Hydrogen peroxide		

1. Hydrolysis

Hydrolysis involves the cleavage of a bond and the addition of the elements of water to the cleaved products. The process may be conducted in water or catalyzed by acids or bases. Hydrolysis is generally considered as a mild degradation procedure which is effective in removing components labily bound to the "core" or "backbone" of HS macromolecules. As such, it has been used primarily as a pretreatment process of HS prior to further treatments. However, some evidence indicates that also some structural units of the "core" may be affected by hydrolysis.

Several important factors influence hydrolytic processes, including the hydrolytic reagents, solubility of the substrate, ratio of reagent solution to substrate, temperature and time of hydrolysis, and presence of oxygen and metal ions.[441] For example, hydrolysis will proceed more rapidly and efficiently if the substrate dissolves in the hydrolytic reagent. For this aspect, no problem is presented by FAs, whereas HAs will dissolve only in alkaline media. Although a ratio of liquid to substrate = 100:1 is desirable to decrease the possibilities of degrading components released during hydrolysis, a ratio of about 10:1 is generally used to reduce difficulties encountered in recovering products from very dilute solutions. In general, high temperatures are necessary for the hydrolytic cleavage of most bonds, and refluxing the mixture at the boiling point of the hydrolyzing medium is often used. The time of hydrolysis is very dependent on the type and strength of reagent used, the bond to be cleaved, and the lability of the products. The presence of oxygen should be avoided in hydrolytic procedures not involving oxidation. Further, interferences of metal ions have been shown in both the actual hydrolytic processes and in the subsequent separation of the products of hydrolysis. In conclusion, the very wide range of variables that have been used and the lack of adoption of standard methods often make comparisons between published results meaningless.

Boiling HAs and FAs with water removed carbohydrates,[445] small amounts (about 1%) of simple phenolic acids and aldehydes,[446] peptides,[320] n-alkanes, n-fatty acids, some furan derivatives, and a benzene carboxylic acid.[447,448] Greater amounts of phenolic compounds (five identified) and benzene carboxylic acids (seven identified) were recovered from FA, amounting to about 4% of the initial weight.[447]

Ether-soluble products isolated from acid hydrolysates of HAs and FAs accounted for from 0.5% to 2.5% of the initial material and contained, besides carbohydrates and protein derivatives, phenolic acids and aldheydes. Protocatechuic acid, p-hydroxybenzoic acid, vanillic acid, and vanillin were identified, possibly originated from lignin impurities in the HS.[446] An increase in the MW was observed for some HAs as a result of hydrolysis in 6 M HCl,[378] whereas other chemical properties of HAs were not altered significantly by the hydrolysis treatment which produced a weight loss estimated at 42%.[449] Boiling in water followed by acid hydrolysis gave a total weight loss of 45% and an increase in C and a decrease in N contents for some HAs.[450] Using ^{13}C NMR spectroscopy, almost complete removal of amino acids and sugars, increase in aromaticity, and loss of substituent carboxyl groups were observed in the HA residues after acid hydrolysis.[451,452] These studies also suggested the possible occurrence of decarbonylation and/or decarboxylation reactions during acid hydrolysis of HS.[441]

Thirty phenolic compounds were identified by chromatographic methods in 5 N NaOH hydrolysates of HAs and FAs.[453] These accounted for 6% of the initial material for reactions at 250°C but only 2% at 170°C, and were suggested to originate from both lignin and microbial products. Various amounts of n-fatty acids, phenolic derivatives, benzenecarboxylic acids and N-containing compounds were obtained by 2 N NaOH hydrolysis of HA and FA at 170°C for 3 h in an autoclave, and identified by mass spectrometry and micro-IR spectrometry after extraction, methylation, and separation by chromatographic techniques.[447] Alkaline hydrolysis was found to be effective in the cleavage of C–O bonds for the liberation of structural phenolic components in HA and FA, but it was ineffective for degrading aromatic structures linked by C–C bonds.[447]

2. Reductive Degradation

The principal mechanisms involved in reductive degradation of HS consist in the saturation of C=C double bonds, reduction of O-containing functional groups, and cleavage of ether linkages.[4]

Reductive degradation procedures used in HS studies include (1) reduction with sodium amalgam, a relatively mild process that mainly causes the cleavage of ether linkages; (2) zinc dust distillation and fusion, conducted in drastic reaction conditions, which result in extensive degradations of O-containing components and production of fused aromatic structures; and (3) use of sodium in liquid ammonia, red phosphorus, and hydriodic acid, and hydrogenation-hydrogenolysis.

The work of several authors who have applied reductive procedures to the degradation of HS has been critically discussed in books and reviews.[45,323,440]

a. Sodium Amalgam Reduction

Sodium amalgam is a relatively mild reductive agent whose action is based on the attack of atomic, or nascent, H on electron-rich sites of the substrate molecule which mainly leads to the cleavage of ether linkages and release of phenol and phenolic acids from the humic macromolecules. The presence of a carboxyl group or more than two OH groups on the aromatic ring generally enhances degradation. Aromatic rings bound by C–C linkages or through a methylene bridge are not attacked, whereas labile OH groups in HS are believed to be lost during the early stages of reduction with Na amalgam.[440]

The typical procedure recommended[454] for the Na amalgam reduction of HS consists in heating to boiling the HS sample dissolved in 0.5 M NaOH and added with 5% Na-amalgam, with continuous stirring and flow of N_2. After 4 h, the solution is cooled, acidified with 6 M HCl to pH 1.0, and filtered. The reaction products are then recovered from the supernatant solution by extraction with diethyl ether. To minimize dehydroxylation and other modifications of phenolic structures during residence in the reaction mixture, methylation of the reduction products soon after recovery is suggested.[455,456] The identification of the degradation products is then attempted by use of chromatographic methods, the most popular one being TLC.

Besides the risk of dehydroxylation of phenolic structures, some of the products are labile and can easily be reoxidized and modified drastically both during release from the macromolecules and during residence in the reactor mixture. Accordingly, the final products may not be all related to those actually occurring as components of the original macromolecules.[457]

Yields of ether-soluble reduction products ranged from 30%–35% to 2%–3%.[454,457-460] In most studies, however, yields of identifiable compounds were less than 5% of the starting material.

The TLC method has been generally used for the identification of the reduction products of Na-amalgam degradation reactions. A number of HAs from different soils released up to 40 different degradation products, with many of these retaining side chains and functional groups.[461] These include compounds of probable lignin origin, such as vanillic and syringic acids, and flavonoid-derived compounds, such as resorcinol-type phenols and hydroxytoluenes, which are believed to have microbial origin. Additional phenolic compounds recovered by reduction of HS with Na-amalgam include catechol, pyrogallol, some hydroxybenzoic acids, ferulic acid, and p-coumaric acid.[4] Aromatic aldheydes were also identified in methylated digest products analyzed by GLC.[455] Substantial amounts of aliphatic substances were also found in the Na-amalgam reduction products from HAs.[460]

Although comparisons between results from different laboratories are difficult to interpret because of variations in analytical procedures, the considerable variations observed in the number and kind of phenolic compounds released by Na-amalgam reduction would indicate the existence of intrinsic differences in the nature of parent HAs.[4] However, the conclusion of Burges et al.[458] that TLC patterns of degradation products of HAs might provide a "fingerprint" for identifying a particular type of HA has not been substantiated. Piper and Posner[454] suggested that the Na-amalgam method could at best provide an estimate of the "degree of transformation" of HAs.

b. Zinc Dust Distillation and Fusion

Zinc dust is a strong reductive agent which is able to cleave C–C and C–N bonds, but not heteroaromatic rings. Zinc dust distillation and fusion are drastic procedures which have been used to obtain information on the central "core" of HS by extensive degradation of O-containing aromatic units, including polyphenols.[462-466]

The typical procedure used for Zn dust distillation of HS[464] consists in heating at 510 to 530°C for about 15 min, in a stream of inert gas, a mixture of the HA sample and Zn dust. Polycyclic hydrocarbons sublimed into the cooler part of the reaction tube are recovered by extraction with benzene, purified by vacuum sublimation, and separated by preparative TLC. Identification of reaction products has been made by UV and fluorescence spectroscopies and GC-MS.

Zn dust fusion is a less drastic procedure, consisting of heating at 300°C for about 15 min a mixture of the HA sample, Zn dust, NaCl, and $ZnCl_2$.[464] Reaction products are recovered by solvent extraction from the reaction mixture and analyzed using the procedures described for Zn dust distillation.

Degradative procedures of HAs and FAs based on Zn dust reduction released as main compounds polycyclic aromatic hydrocarbons containing from two to five condensed rings. Major compounds identified were 1,2,7-trimethylnaphthalene, anthracene and substituted anthracenes, substituted phenanthrenes, pyrene and substituted pyrenes, and perylene.[464,465] Other products included fluoranthrene, 1,2-benzanthracene, benzofluorenes, benzopyrenes, naphthalene, chrysene, and coronene.[462]

The Zn dust distillation and fusion of a soil FA and a soil HA produced the same types of products and almost identical yields, accounting for about 1% of the starting material.[464,465] Assuming a recovery of 10% of products, polycyclic aromatic hydrocarbons were suggested to possibly account for 12% to 25% of the HA and FA "cores".[464]

The drastic conditions associated with Zn-dust procedures have been suggested to produce excessive bond breaking, molecular rearrangements and recombination of fragments, and dehydrogenation, which can give rise to artificial condensed ring structures.[464] However, Cheshire et al.[463] concluded that the majority of polycyclic aromatic hydrocarbons isolated from the digests of HAs were not artifacts, but originated from real structural components of the HA "core". Nevertheless, a portion of polycyclic aromatic hydrocarbons found in HAs and FAs may occur as contaminants produced in soil by natural fires.[4]

c. Other Reductive Procedures

The application of several other reductive degradation procedures to HS has been limitedly successful. Reduction by H_2 (hydrogenation) with cleavage of bonds (hydrogenolysis) in the presence of a catalyst at high temperatures and pressures has been applied to HS.[467-469] A number of phenylcarboxylic acids, phenols, cyclopentanols, and other neutral compounds have been isolated.

However, hydrogenation and hydrogenolysis have provided little if any structural information on HS because the drastic experimental conditions used produced molecular rearrangements and condensation reactions of various types leading to the formation of artifacts.

Yields up to 45% of ether-soluble products were obtained by repeated treatments of HAs with metallic Na in liquid NH_3 at –33°C.[470] Analysis by GLC and MS of the products allowed the identification of hydroxyphenols and phenolic acids amounting to about 3% of the starting material.

Heating a HA at 250°C in the presence of red P and HI gave volatile and nonvolatile oils amounting to 3% and 18%, respectively, of the starting material.[463] Catalytic dehydrogenation of such oils gave fused aromatic ring compounds, which supported the concept of a polynuclear aromatic "core" structure for of HS.[462,463]

3. Oxidative Degradation

Over the years, the application of several oxidation procedures has allowed the identification of more than 70 compounds as oxidative degradation products of HS, and attempts have been made to relate these compounds to structures of the parent macromolecules.

The oxidative degradation of HS has been obtained with variable success by use of: (1) rather drastic procedures, such as alkaline permanganate oxidation; (2) relatively mild oxidants, such as alkaline cupric oxide; (3) acidic conditions, with peracetic acid and nitric acid; and (4) other oxidants, such as alkaline nitrobenzene, sodium hypochlorite, and hydrogen peroxide.

Oxidative degradation methods and results obtained by several authors have been reviewed in some books and chapters.[4,5,323,442]

The separation and purification of oxidative degradation products from digests of HS have been obtained by combinations of chromatographic methods, such as paper, column, and thin-layer chromatographies, and solvents of increasing polarity. Identification of degradation products has been made using computer-aided GC-MS and micro-IR spectrophotometric analysis.

An important approach introduced by Barton and Schnitzer[471] to stabilize the starting materials and oxidation products against production of artifacts consists in the premethylation of HS before and/or after oxidation by CH_3I-Ag_2O, or CH_2N_2, or $(CH_3)_2SO_4$. Advantages of premethylation include stabilization of phenolic constituents against degradation, increase of solubilities of oxidation products in nonpolar organic solvents used in the chromatographic separation processes, and increase of their volatility.[471]

a. Alkaline Permanganate Oxidation

Although rather drastic, alkaline permanganate degradation is considered to be the best of the oxidative procedures since it is able to produce relatively large amounts of identifiable digest products and to provide significant information on the chemical structures of HS.[442]

The general procedure used by Schnitzer[323,439] consists of oxidizing a weighed amount of premethylated HA or FA sample with 4% (w/v) aqueous $KMnO_4$ solution at pH 10, extraction of the products into organic solvents such as ethyl acetate, remethylation, and separation of components by preparative column or thin-layer chromatography. This is followed by further separation of the fractions by preparative GC into well-defined compounds which are then identified by comparing their mass and micro-IR spectra with those of standards.

Major products identified in $KMnO_4$-oxidative degradation digests of both methylated and unmethylated HS from widely different soils generally include from di- to hexa- forms of benzene-carboxylic acids, phenolic acids containing between one and three OH groups and between one and five COOH groups, and aliphatic mono-, di-, and tri-carboxylic acids, which are usually identified in the forms of ethers and esters.[323,442] These structures are considered to constitute the major "building blocks" of HS. The benzenecarboxylic acids are generally thought to derive from oxidation of aliphatic side chains associated with polycyclic aromatic compounds and/or aliphatic

Table 18   Yields of the Major Types of Products (mg) Resulting From the KMnO₄ Oxidation of 1.0 g of Methylated Humic Acids and Fulvic Acids Extracted From Soils From Various Climatic Zones

Type of product	Arctic	Cool, temperate		Warm, temperate	Subtropical	Tropical
		Acid soils	Neutral soils			
Humic acids						
Aliphatic	129.2	9.8–51.6	7.0–15.8	1.0–17.1	0.6–1.0	8.3–116.7
Phenolic	36.8	70.4–79.0	68.1–96.5	5.3–33.6	16.2–31.2	32.3–111.7
Benzenecarboxylic	50.8	80.2–122.4	147.5–173.5	35.6–123.3	99.2–164.1	49.6–183.3
Total identified	224.2	194.4–282.4	248.6–281.3	61.6–133.0	118.1–198.5	100.2–350.4
Fulvic acids						
Aliphatic		9.2	6.8–8.9	2.3–31.3	0–1.6	5.7–27.1
Phenolic		30.5	43.1–46.5	4.6–14.1	7.4–60.0	15.9–26.3
Benzenecarboxylic		46.9	77.6–95.4	27.6–124.9	20.6–86.9	37.9–68.4
Total identified		86.6	129.6–147.8	45.9–174.3	28.0–148.1	64.6–108.0

Data from References 472 and 473.

bridges between aromatic moieties, whereas the aliphatic carboxylic acids may arise from oxidation of straight-chain compounds or labile ring systems.[442]

The yields of benzenecarboxylic acids were almost similar for methylated and unmethylated HAs and FAs, whereas higher proportions of phenolic compounds were found in the methylated samples.[439] The nature and quantities of degradation products are similar in HAs and FAs extracted from soils from various climatic and geographic zones, except for the lower level of phenolic constituents in HAs and FAs from warm temperate soils, and the higher level of aliphatic compounds in HS from Arctic and tropical soils (Table 18).[472,473]

b. Alkaline Cupric Oxide Oxidation

Alkaline CuO oxidation is considered a relatively mild oxidative degradation procedure for HS, which is especially efficient for releasing phenolic structures, whereas aromatic structures bound through C–C bonds are not easily attacked.[474]

About 30% of a premethylated FA was accounted for in degradation products obtained by alkaline CuO oxidation.[474] More than one half of the degradation products were identified. Phenolics, mainly phenolic acids, amounted to two thirds of the identified compounds, and benzenecarboxylic acids accounted for about 15%. Alkanes, fatty acid methyl esters, and aliphatic dicarboxylic acid esters were also detected.

In a previous study,[475] a yield of 20% ether-soluble products was obtained by CuO-NaOH oxidation of HA, of which only 10%, which corresponded to 2% of the original HA, was accounted for, mainly as about equal amounts of resorcinol-derived and guaiacyl-type compounds.

Degradation by CuO-NaOH of HAs and FAs extracted from soils from various climatic zones yielded phenolics as the most abundant compounds and relatively low amounts of benzenecarboxylic acids.[472] Similar to alkaline KMnO₄ oxidation, the Arctic HA produced larger amounts of aliphatic compounds than any other HA, and climate did not appear to affect the yields of phenolic and benzenecarboxylic acid oxidation products.

c. Peracetic Acid Oxidation

Peracetic acid has been used as an oxidizing agent for HS[476,477] with the purpose of minimizing the magnitude and extent of chemical alterations and the formation of possible artifacts which may result in oxidations conducted under alkaline conditions.

Products and yields obtained by peracetic acid oxidation of HA and FA were phenolic acids, 4.3% and 4.1%, respectively, and benzenecarboxylic acids, 15.2% and 7.3%, respectively.[476,477] In addition, small amounts of fatty acids and aliphatic dicarboxylic acids were detected.

The compounds produced by peracetic acid oxidation of unmethylated HS were essentially similar to those obtained by oxidations of methylated HS with $KMnO_4$-NaOH and Cu-NaOH, and by HNO_3 oxidation, except that no nitro compounds were formed in the latter case. Total yields were different, however, particularly for FA, which released greater amounts of aliphatic compounds and smaller amounts of phenolic compounds when oxidized with peracetic acid.

d. Other Oxidative Procedures

Nitric acid oxidation of SOM[478] allowed the identification of a variety of aliphatic dicarboxylic acids, amounting up to 0.7% of the initial material, benzenecarboxylic acid (to 3.8%), hydroxy-benzoic acids (to 0.6%), and nitro compounds (to 5.5%).

Over 50% of the total C was converted to CO_2 by oxidation of a soil HA with sodium hypochlorite, whereas nonvolatile oxidation products consisted of a variety of relatively high amounts of aliphatic mono-, di-, and tricarboxylic acids and benzenepolycarboxylic acids.[479] Phenolic compounds were absent, thus indicating that cleavage of aromatic rings occurred.

Alkaline nitrobenzene oxidation of SOM yielded very low amounts of lignin-derived compounds (about 1%),[480] thus the method was not considered to be efficient for use in HS studies.

Oxidation of HS with H_2O_2 principally produced CO_2 and H_2O, with a maximum of 5% of the original material recovered as nonvolatile material soluble in ether. This approach was thus considered of limited value for degradation studies of HS.[481]

A HA and a FA sample were degraded by a number of oxidation methods or combination of methods to uncover the most efficient ones for producing maximum yields of oxidation products.[323,472] Data obtained showed that HA and FA yielded almost equal proportions of aliphatic structures, but HA were richer in benzenecarboxylic acids, in contrast to FA which produced more phenolic acids. However, calculated aromaticity values were similar (about 70%) for both HA and FA.

4. Degradation with Sodium Sulfide

Degradation with sodium sulfide at elevated temperatures is a promising chemical procedure which has found occasional application for HS studies. Only a few studies of reaction of HS with sodium sulfide have been published,[482-486] and a comprehensive review of these studies has been provided by Hayes and O'Callaghan.[443]

The procedure generally used involves heating of HA in 10% sodium sulfide solution at 250°C under autoclave conditions. At elevated temperature the reactive species are considered to be the nucleophilic and basic HS^- and OH^-. "Acid-boiled" HA recovered after refluxing for 24 h in an excess of 6 M HCl in a stream of N_2 has also been used.[486] Methylation with diazomethane of phenolic OH groups of the different compounds in the digests has been applied prior to separation by GC and identification by GC-MS.

Up to 60% of the original non-"acid-boiled" HAs were solvent-soluble after Na_2S treatment, whereas "acid-boiled" HA residue amounted to 45% of the original HA and gave ether-soluble products amounting to about 40% of the starting "core" material. Acid hydrolysis combined with reaction in Na_2S was estimated to achieve over 80% degradation of HA macromolecules.[486] The yields of ether-soluble products increased when the treatment of HA with Na_2S was conducted in the presence of an hydrodesulfurization catalyst such as cobalt molybdate.[486]

A complete list of compounds identified in the Na_2S digests of HAs and a detailed discussion of their possible structural origin can be found in the review of Hayes and O'Callaghan.[443] The compounds identified are significantly different from those obtained by oxidative procedures, and mostly consist several aliphatic dicarboxylic acids, some long-chain alcohols, fatty acids, and several aromatic structures bearing aliphatic side chains and polar functional groups, whereas only one benzene-polycarboxylic acid was identified in trace amounts.[443]

The origin of these compounds has been discussed in details in the review of Hayes and O'Callaghan.[443] Aliphatic acids were suggested to be formed through cleavage of double bonds and, occasionally, through cleavage of a C–C bond of saturated aliphatic side chains on activated aromatic structures. Ethanoic and lactic acids identified in the Na_2S digest of HA possibly originated from the carbohydrate materials associated with HS. Many of the aromatic compounds identified might originate by cleavage of ether bonds in HA structures bearing hydroxy, methoxy, or other ether substituents. Several of the aromatic structures bearing aliphatic side chains and polar functional groups such as COOH, CO, and OH, could have been derived from lignin-type precursors.

5. Degradation with Phenol

A valuable approach for determining the nature of aliphatic groups linking aromatic structures in HS macromolecules is through depolymerization by refluxing for 24 h in phenol in the presence of a catalyst such as p-toluenesulfonic acid (PTSA).[487] After removal of excess phenol and the catalyst by steam distillation, the light petroleum-soluble components of the methylated residues were fractionated by column chromatography, preparative GLC and TLC, and finally identified by chemical and spectroscopic methods.[487]

The mechanisms involved in depolymerization were suggested to be aromatic replacements by the phenol through reverse Friedel-Crafts reactions, and also involvement of various reactive groups, such as hydroxy, ether, carbonyl, carboxyl, and ethylenic bonds in the addition of phenol. Jackson et al.[487] isolated and identified 11 compounds, including xanthenes, xanthone, phenyl-methanes, -ethanes, and -propanes, and triphenylmethane structures, and proposed relationships between these degradation products and the original linkages and structural features of HAs.

Model studies have been used to make some predictions on the origin of the compounds identified in phenol-PTSA digests of HA.[443] For example, xanthene structures were predicted to be formed from the degradation of 2-hydroxybenzene structures, whereas diphenylalcane structures might have formed from aromatic structures linked by aliphatic bridges. The origin of xanthene could be ascribed to diaromatic ketone structures in the HA macromolecules. The persistence of xanthene and xanthone structures in the digests and studies of model substance reactions suggested that aliphatic-aromatic, and especially aromatic, ether linkages have high resistance to cleavage by phenol-PTSA reagent. Compounds identified in the digests also provided evidence that some of the phenyl to carbonyl C linkages survived the reagents and reaction conditions, but it was evident also that displacements with phenol did occur.

The 11 compounds identified were a small proportion of those contained in the digests of phenol-PTSA degradation of HA. Model studies described in the review of Hayes and O'Callaghan[443] indicated that the phenol-PTSA system can attack several types of functional groups, and cleave C–C bonds activated by unsaturation and by a variety of functional groups. Modern GC-MS instrumentation available nowadays is expected to allow the identification of numerous compounds in the digest mixtures, and provide more appropriate predictions of the origins of phenol-derived digest products in the HA macromolecules. A thorough knowledge of the mechanisms of the interactions based on the information obtained from model studies is, however, considered necessary before attempting to assign the origins of products identified in the reaction digests.[443]

6. Thermal Degradation

Thermogravimetry (TG), differential thermogravimetry (DTG), differential thermal analysis (DTA), differential scanning calorimetry (DSC), isothermal heating, and various pyrolysis techniques have been employed to investigate the reaction kinetics, mechanisms, and products of thermal decomposition of HS. Results provided by these methods have been discussed in books[3,4,323,439] and reviews.[279,444]

a. Thermogravimetry and Differential Thermogravimetry

The TG method allows one to detect and record quantitatively the alterations of sample weights continuously during the course of heating, with the aid of a thermobalance. Because of the possible overlapping of the decomposition temperatures during heating, the DTG is preferred, which allows one to record directly the first derivative of the TG curves by plotting the weight differences per time unit (rate of weight loss) vs. the temperature.

DTG curves of soil HAs generally show two characteristic intense peaks at 280 and 540°C and a weak peak at <100°C, whereas for soil FAs the main peak is at 420°C and two weak peaks appear at 270°C and <100°C.[488,489] Although the assignment of maxima in DTG curves to the decomposition of distinct structural components of HA and FA is troublesome, chemical and IR analyses of the residues of thermal decomposition withdrawn at different temperatures from the reaction system allow one to obtain some indication of the thermal reactions occurring.[488,489]

For both HA and FA samples, the C content of the chars increased with temperature, while the O content simultaneously decreased. Residues of HA and FA heated at 540°C contained identical percentages of C and H but no O. Some of the N and S in the initial samples were recovered in the chars heated to the highest temperature, thus suggesting the partial heterocyclic nature of these elements in HS. For HA, decarboxylation and decomposition of phenolic hydroxyl groups appeared to start above 150°C, reach a maximum at 250°C and 200°C, respectively, and be completed at 400°C. The FA sample appeared to be more heat-resistant than HA since the cleavage of carboxyl groups started at temperatures above 250°C and was completed at 450°C, and the elimination of hydroxyl groups increased up to 300°C and rapidly decreased to the final temperature of 450°C.[488,489] The highest temperature peaks were ascribed to the decomposition of the aromatic "nuclei" of HS. The differences in decomposition rates were related to the different "degrees of aromaticity" of HA and FA.

By applying Van Krevelen's graphical-statistical method to thermal decomposition of HAs and FAs, Schnitzer and Hoffman[488,489] concluded that the main thermal reactions occurring for HA were (1) dehydration (up to 200°C); (2) elimination of functional groups, mainly decarboxylation and dehydration (between 200°C and 250°C); and (3) further dehydration and dehydrogenation (up to the highest temperature). The reactions between 500 and 600°C were attributed to the breaking of aliphatic and alicyclic structures, and those at temperatures above 600°C to decomposition of aromatic structures.[490] The main reactions involved in the thermal decomposition of FA were dehydration and decarboxylation up to 400°C, and dehydration and deoxydation at temperatures >400°C.

b. Differential Thermal Analysis and Differential Scanning Calorimetry

Results of DTA of HS are very similar to those of DTG, and also in this case it is difficult to relate DTA peaks to specific thermal reactions of particular structural units of decomposing HS macromolecules.

In general, DTA curves of FAs and HAs showed:[491] (1) a shallow endothermic peak near 100°C, attributed to loss of water; (2) an exotherm between 300 and 380°C, generally ascribed to decarboxylation and dehydration of phenolic OH; and (3) a prominent exotherm between 440 and 490°C, due to decomposition of aromatic nuclei. Some HAs showed DTA curves with a third exotherm at 530 to 570°C, possibly attributed to particularly stable aromatic structures in HAs and FAs.[416]

The DSC method has been only recently applied to the study of thermal reactions occurring on heating HS.[492] A number of HAs and FAs were heated from 50 to 550°C at a constant rate and the residues of heating at different temperatures were analyzed by FT-IR spectroscopy. The DSC curves generally showed three endothermic peaks at (1) about 160°C, possibly ascribed to loss of polysaccharide chains; (2) about 300°C, possibly due to decarboxylation processes; and (3) about 500°C, possibly associated with recombination and polycondensation of aromatic nuclei. However, the authors concluded that more studies are necessary to understand the thermal reactions associated with DSC curves of HS.

c. Pyrolysis Techniques

Analytical pyrolysis is a thermal degradation method in which thermal energy is applied to and absorbed by the sample, giving rise to excitation of the bond vibrational modes in the molecule. Relaxation then occurs through both heterolytic and homolytic cleavage of weaker bonds with release of volatile products characteristic of the structure of the original macromolecule. The products must be removed rapidly from the reaction system to prevent secondary reactions between the products, and between the products and the substrate. Secondary product formation is generally avoided by employing pyrolysis in vacuum, or in a rapid stream of inert gas, and small sample amounts (<100 μg).[444]

The pyrolysis process is generally combined with analytical methods of product assessment, such as mass spectrometry (Py-MS), gas chromatography (Py-GC), or a combination of the two (Py-GC-MS). Although Py-MS has the advantage of allowing pyrolysis products to pass directly into the mass spectrometer, it does not distinguish between fragments of similar MW and thereby fails to provide identification of all pyrolysis products. Initial separation of pyrolysis products by GC followed by the flow of the column effluent directly into the mass spectrometer (Py-GC-MS) is thus preferred.

Ion fragmentation in the MS should be minimal, as otherwise the original thermal fragmentation of the sample can be obscured. Low-energy electron impact ionization (EI) is the most commonly used method, in which a composite mass spectrum is obtained consisting essentially of molecular ions up to about m/z = 300.[444] To obtain molecular ions up to mass 3000, more advanced methods of soft (i.e., low-energy) ionization are adopted, including field ionization (FI) and field desorption (FD).[493]

Details of pyrolysis equipment and techniques, and related methods for automation and multivariate statistical analysis can be found in specialized reviews.[444,494-496] A comprehensive review of methods of computerized data analysis, which is now an integral part of analytical pyrolysis techniques, can be found in Howarth and Sinding-Larsen.[497] The pyrolysis-methylation technique applied to soil FAs has been reviewed by Saiz-Jimenez.[498] Modern methods of analytical pyrolysis, which include off-line reactor pyrolysis, direct, in-source Py-MS using soft ionization techniques (Py-FIMS and Py-FDMS), derivatization Py-MS, and flash pyrolysis or Curie-point Py-GC-MS, in combination with library searches, have been recently reviewed.[499] Pitfalls, limitations, and possible solutions of analytical pyrolysis of HS have been recently reconsidered.[500]

Pioneering application of Py-GC and Py-GC-MS to the study of soil HAs and FAs were reported by Nagar,[501] Wershaw and Bohner,[502] and Kimber and Searle.[503,504] Nineteen compounds, including alkanes, olefins, and aromatics, were identified by GC analysis of pyrolysis products of soil HAs and FAs.[505,507] In a successive, comprehensive pyrolysis study, Martin et al.[508,509] identified by MS analysis 132 compounds in the pyrolysates of soil HAs and FAs. By comparing the products of pyrolysis obtained from various known biopolymers with those found in pyrolysates of HAs and FAs, Martin et al.[508,509] suggested that the vast majority of low-boiling-point compounds detected in the latter could arise from polysaccharides, proteins, and aliphatic hydrocarbons. As regards the high-boiling-point compounds, a relatively large number of lignin fragments was identified, indicating the presence of lignins or modified lignins in HAs, in addition to compounds derived from polysaccharides and proteins. The composition of the pyrolysate obtained from acid-hydrolysed HA was constituted of a series of alkanes, alkenes, phenols, benzofurans, alkylbenzenes, and alkylpolycyclic aromatic hydrocarbons, which indicated a prevalent aromatic core for HS macromolecules.[508,509]

The identification of various cyclopentenones and dimethyl maleic anhydride in pyrolysates of soil FAs and HAs was related to the presence of aliphatic polycarboxylic acids.[510,511] However, the finding of cyclopentenones in pyrolysates of simple polysaccharides led Saiz Jimenez and de Leeuw[512] to the conclusion that the presence of such compounds in pyrolysates of HS could not be exclusively ascribed to polycarboxylic structures.

In a vast study of soil HS,[513-516] more than 300 pyrolysis products have been identified, and a possible origin for each compound has been assigned. The major discovery of these authors was that HAs, from which lipids, polysaccharides, proteins, lignins, etc., were removed by various pretreatments, yielded upon pyrolysis homologous series of n-alkanes, n-alkenes, and alkadienes ranging from C_5 to C_{35}. Evidence was obtained that such a characteristic pattern of aliphatic hydrocarbons in pyrolysates of HAs had its origin in aliphatic, resistant biopolymeric structures present in HAs and originated from algal and plant cell walls and/or plant cuticles and barks.[517] Based on these findings, the prevalent aromatic nature of HAs was questioned, and a significant contribution of aliphatic biopolymers to a highly refractory, polymethylenic moiety in HAs was proposed.[518,519]

It should be noted that yields from pyrolysis techniques are frequently 50% or less, owing to competing char-forming reactions, thus data obtained are selective for the more volatile components. With HAs and FAs, less than one half of the total C is usually recovered as volatile products. Accordingly, although the volatile products evolved bear the expected relationships to the parent structures, pyrolysis data cannot be regarded as being representative of the entire sample under examination. Further, pyrolysis does not provide quantitative data for structural units or components of HS, and some of the products may be artifacts of the thermal degradation process.

Pyrolysates of HS contain a rich mixture of products that can be related to their constituent biopolymers and building blocks. A complete array of products identified in pyrolysates of HS and their parent biopolymers are listed in Table 19.[444,520] From this list, it appears that all HS give, in various proportions, products highly characteristic of polypeptide and polysaccharide moieties, lignin units, microbially synthesized polyphenols, substituted polycarboxylic acids, lipids and other aliphatic constituents, amino sugars, and aromatic hydrocarbons. Broad differences were found in the proportions of pyrolysis products between FAs, HAs, and humins.[521,522] Pyrolysates of HAs were enriched in polypeptide products, lignin derivates, and phenols, but contained little polysaccharide products, whereas FA pyrolysates contained less polypeptide products and higher levels of polysaccharides. Similar results were obtained in another study,[523] where carbohydrate and phenolic constituents were found to be more pronounced in pyrolysis products from FAs, whereas saturated and unsaturated hydrocarbons were the most relevant products obtained from HAs.

In addition to the list of compounds shown in Table 19, a successive study conducted by Py-FIMS and Py-FIDS techniques[524-526] allowed one to extend the MW range to close to m/z = 1500, and indicated the presence of novel signals and fragments in pyrolysates of several soil HAs and FAs. Two typical Py-FIMS of HAs are shown in Figure 14.[271] Major novel components identified in HA extracts were n-alkanes with C_{17} to C_{101}, n-fatty acids with C_{15} to C_{33}, alkanes with C_{22} to C_{51}, sterols with C_{28} to C_{29}, diols with C_{16}, C_{24}, C_{31}, and C_{34}, monoesters with C_{40} to C_{68}, diesters with C_{65} and C_{66}, and triesters with C_{75} to C_{93}. Major novel constituents of the FA extracts were n-alkanes with C_{20} to C_{74}, fatty acids with C_{16} to C_{34}, the C_{24} diol, sterols with C_{27} to C_{29}, alkyl monoesters with C_{44} to C_{68}, and alkyl diesters with C_{56} to C_{66}. These results showed that the components in all extracts are similar although they are present in different proportions.

Results of time-resolved Py-FIMS also showed the presence of a homologous series of bound aromatic esters from suberin and pyrolysis products of intact and modified "condensed" lignin subunits in HA hydrolysates, and of unsaturated ethylcolestanes and aromatic esters in humin hydrolysates.[527-530]

Results from pyrolysis studies provided evidence of important processes occurring during the humification phenomena.[444] Primary polysaccharides of plant origin, mainly xylans and cellulose, are suggested to be consumed and secondary, microbially produced polysaccharides, which give pyrolysis products containing the furan ring, are formed and accumulate in the FA fraction of SOM.

Table 19   Soil Biopolymers and their Pyrolysis Products

Polymer	Molecular ion, *m/z*	Compounds
Primary polysaccharide	60	Acetic acid
	84	Hydroxyfuran
	96	Furfural
	98	Dihydropyrone, furfuryl alcohol
	112	—
	114	3-Hydroxy-2-penteno-1,5-lactone (pentose)
	126	Levoglucosenone (hexose)
	128	Anhydro-monomer (deoxyhexose)
	144	Anhydro-monomer (hexose)
	162	Levoglucosan (hexose "monomer")
Secondary polysaccharide	68, 82	Furan and methylfuran
	96	Furfural, dimethylfuran
	110	Methylfurfural
N-Acetyl-amino sugars	59, 73, 125, 135, 137, 151	Acetamide, methylacetamide
Polypeptide and protein	17	Ammonia
	34, 48	Sulfides (Cys, Met)
	67, 81, 95	Pyrroles (Pro, Hpro, Glu)
	78, 92, 106	Benzenes (Phe)
	94, 108	Phenols (Tyr)
	117, 131, 145	Indoles (Trp)
	79, 93, 107	Pyridines
Lignins	94, 108, 120, 122, 134, 136, 148, 150, 152, 164, 166	*p*-Hydroxyphenyl compounds
	124, 138, 150, 152, 164, 166, 178, 180, 182, 192, 194, 196	Guaiacyl compounds
	154, 168, 180, 182, 194, 196, 208, 210, 212, 224, 226	Syringyl compounds
Lipids	42, 56, 70, 84, 98, 112	Alkenes
	54, 68, 82, 96, 110	Alkadienes
Polycarboxylic acids	82	2-Cyclopenten-1-one
	128	Dimethylmaleic anhydride
Aromatic hydrocarbon precursors	78, 92, 106, 120, 134, 148, 162	Alkylbenzenes
	104, 118, 132, 146, 160	Alkennylbenzenes
	116, 130, 144, 158	Alkylindenes
	142, 156	Alkenylindenes
Phenol precursors	94, 108, 122, 136, 150	Alkylphenols
	110, 124	Dihydroxybenzenes

From Bracewell, J. M. et al., in *Humic Substances, Vol. 2, In Search of Structure,* Hayes, M. H. B. et al., Eds., John Wiley & Sons, 1989, 181. With permission.

During humification, the typical pyrolysis pattern of native lignin is lost, possibly by release of the methoxyl groups and oxidative degradation of the propanoid side chains. The residual aromatic rings may then contribute to the substituted benzenes and phenols obtained on pyrolysis, and be oxidatively opened giving rise to saturated or partly unsaturated aliphatic chains bearing COOH, CO, and OH substituents, which would give cyclic ketones, lactones, and furans as pyrolysis products. Further, the pyrolytic elimination of OH groups to give C=C bonds and decarboxylation of COOH groups could be responsible for alkanes and alkylbenzenes, also obtained on pyrolysis of HS.

Methods of analytical pyrolysis, especially Curie-point Py-GC-MS and Py-FIMS made possible the identification of chemical building blocks in HS (Table 20),[271] and provided a molecular-chemical basis for modeling a structural network for HS in which aromatic rings are joined by long-chain alkyl structures.[530,531] Results of extensive, long-term investigations by analytical pyrolysis have contributed, in combination with other chemical, microscopic, and spectroscopic data available in the literature to the state-of-the-art proposal of a structural model for HS (Figure 6).[271,529,530]

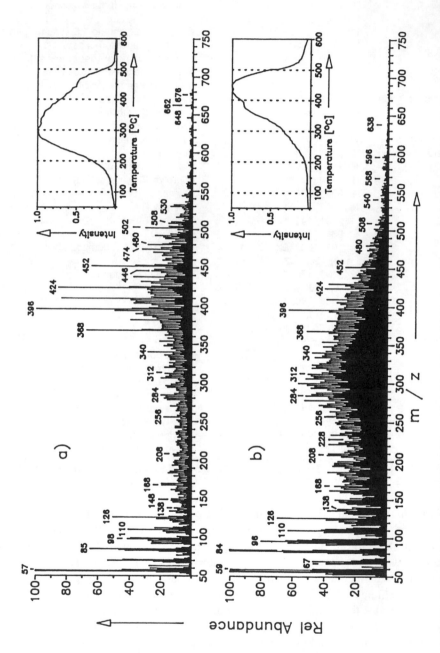

Figure 14 Pyrolysis-field ionization mass spectra of (a) Haplaquod humic acid and (b) Haplaquoll humic acid. (From Schulten,, H.-R., in *Humic Substances in the Global Environment and Implications on Human Health*, Senesi, N. and Miano, T.M., Eds., Elsevier, Amsterdam, 1994, 43. With permission.)

Table 20   Principal Chemical Structures of Soil HAs Identified by Curie-Point Pyrolysis-Gas Chromatography/Mass Spectrometry

Structure	Name
$CH_3-(CH_2)_n-CH_3$	Alkanes (n = 2–29)
$CH_2=CH-(CH_2)_n-CH_3$	Olefins (n = 1–28)
⬡$-(CH_2)_n-CH_3$	Alkylbenzenes (n = 0–17)
CH_3-⬡$-(CH_2)_n-CH_3$	Methyl alkylbenzenes (n = 1–11)
CH_3-⬡$-CH_3$ (with CH_3)	Methyl-substituted benzenes
$R_{1-2}-$⬡⬠	Methyl-substituted indenes
⬡$-R_{1-5}$ (with OH)	Methyl-substituted phenols
$R_{1-4}-$⬡$-O-CH_3$ (with OH)	Methyl-substituted methoxyphenols
⬠$-R_{1-2}$ (furan)	Methyl-substituted furans
⬠$-R_{1-2}$ (pyrrole)	Methyl-substituted pyrroles
⬡$-R_{1-2}$ (pyridine)	Methyl-substituted pyridine

Note: R = H; CH_3

From Schulten, H.-R., in *Humic Substances in the Global Environment and Implications on Human Health,* Senesi, N. and Miano, T. M., Eds., Elsevier, Amsterdam, 1994, 43. With permission.

H. Structural and Chemical Information by Spectroscopic Methods

Spectroscopic methods have had wide application to the study of HS and have added considerably to the knowledge of their chemical structure and properties. Spectroscopic techniques that have been used comprise both techniques employing high-energy radiation, including UV-Vis absorption and fluorescence spectroscopies for studying electronic transitions of bonding electrons, and IR spectroscopy for studying molecular vibrations, and techniques employing low-energy radiation, such as NMR and electron spin (or paramagnetic) resonance (ESR or EPR) spectroscopies.

Spectroscopic methods are severely limited when applied to HS, which yield spectra that generally represent the summation of the responses of several different species of which HS are composed. Unlike NMR and IR spectroscopies, which can provide direct qualitative identification and, sometimes, semiquantitative estimation of structural components of HS, the other spectroscopic techniques generally do not provide direct identification of functional groups and structural entities of HS. However, the latter techniques may provide additional information that may be valuable in elucidating some structural and functional aspects of the chemistry of HS and in differentiating HS of different origin and nature.

1. Ultraviolet Visible Spectroscopy

The absorption of radiation in the UV-Vis region of the electromagnetic spectrum arises from electronic transitions from bound states (outer valence orbitals) to excited electron states.[532] The wavelength limits of these regions are 200 to 400 nm for the UV and 400 to ~800 nm for the visible. In organic molecules, such as HS, these exceptionally low-energy transitions are associated with the presence of chromophores, i.e., functional groups containing conjugated double bonds and sulfur, nitrogen, or oxygen atoms with delocalized electronic orbitals. The strongest UV-Vis absorption bands are associated with $\pi \rightarrow \pi^*$ transitions, and weaker ones with $n \rightarrow \pi^*$ transitions, where π and π^* refer to bonding and antibonding p-type orbitals and n refers to lone-pair orbitals.[532] Electronic transitions can occur within the molecular orbitals of chromophores or involve the transfer of an electron from one chromophore to another chromophore or to a nonchromophore (electron or charge-transfer excitation). Groups that are not chromophores but affect absorption of chromophores are called auxochromes, which typically include hydroxyl and amine groups.

The two most important parameters measured in UV-Vis spectrometry are the wavelength(s) of maximum absorption, λ_{max}, and the absorptivity, ε. A change in both parameters of chromophores may result by introduction of an auxochrome to various positions of an aromatic ring and/or by ionization of carboxyl and hydroxyl groups.[533] Generally, increasing substitution is accompanied by an increase of λ_{max}.

The UV-Vis spectra of HAs and FAs are generally featureless, showing no well-defined maxima or minima, with absorptivities increasing as the wavelength decreases.[3-6,279,323,534-538] However, a slight maximum is occasionally present in the 260- to 300-nm region of the UV, especially for FA. The little structure of UV-Vis spectra of HS feasibly results from extended overlap of the absorbances of a wide variety of chromophores affected by various substitution.[6] Thus, it can be concluded that UV-Vis spectroscopy has little value for studying functional groups in HS, being severely limited by the inherent nature of HS.[537]

Despite these limitations, UV-Vis spectroscopy has found some other useful applications in HS research. UV-Vis light absorption of HS appears to increase with an increase in: (1) degree of condensation of aromatic rings;[9] (2) total C content; (3) molecular weight; and (4) ratio of C in aromatic rings to C in aliphatic side chains.[4,323] Absorptivity of HAs has been shown to vary somewhat according to the soil from which they originate, and to be related to differences in degree of humification.[9]

The UV-Vis spectra of HS are affected by solution pH, which causes a change in λ_{max} and absorptivity of the chromophores.[539,540] These changes are possibly attributed to variations in structural properties, such as degree of dissociation or protonation of carboxyl and phenolic hydroxyl groups, and to conformational changes in the macromolecular structure resulting in greater or lesser exposure of chromophores to the light.[538] Tsutsuki and Kuwatsuka[540] used the technique of difference spectroscopy to study the spectral changes in HAs and FAs due to changes in pH. The low pH difference spectra obtained by these authors showed that absorbance was greater at higher pH over the complete range of the UV-Vis spectrum, and that peaks were apparent in the 280- to 285-nm region for all samples, and at 360 nm for some samples. The pH-dependent absorbance difference observed at 280 to 285 nm was ascribed to the effect of ionization of hydroxybenzenecarboxylic acid chromophores and, possibly, of phenolic chromophores.[538] Trihydroxybenzenes were considered as possibly responsible for the high-pH difference peak observed at 350 to 360 nm.[539,540]

The ratio of the absorbance at 465 nm to that at 665 nm, referred to as the E_4/E_6 ratio, has been widely used to characterize HS.[3-5,9,279,323,535] The E_4/E_6 ratios for HAs are usually <5, whereas those for FAs are generally >5. The E_4/E_6 ratio of HAs and FAs varies with pH and salt concentration. With decreasing pH, the absorbances decrease but the variation is not the same at 465 and 665 nm.[536,540,541] According to Tsutsuki and Kuwatsuka,[540] the effect of pH on absorption in the visible region and on the E_4/E_6 ratio can be due to changes in λ_{max} and absorptivity with ionization of phenolic OH contained in complex quinonic unsaturated structures. However, in a recent study on solvent and pH effects on the UV-Vis spectra of FAs, it was shown that absorbances at variable wavelengths were not consistent with quinone chromophores but might be due to chromophores in extensively conjugated polyaromatic structures.[542] The E_4/E_6 ratios also decrease with increasing salt (NaCl) concentration.[536] Thus, the best procedure recommended [541] for determining the E_4/E_6 is to dissolve from 2 to 4 mg of HA or FA in 10 ml of 0.05 M NaHCO$_3$, which results in a pH of approximately 8 and a fixed salt concentration.

The E_4/E_6 ratio has been found to vary with the origin of HS to which it has been correlated.[9] According to Kononova,[9] the E_4/E_6 ratio is inversely related to the degree of condensation of the aromatic network in HS, and can serve as an index of humification. Thus, a low E_4/E_6 ratio would be indicative of a relatively high degree of condensation of aromatic constituents in HS, whereas a high ratio would reflect a low degree of aromatic condensation and the presence of relatively large proportions of aliphatic structures.

Differently, Chen et al.[541] observed that: (1) much of the observed visible absorption by HS may be due to light scattering, which thus may contribute to the lowering of E_4/E_6 in high-MW HAs compared with low-MW FAs; (2) the E_4/E_6 ratio of HAs and FAs is governed primarily by particle sizes and weights, and is not apparently related to the relative amount of aromatic condensed rings in the structure; and (3) the effect of pH on absorption and E_4/E_6 ratios is due to changes in particle sizes possibly caused by folding or unfolding, or aggregation/dispersion of the HS macro-molecules.

Quantitative applications of UV-Vis spectrometry to HS studies, e.g., for determining the concentration of dissolved HS based on Beer's law do not appear to be relevant, and are described elsewhere.[3-5]

2. Fluorescence Spectroscopy

Molecular fluorescence consists in a radiative photoprocess based on the emission of a photon when an excited electron undergoes a transition from the first excited-singlet state to the singlet ground state. This implies that electrons in a molecule of interest be previously excited, e.g., by absorbing incident electromagnetic radiation, and transferred from ground-state molecular orbitals to excited-state nonbonding or antibonding orbitals. Typical transitions of this kind involve the promotion of available n or π electrons to π^* orbitals. These processes are highly probable in complex molecular systems, such as HS, which contain atoms with lone pairs of electrons, such as O and N, and aromatic and/or aliphatic conjugated unsaturated systems capable of a high degree of resonance, i.e., of electron delocalization. A complete treatment of the theory, methods, and applications of fluorescence spectroscopy can be found in several books,[543-545] whereas a succint treatment oriented to applications to HS studies has been provided by Senesi.[546]

Molecular parameters that affect fluorescence intensity and wavelength include the extension of the π-electron system, the level of heteroatom substitution, and the type and number of substituent groups on the aromatic rings. The overall fluorescence behavior of a molecule results, therefore, from a cumulative effect which primarily depends on the various structural components of the molecule, and the fluorescence spectrum obtained consists of the sum of all the individual spectra of the different contributing fluorophores in the molecule. The well-known molecular heterogeneity of HS macromolecules thus represents a great obstacle for the direct identification of the individual structural components responsible for fluorescence.

Fluorescence spectra can be obtained in the three modes of emission, excitation, and synchro-nous-scan excitation on HA or FA samples dissolved in aqueous media at a suitable concentration and pH. The emission spectrum is recorded by measuring the relative intensity of radiation emitted as a function of the wavelength at a constant excitation wavelength. The excitation spectrum is recorded by measuring the emission intensity at a fixed wavelength while varying the excitation wavelength. Synchronous-scan excitation spectra are obtained by measuring the fluorescence intensity while simultaneously scanning over both the excitation and emission wavelengths, and keeping a constant, optimized wavelength difference, $\Delta\lambda = \lambda_{em} - \lambda_{exc}$, between them.[545] When a suitable $\Delta\lambda$ is used, the synchronous-scan technique can selectively increase the intensity of specific peaks, thus improving the general sensitivity of the method. In multicomponent samples, interferences resulting from spectral superposition will be greatly reduced, and some sort of "spectral" separation into individual components will be achieved without requiring any actual physical separation process.

Several limitations and problems, both theoretical and instrumental, are, however, associated with fluorescence spectroscopy analysis. For example, a number of processes may compete with the underlying fluorescence emission of energy from the lowest excited singlet to the ground state. These include internal conversion, or collision deactivation, intersystem crossing, and photodecomposition processes. The shape of the excitation spectrum should be identical with that of the absorption spectrum of the molecule and independent of the wavelength of the exciting radiation and the wavelength at which fluorescence is measured. However, differences may be observed due to instrumental artifacts arising from variations in source intensity with wavelength and other causes.[543]

A number of environmental factors, such as temperature, nature of solvent, presence of other solutes, hydrogen bonding, pH, and interactions with metal ions, can markedly influence the fluorescence behavior of a compound. For example, the so-called "inner-filter" effect, which arises generally from light absorption by the solvent or high concentration of an absorbing solute, can cause quenching of the fluorescence of the molecule of interest. Dissolved molecular O is considered to be the most ubiquitous quencher of fluorescence, especially in polar solvents. Commonly encountered quenchers are also metal ions, even when they do not form complexes with the fluorescent molecule. Inner-filter effect and quenching can be strongly reduced, e.g., by diluting the sample.

Fluctuations in intensity of the exciting light source, variations in instrumental sensitivity of the detection system and other instrumental factors, as well as various light scattering effects, including Rayleigh, Tyndall, and Raman effects, also markedly influence the fluorescence spectra recorded. Corrections accounting for instrumental artifacts and for scattering effects should be applied to fluorescence spectra of HS, particularly when quantitative comparisons are made between spectra measured on a variety of spectrofluorimeters and on HS from different sources.[543] However, a comparative discussion of uncorrected fluorescence spectra may be allowed on a qualitative basis when they are recorded on the same instrument using the same experimental conditions.

The potential utility and applications of fluorescence spectroscopy as a sensitive, nonseparative, noninvasive, and relatively simple means for studying some molecular and quantitative aspects of the structural and functional chemistry of natural organic matter, including HS, and their interactions with metal ions and organic chemicals have been recently reviewed.[546-548] Although "native" or indigenous fluorescent structures generally constitute only minor components of the HS macromolecules, their variety and the dependence of their properties on several molecular and environmental parameters, including MW, concentration, and pH, render their investigation particularly useful to obtain unique information, not only about the fluorescence behavior but also about the general nature and chemistry of HS.

The fluorescence technique has been mostly applied to the study of HS and natural organic matter of aquatic origin, whereas applications to soil HS are relatively limited.[546,548] However, the few fluorescence studies available at present on soil HS have provided results that appear promising in their potentiality for attempting an interpretation at the molecular level of fluorescence properties of HS, for distinguishing between FA and HA fractions from the same source, and for differentiating FA or HA of various origin.[549-551]

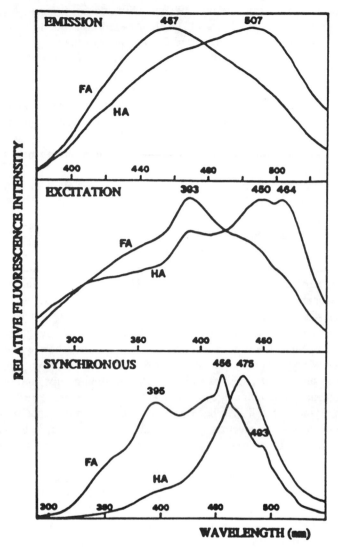

Figure 15   Fluorescence emission, excitation, and synchronous-scan excitation spectra of the IHSS reference humic acid (HA) and fulvic acid (FA) from a Mollic Epipedon soil. (From Senesi, N. et al., *Soil Sci.*, 152, 259, 1991. With permission.)

The fluorescence emission spectra of soil HAs and FAs generally consist of a unique broad band with a maximum wavelength which ranges from 500 to 520 nm for HA and from 445 to 465 nm for FA (Figure 15).[549-556] However, longer wavelengths for the emission maximum have been reported for some soil FAs and shorter wavelength for some soil HAs.[546,550] The overall fluorescence intensity of FA is generally higher than that of HA.[550]

Fluorescence excitation spectra of HS are generally better resolved than emission spectra and characterized by a number of peaks and/or shoulders that are generally localized in three wavelength regions, long (480 to 440 nm), intermediate (400 to 380 nm), and short (360 to 300 nm).[550,551] Most soil HAs feature two closely spaced major peaks in the long wavelength region (around 465 and 450 nm), often accompanied by a minor peak or shoulder in the intermediate wavelength range (at 395 to 390 nm) (Figure 15).[550] However, the relative intensity of these three peaks may be different in some soil HAs.[550] Differently, soil FAs generally feature one main excitation peak in the intermediate region of the spectrum (around 390 nm) with additional minor peaks and shoulders at longer and/or shorter wavelengths (Figure 15).[550] However, lower (360 to 345 nm)[549,552]or higher (455 to 470 nm) wavelengths have been reported for the major excitation peak of some soil FA.[546]

Although fluorescence spectroscopy in the synchronous-scan mode is expected, by principle, to yield a peak resolution better than that obtained by conventional emission and excitation spectra, soil HAs generally feature only one main, relatively broad peak in the long wavelength region of the synchronous-scan spectrum, often accompanied by faint shoulder(s) at longer and shorter wavelengths (Figure 15).[550,551,555,556] Soil FAs generally exhibit synchronous-scan spectra more structured than their HA counterparts, featuring two main peaks at long (450 to 460 nm) and intermediate (390 to 400 nm) wavelengths, often with some less intense peaks and shoulders at both sides (Figure 15).[550,551,555,556]

Based on the above studies, the possible use of fluorescence properties of HS has been proposed as an adequate diagnostic criterion for the differentiation and classification of HS according to their origin, genesis, and nature.[550]

Because of the molecular complexity and chemical heterogeneity of HS, the observed fluorescence spectrum is often broad and ill-resolved, probably representing the sum of the spectra of many different fluorophores present in the HS macromolecule. Hence, the identification of the molecular components responsible for fluorescence of HAs and FAs is still far from being clarified. However, some reasonable hypotheses have been suggested on the possible chemical nature of relevant fluorescing structures in HS on the basis of available data and trends of fluorescing properties.[546,548,550,551] The low fluorescence intensities and long wavelengths measured for the principal fluorescence peaks of soil HAs have been associated with the presence in these macromolecules of high-MW components possessing linearly condensed aromatic ring systems bearing electron-withdrawing substituents, such as carbonyl and carboxyl groups, and/or of other unsaturated bond systems capable of a high degree of conjugation. The higher fluorescence intensities and shorter wavelengths typical of soil FAs have been ascribed to the presence of simpler structural components of low-MW-bearing, electron-donating substituents, such as hydroxyl, methoxyl, and amino groups, low degree of aromatic polycondensation, and low levels of conjugated chromophores.

Several structural components that have been identified in HS macromolecules by degradative methods or by other physicochemical or spectroscopic means such as NMR and IR, and which are known to be fluorescent as simple compounds, have been suggested as potential contributors to the fluorescence of HS.[546,548,550,551] For example, hydroxy- and methoxy-coumarin-like structures, such as esculetin and scopoletin originated from lignin, chromone, and xanthone derivatives of plant origin, and fluorophores of the Shiff-base type derived from polycondensation reactions of carbonyls with amino groups, have all been indicated as possibly responsible for fluorescence of HS at intermediate wavelengths (emission, 470 to 440 nm and excitation, 390 nm). Excitation peaks at low wavelengths (450 to 435 nm) have been accounted for by simpler fluorescent units, such as a benzene ring bearing an hydroxyl conjugated to a carbonyl, methylsalicylate units, and dihydroxybenzoic acid moieties, such as protocatechuic, caffeic, and ferulic acids. Further, flavonoid moieties, as well as naftols and hydroxy-quinoline moieties, have been suggested to contribute to long emission wavelengths (up to 520 nm) of HS.

Fluorescence properties of HS, i.e., the intensity and shape of the spectra and peak position, have been shown to be highly affected by some molecular parameters and conditions of the medium.[546,548,551] With increasing MW, a decrease in fluorescence emission and excitation intensities and a distinct shift to longer wavelengths, accompanied by a broadening of the emission maximum, have been observed for several soil HAs and FAs.[547,557] This was related to the lesser degree of polymerization and substitution of the lighter HS fractions in comparison to the more saturated and more highly substituted aromatic network which characterize the heavier HS fractions.

Fluorescence intensity of soil FAs increased with increasing concentration from 10 to 120 mg l^{-1} at pH = 7.0, whereas soil HAs fluoresced more intensively at intermediate concentration values

(16 to 70 mg l^{-1}).[558] As the concentration increased, a shift of the fluorescence emission to longer wavelengths and modifications of the relative intensities of excitation peaks were also observed.[546] These effects were ascribed to modifications of the degree of association and to configuration rearrangements affecting fluorophore structures. In particular, the decrease of fluorescence intensity at the highest concentration of HA has been ascribed to collisions occurring among excited HA molecules and consequent excimer formation, made possible by the molecular association in local, ordered clusters.[546,548] Excimers could also be in part responsible for the observed increase of the maximum emission wavelength.

Fluorescence properties of HS also varied markedly with pH, although contrasting results of this function have been obtained by various authors working with HAs and FAs of different origin and nature.[546,548] The emission intensity generally decreased with raising the pH up to 7, and then increased up to pH 10, whereas excitation intensity increased over all the pH range.[552,555,558] The relative intensity and the number and position of excitation and synchronous-scan peaks were also variously affected by pH changes and the nature of the HS.[555,558] These effects have been attributed to variations of the ionization of the acidic functional groups, changes of particle association, and molecular rearrangements, such as decoiling of macromolecular structures, e.g., by disruption of hydrogen bonds.

A gradual decrease of excitation fluorescence intensity of soil HAs and FAs has also been measured upon increasing salt (NaCl) concentration (ionic strength) from 0.001 to 0.01 M.[552] This effect has been ascribed to both the gradual coiling up of the HS macromolecular structure and the salt-depressing ionization of functional groups.

3. Infrared Spectroscopy

The most interesting portion of the IR region of the electromagnetic spectrum for the structural and analytical study of organic molecules is the medium IR region, between 4000 and 400 cm^{-1} (2.5 and 25 nm). Energy absorbed by an organic molecule in this region is converted into energy of molecular vibration, and the IR spectrum of the organic molecule consists of vibrational bands. There are two general types of bond vibrational modes: stretching (symmetric and asymmetric), which involves changes in the bond length between the atoms along the bond axis; and bending or deformation (in-plane, such as scissoring and rocking, and out-of-plane, i.e., wagging and twisting), which involves a change of bond angles.

Infrared spectroscopy is a very powerful analytical tool because it can provide information on the presence of specific functional groups or other structural entities within a molecule, on the basis that, within limits, each absorption band corresponding to a particular vibration of a given bond occurs at a given frequency. The characteristic frequency (or wavenumber) of absorption of a specific functional group is dependent on the vibrational mode, the strength of the bonds involved, and the masses of the atoms. Within the IR spectral region from 4000 to ~1250 cm^{-1}, this absorption is relatively unaffected by the remainder of the molecule, whereas absorption bands occurring at frequencies less than ~1250 cm^{-1} (the so-called "fingerprint" region) are profoundly affected by the entire molecular structure.[559]

The intensity of absorption mostly depends on dipole-moment change involved in the vibration, and so the most polar bonds and groups, such as those involving O atoms, frequently give the strongest absorptions in IR. Thus, the position and intensity of an absorption band can be used to determine the presence of a particular functional group or structural entity in the molecule, whereas the absence of absorption is often indicative of the absence of that group in the molecule.

To obtain an IR spectrum, the experimental sample is irradiated with IR radiation generally from 4000 to 400 cm^{-1} (2.5 to 25 nm) and the amount of radiation transmitted, or absorbed, by the sample is measured vs. wavenumber, or wavelength, by a recording spectrometer. Generally, two methods are mostly applied for preparing solid HS samples for IR spectroscopy, the alkali halide pressed-pellet method and the mull method. In the first method, about 1 mg of the HS sample is thoroughly mixed with 100 to 400 mg KBr and pressed in a suitable die under vacuum at elevated pressure, to obtain a small pellet. The advantage of this method is that KBr does not absorb IR radiation, thus only the spectrum of the sample is obtained. However, a limitation is represented by the presence of moisture retained in KBr, which can affect OH absorptions of the sample. The mull technique involves thoroughly mixing the sample with a mulling agent, generally nujol, which is a mixture of high-MW paraffinic hydrocarbons, and placing the dispersion between two NaCl discs. Since the nujol itself absorbs IR radiation, interference in the C–H stretching region is a limiting factor of this technique.

IR spectra of HS have also been recorded in aqueous media either on concentrated solutions or slurries of HS in D_2O using a Fourier transform IR (FTIR) spectrometer.[560,561] In FTIR spectrometry dispersive, diffraction grating is replaced by interferometers, together with the necessary computing facilities for Fourier transform, which allow conversion of the interferogram into a normal (intensity vs. wavenumber) spectrum. FTIR spectroscopy has a number of advantages compared to dispersive IR spectroscopy,[562] including enhanced resolution, improved signal-to-noise ratio, and higher energy output.

Nontransmission IR techniques based on FT methods have also been developed, including diffuse reflectance infrared Fourier transform (DRIFT) spectroscopy.[563] Besides its high sensitivity, the great practical advantages of the DRIFT method are that solid samples can be studied directly with no or little preparation, and interference due to water absorption can be eliminated by digital subtraction. DRIFT spectra of HS have been obtained by some authors.[564,565]

Band frequencies occurring in complex organic molecules, with use of correlation tables and empirical rules of comparison, have been successfully applied to the interpretation of IR spectra for the assignment of experimentally observed absorption bands to specific functional groups.[566,567] The theory of IR spectroscopy has been outlined briefly by MacCarthy and Rice[537] insofar as it is required for interpreting the spectra of HS.

Infrared spectroscopy has been found useful in HS research for several reasons, including (1) the gross characterization of important functional groups and structural entities in HS of diverse nature and origin; (2) the evaluation of the effects of chemical modifications such as methylation, acetylation, saponification, and the formation of derivatives; (3) the detection of changes in the chemical structure of HS following hydrolysis, oxidation, thermal degradation, and similar treatments; (4) the ascertainment of the presence or absence of inorganic impurities, such as metal ions and clays; and (5) the characterization of metal-HS and pesticide-HS interactions. A great number of IR studies on HS can be found in the literature, and they have been reviewed by several authors.[3-5,323,328,535,537,538,568]

Typical IR spectra of soil HAs and FAs are shown in Figure 16.[556] The most evident features of the IR spectra of HS are their overall "simplicity" and "similarity", but these are more apparent than real. For complex macromolecules such as those of HS, the observed broadness of the bands feasibly results from the extended overlapping of very similar absorptions arising from individual functional groups of the same type, with different chemical environments. Further, the fact that most groups of atoms vibrate with almost the same frequency, irrespective of the molecule to which they are attached, does not mean that HS displaying similar IR spectra must have similar overall structures, but only that the net functional group and structural entities content may be similar.[537]

Main IR absorption bands of HS and assignments are listed in Table 21.[3-5,535] Different researchers report the bands at slightly different frequencies, and the listed values should be regarded as approximate. All HS show a broad absorption in the 3450 to 3300 cm^{-1} region, which is usually attributed to O–H stretching of H-bonded OH groups.[569,570] In the free state, OH groups absorb near 3600 cm^{-1}, whereas the absorption frequency is reduced somewhat upon association, e.g., through H-bonding. This assignment is consistent with the decrease in the absorption intensity observed upon methylation and acetylation and the increase to its original value when the methylated sample is saponified.[569] However, the fact that spectra of methylated or acetylated derivatives show a prominent residual absorption in the 3450 to 3300 cm^{-1} region has been attributed, in part, to

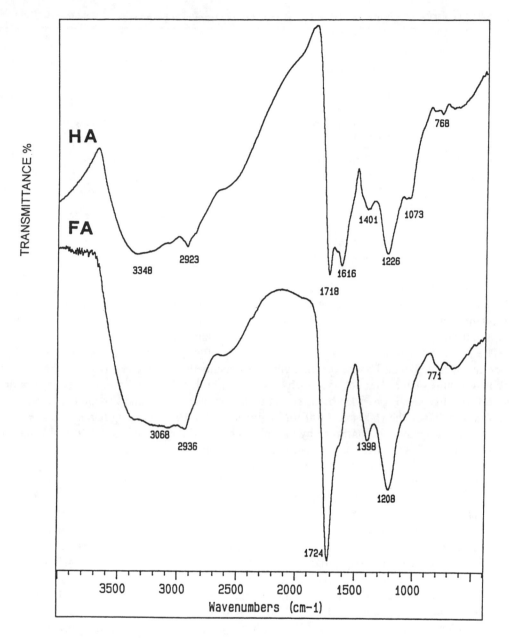

Figure 16 FT-IR spectra of the IHSS reference humic acid (HA) and fulvic acid (FA) from a Mollic Epipedon soil. (From Senesi, N. et al., *Sci. Total Environ.*, 81/82, 143, 1989. With permission.)

N–H stretching and/or to OH groups resistant to these treatments. It should be noted also that moisture contamination can contribute to absorption in the OH region.

Very rarely absorption due to aromatic C–H stretching occurring at frequencies slightly higher than 3000 cm⁻¹ is detected in the IR spectra of HS. The absence of an expected band in this region may be due to extensive substitution of the aromatic rings in HS macromolecules and/or to masking from the broad band due to the OH stretching.

Two absorption bands in the 2900 cm⁻¹ region are evident in the spectra of most HS, usually superimposed on the shoulder of the broad O–H stretching band. These bands are attributed to the asymmetric and symmetric stretching vibrations of aliphatic C–H bonds in CH_3 and CH_2 groups.[570]

Table 21   Main IR Adsorption Bands and Assignments for Humic Substances

Frequency (cm^{-1})	Assignment
3450–3300	O–H stretching, N–H stretching (trace), hydrogen-bonded OH
3080–3030	Aromatic C–H stretching
2950–2840	Aliphatic C–H stretching
1725–1710	C=O stretching of COOH, aldheydes and ketones
1660–1630	C=O stretching of amide groups (amide I band), quinone C=O and/or C=O of H-bonded conjugated ketones
1620–1600	Aromatic C = C stretching, COO$^-$ symmetric stretching
1540–1510	N–H deformation and C = N stretching (amide II band), Aromatic C=C stretching
1460–1440	Aliphatic C–H deformation
1400–1380	OH deformation and C–O stretching of phenolic OH, C–H deformation of CH$_2$ and CH$_3$ groups, COO$^-$ antisymmetric stretching
1260–1200	C–O stretching and OH deformation of COOH, C–O stretching of aryl ethers and phenols
1170	C–OH stretching of aliphatic O–H
1080–1030	C–O stretching of polysaccharide or polysaccharide-like substances, Si–O of silicate impurities
975–775	Out-of-plane bending of aromatic C–H

This assignment is substantiated by the increase in absorption of these bands observed upon methylation or acetylation.[569] Absorption in this region varies considerably among different HS samples, although it is often more pronounced for HAs than for FAs.

The pronounced band at 1725 to 1710 cm^{-1}, very often present in the IR spectra of HS, is generally attributed to the C=O stretching vibration, mainly due to COOH groups, but also to other carbonyl groups such as ketones and aldehydes. Evidence that the major part of this absorption is due to COOH groups is provided by its strong reduction in intensity, and concomitant appearance of two new bands in the 1600 and 1400 cm^{-1} regions, due to the COO$^-$ ion upon conversion of HS to salts. This assignment is confirmed by the shift to higher frequency of the 1720 cm^{-1} band upon methylation of HS, which can be attributed to the formation of esters, and by the appearance of IR bands typical of anhydrides upon acetylation of HS.[569]

A broad and intense band in the region 1660 to 1600 cm^{-1} is often present in IR spectra of HS, which is probably formed by a combination of several absorptions due to several groups. The preferential assignment of this band is to C=C vibrations of aromatic structures, possibly conjugated with C=O. However, several other groups are thought to contribute to this broad band, including C=O stretching of amide groups (the so-called amide I band), nonaromatic double bonds, H-bonded C=O of conjugated ketones and quinones, and COO$^-$ symmetric stretching. The bending vibration of water is centered at 1640 cm^{-1} and may contribute to the IR absorption in this region if the sample is not well dried. Quinone and conjugated ketone groups were observed at 1670 cm^{-1} by methylation and/or acetylation of HS,[334,571,572] whereas the contribution of H-bonded unsaturated conjugated ketones to the band near 1610 cm^{-1} has been stressed by Theng and Posner.[573]

Other absorption bands shown in the IR spectra of HS are (1) in the 1540 to 1510 cm^{-1} region, attributed to stretching vibrations of aromatic C=C bonds and/or to N–H deformation and C=N stretching (the so-called amide II band); (2) at 1460 to 1440 cm^{-1}, assigned to bending vibration of aliphatic C–H groups; (3) in the region 1400 to 1380 cm^{-1}, due to OH bending and C–O stretching of phenolic groups, COO$^-$ antisymmetric stretching, and aliphatic C-H deformation; (4) 1260 to 1200, preferentially assigned to C–O stretching and OH deformation of COOH groups and C–O stretching of aryl ethers and phenols; (5) around 1170 cm^{-1}, due to alcoholic groups; (6) in the region 1080 to 1030 cm^{-1}, due to C–O stretching of polysaccharides and Si–O of silicate impurities; and (7) several weak bands and shoulders below 1000 cm^{-1}, some of which are due to out-of-plane bending of CH of variously substituted aromatic groups, and CH of long-chain aliphatics.

Stevenson and Goh[574] classified IR spectra of HS into three general types on the basis of typical bands diagnostic of specific functional groups and structural entities. Type I spectra are typical of

HAs and are characterized by strong bands near 3400, 1720, 1600, and 1220 cm^{-1}. Type II spectra are typically shown by FAs and are characterized by strong absorptions at 1720 and 1200 cm^{-1}, whereas absorption in the 1600 cm^{-1} region is weak and centered near 1640 cm^{-1}. Type III spectra show typical absorptions at near 1540 cm^{-1}, in the 1900 cm^{-1} region, and at 1050 cm^{-1}. A more elaborate classification of IR spectra of HS into four major types has been made by Kumada,[575] based on the relative intensities of specific absorption bands.

The utility of IR spectroscopy has been expanded considerably when used in conjuction with chemical derivatization. As discussed previously, confirmatory evidence for the assignment of various IR absorption bands to various groups in HS has been obtained from spectra of derivatives produced by various chemical treatments, such as methylation and acetylation.

4. Nuclear Magnetic Resonance

The new and powerful NMR techniques are among the most useful tools currently available for the study of SOM and structural and functional groups of HS.[576] In the last two decades, NMR spectroscopy of HS has received considerable attention, and there has been a dramatic increase in the number of papers published on this subject in the last few years. Rapid advances in instrumentation have enabled one to apply NMR techniques to complex macromolecules such as HS. The technique of pulse FT-NMR spectroscopy using multichannel excitation has allowed for the rapid acquisition and averaging of many individual scans to produce an average spectrum with a high signal-to-noise ratio. As well, the development of signal-averaging and cross-polarization techniques has increased the sensitivity of NMR experiments, thus allowing otherwise unobtainable signals from such complex mixtures as HS to be recorded relatively easily. Recent advances in NMR spectroscopy have shown this technique to be the most definitive characterization tool compared to other characterization methods such as IR spectroscopy. Further, for relatively ash-free samples of HS, the NMR spectra are semiquantitative to quantitative, thus allowing for the calculation of the proportions of different structural entities and functional groups in HS samples.

The subject of NMR spectroscopy is very complex and requires a knowledge of advanced physics, chemistry, computer science, and mathematics for a thorough comprehension of the theory, experimental design, instrumental operation, computer control, and data interpretation. However, according to Malcolm[577] the acquisition and/or use of NMR data for application to HS is possible without having such a detailed understanding of all the associated theory and technology.

The basic theory of liquid-state and solid-state NMR spectroscopy with particular reference to its applicability to HS has been reviewed in a nonmathematical way by several authors,[578-580] whereas more detailed discussions are provided elsewhere.[155,581,582]

Nuclei which are active in NMR spectroscopy are those which possess a nuclear spin that is an odd integer multiple of ½, such as ^{1}H, ^{13}C, ^{15}N, and ^{31}P. In an NMR experiment, the sample is placed in a uniform magnetic field and an oscillating magnetic field is applied perpendicular to it. Either the steady field or the frequency of the oscillating field is varied until the condition of resonance is met. At resonance, the nuclei will absorb energy from the oscillating field and a spectrum is produced. When a nucleus that possesses a spin, such as ^{1}H or ^{13}C, exists in a chemical compound, the spinning nucleus is partially shielded by the surrounding electrons from the external magnetic field. Therefore, the effective magnetic field impinging upon the nuclear spin is altered and this in turn requires that the frequency of the oscillating field be changed in order to obtain resonance, if the stationary field is fixed, as is the case in all FT-NMR spectrometers.

The basis for the analysis is that the frequency at which nuclei such as ^{13}C and ^{1}H resonate is governed by the chemical environment of the C and H nuclei in the sample. Different functional groups, or the same functional group in different chemical environments in organic molecules, have different electron distributions and thus the constituent nuclei resonate at different frequencies. Consequently, from the resonance signals in the NMR spectrum, information is provided regarding

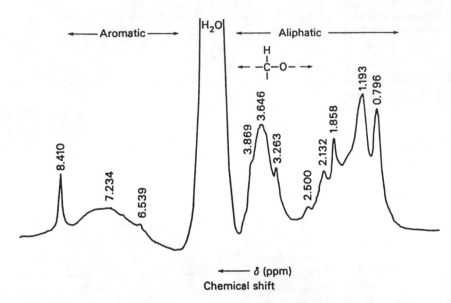

Figure 17 ¹H NMR spectrum of humic acid from Wakanui silt loam soil. (From Wilson, M. A. et al., *Nature,* 276, 487, 1978. With permission.

the chemical environments of C or H atoms in the sample, and identification of a particular functional group is possible.

The NMR frequency of a given nucleus is generally measured relative to a suitable standard, usually tetramethylsilane (TMS) for ¹H and ¹³C nuclei. The difference in resonance frequency with respect to a reference is termed "chemical shift", δ, which is usually quoted in ppm. In conclusion, results of NMR experiments are expressed in terms of chemical shift, which is the parameter from which structural information is obtained.

a. ¹H NMR

The ¹H nucleus is the most easily observed nucleus in NMR spectroscopy; however, resolution of resonance groups has been generally poor, as applied to HS. Solution ¹H NMR is the preferred method, but the interference from proton signals is a major problem, although unprotonated solvents, such as sodium deuteroxide (NaOD) or dimethylsulfoxide (DMSO), have been used to dissolve HS.

Although generally lacking hyperfine structure, the ¹H NMR spectra of soil HS may have features indicating a wide range of chemical structures, as is shown in the representative spectrum in Figure 17.[583] In Table 22,[583] the chemical shifts and corresponding assignments which have been obtained from ¹H NMR spectra of HS are summarized.

Early proton spectra of soil HAs dissolved in DOH obtained by FT-NMR[584] showed several broad bands in the aliphatic region from 1 to 2.5 ppm, a single methoxyl band at 3.8 ppm, and a single very broad band in the aromatic region extending from 6.5 to 8.5 ppm. Percentages of aromatic protons to the total protons from 35% to 19% were calculated by integration of the areas under the peaks. Using high-field (270 MHz) techniques, Wilson et al.[583] obtained spectra for soil HAs which contained considerable detail (Figure 17) and showed resonances from protons attached α to oxygen, which were assigned to carbohydrates. These spectra also showed a much stronger aromatic band from which it was estimated that 65% of C in this soil HS was in aromatic and carboxylic acid groups. In addition to the broad band in the aromatic region from 6 to 8.5 ppm, several well-defined peaks attributed to methyl and methylene groups indicated the presence of alkyl chains in the HS sample. The presence of polymethylene groups in different SOM extracts was also detected by Sciacovelli et al.[585] The presence of aromatic bands in terrestrial HAs has

Table 22   Chemical Shift Ranges and Tentative Assignments of Major ¹H Resonances Observed From Humic Substances, Using a DMSO Standard

Chemical shift range (ppm)	Tentative assignments
0.8–1.0	Terminal methyl groups of methylene chains
1.0–1.4	Methylene of methylene chains; CH_2CH at least two carbons or further from aromatic rings, or polar functional groups
1.4–1.7	Methylene of alicyclic compounds
1.7–2.0	Protons of methyl and methylene groups α to aromatic rings
2.0–3.3	Protons of methyl groups and methylene groups α to aromatic rings; protons α to carboxylic acid groups
3.3–5.0	Protons α to carbon attached to oxygen groups; carbohydrates
5.0–6.5	Olefins
6.5–8.1	Aromatic protons including phenols
8.1–9.0	Sterically hindered protons of aromatics
8.0–13.0	Acidic protons of phenols and carboxylic acids

From Wilson, M.A. et al., *Nature*, 276, 487, 1978. With permission.

been shown also by Hatcher et al.,[586] whereas Ruggiero et al.[587,588] have found that ¹H NMR spectra of different MW fractions of HAs and FAs showed differences in the aliphatic and aromatic regions. In another study, Wilson et al.[589] found that about 35% of the protons in a soil HA was associated with carbohydrates, whereas proton aromaticity was as low as 17%. Estimates for proton aromaticities based on proton NMR spectra are, however, subject to considerable error both because resonance due to aromatic protons often appears as a broad, slight rise of the baseline and because of the extensive substitution of the aromatic ring.

Proton NMR spectroscopy has been also used in several experiments with HS, including investigations of oxidation and other degradation methods, hydrogenation, metal interactions, and deuterium exchange. The work on ¹H NMR of HS has been reviewed by Wilson [590] and Wershaw,[578] where more detailed discussion may be found.

b. ¹³C NMR

Two general approaches, liquid-state and solid-state, have been used for ¹³C NMR spectroscopy of HS. The ¹³C NMR spectroscopy of HS presents some intrinsic limitations that are not encountered in proton NMR, which are (1) the low abundance of ¹³C; (2) the low sensitivity of ¹³C with respect to protons; (3) the highly variable relaxation times of ¹³C in different environments; and (4) the variable nuclear Overhauser effect (NOE).[590]

One advantage of liquid-state ¹³C NMR is the low line broadening due to chemical shift anisotropy and to dipolar interactions, which has allowed one to obtain spectra consisting of very sharp, well-defined lines. Most liquid-state ¹³C NMR spectra of HS have been obtained in an alkaline (i.e., 0.1 *M* NaOH) solution with sample concentrations of 50 to 100 mg ml⁻¹ on 200 to 400 MHz NMR spectrometers, with a required spectrometer time of 6 to 12 h.

Some disadvantages of liquid-state ¹³C NMR are (1) the low solubility of HS in suitable NMR solvents; (2) the possible formation of aggregates and colloidal particles; (3) long relaxation times and proton coupling effects; (4) solvent effects on chemical shifts; and (5) inability to maintain a liquid phase by cooling the sample to the low temperatures required.[4]

Because of the low solubility of several organic materials, such as HS, in organic solvents, a great interest has been developed in obtaining solid-state ¹³C NMR spectra. In the past, it has not been possible to obtain high-resolution ¹³C NMR spectra of solids, which generally consisted of much broader lines than those of liquid samples, but successive advances in solid-state NMR technology have greatly improved the situation. Line-broadening in solid-state ¹³C NMR is due to:

(1) dipolar interactions between ^{13}C and adjacent ^{1}H nuclei; (2) chemical-shift anisotropy of highly anisotropic carbons; and (3) poor signal-to-noise ratio.

Spectral quality of solid-state ^{13}C NMR has been greatly enhanced by using a combination of the following techniques: (1) high-power proton decoupling, which eliminates or minimizes dipole-dipole coupling between ^{1}H and ^{13}C interacting nuclei, i.e., the magnetic influence of the proton nuclei on a neighboring ^{13}C nucleus, thereby reducing line broadening; (2) cross-polarization (CP), or proton-enhanced nuclear induction, which results in the transfer of net magnetization from the abundant ^{1}H spins to the less abundant ^{13}C spins, thus overcoming the problem of dipolar line broadening and enhancing resolution; and (3) magic angle spinning (MAS), which eliminates the remaining dipolar ^{13}C-^{1}H interactions and chemical shift anisotropy effects by rapidly rotating the solid sample at the so-called magic angle (54.7°), with respect to the applied magnetic field.

A combination of the three above techniques into the so-called cross-polarization magic-angle spinning (CP-MAS) experiment usually overcomes the undesirable line-broadening factors in solids and yields a spectrum that is relatively well resolved and similar to that of a liquid sample.[577] A further improvement can be obtained by using the "dephasing" technique in which an extra delay period is used to enhance signal responses.

Many of the early ^{13}C NMR spectra of HS, obtained at 20 to 25 MHz, contained little information due to the poor sensitivity, low stability, and limited data processing capabilities of the older instruments based on electromagnets.[591] Detailed liquid-state ^{13}C NMR spectra of a soil HA and FA (Figure 18)[592] were obtained at 62.8 MHz, including variation of acquisition conditions, quantitative measurements of spin-lattice relaxation time (T_1) and NOE, and the use of multiple sequences to enhance resolution and assign carbon multiplicity. Detailed discussion of ^{13}C chemical shift assignments may be found in Preston and Blackwell.[592]

A typical solid state ^{13}C NMR spectrum for a soil HA is shown in Figure 19.[593] This spectrum is broader and contains fewer details than the liquid-state spectra but the major types of C in the HA are well resolved. The spectra shown in Figures 18 and 19 have features similar to those widely reported for HS samples from soils and other sources.

A common practice has been to divide the spectrum into regions corresponding to specific chemical classes. Typical chemical shift regions observed for HS in reference to the commonly used TMS standard, and corresponding assignments of ^{13}C resonances are summarized in Table 23. The chemical shifts are essentially the same for liquid- and solid-state ^{13}C NMR, but may vary by a few units owing to solvent and temperature effects. Specific signals observed in each of these regions in spectra published by various authors are, however, somewhat variable.

Almost all the best resolved ^{13}C NMR spectra of soil HS in the literature are of the CP MAS type.[577,586,594-596] Soil HAs obtained by alkaline extraction all yield the same general CP MAS ^{13}C NMR spectra showing five well-resolved and three poorly resolved peaks.[577,596] The chemical shifts for the five well-resolved peaks have been assigned to: (1) unsubstituted aliphatic C comprising methyl, methylene, and methine groups (0 to 50 ppm); (2) C in C–O of methoxyl groups (50 to 60 ppm); (3) C in all other aliphatic C–O and C–N groups (60 to 95 ppm); (4) aromatic C (110 to 160 ppm); and (5) carbonyl C in carboxyl, ester, and amide groups (160 to 190 ppm). The chemical shifts and assignments for the three poorly resolved peaks are (1) anomeric C (95 to 110 ppm); (2) aromatic C in phenolic groups, aromatic NH_2 groups, and aromatic ethers (142 to 160 ppm); and (3) carbonyl C in ketonic groups (190 to 230 ppm). Primarily C–O bonds are believed to contribute to resonances in the spectral region 60 to 90 ppm, because of the low N content in HAs. The region corresponding to 60 to 110 (C–O resonances) is commonly attributed to carbohydrates but other C–O containing compounds, such as ethers, may contribute to resonance in this region. Alkene double bonds may also contribute to resonance in the 110 to 160 ppm region and amino acid C to resonance in the 50 to 60 ppm region. The results of these studies indicated that soil HAs are comprised predominantly of aliphatic C (0 to 110 ppm) with a large component of aromatic C (110 to 160 ppm).

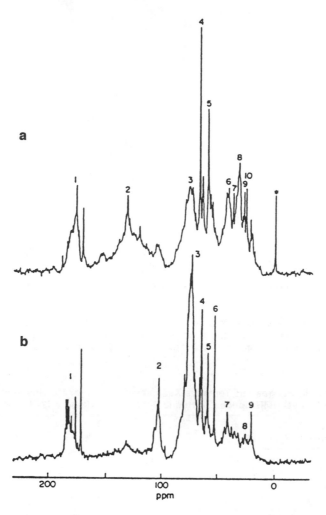

Figure 18   Liquid-state ^{13}C NMR spectra of humic and fulvic acids from Bainsville soil, showing discrete resonances. Peaks (ppm) humic acid: **1**, 176; **2**, 132; **3**, 76; **4**, 65; **5**, 59; **6**, 41; **7**, 37; **8**, 32; **9**, 27; **10**, 25; fulvic acid: **1**, 180; **2**, 103; **3**, 74; **4**, 64; **5**, 58; **6**, 52; **7**, 41; **8**, 25; 9, 20; * = tetramethylsilane. (From Preston, C. M. and Blackwell, B. A., *Soil Sci.*, 139, 88, 1985. With permission.)

The CP MAS ^{13}C NMR spectra of soil FAs are not only very different from those of the corresponding soil HAs in the number of resolvable peaks, but also exhibit a broad range of variation of chemical shift maxima in the same respective C functionalities.[577] In general, the FA spectra show the same well-resolved carbon chemical shift regions as previously described for soil HAs. An additional well-resolved peak is in the anomeric C region at 95 to 108 ppm. The dominant peak area in all the CP MAS ^{13}C NMR spectra of soil FA samples is the C–O chemical shift region between 60 and 110 ppm, which is believed to be due primarily to soil polysaccharides.[577,596] Spectra of FAs obtained after hydrolysis or passage through a XAD-8 resin column, which remove a large amount of polysaccharides, showed that the large C–O and anomeric C peaks are reduced and the aromatic C peak is increased.[577,596]

Most of the resonances in the 40 to 105 ppm region, arising from proteinaceous materials and carbohydrates, were no longer observed after acid hydrolysis of soil HAs and FAs.[451,597] Also, the intensity of signals between 170 to 180 ppm, due largely to C in COOH groups, were reduced because of partial decarboxylation occurring during hydrolysis. Intensities of the two principal components, aliphatic C (0 to 40 ppm) and aromatic C (106 to 150 ppm) were unaffected by acid

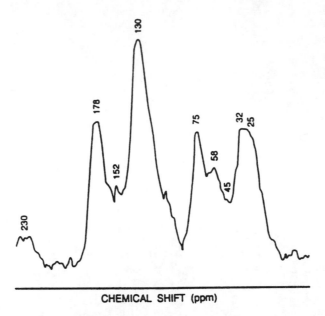

CHEMICAL SHIFT (ppm)

Figure 19 Solid-state ^{13}C NMR spectrum of a humic acid from a Mollisol A_h horizon. (From Schnitzer, M., in *Humic Substances in Soil and Crop Sciences: Selected Readings,* MacCarthy, P. et al., Eds., American Society of Agronomy, Madison, WI, 1990, 65. With permission.)

Table 23   Chemical Shift Ranges and Tentative Assignments of Major ^{13}C Resonances Observed in the CPMAS ^{13}C NMR Spectra of Humic Substances, Using a TMS Standard

Chemical shift (ppm)	Tentative assignments
0–50	Unsubstituted saturated aliphatic carbons
10–20	Terminal methyl groups
15–50	Methylene groups in alkyl chains
25–50	Methine groups in alkyl chains
50–95	Aliphatic carbon singly bonded to one oxygen or nitrogen atom
51–61	Aliphatic esters and ethers; methoxy, ethoxy
57–65	Carbon in CH_2OH groups; C_6 in polysaccharides
65–85	Carbon in CH(OH) groups; ring carbons of polysaccharides; ether-bonded aliphatic carbon
90–110	Carbon singly bonded to two oxygen atoms; anomeric carbon in polysaccharides, acetal or ketal
110–160	Aromatic and unsaturated carbon
110–120	Protonated aromatic carbon, aryl H
120–140	Unsubstituted and alkyl-substituted aromatic carbon
140–160	Aromatic carbon substituted by oxygen and nitrogen; aromatic ether, phenol, aromatic amines
160–230	Carbonyl, carboxyl, amide, ester carbons
160–190	Largely carboxyl carbons
190–230	Carbonyl carbons

From Malcolm, R. L., in *Humic Substances, Vol. 2, In Search of Structure,* Hayes, M. H. B. et al., John Wiley & Sons, New York, 1989, chap. 12. With permission.

hydrolysis and could be examined in greater detail. Gonzales-Vila et al.[598] obtained similar results in extensive studies of acid hydrolysis of HS.

The few ^{13}C NMR spectra that have been reported for soil humin fractions show moderate resolution with large variations in the type and quantity of C in each humin.[599]

For quantitative analyses, peak areas of the spectrum corresponding to the various chemical shift zones are measured often by integration, thus providing the distribution of various C types in HA and FA. The basic problem for quantitation is that the integrated area under a given band in the ^{13}C NMR spectrum is not only a function of the number of C atoms resonating at that frequency, but is also a function of the relaxation time and the NOE of the C atoms. Difficulties also arise when comparison of quantitative results from different laboratories are attempted, because of a lack of standardized procedures for sample treatment and instrumental operation, e.g., acquisition parameters optimized for quantitative analysis.[577] Another major problem is that HA and FA samples analyzed are not completely ash-free, and the presence of paramagnetic species can mask responses of the ^{13}C nuclei. However, visual interpretations of multipeak spectra should be avoided because they can be misleading, and semiquantitative C distributions calculated in several cases from ^{13}C NMR spectra of soil HAs and FAs are useful for comparisons among samples of diverse origin and nature, or submitted to different chemical treatments.[577,594,596]

An examination of the published literature on the application of ^{13}C NMR spectroscopy to soil HAs and FAs shows that there are wide differences in the chemical composition of HS depending on the soil type and geographic location. Aromaticities calculated by Malcolm[577,596] for a number of soil HAs ranged from 25% to 42%, with an average of 29%, whereas aromaticity calculated with exclusion of the C of COOH averaged 37%. These ranges were of the same order as those generally recorded for soil HAs.[600-605] Somewhat higher aromaticities were obtained by Hatcher et al.,[586] ranging from 49% to 54%, with exclusion of the C in COOH groups, and by Schnitzer,[2] ranging from 21% to 60%. Somewhat lower aromaticities, up to 26% and 37%, respectively, with or without C in COOH, were calculated for soil FAs by Malcolm.[577,596] An exception is provided by spectra of Saiz Jimenez et al.,[595] from which aromaticities <20% were calculated for soil HAs, and consistently higher values were obtained for the corresponding FAs.

Aliphatic structures are generally found to be the major components of most HS and to predominate in HS from submerged or poorly drained soils. Soil HAs contain approximately 20% C in the C–O region (60 to 110 ppm) of the spectra, a large portion of which has been assigned to covalently bonded saccharide moieties within the HA structure, which could not be removed by separation on XAD-8 resin but could be partially removed by hydrolysis.[597] The carboxyl and ester C peak from 160 to 190 ppm was intense in the spectra of soil HAs, and accounted for approximately 15% of the total C, being attributed to about two thirds COOH carbon and one-third ester C.[596] In general, ^{13}C NMR spectra showed that HAs are slightly more aromatic and contain more paraffinic C and methoxyl C than FAs, whereas FAs are richer in COOH groups and carbohydrates than HAs.

A different approach has been used by Wershaw et al.[606] and Mikita et al.[607] in measuring the relative amounts of various hydroxyl-containing groups, including aliphatic and aromatic carboxyls, phenolic, alcoholic, and carbohydrate OH, and amino N in HAs and FAs by ^{13}C NMR. In these experiments, HAs and FAs have been permethylated with ^{13}C-enriched reagents and then the areas of the detailed ^{13}C NMR peaks of the samples in the region 50 to 62 ppm have been measured by integration. An accurate representation of the relative abundance of functional groups has been obtained as the NOE effect was found uniform for all OCH_3 groups in this region, and no distortion due to relaxation effects was expected for these groups.

CP MAS ^{13}C NMR spectroscopy has also been used for evaluating purification procedures for HAs, for examining pathways of HS synthesis, to determine cultivation effects on SOM, and to monitor chemical changes associated with incorporation of ^{13}C-labeled substrates into components of SOM.[4] The changes in structural composition of forest HAs during humification have also been studied by a combination of CP MAS ^{13}C NMR spectroscopy and various degradation methods.[608-610] Valuable information on the chemical structure of HS has been obtained by combining ^{13}C NMR with chemical methods, including methylation, hydrolysis, oxidation, reduction, and use of different extractants.[155,578,591]

c. ^{15}N NMR

In the past decade, ^{15}N NMR spectroscopy has developed into a powerful new probe for examining the structures and reactions of soil HS. For a long time, insufficient sensitivity and the low natural abundance of ^{15}N isotope hampered the detection of the ^{15}N nucleus, especially in natural biopolymers. The development of new instrumental techniques and more powerful spectrometers have solved many prior detection problems and made possible the acquisition of routinely well-defined ^{15}N NMR spectra, which have allowed the examination of the molecular structures of N-containing compounds in great detail. The application of ^{15}N NMR to the study of HS is quite recent, but has already provided new insights and new opportunities. Two reviews on this subject have been provided recently.[576,611]

The N nucleus is the third most important probe (after ^{1}H and ^{13}C) for investigating the structures of organic molecules by NMR spectroscopy. Both ^{14}N and ^{15}N isotopes are suitable for NMR spectroscopy. The most abundant (99.63%) ^{14}N nucleus was usually preferred by NMR spectroscopists, but it had two disadvantages, line broadening and inability to discriminate among nonequivalent N nuclei in macromolecules such as HS. On the other hand, for a long time detection of the ^{15}N nucleus was hampered by insufficient sensitivity and low natural abundance. New techniques, such as new pulse sequences and polarization transfer, and the use of high-field magnets and large sample probe heads, have solved the detection problems to a large extent, rendering ^{15}N NMR a versatile method for studying the molecular structure of organic macromolecules, both in the liquid and solid states.[611] Two common standards are used in the literature for ^{15}N NMR, $CH_3{}^{15}NO_2$ and [15]NH_3, with the latter giving positive values to all ^{15}N chemical shifts. A wide array of pulse sequences, each having their unique advantages and limitations, are used in modern ^{15}N NMR spectroscopy. The recent article of Thorn et al.[612] provides an excellent brief review of the application of pulse sequences to the study of HS.

Some disadvantages inherent in the ^{15}N technology itself are: (1) long relaxation times and unfavorable NOE effects, which may make difficult the acquisition of reliable spectra; (2) acoustic ringing, an instrumental phenomenon associated with low-frequency nuclei like ^{15}N, which distorts the spectral baseline; (3) rapid exchange of attached protons, which may reduce polarization transfer and thus reduce the intensity of the N resonance; (4) the low natural abundance of ^{15}N, which makes the acquisition of ^{15}N spectra of non-labeled natural compounds very difficult; and (5) the high sensitivity of the N nucleus to changes in solvent polarity and pH.[611]

However, solid-state ^{15}N-CP MAS-NMR technology offers many attractive features for the study of macromolecules such as HS. In contrast to ^{13}C and ^{1}H, ^{15}N nuclei are favored for solid-state NMR studies of HS, since there are relatively few nonequivalent N atoms and the chemical shift differences are usually large. Solid-state studies of N incorporation into HS and of SOM have been reported.[613,614]

Four interesting studies were published in 1992 by Thorn and collaborators on the application of liquid-state ^{15}N NMR to HA and FA chemistry. Thorn et al.[612] identified by liquid-state ^{15}N and ^{13}C NMR the chemical nature of carbonyl functionality in a number of FA and HA samples of diverse origin, including a soil HA and a soil FA, which were previously derivatized with ^{15}N-labeled hydroxylamine. Unambiguous evidence was provided by ^{15}N NMR spectra for the presence of quinones, ketones, and esters. The results of the ^{13}C and ^{15}N examinations also indicated that all samples had significant ketone plus quinone contents, amounting to 6%–7% of total O, and that a significant fraction of the ketone and quinone groups did not react with hydroxylamine.

In another study, Thorn and Mikita [615] demonstrated by ^{15}N NMR spectroscopy that pyridine structures were formed when ^{15}N-labeled ammonia was fixed by HAs and FAs. A possible mechanism suggested for the formation of pyridine moieties in FA was a reaction between ammonia and 1,5-diketone groups, which could reveal useful information about the structure of the 1,5-diketone precursor.

Weber et al.[616] have shown that aniline, a degradation product of many pesticides, was incorporated into heterocyclic structures in FA by formation of covalent bonds. Thus, aromatic amines, like ammonia, could bind to HS to produce heterocyclic compounds, and the ^{15}N probe appeared to be well suited for elucidating the nature of interaction of N-containing herbicides and HS.

Thorn and Mikita[617] reported that HS could fix nitrite ion under acidic conditions, with ^{15}N NMR spectra revealing major oxime and nitrosophenol groups in the reaction product. Thus, direct examination of HA and FA adducts with ^{15}N-labeled compounds is now possible with ^{15}N NMR spectroscopy.

Other ^{15}N incorporation studies carried out since 1986 and relevant to SOM studies, as well as studies on nitrification and denitrification processes, based on liquid- and solid-state ^{15}N NMR spectroscopy, have been reviewed by Steelink.[611]

Results of CP MAS ^{15}N NMR analysis of ^{15}N-labeled HAs and FAs indicated that most of the N (80% and 86%, respectively, for HA and FA) occurred in amide forms, from 9% to 12% in amine forms, and from 4% to 9% in heterocyclic, mainly pyrrole forms.[618-620] Data collected in an extensive ^{15}N NMR study on the nitrogen metabolism in SOM have confirmed these proportions of N structures.[621,622]

d. ^{31}P NMR

Although present in low concentrations in SOM, ^{31}P is a sensitive nucleus and has proved to be very useful for examining the proportions of phosphorus compounds in alkaline soil extracts and soil HAs and FAs.[145,157-159,162,623-626] The proportions of orthophosphate, orthophosphate mono- and diesters, pyrophosphate, polyphospate, and phosphonate can be detected. Generally, SOM from undisturbed soils contained higher proportions of organic P forms, whereas phosphate fertilization resulted in a much higher proportion of orthophosphate. More information on SOM phosphorus by NMR spectroscopy has been provided before in this chapter. A recent review on ^{31}P NMR of SOM and HS has been provided by Preston.[576]

5. Electron Spin (or Paramagnetic) Resonance

Electron spin resonance (ESR) otherwise known as electron paramagnetic resonance (EPR) spectroscopy is a nondestructive, noninvasive, highly sensitive and accurate analytical technique that can detect and characterize species containing unpaired electrons. These include organic and inorganic free radicals and paramagnetic transition metal ions in free or complexed forms and in solid, colloidal, or solution states. Many important processes occurring in soils involve unpaired electrons amenable to ESR measurement.

The basic phenomenon underlying ESR spectroscopy is the Zeeman effect, which involves an interaction between the spin of an unpaired electron and an external magnetic field. The energy of an unpaired electron in a magnetic field of strength H, applied parallel to the z-axis, is given by

$$E = -g\beta HM_z \qquad (17)$$

where E is the energy of the unpaired electron, g is the spectroscopic splitting factor (g-value), β is the Bohr magneton, and M_z is the component of spin angular momentum of the electron in the direction of the applied magnetic field, H. The values that M_z may assume are $+\frac{1}{2}$ or $-\frac{1}{2}$, depending on the alignment of the spin magnetic moment either with the magnetic field direction (high energy) or against it (low energy). For any given value of H, H_0, the energy difference, ΔE, between the two spin states is

$$\Delta E = g\beta H_0 = h\nu \qquad (18)$$

If an incident electromagnetic radiation of frequency ν, applied perpendicular to the static magnetic field H_0, is supplied to the sample, absorption occurs provided that ν satisfies Equation 18. This is known as the "resonance condition" and the measurement of this absorption forms the basis of ESR spectroscopy.

When Equation 18 is satisfied, electrons in the lower level absorb radiation energy and are excited to the upper level, which is immediately followed by emission of energy of the same frequency ν by electrons in the upper level that fall to the lower level. A net absorption of radiation occurs, however, only when the population of the lower level is maintained greater than that of the upper level, which occurs by a mechanism known as "spin-lattice relaxation" that involves the dissipation of energy from electrons of the upper level to the surrounding lattice, and their return to the lower level.

ESR spectrometers operate in the microwave region of the electromagnetic spectrum (9 to 35 GHz), the X-band spectrometers (typical microwave frequency around 9.5 GHz) being the most common. The static, or DC magnetic field (up to 10 KG for most commercial spectrometers) is generated by large electromagnets flanking a chamber, called "microwave cavity" where the sample is placed and irradiated with microwaves supplied by a source called "Klystron". In the practical ESR experiment, the microwave frequency is usually held constant at about 9.5 GHz and the external magnetic field is swept continuously until the resonance condition is satisfied at H_0 (Equation 18), and absorption of radiation by the sample occurs. The signal detection is improved by greatly increasing the intensity of the absorption by modulating the magnetic field, usually at 100 KHz. The output of the detector-amplifier system thus appears as the first derivative of the absorption, which is usually recorded as a function of the external magnetic field, H.

The sensitivity of the ESR technique depends on several factors, including type of cavity, sample position in the cavity, nature of radical centers, spin concentration in the sample, microwave power level, modulation amplitude used, and temperature at which the measurement is made. In particular, according to Curie's law, maximum sensitivity for paramagnetic species will be attained at the lowest possible temperature; thus ESR spectra are often recorded at liquid N_2 (bp 77 K) temperature. A major disadvantage of working at room temperature in aqueous systems is that water has an extremely high dielectric loss at microwave frequencies, thus small samples are used to maintain sensitivity, and the sample must be placed in a capillary tube or a specially constructed flat solution cell, to minimize dielectric losses sufficiently.

One of the major limitations of an ESR experiment is the loss of resolution of the signal, which is determined by the overlap of the component lines to such an extent that information is lost. Line-width broadening may arise from "microwave power saturation" effects, thus the choice of power is critical because, if spin-lattice relaxation times are long, a high microwave power will tend to equalize the electron populations of the two levels, i.e., to "saturate" the spin system. Consequently, the net energy absorption, i.e., signal intensity, will be reduced, causing "homogeneous" broadening of signal line widths with loss of line resolution. Another mechanism of line-width broadening is the so-called "inhomogeneous" broadening, which may arise from nonuniformities in the magnetic field throughout the sample which can be caused by neighboring nuclei, that is, unresolved fine, hyperfine, or superhyperfine structure, or from dipolar interactions between unlike spins. These effects result in merging of the individual resonant lines or spin packets into a single overall line or envelope, with loss of resolution and related information. Two more sophisticated magnetic resonance techniques, namely, electron nuclear double resonance (ENDOR) and electron spin echo envelope modulation (ESEEM) spectrometries, can be used to overcome some difficulties caused by inhomogeneous broadening.

A detailed description of the theory and practice of ESR spectroscopy is beyond the scope of this chapter, and may be found elsewhere.[627-633]

Humic substances of any origin and nature are known to contain organic free radicals, which are indigenous to their structure, and that may be involved to a various extent in several chemical, biochemical, and photochemical processes occurring in soil and water systems. By definition, organic free radicals are characterized by the presence in their structure of one or more unpaired electrons, i.e., they are paramagnetic, and can, therefore, be measured and studied by ESR spectroscopy. Application of ESR spectroscopy can thus provide unique information on the nature and concentration of organic free radicals present in HS and on their changes as a function of various environmental factors, also including interaction with organic chemicals and paramagnetic metal ions. This work has been widely reviewed in the past[634] and more recently.[269,635-640]

In the usual experiment, the ESR spectrum of organic free radicals in HAs and FAs is obtained at room temperature by an ESR spectrometer operating at X-band frequency with a 100-KHz magnetic field modulation. Solid samples are usually placed in the resonant cavity packed in columns in suitable ESR quartz tubes; for water-dissolved samples, a special ESR flat cell is commonly used. The magnetic field is swept over a relatively narrow scan range (generally ≤ 100 G), through the field at which the free electron resonates ($g = 2.00232$). Most organic free radicals resonate, in fact, at a field corresponding to g-values close to this number.

The typical ESR spectra of organic free radicals in soil HAs and FAs are featured by a single-line resonance, devoid of any hyperfine structure, while a partially resolved hyperfine structure is rarely observed.[637,638] Four types of information are available from these spectra: (1) the spectroscopic splitting factor, that is, the g-value, which can be accurately approximated from the magnitudes of the magnetic field at which resonance occurs for the sample and for a standard of known g-value, usually N,N-diphenylpycrylhydrazyl (DPPH) diluted in powdered KCl ($g_{DPPH} = 2.0036$); (2) the width of the adsorption line (line width), which is generally measured in gauss or tesla (1 gauss $= 10^{-4}$ tesla), as the peak-to-peak separation of the first derivative line; (3) the hyperfine splitting, where observed, which is measured as the separation (in gauss or in tesla) between the hyperfine lines of the hyperfine structure; and (4) the concentration of unpaired electrons (free radicals) expressed in spins g^{-1}, usually estimated by comparing the area of the ESR signal of the sample with that of a standard char containing a known number of paramagnetic centers, e.g., the so-called "strong pitch", supplied by the manufacturer. The calculation is based on the assumption that the area, calculated by double integration or, more commonly, by multiplying the height and square of the width of the first derivative signal, is directly proportional to the number of paramagnetic centers contributing to the resonance. However, for a valid comparison of signal areas, the procedure requires that a number of factors related to the sample, standard, and instrumental setting be strictly controlled.[637] Despite these constraints, a reasonably reliable comparison can be expected, especially when measuring the relative changes in spin concentrations in a particular sample subjected to variations in physical and chemical factors.

The g-value often can supply useful information concerning the chemical nature of the radical. The g-values commonly measured for soil HAs and FAs do not differ significantly one from another, generally ranging from 2.0023 to 2.0050.[637,638] They are consistent with semiquinone radical units possibly conjugated to aromatic rings ($g = 2.0041$ for anthraquinones), although a contribution from methoxybenzene radicals (g-values range from 2.0035 to 2.0040) and N-associated radicals ($g = 2.0031$ to 2.0037) cannot be excluded.[637,638,641] Extended aromatic network conjugated to the semiquinone moiety may cause partial delocalization of the free electron from the O atom of the semiquinone to aromatic C atoms, with the consequent lowering of g-values.[642,643]

The line widths of the ESR signal also do not show any particular trend in relation to the origin and nature of HS.[637,638] Water solutions of HS always exhibit spectral line widths narrower (2.0 to 2.5 G) than those of solid-state samples (4.0 to 7.5 G). This behavior has been ascribed to rapid tumbling of molecules in liquids, great freedom of rotation, and low association with neighboring molecules. FAs usually show line widths slightly greater than HAs of the same source. Small variations in the line widths of ESR signals may be ascribed to a different number of superimposed resonances at slightly differing field values, all contributing to the signal. Line broadening may thus reflect either a partial delocalization of the unpaired electron from the semiquinone onto the conjugated aromatic network, or unresolved superhyperfine interactions of the free electron with a number of neighboring aromatic and aliphatic hydrogen nuclei. Additional factors that can influence the line width are the free radical concentration of the sample, its state of aggregation

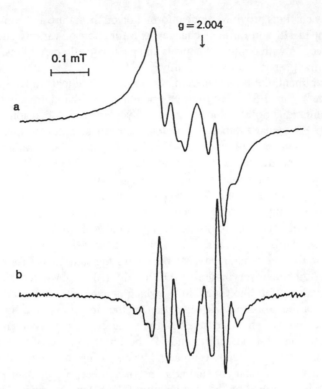

g = 2.004
↓

0.1 mT
├─────┤

a

b

Figure 20   Comparison of (a) first- and (b) second-derivative ESR spectra at a microwave of 1.01 mW and modulation amplitude of 0.0189 mT for solution of a Countesswells soil humic acid. (From Cheshire, M. V. and McPhail, D. B., *Eur. J. Soil Sci.*, 47, 205, 1996. With permission.)

(solid state or solution), temperature, solvent-solute interactions, interactions with metal ions, and power-saturation effects.

In most cases, the ESR spectra of HS lack hyperfine structure, which renders impossible any further description of the chemical and structural environment around the radical itself. Four-line and three-line hyperfine patterns, attributed to the interaction of the unpaired electron on the semiquinone O atom with two adjacent nonequivalent or equivalent H nuclei, were observed, respectively, for acid-boiled HAs originated from acid soils[644,645] and for an oxidized soil FA.[646] Steelink[635] observed that the ESR spectra of 3,4-dihydroxybenzoic acid semiquinone anion, of some flavonoids and quercetin and rutin, both of which possess a 3,4-dihydroxyphenyl structural unit, obtained after standing for few hours in alkaline solution, resembled the four-line spectrum obtained for acid-boiled HAs from Atherton et al.[644] A relation was found between dihydroxybenzoic acid groups in HS and their spin content.[647] These findings suggested that organic free radicals in HS may be at least partly associated with a catechol moiety. More recently, Cheshire and McPhail[648] have observed hyperfine splitting of the free radical signal in solution spectra of soil HAs (Figure 20). These authors have concluded that the measurement of hyperfine splitting depends on appropriate instrument settings, such as modulation amplitude and microwave power of the spectrometer and is facilitated by recording spectra in the second derivative mode.

The concentration of free radicals is probably the most important piece of data that can be obtained from the ESR spectrum of HS and the most frequently cited. Data from literature survey shows that the free radical concentration of HS ranges between about 10^6 and 10^8 spins g^{-1} and that FA usually contain one third to one fifth the spin content of HA from the same source.[637,638] Although some general relationships have been suggested to occur between soil type and spin content of HAs,[416,643] no specific correlation has been validated. Although the absolute number of unpaired electrons per mole of HS is generally small, it may contribute significantly to chemical

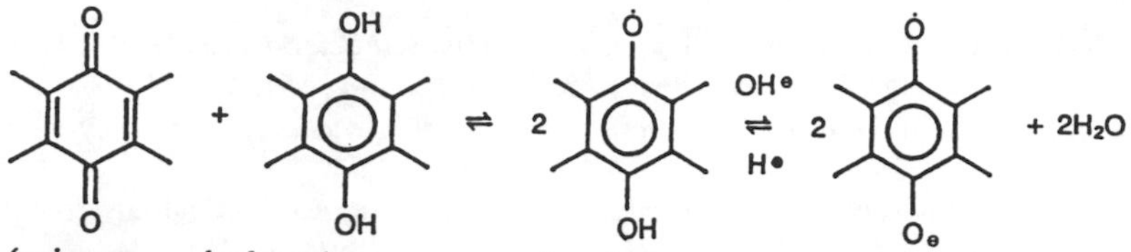

Figure 21   Quinone-hydroquinone model for the production of semiquinone free radicals in humic substances. (From Senesi, N., *Adv. Soil Sci.*, 14, 77, 1990. With permission.)

reactivity and functions exerted by HS. For example, a concentration ranging between 10^{17} and 10^{18} spins g^{-1} in compounds of MW 10,000 Da represents about one radical per 600 to 60 molecules, whereas in the case of higher MW, e.g., with MW = 100,000 Da, the value increases considerably to one radical over 60 to 6 molecules.

The free radical concentration of HS has been shown to depend on numerous environmental and laboratory factors which, however, left almost unaltered both the g-value and the line width of the ESR signal. This work has been amply reviewed elsewhere.[636-640]

Raising pH, chemical reduction, UV-Vis light irradiation, acid hydrolysis, and temperature increases produce a marked increase in free radical concentrations of HAs and FAs. In most of the cited experiments, however, the increase of free radical concentration was not sustained in time, but was followed by a gradual decrease soon after the maximum value was attained. A direct quantitative relationship was developed between pH and free radical content of FA by Wilson and Weber,[649] who concluded that a group of closely related semiquinone species, and not a single entity, was responsible for the free radical variation with pH. Photoinduced, charge-transfer mechanisms occurring between donor groups containing mobile H atoms and acceptor dark pigments, both present in HA, were suggested to be the main factors responsible for the increase of free radical content upon irradiation.[650]

Although contradictory data are reported, mild chemical or electrochemical oxidation, methylation, and increase in neutral electrolyte concentration often produced a time- and pH-dependent decrease of free radical concentration in HAs and FAs. The effect of oxidation could be reversed, however, by treatment with a reductant or by light-irradiation of the HS sample. The tenfold decrease of free radical concentrations measured for some soil HAs and FAs seemed to confirm that the OH groups are the most important electron donors responsible for the formation and existence of free radicals in HS.[642]

The accumulated ESR evidence supports the existence of a quinone-hydroquinone electron donor-acceptor (or charge-transfer) system for the reversible generation and maintainance of free radicals of semiquinonic nature in HS (Figure 21).[636-638] Two classes of free radicals of similar nature, but different stability, were suggested to exist in HS: (1) indigenous or "native" free radicals, stable over long time spans, which survive in any condition of the system; and (2) "transient" or short-lived radicals, which can be generated by a shift to the right of the equilibrium in Figure 21 caused by changes in the factors discussed above. The latter can persist only over relatively short time spans, since the equilibrium in Figure 21 can be easily reversed.

High spin contents in HS were usually associated with low H%, low H/C and O/C atomic ratios, low absorbance at 465 nm, high phenolic content, high E_4/E_6 ratios, and intense IR absorptions due to aromatic structures, whereas low free radical contents corresponded to a prevalence of aliphatic and olephinic bands in the IR.[541,642,643] These results provided evidence that the free radical concentration in HS is directly related to the dark color, degree of aromaticity, and molecular size and complexity of HS.

REFERENCES

1. Bohn, H. L., Estimate of organic carbon in world soils, *Soil Sci. Soc. Am. J.,* 46, 1118, 1982.
2. Schnitzer, M., Soil organic matter. The next 75 years, *Soil Sci.,* 151, 41, 1991.
3. Stevenson, F. J., *Humus Chemistry. Genesis, Composition, Reactions,* John Wiley & Sons, New York, 1982.
4. Stevenson, F. J., *Humus Chemistry. Genesis, Composition, Reactions,* 2nd ed., John Wiley & Sons, New York, 1994.
5. Schnitzer, M. and Khan, S. U., *Humic Substances in the Environment,* Marcel Dekker, New York, 1972.
6. Hayes, M. H. B. and Swift, R. S., The chemistry of soil organic colloids, in *The Chemistry of Soil Constituents,* Greenland, D. J. and Hayes, M. H. B., Eds., John Wiley & Sons, Chichester, 1978, 179.
7. Waksman, S. A., *Humus. Origin, Chemical Composition and Importance in Nature,* Balliere, Tindall and Cox, London, 1936.
8 Scheffer, F. and Ulrich, B., Lehrbuch der Agrikulturchemie und Bodenkunde. 111 Teil, *Humus and Humus/Düngung,* Bd 1. Stuttgart, 1960.
9. Kononova, M. M., *Soil Organic Matter,* 2nd ed., Pergamon Press, Oxford, 1966.
10. Felbeck, G. T., Structural chemistry of soil humic substances, *Adv. Agron.,* 17, 327, 1965.
11. Hempfling, R. and Schulten, H.-R., Pyrolysis-(gas chromatography)/mass spectrometry of agricultural soils and their humic fractions, *Z. Pflanzenernähr. Bodenk.,* 154, 425, 1991.
12. Tan, K. H., Himmelsbach, D. S., and Lobartini, J. C., The significance of solid-state ^{13}C NMR spectroscopy of whole soil in the characterization of humic matter, *Commun. Soil Sci. Plant Anal.,* 23, 1513, 1992.
13. Sorge, C., Schnitzer, M., and Schulten, H.-R., In-source pyrolysis-field ionization mass spectrometry and Curie-point pyrolysis-gas chromatography/mass spectrometry of amino acids in humic substances and soils, *Biol. Fertil. Soils,* 16, 100, 1993.
14. Sorge, C., Schnitzer, M., Leinweber, P., and Schulten, H.-R., Molecular-chemical characterization of organic matter in whole soil and particle-size fractions of a spodosol by pyrolysis-field ionization mass spectrometry, *Soil Sci.,* 158, 189, 1994.
15. Schulten, H.-R., The three-dimensional structure of humic substances and soil organic matter studied by computational analytical chemistry, *Fresenius J. Anal. Chem.,* 351, 62, 1995.
16. Golchin, A., Oades, J. M., Skjemstad, J. O., and Clarke, P., Structural and dynamic properties of soil organic matter as reflected by ^{13}C natural abundance, pyrolysis mass spectrometry and solid-state ^{13}C NMR spectroscopy in density fractions of an oxisol under forest and pasture, *Aust. J. Soil Res.,* 33, 59, 1995.
17. Kinchesh, P., Powlson, D. S., and Randall, E. W., ^{13}C NMR studies of organic matter in whole soils: I. Quantitation possibilities, *Eur. J. Soil Sci.,* 46, 125, 1995.
18. Kinchesh, P., Powlson, D. S., and Randall, E. W., ^{13}C NMR studies of organic matter in whole soils: II. A case study of some Rothamsted soils, *Eur. J. Soil Sci.,* 46, 139, 1995.
19. Lowe, L. E., Carbohydrates in soil, in *Soil Organic Matter,* Schnitzer, M. and Khan, S. U., Eds., Elsevier, New York, 1978, 65.
20. Gupta, U. C., Carbohydrates, in *Soil Biochemistry,* Vol. 1, McLaren, A. D. and Peterson, G. H., Eds., Arnold, London, 1967, 91.
21. Swincer, G. D., Oades, J. M. and Greenland, D. J., Extraction, characterization, and significance of soil polysaccharides, *Adv. Agron.,* 21, 195, 1969.
22. Greenland, D. J. and Oades, J. M., *Saccharides,* in *Soil Components, Organic Components,* Vol. 1, Gieseking, J. E., Ed., Springer-Verlag, New York, 1975, 213.
23. Cheshire, M. V., *Nature and Origin of Carbohydrates in Soil,* Academic Press, New York, 1979.
24. Doutre, D. A., Hay, G. W., Hood, A., and VanLoon, G. W., Spectrophotometric methods to determine carbohydrates in soil, *Soil Biol. Biochem.,* 10, 457, 1978.
25. Martens, D. A. and Frankenberger, W. T., Jr., Quantification of soil saccharides by spectrophotometric methods, *Soil Biol. Biochem.,* 22, 1173, 1990.
26. Whitehead, D. C., Buchan, H., and Hartley, R. D., Components of soil organic matter under grass and arable cropping, *Soil Biol. Biochem.,* 7, 65, 1975.
27. Folsom, B. L., Wagner, G. H., and Scrivner, C. L., Comparison of soil carbohydrate in several prairie and forest soils by gas-liquid chromatography, *Soil Sci. Soc. Am. Proc.,* 38, 305, 1974.

28. Swincer, G. D., Oades, J. M., and Greenland, D. J., Studies on soil polysaccharides. II. The composition and properties of polysaccharides in soils under pasture and under fallow-wheat rotation, *Aust. J. Soil Res.*, 6, 225, 1968.
29. Gupta, U. C. and Sowden, F. J., Occurrence of free sugars in soil organic matter, *Soil Sci.*, 96, 217, 1963.
30. Martin, J. P., Decomposition and binding action of polysaccharides in soil, *Soil Biol. Biochem.*, 3, 33, 1971.
31. Cheshire, M. V., Greaves, M. P., and Mundie, C. M., Decomposition of soil polysaccharide, *J. Soil Sci.*, 25, 483, 1974.
32. Cheshire, M. V., Origins and stability of soil polysaccharide, *J. Soil Sci.*, 28, 1, 1977.
33. Olness, A. and Clapp, C. E., Influence of polysaccharide structure on dextran adsorption by montmorillonite, *Soil Biol. Biochem.*, 7, 113, 1975.
34. Clapp, C. E., Dawson, J. E., and Hayes, M. H. B., Composition and properties of a purified polysaccharide isolated from an organic soil, in *Proc. Int. Symp. Peat in Agriculture and Horticulture*, Bet Dagan, Israel, 1979, 153.
35. Parsons, J. W. and Tinsley, J., Chemical studies of polysaccharide material in soils and composts based on extraction with anhydrous formic acid, *Soil Sci.*, 92, 46, 1961.
36. Barker, S. A., Finch, P., Hayes, M. H. B., Simmonds, R. G., and Stacey, M., Isolation and preliminary characterization of soil polysaccharides *Nature*, 205, 68, 1965.
37. Black, W. A. P., Cornhill, W. J., and Woodward, F. N., A preliminary investigation on the chemical composition of sphagnum moss and peat, *J. Appl. Chem.*, London, 5, 484, 1955.
38. Swincer, G. D., Oades, J. M., and Greenland, D. J., Studies on soil polysaccharides. I. The isolation of polysaccharides from soil, *Aust. J. Soil Res.*, 6, 211, 1968.
39. Hayes, M. H. B., Stacey, M., and Swift, R. S., Techniques for fractionating soil polysaccharides, *Trans. 10th Int. Congr. Soil Sci. Suppl.*, 1, 75, 1975.
40. Bernier, B., Characterisation of polysaccharides isolated from forest soils, *Biochem. J.*, 70, 590, 1958.
41. Forsyth, W. G. C., Studies on the more soluble complexes of soil organic matter. II. The composition of the soluble polysaccharide fraction, *Biochem. J.*, 46, 141, 1950.
42. Acton, C. J., Paul, E. A., and Rennie, D. A., Measurement of the polysaccharide content of soils, *Can. J. Soil Sci.*, 43, 141, 1963.
43. Swift, R. S. and Posner, A. M., Gel chromatography of humic acids, *J. Soil Sci.*, 22, 237, 1971.
44. Mehta, N. C., Strenli, H., Muller, M., and Denel, H., Role of polysaccharides in soil aggregation, *J. Sci. Food Agric.*, 11, 40, 1960.
45. Barker, S. A., Hayes, M. H. B., Simmonds, R. G., and Stacey, M., Studies on soil polysaccharides. I, *Carbohydrate Res.*, 5, 13, 1967.
46. Finch, P., Hayes, M. H. B., and Stacey, M., Studies on soil polysaccharides and their interactions with clay preparations, *Int. Soil Sci. Soc., Trans. Commun. II and IV, 1966*, Aberdeen, 1967, 19.
47. Cheshire, M. V., Russel, J. D., Fraser, A. R., Bracewell, J. M., Robertson, G. W., Benzingpurdie, L. M., Ratcliffe, C. I., Ripmeester, J. A., and Goodman, B. A., Nature of soil carbohydrate and its association with soil humic substancesm, *J. Soil Sci.*, 43, 359, 1992.
48. Cheshire, M. V., Mundie, C. M., Bracewell, J. M., Robertson, G. W., Russel, J. D., and Fraser, A. R., The extraction and characterization of soil polysaccharide by whole soil methylation, *J. Soil Sci.*, 34, 539, 1983.
49. Saini, G. R. and Salonius, P. O., Relation between Staudinger indices and precipitabilities of polysaccharides extracted from forest soils, *Soil Sci. Soc. Am. Proc.*, 33, 693, 1969.
50. Whistler, R. L. and Smart, C. L., *Polysaccharide Chemistry*, Academic Press, New York, 1953.
51. Hayes, M. H. B., Soil organic matter extraction, fractionation, structure and effects on soil structure, in *The Role of Organic Matter in Modern Agriculture*, Chen, Y. and Awimelech, Y., Eds., Nijhoff Publishing, Dordrecht, 1986.
52. Mortensen, J. L., Physico-chemical properties of a soil polysaccharide, *Trans. 7th Int. Congr. Soil Sci.*, 2, 98, 1960.
53. Parfitt, R. L. and Greenland, D. J., Adsorption of polysaccharides by montmorillonite, *Soil Sci. Soc. Am. Proc.*, 34, 862, 1970.
54. Lynch, D. L. and Cotnoir, L. J., The influence of clay minerals on the breakdown of certain organic substances, *Soil Sci. Soc. Am. Proc.*, 20, 367, 1956.

55. Bradley, D. B. and Sieling, D. H., Effect of organic anions and sugars on phosphate precipitation by iron and aluminium as influenced by pH, *Soil Sci.*, 76, 175, 1953.

56. Duff, R. B. and Webley, D. M., 2-ketogluconic acid as a natural chelator produced by soil bacteria, *Chem. Ind. (London)*, 1376, 1959.

57. Parsons, J. W. and Tinsley, J., Nitrogenous substances, in *Soil Components, Organic Components*, Vol. 1, Gieseking, J. E., Ed., Springer-Verlag, New York, 1975, 263.

58. Kelley, K. R. and Stevenson, F. J., Organic forms of N in soils, in *Humic Substances in Terrestrial Ecosystems*, Piccolo, A., Ed., Elsevier, Amsterdam, 1996, 407.

59. Jansson, S. L., Use of ^{15}N in studies of soil nitrogen, in *Soil Biochemistry*, Vol. 2, McLaren, A. D. and Skujins J., Eds., Marcel Dekker, New York, 1971, 129.

60. Kowalenko, C. G., Organic nitrogen, phosphorus and sulfur in soils, in *Soil Organic Matter*, Schnitzer, M. and Khan S. U., Eds., Elsevier, New York, 1978, 95.

61. Stevenson, F. J., *Cycles of Soil. Carbon, Nitrogen, Phosphorus, Sulfur and Micronutrients,* John Wiley & Sons, New York, 1986.

62. Bremner, J. M., Total nitrogen, in *Methods of Soil Analysis, Pt. 2, Chemical and Microbiological Properties*, Black, C. A., Evans, D. D., White, J. L., Ensminger, L. E., and Clark, F. E., Eds., American Society of Agronomy, Madison, WI, 1965, 1149.

63. Stevenson, F. J., Nitrogen — organic forms, in *Methods of Soil Analysis, Pt. 2, Chemical and Microbiological Properties*, 2nd ed., Page, A. L., Miller, R. H., and Keeney, D. R., Eds., American Society of Agronomy, Madison, WI, 1982, 625.

64. Sowden, F. J., Nature of the amino acid compounds of soil. II. Amino acids and peptides produced by partial hydrolysis, *Soil Sci.*, 102, 264, 1966.

65. Bremner, J. M., Studies on soil organic matter. I. The chemical nature of soil organic nitrogen, *J. Agric. Sci.*, 39, 183, 1949.

66. Cheng, C.-N. and Ponnamperuma, C., Note-extraction of amino acids from soils and sediments with superheated water, *Geochim. Cosmochim. Acta*, 38, 1843, 1974.

67. Bremner, J. M., Organic forms of nitrogen, in *Methods of Soil Analysis, Pt. 2, Chemical and Microbiological Properties*, Black, C. A., Evans, D. D., White, J. L., Ensminger, L. E., and Clark, F. E., Eds., American Society of Agronomy, Madison, WI, 1965, 1238.

68. Bremner, J. M., Nitrogenous compounds, in *Soil Biochemistry*, Vol. 1, McLaren A. D. and Peterson, G.H., Eds., Marcel Dekker, NewYork, 1967, 19.

69. Lowe, L. E., Amino acid distribution in forest humus layers in British Columbia, *Soil Sci. Soc. Am. Proc.*, 37, 569, 1973.

70. Sowden, F. J., Estimation of amino acids in soil hydrolysates by the Moore and Stein method, *Soil Sci.*, 80, 181, 1955.

71. Freney, J. R., Determination of water-soluble sulfate in soils, *Soil Sci.*, 86, 241, 1958.

72. Cheng, C.-N., Extracting and desalting amino acids from soils and sediments: evaluation of methods, *Soil Biol. Biochem.*, 7, 319, 1975.

73. Yonebayashi, K., Kyuma, K., and Kawaguchi, K., The relationship of soil humus fraction to acid hydrolyzable organic matter. I. Studies on readily decomposable organic matter, *Soil Sci. Plant Nutr.*, 20, 421, 1974.

74. Sowden, F. J., Chen, Y., and Schnitzer, M., The nitrogen distribution in soils formed under widely differing climatic conditions, *Geochim. Cosmochim. Acta*, 41, 1524, 1977.

75. Cheng, C.-N., Shufeldt, R. C., and Stevenson, F. J., Amino acid analysis of soil and sediments: extraction and desalting, *Soil Biol. Biochem.*, 7, 143, 1975.

76. Griffith, S. M., Sowden, F. J., and Schnitzer, M., The alkaline hydrolysis of acid-resistant soil and humic acid residues, *Soil Biol. Biochem.*, 8, 529, 1976.

77. Goh, K. M. and Edmeades, D. C., Distribution and partial characterisation of acid hydrolyzable organic nitrogen in six New Zealand soils, *Soil Biol. Biochem.*, 11, 127, 1979.

78. Stevenson, F. J., Organic forms of soil nitrogen, in *Nitrogen in Agricultural Soils*, Stevenson, F. J., Ed., American Society of Agronomy, Madison, WI, 1982, 67.

79. Stevenson, F. J. Distribution of the forms of nitrogen in some soil profiles, *Soil. Sci. Soc. Am. Proc.*, 21, 283, 1957.

80. Stevenson, F. J. and Cheng, C.-N., Amino acids in sediments: recovery by acid hydrolysis and quantitative estimation by a colorimetric procedure, *Geochim. Cosmochim. Acta*, 34, 77, 1970.

81. Stevenson, F. J. and Cheng, C.-N., Organic geochemistry of the Argentine Basin sediments: carbon nitrogen relationships and quaternary correlations, *Geochim. Cosmochim. Acta,* 36, 653, 1972.
82. Paul, E. A. and Schmidt, E. L., Extraction of free amino acids from soil, *Soil Sci. Soc. Am. Proc.,* 24, 195, 1960.
83. Grov, A., Amino acids in soil. II. Distribution of water-soluble amino acids in a pine forest soil profile, *Acta Chem. Scand.,* 17, 2316, 1963.
84. Sowden F. J. and Ivarson, K. C., The "free" amino acids of soil, *Can J. Soil Sci.,* 46, 109, 1966.
85. Rovira, A. D. and McDougall, B. M., Microbiological and biochemical aspects of the rizosphere, in *Soil Biochemistry,* Vol. 1, McLaren, A. D. and Peterson, G. H., Eds., Marcel Dekker, New York, 1967, 417.
86. Bremner, J.M., The amino acid composition of the protein materials in soil, *Biochem. J.,* 47, 538, 1950.
87. Stevenson, F. J., Ion exchange chromatography of the amino acids in soil hydrolysates, *Soil Sci. Soc. Am. Proc.,* 18, 373, 1954.
88. Stevenson, F. J., Isolation and identification of some amino compounds in soils, *Soil Sci. Soc. Am. Proc.,* 20, 201, 1956.
89. Pollock, G. E., Cheng, C.-N., and Cronin S. E., Determination of the D and L isomers of some protein amino acids present in soils, *Anal. Chem.,* 49, 2, 1977.
90. Young, J. L. and Mortensen, J. L., Soil nitrogen complexes. I. Chromatography of amino compounds in soil hydrolysates, *Ohio Agric. Exp. Stat. Res. Circular 61,* 1958, 1.
91. Hayashi, R. and Harada, T., Characterization of the organic nitrogen becoming decomposable through the effect of drying of a soil, *Soil Sci. Plant Nutr.,* 15, 226, 1969.
92. Kai, H., Ahmad, Z., and Harada, T., Factors affecting immobilization and release of nitrogen in soil and chemical characteristics of the nitrogen newly immobilized. III, *Soil Sci. Plant Nutr.,* 19, 275, 1973.
93. Stevenson, F. J. and Braids, O.C., Variation in the relative distribution of amino sugars with depth in some soil profiles, *Soil. Sci. Soc. Am. Proc.,* 32, 598, 1968.
94. Stevenson, F. J., Isolation and identification of amino sugars in soil, *Soil. Sci. Soc. Am. J.,* 47, 61, 1983.
95. Benzing-Purdie, L., Glucosamine and galactosamine distribution in a soil as determined by gas liquid chromatography of soil hydrolysates: effect of acid strength and cations, *Soil Sci. Soc. Am. J.,* 45, 66, 1981.
96. Benzing-Purdie, L., Amino sugar distribution in four soils as determined by high resolution gas chromatography, *Soil Sci. Soc. Am. J.,* 48, 219, 1984.
97. Millar, W. N. and Casida, L. E., Evidence for muramic acid in soil, *Can. J. Microbiol.,* 16, 299, 1970.
98. Sowden, F. J., Investigations on the amounts of hexosamine found in various soils and methods for their determination, *Soil Sci.,* 88, 138, 1959.
99. Stevenson, F. J., Investigations of aminopolysaccharides in soils. I. Colorimetric determination of hexosamines in soil hydrolysates, *Soil Sci.,* 83, 113, 1957.
100. Stevenson, F. J., Investigations of aminopolysaccharides in soils. II. Distribution of hexosamines in some soil profiles, *Soil Sci.,* 84, 99, 1957.
101. Anderson, G., Nucleic acids, derivatives, and organic phosphates, in *Soil Biochemistry,* Vol. 1, McLaren, A. D. and. Peterson, G. H., Eds., Marcel Dekker, New York, 1967, 67.
102. Cortez, J. and Schnitzer, M., Nucleic acid bases in soils and their association with organic and inorganic soil components, *Can. J. Soil Sci.,* 59, 277, 1979.
103. Hoyt, P. B., Chlorophyll-type compounds in soil. I. Their origin, *Plant Soil,* 25, 167, 1966.
104. Hoyt, P. B. Chlorophyll-type compounds in soil. II. Their composition, *Plant Soil,* 25, 313, 1966.
105. Hoyt, P. B., Fate of chlorophyll in soil, *Soil Sci.,* 111, 49, 1971.
106. Cornforth, I. S., The potential availability of organic nitrogen fractions in some West Indian soils, *Exper. Agric.,* 4, 193, 1968.
107. Gorham, E., Chlorophyll derivatives in Woodland soils, *Soil Sci.,* 87, 258, 1959.
108. Kowalenko, C. G. and McKercher, R. B., An examination of methods for extraction of soil phospholipids, *Soil Biol. Biochem.,* 2, 269, 1971.
109. Kowalenko, C. G. and McKercher, R. B., Phospholipid components extracted from Saskatchewan soils, *Can. J. Soil Sci.,* 51, 19, 1971.
110. Hance, R. J. and Anderson, G., Identification of hydrolysis products of soil phospholipids, *Soil Sci.,* 96, 157, 1963.
111. Fujii, K., Kobayashi, M., and Takahashi, E., On the alteration of microflora during the decomposition of plant residue. IV. Studies on the alteration of soil microflora and their metabolism, *Soil Sci. Plant Nutr.,* 20, 101, 1974.

112. Senesi, N. and Miano, T. M., The role of abiotic interactions with humic substances on the environ-mental impact of organic pollutants, in *Environmental Impact of Soil Component Interactions. Natural and Anthropogenic Organics*, Vol. 1, Huang, P. M., Berthelin, J., Bollag, J. M., McGill, W. B., and Page, A. L., Eds., CRC-Lewis, Boca Raton, FL, 1995, 311.

113. Cosgrove, D. J., Metabolism of organic phosphates in soil, in *Soil Biochemistry*, Vol. 1, McLaren, A. D. and Peterson, G. H., Eds., Marcel Dekker, New York, 1967, 216.

114. Halstead, R. L. and McKercher, R. B., Biochemistry and cycling of phosphorus, in *Soil Biochemistry*, Vol. 4, Paul, E. A. and McLaren, A. D., Eds., Marcel Dekker, New York, 1975, 31.

115. John, M. K., Sprout, N., and Kelley, C. C., The distribution of organic phosphorus in British Columbia soils and its relationship to soil characteristics, *Can. J. Soil Sci.*, 45, 87, 1965.

116. Elliott, E. T., Aggregate structure and carbon, nitrogen, and phosphorus in native and cultivated soils, *Soil Sci. Soc. Am. J.*, 50, 627, 1986.

117. Goh, K. M. and Williams, M. R., Distribution of carbon, nitrogen, phosphorus, sulfur, and acidity in two molecular weight fractions of organic matter in soil chronosequences, *J. Soil Sci.*, 33, 73, 1982.

118. Anderson, C. A. and Black, C. A., Separation of organic phosphorus in soil extracts by mechanical and chromatographic filtration, *Soil Sci. Soc. Am. Proc.*, 29, 255, 1965.

119. Steward, J. H. and Oades, J. M., The determination of organic phosphorus in soil, *J. Soil Sci.*, 23, 38, 1972.

120. Anderson, G., Other phosphorus compounds, in *Soil Components, Organic Components*, Vol. 1, Gieseking, J. E., Ed., Springer-Verlag, New York, 1975, 305.

121. Dick, W. A. and Tabatabai, M. A., An alkaline oxidation method for determination of total phosphorus in soils, *Soil Sci. Soc. Am. J.*, 41, 511, 1977.

122. Dalal, R. C., Soil organic phosphorus, *Adv. Agron.*, 29, 83, 1977.

123. Anderson, G., Assessing organic phosphorus in soils, in *The Role of Phosphorus in Agriculture*, Khasawneh, F. E., Sample, E. C., and Kamprath, E. J., Eds., American Society of Agronomy, Madison, WI, 1980, 411.

124. Islam, A. and Ahmed, A., Distribution of inositol phosphates, phospholipids, and nucleic acids and mineralization of inositol phosphates in some Bangladesh soils, *J. Soil Sci.*, 24, 193, 1973.

125. Magid, J., Tiessen, H., and Condron, L. M., Dynamics of organic phosphorus in soils under natural and agricultural ecosystems, in *Humic Substances in Terrestrial Ecosystems*, Piccolo, A., Ed., Elsevier, Amsterdam, 1996, 429.

126. Tate, K. R., Soil phosphorus, in *Soil Organic Matter and Biological Activity*, Vaughan, D. and Malcolm, R. E., Eds., Nijhof-Junk, Dordrecht, 1985, 329.

127. Harrison, A. F. *Soil Organic Phosphorus: A Review of World Literature*, C.A.B. International, Wall-ingsford, Oxon, U.K., 1987.

128. Stewart, J. W. B. and Tiessen, H., Dynamics of soil organic phosphorus, *Biogeochemistry*, 4, 41, 1987.

129. Sanyal, S. K. and De Datta, S. K., Chemistry of phosphorus transformations in Soil, *Adv. Soil Sci.*, 16, 1, 1992.

130. Smith, D. H. and Clark, F. E., Chromatographic separation of inositol phosphorus compounds, *Soil Sci. Soc. Am. Proc.*, 16, 170, 1952.

131. Cosgrove, D. J., *Inositol Phosphates*, Elsevier, New York, 1980.

132. Cosgrove, D. J., The chemical nature of soil organic phosphorus. I. Inositol phosphates, *Austr. J. Soil Res.*, 1, 203, 1963.

133. Cosgrove, D. J., Detection of isomers of phytic acid in some Scottish and Californian soils, *Soil Sci.*, 102, 42, 1966.

134. Cosgrove, D. J., The chemical nature of soil organic phosphorus. II. Characterization of the supposed DL-*chiro*-inositol hexaphosphate component of soil phytate as D-*chiro*-inositol hexaphosphate, *Soil Biol. Biochem.*, 1, 325, 1969.

135. Anderson, G., The identification and estimation of soil inositol phosphates, *J. Sci. Food Agric.*, 7, 437, 1956.

136. McKercher, L. B. and Anderson, G., Characterization of the inositol penta- and hexa-phosphate fractions of a number of Canadian and Scottish soils, *J. Soil Sci.*, 19, 302, 1968.

137. Omotoso, T. I. and Wild, J., Content of inositol phosphates in some English and Nigerian soils, *J. Soil Sci.*, 21, 216, 1970.

138. Omotoso, T. I. and Wild, A., Occurrence of inositol phosphates and other organic phosphate components in an organic complex, *J. Soil Sci.*, 21, 224, 1970.
139. Steward, J. H. and Tate, M. E., Gel chromatography of soil organic phosphorus, *J. Chromatogr.*, 60, 75, 1971.
140. Baker, R. T., A new method for estimating the phospholipid content of soils, *J. Soil Sci.*, 26, 432, 1975.
141. Anderson, G., The isolation of nucleoside diphosphates from alkaline extracts of soil, *J. Soil Sci.*, 21, 96, 1970.
142. Torsvik, V. L. and Goksoir, J., Determination of bacterial DNA in soil, *Soil Biol. Biochem.*, 10, 7, 1978.
143. Torsvik, V. L., Isolation of bacterial DNA from soil, *Soil Biol. Biochem.*, 12, 15, 1980.
144. Cheshire, M. V. and Anderson, G., Soil polysaccharides and carbohydrate phosphates, *Soil Sci.*, 119, 356, 1975.
145. Newman, R. H. and Tate, K. R., Soil phosphorus characterization by ^{31}P NMR, *Commun. Soil Sci. Plant Anal.*, 11, 835, 1980.
146. McKercher, L. B., Studies on soil organic phosphorus, in *Trans. 9th Int. Congr. Soil Sci.*, 3, 547, 1968.
147. Anderson, G. and Malcolm, R. E., The nature of alkali-soluble soil organic phosphates, *J. Soil Sci.*, 25, 282, 1974.
148. Robertson, G., Determination of phosphate in citric acid extracts, *J. Sci. Food Agric.*, 9, 288, 1958.
149. Kaila, A., Organic phosphorus in Finnish soils, *Soil Sci.*, 95, 38, 1963.
150. Moyer, J. R. and Thomas, R. L., Organic phosphorus and inositol phosphates in molecular size fractions of a soil organic matter extract, *Soil Sci. Soc. Am. Proc.*, 34, 80, 1970.
151. Swift, R. S. and Posner, A. M., Nitrogen, phosphorus and sulfur contents of humic acids fractionated with respect to molecular weight, *J. Soil Sci.*, 23, 50, 1972.
152. Veinot, R. L. and Thomas, R. L., High molecular weight organic phosphorus complexes in soil organic matter: inositol and metal content of various fractions, *Soil Sci. Soc. Am. Proc.*, 36, 71, 1972.
153. Tate, K. R., Fractionation of soil organic phosphorus in two New Zealand soils by use of sodium borate, *NZ J. Soil Sci.*, 22, 137, 1979.
154. Gerritse, R. G., Assessment of a procedure for fractionating organic phosphates in soil and organic materials using gel filtration and H.P.L.C., *J. Sci. Food. Agric*, 29, 577, 1978.
155. Wilson, M. A., *Techniques and Applications of Nuclear Magnetic Resonance Spectroscopy in Geochemistry and Soil Science*, Pergamon Press, Oxford, 1987.
156. Tate, K. R. and Newman, R. H., Phosphorus fractions of a climosequence of soils in New Zealand tussock grassland, *Soil Biol. Biochem.*, 14, 191, 1982.
157. Hawkes, G. E., Powlson, D. S., Randall, E. W., and Tate, K. R., A ^{31}P nuclear magnetic resonance study of the phosphorus species in alkali extracts of soils from long-term field experiments, *J. Soil Sci.*, 35, 35, 1984.
158. Condron, L. M., Goh, K. M., and Newman, R. H., Nature and distribution of soil phosphorus as revealed by a sequential extraction method followed by ^{31}P nuclear magnetic resonance analysis, *J. Soil Sci.*, 36, 199, 1985.
159. Condron, L. M., Frossard, E., Tiessen, H., Newman, R. H., and Stewart, J. W. B., Chemical nature of organic phosphorus in cultivated and uncultivated soils under different environmental conditions, *J. Soil Sci.*, 41, 41, 1990.
160. Condron, L. M., Moir, J. O., Tiessen, H., and Stewart, J. W. B., Critical evaluation of methods for determining total organic phosphorus in tropical soils, *Soil Sci. Soc. Am. J.*, 54, 1261, 1990.
161. Gil-Sotres, F., Zech, W., and Alt, H. G., Characterization of phosphorus fractions in surface horizons of soils from Galicia (N.W. Spain) by ^{31}P NMR spectroscopy, *Soil Biol. Biochem.*, 22, 75, 1990.
162. Hinedi, Z. R., Chang, A. C., and Lee, R. W. K., Mineralization of phosphorus in sludge-amended soils monitored by ^{31}P nuclear magnetic resonance spectroscopy, *Soil Sci. Soc. Am. J.*, 52, 1593, 1988.
163. Williams, R. J. P., Giles, R. G. F., and Posner, A. M., Solid state phosphorus NMR spectroscopy of minerals and soils, *J. Chem. Soc. Chem. Commun.*, 20, 1051, 1981.
164. Emsley, J. and Niazi, S. B., Chemical phosphorylation of myo-inositol, *Soil Biol. Biochem.*, 16, 73, 1983.
165. Turner, G. L., Smith, K. A., Kirkpatrick, R. J., and Oldfield, E., Structure and cation effect on ^{31}P NMR chemical shifts and chemical shift anisotropies of orthophosphates, *J. Magnetic Resonance*, 70, 408, 1986.

166. Vassallo, A. M., Wilson, M. A., Collin, P. J., Oades, J. M., Waters, A. G., and Malcolm, R. L., Structural analysis of geochemical samples by solid-state nuclear magnetic resonance spectroscopy. Role of paramagnetic material, *Anal. Chem.*, 59, 558, 1987.

167. Whitehead, D. C., Soil and plant-nutrition aspect of the sulfur cycle, *Soils Fertil.*, 27, 1, 1964.

168. Freney, J. R. and Williams, C. H., The sulfur cycle in soil, in *The Global Biogeochemical Sulfur Cycle, SCOPE 19*, Ivanov, M. V. and Freney, J. R., Eds., John Wiley & Sons, Chichester, UK, 1983, 129.

169. Mitchell, M. J., David, M. B., and Harrison, R. B., Sulfur dynamics of forest ecosystems, in *Sulfur Cycling on the Continents*, SCOPE 48, Howarth, R.W., Stewart, J.W., and Ivanov, M.V., Eds., John Wiley & Sons, Chichester, UK, 1992, 215.

170. Zhao, F. J., Wu, J., and McGrath, S. P., Soil organic sulfur and its turnover, in *Humic Substances in Terrestrial Ecosystems*, Piccolo, A., Ed., Elsevier, Amsterdam, 1996, 467.

171. Freney, J. R. Some observations on the nature of organic sulfur compounds in soil, *Aust. J. Soil Res.*, 11, 424, 1961.

172. Freney, J. R. Forms and reactions of the organic sulfur compounds in soils, in *Sulfur in Agriculture*, Tabatabai, M. A., Ed., American Society of Agronomy, Madison, WI, 1986, 207.

173. Jones, L. H. P., Cowling, D. W., and Lockyer, D. R., Plant-available and extractable sulfur in some soils of England and Wales, *Soil Sci.*, 114, 104, 1972.

174. Tabatabai, M. A. and Bremner, J. M., Distribution of total and available sulfur in selected soils and soil profiles, *Agron. J.*, 64, 40, 1972.

175. Tabatabai, M. A. and Bremner, J. M., Forms of sulfur, and carbon, nitrogen and sulfur relationships, in Iowa soils, *Soil Sci.*, 114, 380, 1972.

176. Bettany, J. R., Stewart, J. W. B., and Halstead, E. H., Sulfur fractions and carbon, nitrogen, and sulfur relationships in grassland, forest and associated transitional soils, *Soil Sci. Soc. Am. Proc.*, 37, 915, 1973.

177. Fitzgerald, J. W., Sulfate ester formation and hydrolysis: a potentially important yet often ignored aspect of the sulfur cycle of aerobic soils, *Bacteriol. Rev.*, 40, 698, 1976.

178. Thompson, J. F., Smith, I. K., and Madison, J. T., Sulfur metabolism in plants, in *Sulfur in Agriculture*, Tabatabai, M. A., Ed., American Society of Agronomy, Madison, WI, 1986, 57.

179. Saggar, S., Bettany, J. R., and Stewart, J. W. B., Measurement of microbial sulfur in soil, *Soil Biol. Biochem.*, 13, 493, 1981.

180. Strick, J. E. and Nakas, J. P., Calibration of a microbial sulfur technique for use in forest soils, *Soil Biol. Biochem.*, 16, 289, 1984.

181. Chapman, S. J., Microbial sulfur in some Scottish soils, *Soil Biol. Biochem.*, 19, 301, 1987.

182. Wu, J., O'Donnel, A. G., He, Z. L., and Syers, J. K., Fumigation-extraction method for the measurement of soil microbial biomass-S, *Soil Biol. Biochem.*, 26, 117, 1994.

183. Blanchar, R. W., Measurement of sulfur in soils and plants, in *Sulfur in Agriculture*, Tabatabai, M. A., Ed., American Society of Agronomy, Madison., WI, 1986, 455.

184. Johnson, C. M. and Nishita, H., Microestimation of sulfur in plant materials, soils, and irrigation waters, *Anal. Chem.*, 24, 736, 1952.

185. Elliott, L. F. and Travis, T. A., Detection of carbonyl sulfide and other gases emanating from beef cattle manure, *Soil Sci. Soc. Am. Proc.*, 37, 700, 1973.

186. Banwart, W. L. and Bremner, J. M., Volatilization of sulfur from unamended and sulfate-treated soils, *Soil Biol. Biochem.*, 8, 19, 1976.

187. Freney, J. R., Stevenson, F. J., and Beavers, A .H., Sulfur-containing amino acids in soil hydrolysates, *Soil Sci.*, 114, 468, 1972.

188. Scott, N. M., Bick, W., and Anderson, H. A., The measurement of sulfur-containing amino acids in some Scottish soils, *J. Sci. Food Agric.*, 13, 21, 1981.

189. Chae, Y. M. and Tabatabai, M. A., Sulfolipid and phospholipid in soils and sewage sludges in Iowa, *Soil Sci. Soc. Am. J.*, 45, 20, 1981.

190. Chae, Y. M. and Lowe, L. E., Distribution of lipid sulfur and total lipids in soils of British Columbia, *Can. J. Soil Sci.*, 60, 633, 1980.

191. Chae, Y. M. and Lowe, L. E., Fractionation by column chromatography of lipid and lipid sulfur extracted from soils, *Soil Biol. Biochem.*, 13, 257, 1981.

192. Lowe, L. E., Soluble polysaccharides fractions in selected Alberta soils, *Can. J. Soil Sci.*, 48, 215, 1968.

193. Freney, J. R. Sulfur-containing organics, in *Soil Biochemistry*, Vol. 1, McLaren, A. D. and Peterson, G. H., Eds., Marcel Dekker, New York, 1967, 229.

194. De Long, W. A. and Lowe, L. E., Carbon bonded sulfur in soil, *Can. J. Soil Sci.*, 42, 223, 1962.

195. Freney, J. R., Melville, G. E., and Williams, C. H., The determination of carbon bonded sulfur in soil, *Soil Sci.*, 109, 310, 1970.

196. Freney, J. R., Melville, G. E., and Williams, C. H., Soil organic matter fractions as sources of plant-available sulfur, *Soil Biol. Biochem.*, 7, 217, 1975.

197. Lowe, L. E., Sulfur fractions of selected Alberta profiles of the gleysolic order, *Can. J. Soil Sci.*, 49, 375, 1969.

198. Scott, N. M. and Anderson, G., Organic sulfur fractions in Scottish soils, *J. Sci. Food Agric.*, 27, 357, 1976.

199. Fitzgerald, J. W., Naturally occurring organosulfur compounds in soil, in *Sulfur in the Environment, Pt. II, Ecological Impacts*, Nriagu, J. O., Ed., John Wiley & Sons, New York, 1978, 391.

200. Lowe, L. E., An approach to the study of the sulfur status of soils and its application to selected Quebec soils, *Can. J. Soil Sci.*, 44, 176, 1964.

201. Lowe, L. E., Sulfur fractions of selected Alberta profiles of the chernozemic and podzolic orders, *Can. J. Soil Sci.*, 45, 293, 1965.

202. Neptune, A. M. L., Tabatabai, M.A., and Hanway, J. J., Sulfur fractions and carbon-nitrogen-phosphorus-sulfur relationships in some Brazilian and Iowa soils, *Soil Sci. Soc. Am. Proc.*, 37, 51, 1975.

203. David, M. B., Mitchell, M. J., and Nakas, J. P., Organic and inorganic sulfur constituents of a forest soil and their relationship to microbial activity, *Soil Sci. Soc. Am. J.*, 46, 847, 1982.

204. Zucker, A. and Zech, W., Sulfur status of four uncultivated soil profiles in northern Bavaria, *Geoderma*, 36, 229, 1985.

205. Mitchell, M. J., David, M. B., Maynard, D. G., and Telang, S. A., Sulfur constituents in soils and streams of a watershed in the Rocky Mountains of Alberta, *Can. J. Forest Res.*, 16, 315, 1986.

206. Strickland, T. C., Fitzgerald, J. W., Ash, J. T., and Swank, W. T., Organic sulfur transformations and sulfur pool sizes in soil and litter from a southern Appalachian hardwood forest, *Soil Sci.*, 143, 453, 1987.

207. Fitzgerald, J. W., Hale, D. D., and Swank, W. T., Sulfur-containing amino acid metabolism in surface horizons of a hardwood forest, *Soil Biol. Biochem.*, 20, 825, 1988.

208. Vannier, C., Didon-Lescot, J. F., Lelong, F., and Guillet, B., Distribution of sulfur forms in soils from beech and spruce forests of Mont Lozére (France), *Plant Soil*, 154, 197, 1993.

209. Acquaye, D. K. and Beringer, H., Sulfur in Ghanaian soils. I. Status and distribution of different forms of sulfur in some typical profiles, *Plant Soil*, 113, 197, 1989.

210. Nguyen, M. L. and Goh, K. M., Accumulation of soil sulfur fractions in grazed pastures receiving long-term superphosphate applications, *NZ J. Agric. Res.*, 33, 111, 1990.

211. Anderson, D. W., Saggar, S., Bettany, J. R., and Stewart, J. W. B., Particle size fractions and their use in studies of soil organic matter. I. The nature and distribution of forms of carbon, nitrogen, and sulfur, *Soil Sci. Soc. Am. J.*, 45, 767, 1981.

212. Bettany, J. R., Stewart, J. W. B., and Saggar, S., The nature and forms of sulfur in organic matter fractions in soils selected along an environmental gradient, *Soil Sci. Soc. Am. J.*, 43, 981, 1979.

213. Lowe, L. E. and Delong, W. A., Carbon-bonded sulfur in selected Quebec soils, *Soil Sci.*, 43, 151, 1963.

214. Scott, N. M. and Anderson, G., Sulfur, carbon, and nitrogen contents of organic fractions from acetylacetone extracts of soils, *J. Soil Sci.*, 27, 324, 1976.

215. Keer, J. I., McLaren, R. G., and Swift, R. S., Acetylacetone extraction of soil organic sulfur and fractionation using gel chromatography, *Soil Biol. Biochem.*, 22, 97, 1990.

216. Freney, J. R., Melville, G. E., and Williams, C. H., Extraction, chemical nature, and properties of soil organic sulfur, *J. Sci. Food Agric.*, 20, 440, 1969.

217. Schoenau, J. J. and Bettany, J. R., Organic matter leaching as a component of carbon, nitrogen, phosphorus, and sulfur cycles in a forest, grassland, and gleyed soil, *Soil Sci. Soc. Am. J.*, 5, 646, 1987.

218. Bettany, J. R., Saggar, S., and Stewart, J. W. B., Comparisons of the amounts and forms of sulfur in soil organic matter fractions after 65 years of cultivation, *Soil Sci. Soc. Am. J.*, 44, 70, 1980.

219. Freney, J. R., Melville, G. E., and Williams, C. H., Organic sulfur fractions labeled by addition of ^{35}S-sulfate to soil, *Soil Biol. Biochem.*, 3, 133, 1971.

220. MacLaren, R. G. and Swift, R. S., Changes in soil organic sulfur fractions due to long term cultivation of soils, *J. Soil Sci.*, 28, 445, 1977.

221. McLachlan, K. D. and De Marco, D. G., Changes in soil sulfur fractions with fertilizer additions and cropping treatments, *Aust. J. Soil Res.*, 13, 169, 1975.

222. Castellano, S. D. and Dick, R. P., Cropping and sulfur fertilization influence on sulfur transformations in soil, *Soil Sci. Soc. Am. J.*, 54, 114, 1990.

223. Ghani, A., McLaren, R. G., and Swift, R. S., Sulfur mineralization in some New Zealand soils, *Biol. Fertil. Soils*, 11, 68, 1991.

224. Haynes, R. J. and Williams, P. H., Accumulation of soil organic matter and the forms, mineralization potential and plant-availability of accumulated organic sulfur: effects of pasture improvement and intensive cultivation, *Soil Biol. Biochem.*, 24, 209, 1992.

225. Krouse, H. R. and Tabatabai, M. A., Stable sulfur isotopes, in *Sulfur in Agriculture*, Tabatabai, M. A., Ed., American Society of Agronomy, Madison, WI, 1986, 169.

226. Schoenau, J. J. and Bettany, J. R., ^{34}S natural abundance variations in prairie and boreal forest soils, *J. Soil Sci.*, 40, 397, 1989.

227. Howard, A. J. and Hamer, D., The extraction and constitution of peat wax. Review of peat wax chemistry, *J. Am. Oil Chem. Soc.*, 37, 478, 1960.

228. Stevenson, F. J., Lipids in soil, *J. Am. Oil Chem. Soc.*, 43, 203, 1966.

229. Morrison, R. I., Soil lipids, in *Organic Geochemistry*, Eglinton, G. and Murphy, M. T. J., Eds., Pergamon, New York, 1969, 559.

230. Braids, O. C. and Miller, R. H., Fats, waxes and resins in soils, in *Soil Components, Organic Components*, Vol. 1, Gieseking, J. E., Ed., Springer-Verlag, Berlin, 1975, 343.

231. Robinson, T., *The Organic Constituents of Higher Plants*, Burgess Publishing, Minneapolis, 1963.

232. Turfitt, G. E., The microbiological degradation of steroids. I. The sterol content of soils, *Biochem. J.*, 37, 115, 1943.

233. Wang, Yu-Cheng Liang and Wei-Chiang Shen, Method of extraction and analysis of higher fatty acids and triglycerides in soils, *Soil Sci.*, 107, 181, 1969.

234. Meinschein, W. G. and Kenny, G. S., Analyses of a chromatographic fraction of organic extracts of soils, *Anal. Chem.*, 29, 1153, 1957.

235. Eglinton, G. and Murphy, M. T. J., *Organic Geochemistry: Methods and Results*, Springer-Verlag, New York, 1969.

236. Schnitzer, M. and Schulten, H.-R., Pyrolysis-soft ionization mass spectrometry of aliphatics extracted from a soil clay and humic substances, *Sci. Total Environ.*, 81/82, 19, 1989.

237. Schulten, H.-R. and Schnitzer, M., Aliphatics in soil organic matter in fine-clay fractions, *Soil Sci. Soc. Am. J.*, 54, 98, 1990.

238. Kolattukudy, P. E., *Chemistry and Biochemistry of Natural Waxes*, Elsevier, Amsterdam, 1976.

239. Butler, J. H. A., Downing, D. T., and Swaby, R. S., Isolation of chlorinated pigment from green soil, *Aust. J. Chem.*, 17, 717, 1964.

240. Morrison, R. I. and Bick W., Long-chain methyl ketones in soils, *Chem. Ind. (London)*, p. 596, 1966.

241. Galoppini, C. and Riffaldi, R., Composition of ether extracts of soil, *Agrochimica*, 13, 207, 1969.

242. Schnitzer, M., Hindle, C. A., and Meglic, M., Supercritical gas extraction of alkanes and alkanoic acids from soils and humic materials, *Soil Sci. Soc. Am. J.*, 50, 913, 1986.

243. Schnitzer, M. and Preston, C. M., Supercritical gas extraction of a soil with solvents of increasing polarities, *Soil Sci. Soc. Am. J.*, 51, 639, 1987.

244. Fox, T. R. and Comerford, N. B., Low-molecular-weight organic acids in selected forest soils of the Southeastern U.S.A., *Soil Sci. Soc. Am. J.*, 54, 1139, 1990.

245. Jones, J. G., Studies on lipids of soil micro-organisms with particular reference to hydrocarbons, *J. Gen. Microbiol.*, 59, 145, 1969.

246. Simonart, P. and Batistic, L., Aromatic hydrocarbons in soils *Nature*, 212, 1461, 1966.

247. Blumer, M., Benzypyrenes in soil, *Science*, 134, 474, 1961.

248. Hites, R. A., Laflamme, R. E., and Farrington, J. W., Polycyclic aromatic hydrocarbons in recent sediments: the historical record, *Science*, 198, 829, 1977

249. Laflamme, R. E. and Hites, R. A., The global distribution of polycyclic aromatic hydrocarbons in recent sediments, *Geochim. Cosmochim. Acta*, 42, 289, 1978.

250. Swan, E. P., Identity of a hydrocarbon found in a forest soil, *Forest Prod. J.*, 15, 272, 1965.

251. Chopra, M. N., Investigations into the fate of plant pigments in some Canadian soils, *Soil Sci.,* 121, 103, 1976.
252. Wang, T. S. C. and Chuang, T-T., Soil alcohols, their dynamics and their effect upon plant growth, *Soil Sci.,* 104, 40, 1967.
253. Wang, T. S. C., Pau-Tsung Hwang and Chung-Yi Chen, Soil lipids under various crops, *Soil Sci. Soc. Am. Proc.,* 35, 584, 1971.
254. Lambert, E. N., Seaforth, C. E., and Ahmed, N., The occurrence of 2-methoxy-1,4-naphthoquinone in Carribean Vertisols, *Soil Sci. Soc. Am. Proc.,* 35, 463, 1971.
255. Cifrulak, C. F., Spectroscopic evidence of phthalates in soil organic matter, *Soil Sci.,* 107, 63, 1969.
256. Khan, S. U. and Schnitzer, M., Sephadex gel filtration of fulvic acid: the identification of major components in two low-molecular weight fractions, *Soil Sci.,* 112, 231, 1971.
257. Smith, K. A. and Russell, R. S., Occurrence of ethylene, and its significance, in anaerobic soil, *Nature,* 222, 769, 1969.
258. Smith, K. A. and Restall, S. W. F., The occurrence of ethylene in anaerobic soil, *J. Soil Sci.,* 22, 430, 1971.
259. Goodlass, G. and Smith, K. A., Effects of organic amendments on evolution of ethylene and other hydrocarbons from soil, *Soil Biol. Biochem.,* 10, 201, 1978.
260. Sexstone, A. J. and Mains, C. N., Production of methane and ethylene in organic horizons of spruce forest soils, *Soil Biol. Biochem.,* 22, 1315, 1990.
261. Whitehead, D. C., Identification of *p*-hydroxybenzoic, vanillic, *p*-coumaric and ferulic acids in soils, *Nature,* 202, 417, 1963.
262. McCalla, T. M. and Haskins, F. A., Phytotoxic substances from soil microorganisms and crop residues, *Bacteriol. Rev.,* 28, 181, 1964.
263. Mojé, W., Organic soil toxins, in *Diagnostic Criterion for Plants and Soils,* Chapman, H. D., Ed., University of California Press, Berkeley, 1971, 533.
264. Patrick, Z. A., Phytotoxic substances associated with the decomposition in soil of plant residues, *Soil Sci.,* 111, 13, 1971.
265. Flaig, W., Effect of lignin degradation products on plant growth, *in Isotopes and Radiation in Soil Organic Matter Studies,* International Atomic Energy Agency (IAEA), Vienna, 1965, 3.
266. Jambu, P., Fustec, E., and Jacquesy, R., Les lipids des sols: nature, origine, evolution, propriétes, *Sci. Sol,* 4, 229, 1978.
267. Thurman, E. M., *Organic Geochemistry of Natural Waters,* Nijhoff, Dordrecht, 1986.
268. Langford, C. H., Gamble, D. S., Underdown, A. W., and Lee, S., Interaction of metal ions with a well characterized fulvic acid, in *Aquatic and Terrestrial Humic Materials,* Christman, R. F. and Gjessing, E. T., Eds., Ann Arbor Science, Ann Arbor, MI, 1983, 219.
269. Senesi, N., Metal-humic substance complexes in the environment. Molecular and mechanistic aspects by multiple spectroscopic approach, in *Biogeochemistry of Trace Metals,* Adriano, D. C., Ed., Lewis, Boca Raton, FL, 1992, 429.
270. Schulten, H.-R. and Schnitzer, M., A state of the art structural concept for humic substances, *Naturwissenschaften,* 80, 29, 1993.
271. Schulten, H.-R., A chemical structure for humic acid. Pyrolysis-gas chromatography/mass spectrometry and pyrolysis-soft ionization mass spectrometry evidence, in *Humic Substances in the Global Environment and Implications on Human Health,* Senesi, N. and Miano, T. M., Eds., Elsevier, Amsterdam, 1994, 43.
272. Schulten, H.-R., Three-dimensional molecular structures of humic acids and their interactions with water and dissolved contaminants, *Int. J. Environ. Anal. Chem.,* 64, 147, 1996.
273. Waksman, S. A., *Humus,* Williams & Wilkins, Baltimore, 1932.
274. Martin, J. P. and Haider, K., Microbial activity in relation to soil humus formation, *Soil Sci.,* 111, 54, 1971.
275. Haider, K., Martin, J. P., and Filip, Z., Humus biochemistry, in *Soil Biochemistry,* Vol. 4, Paul, E. A. and McLaren, A. D., Eds., Marcel Dekker, New York, 1975, 195.
276. Flaig, W., Generation of model chemical precursors, in *Humic Substances and their Role in the Environment,* Frimmel, F. H. and Christman, R. F., Eds., John Wiley & Sons, New York, 1988, 75.
277. Hatcher, P. G. and Spiker, E. C., Selective degradation of plant biomolecules, in *Humic Substances and their Role in the Environment,* Frimmel, F. H. and Christman, R. F., Eds., John Wiley & Sons, New York, 1988, 59.

278. Hedges, J. I., Polymerization of humic substances in natural environments, in *Humic Substances and their Role in the Environment*, Frimmel, F. H. and Christman, R. F., Eds., John Wiley & Sons, New York, 1988, 45.

279. Flaig, W., Beutelspacher, H., and Rietz, E., Chemical composition and physical properties of humic substances, in *Soil Components: Organic Components*, Vol. 1, Gieseking, J. E., Ed., Springer-Verlag, New York, 1975, 1.

280. Hedges, J. I., The formation and clay mineral reactions of melanoidins, *Geochim. Cosmochim. Acta*, 42, 69, 1978.

281. Ikan, R., Rubinsztain, Y., Ioselis, P., Aizenshtat, Z., Pugmire, R., Anderson, L. L., and Woolfenden, W. R., Carbon-13 cross polarized magic-angle samples spinning nuclear magnetic resonance of melanoidins, *Org. Geochem.*, 9, 199, 1986.

282. Ishiwatari, R., Morinaga, S., Yamamoto, S., Machihara, T., Rubinsztain, Y., Ioselis, P., Aizenshtat, Z., and Ikan, R., A study of formation mechanism of sedimentary humic substances. I. Characterization of synthetic humic substances (melanoidins) by alkaline potassium permanganate oxidation, *Org. Geochem.*, 9, 11, 1986.

283. Hodge, J. E., Dehydrated foods, chemistry of browning reactions in model systems, *J. Agric. Food Chem.*, 1, 928, 1953.

284. Popoff, T. and Theander, O., Formation of aromatic compounds from carbohydrates. IV. Chromones from reaction of hexuronic acids in slightly acidic, aqueous solution, *Acta Chem. Scand.*, B30, 705, 1976.

285. Saiz-Jimenez, C., Haider, K., and Martin, J. P., Anthraquinones and phenols as intermediates in the formation of dark-colored, humic acid-like pigments by *Eurotium echinulatum*, *Soil Sci. Soc. Am. Proc.*, 39, 649, 1975.

286. Hayes, M. H. B., Extraction of humic substances from soil, in *Humic Substances in Soil, Sediment, and Water. Geochemistry, Isolation, and Characterization*, Aiken, G. R., McKnight, D. M., Wershaw, R. L., and MacCarthy, P., Eds., John Wiley & Sons, New York, 1985, 329.

287. Swift, R.S., Fractionation of soil humic substances, in *Humic Substances in Soil, Sediment, and Water. Geochemistry, Isolation, and Characterization*, Aiken, G. R., McKnight, D. M., Wershaw, R. L., and MacCarthy, P., Eds., John Wiley & Sons, New York, 1985, 387.

288. Senesi, N., Miano, T. M., and Brunetti, G., Methods and related problems for sampling soil and sediment organic matter. Extraction, fractionation and purification of humic substances, *Quim. Anal. Acta*, 13, S26, 1994.

289. Whitehead, D. C. and Tinsley, J., Extraction of soil organic matter with dimethylformamide, *Soil Sci.*, 97, 34, 1964.

290. Hayes, M. H. B., Swift, R. S., Wardle, R. E., and Brown, J. K., Humic materials from an organic soil: a comparison of extractants and of properties of extracts, *Geoderma*, 13, 231, 1975.

291. Bremner, J. M., Some observations on the oxidation of soil organic matter in the presence of alkali, *J. Soil Sci.*, 1, 198, 1950.

292. Porter, L. K., Factors affecting the solubility and possible fractionation of organic colloids extracted from soil and leonardite with an acetone-H_2O-HCl solvent, *J. Agric. Food Chem.*, 15, 807, 1967.

293. Gasho, G. J. and Stevenson, F. J., An improved method for extracting organic matter from soil, *Soil Sci. Soc. Am. Proc.*, 32, 117, 1968.

294. Oden, S., Die Humiansäuren, chemische, physikalische und bodenkundliche Forschung, *Kolloidchem. Beih.*, 11, 75, 1919.

295. Springer, U., Der heutige Stand der Humusuntersuchungsmethodik mit besonderer Berücksichtigung der Trennung, Bestimmung und Charakterisierung der Huminsäuretypen und ihre Anwendung auf charakteristische Humusformen, *Bodenk. Pflanzanernahr.*, 6, 312, 1938.

296. Theng, B. K. G., Wake, J. R. H., and Posner, A. M., The fractional precipitation of soil humic acid by ammonium sulfate, *Plant Soil*, 29, 305, 1968.

297. Berzelius, J. J., *Lehrbuch der Chemie*, 3rd ed. (translation by Wöhler). Dresden and Leipzig, 1839.

298. Posner, A. M., The humic acids extracted by various reagents from a soil. I. Yield, inorganic components, and titration curves, *J. Soil Sci.*, 17, 65, 1966.

299. Scheffer, F., Ziechmann, W., and Pawelke, G., Über die schonende Gewinnung natürlicher Huminstoffe mit Hilfe milder organischer Lösungsmittel, *Zeitsch. Pflanzen.Bodenk.*, 90, 58, 1960.

300. Felbeck, G. T., Jr., Chemical and biological characterization of humic matter, in *Soil Biochemistry*, Vol. 2, McLaren, A. D. and Skujins, J., Eds., Marcel Dekker, New York, 1971, 36.

301. Lisanti, L. E., Senesi, N., and Testini, C., Ricerche sulle proprietà paramagnetiche dei composti umici. III. Indagini su frazioni ottenute mediante l'impiego di estraenti organici blandi, *Agrochimica*, 17, 282, 1973.
302. Senesi, N., Testini, C., and Polemio, M., Chemical and spectroscopic characterization of soil organic matter fractions isolated by sequential extraction procedure, *J. Soil Sci.*, 34, 801, 1983.
303. Yau, W. W., Kirkland, J. J., and Bly, D. D., *Modern Size-Exclusion Liquid Chromatography: Practice of Gel Permeation and Gel Filtration Chromatography*, John Wiley & Sons, New York, 1979.
304. Schnitzer, M. and Skinner, S. I. M., Gel filtration of fulvic acid, a soil humic compound, *Isotopes and Radiation in Soil Organic Matter Studies, International Atomic Energy Agency*, Vienna, 1968, 41.
305. Posner, A. M., Importance of electrolyte in the determination of molecular weights by Sephadex gel with special reference to humic acids, *Nature (London)*, 198, 1161, 1963.
306. Lindqvist, I., Adsorption effects in gel filtration of humic acid, *Acta Chem. Scand.*, 21, 2564, 1967.
307. Ladd, J. N., The extinction coefficients of soil humic acids fractionated by Sephadex gel filtration, *Soil Sci.*, 107, 303, 1969.
308. De Nobili, M., Gjessing, E., and Sequi P., Sizes and shapes of humic substances by gel chromatography, in *Humic Substances, Vol. 2 In Search of Structure*, Hayes, M. H. B., MacCarthy, P., Malcolm, R. L., and Swift, R. S., Eds., John Wiley & Sons, New York, 1989, chap. 20.
309. Wershaw, R. L. and Pinckney, D. J., The fractionation of humic acids from natural water systems, *J. Res. U.S. Geol. Surv.*, 1, 361, 1973.
310. Sapek, A., Effect of pH value of sample solution on separation of humus substances by gel filtration (in Polish), *Rocz. Glebozn.*, 24, 519, 1973.
311. Wershaw, R. L. and Aiken, G. R., Molecular size and weight measurements of humic substances, in *Humic Substances in Soil, Sediment and Water. Geochemistry, Isolation, and Characterization*, Aiken, G. R., McKnight, D. M., Wershaw, R. L., and MacCarthy, P., Eds., Wiley-Interscience, New York, 1985, chap. 19.
312. Rickwood, D., Centrifugation: A Practical Approach. Information Retrieval. 1978.
313. Deyl, Z., Everaerts, F. M., Prusik, Z., and Svenden, P. J., *Electrophoresis: a Survey of Techniques and Applications, Journal of Chromatography Library*, Vol. 18, Elsevier, Amsterdam, 1979.
314. Everaerts, F. M., Beckers, J. L., and Verhaggen, Th. P. E. M., *Isotachophoresis, Journal of Chromatography Library*, Vol. 6, Elsevier, Amsterdam, 1976.
315. Duxbury, J. M., Studies of the molecular size and charge of humic substances by electrophoresis, in *Humic Substances Vol. 2, In Search of Structure*, Hayes, M. H. B., MacCarthy, P., Malcolm, R. L., and Swift, R. S., Eds., John Wiley & Sons, New York, 1989, chap. 21.
316. Wright, J. R. and Schnitzer, M., Oxygen containing functional groups in the organic matter of the Ao and Bh horizon of a podzol, in *Proc. 7th Int. Congr. Soil Sci.*, Vol. 2, Madison, WI, 1960, 120.
317. Roulet, N., Metha, N. C., Dubach, P., Deuel, H., Abtrennung von Kohlehydraten und Stickstoffverbindungen aus Huminstoffen durch Gelfiltration und Ionenaustausch-Chromatographie, *Z. Pflanzenernahr. Dung. Bodenk.*, 103, 1, 1963.
318. Dormaar, J. F., Metche, M., and Jacquin, J., *Soil Biol. Biochem.*, 2, 285, 1970.
319. Khan, S. U., Distribution and characteristics of organic matter extracted from black solonetzic and black chernozemic soils of Alberta: the humic acid fraction, *Soil Sci.*, 112, 401, 1971.
320. Haworth, R. D., The chemical nature of humic acid, *Soil Sci.*, 111, 71, 1971.
321. Thurman, E. M., Isolation of soil and aquatic humic substances, Group Report, in *Humic Substances and their Role in the Environment*, Frimmel, F. H. and Christman, R. F., Eds., Wiley-Interscience, Chichester, 1988, 31.
322. Huffman, E. W. D., Jr. and Stuber, H. A., Analytical methodology for elemental analysis of humic substances, in *Humic Substances in Soil, Sediment and Water. Geochemistry, Isolation, and Characterization*, Aiken, G. R., McKnight, D. M., Wershaw, R. L., and MacCarthy, P., Eds., Wiley-Interscience, New York, 1985, chap. 17.
323. Schnitzer, M., Humic substances: chemistry and reactions, in *Soil Organic Matter*, Schnitzer, M. and Khan, S. U., Eds., Elsevier, Amsterdam, 1978, chap. 1.
324. Steelink, C., Implications of elemental characteristics of humic substances, in *Humic Substances in Soil, Sediment and Water. Geochemistry, Isolation, and Characterization*, Aiken, G. R., McKnight, D. M., Wershaw, R. L., and MacCarthy, P., Eds., Wiley-Interscience, New York, 1985, chap. 18.

325. Perdue, E. M., Acidic functional groups in humic substances, in *Humic Substances in Soil, Sediment and Water. Geochemistry, Isolation, and Characterization*, Aiken, G. R., McKnight, D. M., Wershaw, R. L., and MacCarthy, P., Eds., Wiley-Interscience, New York, 1985, 493.

326. Tsutsuki, K. and Kuwatsuka, S., Chemical studies on soil humic acids. II. Composition of oxygen-containing functional groups of humic acids, *Soil Sci. Plant Nutr.*, 24, 547, 1978.

327. Schnitzer, M. and Desjardins, J. G., Oxygen-containing functional groups in organic soils and their relation to the degree of humification as determined by solubility in sodium pyrophosphate solution, *Can. J. Soil Sci.*, 46, 237, 1966.

328. Stevenson, F. J. and Butler, J. H. A., Chemistry of humic acids and related pigments, in *Organic Geochemistry*, Eglinton, G. and Murphy, M. T. J., Eds., Springer-Verlag, New York, 1969, 534.

329. Wershaw, R. L. and Pinckney, D. J., Methylation of humic acid fractions, *Science*, 199, 906, 1978.

330. Dubach, P., Mehta, N. C., Jakab, T., Martin, F., and Roulet, N., Chemical investigations on soil humic substances, *Geochim. Cosmochim. Acta*, 28, 1567, 1964.

331. Martin, F., Dubach, P., Mehta, N. C., and Deuel, H., Bestimmung der funktionellen Gruppen von Huminstoffen, *Z. Pfl. Dung. Bodenk*, 103, 27, 1963.

332. Perdue, E. M., Reuter, J. H., and Ghosal, M., The operational nature of acidic functional group analyses and its impact on mathematical descriptions of acid-base equilibria in humic substances, *Geochim. Cosmochim. Acta*, 44, 1841, 1980.

333. Holtzclaw, K. M. and Sposito, G., Analytical properties of the soluble, metal-complexing fractions in sludge-soil mixtures. IV. Determination of carboxyl groups in fulvic acids, *Soil Sci. Soc. Am. J.*, 43, 318, 1979.

334. Schnitzer, M., The methylation of humic substances, *Soil Sci.*, 117, 94, 1974.

335. Leehneer, J. A. and Noyes, T. I., Derivatization of humic substances for structural studies, in *Humic Substances Vol. 2, In Search of Structure*, Hayes, M. H. B., MacCarthy, P., Malcolm, R. L., and Swift, R. S., Eds., John Wiley & Sons, New York, 1989, 257.

336. Schnitzer, M. and Skinner, S. I. M., A polarographic method for the determination of carbonyl groups in soil humic compounds, *Soil Sci.*, 101, 120, 1966.

337. Kukharenko, T. A. and Yekaterinina, L. N., Method of determining quinoid groups in humic acids, *Soviet Soil Sci.*, 7, 933, 1967.

338. Vasilyevskaya, N. A., Glebko, L. I., and Maximov, O. B., Determination of quinoid groups in humic acids, *Soviet Soil Sci.*, 3, 224, 1971.

339. Schnitzer, M. and Riffaldi, R., The determination of quinone groups in humic substances, *Soil Sci. Soc. Am. Proc.*, 36, 772, 1972.

340. Maximov, O. B. and Glebko, L. I., Quinoid groups in humic acids, *Geoderma*, 11, 17, 1974.

341. Glebko, L. I., Ulkina, Zh. I., and Maximov, O. B., Semimicro-method for the determination of quinoid groups in humic acids, *Mikrochim. Acta*, 1970, 1247, 1970.

342. Arp, P. A., Potentiometric analysis of humic substances and other colloids, *Can. J. Chem.*, 61, 1671, 1983.

343. Sposito, G., Holtzclaw, K. M., and Keech, D. A., Proton binding in fulvic acid extracted from sewage sludge-soil mixture, *Soil Sci. Soc. Am. Proc.*, 41, 1119, 1977.

344. Posner, A. M., Titration curves of humic acid, *Trans. 8th Int. Congr. Soil Sci.*, 3, 161, 1964.

345. Davis, H. and Mott, C. J. B., Titrations of fulvic acid fractions. I. Interactions influencing the dissociation/reprotonation equilibria, *J. Soil Sci.*, 32, 379, 1982.

346. Gran, G., Determination of the equivalent point in potentiometric titrations, *Acta Chem. Scand.*, 1, 559, 1950.

347. Gran, G., Determination of the equivalence point in potentiometric titrations. II, *Analyst*, 77, 661, 1952.

348. Tan, K. H., *Principles of Soil Chemistry*, 2nd ed., Marcel Dekker, New York, 1993

349. Perdue, E. M., Reuter, J. H., and Parrish, R. S., A statistical model of proton binding by humus, *Geochim. Cosmochim. Acta*, 48, 1257, 1984.

350. Gamble, D. S., Titration curves of fulvic acid: the analytical chemistry of a weak acid polyelectrolyte, *Can. J. Chem.*, 48, 2662, 1970.

351. Schuman, M. S., Collins, G. J., Fitzgerald, P. J., and Olson, D. L., Distribution of stability constants and dissociation rate constants among binding sites on estaurine copper-organic complexes: rotated disk electrode studies and an affinity spectrum analysis of ion-selective electrode and photometric data, in *Aquatic and Terrestrial Humic Materials*, Christman, F. R. and Gjessing, E. T., Eds., Ann Arbor Science Press, Ann Arbor, 1983, 349.

352. van Dijk, H., Electrometric titration of humic acids, *Sci. Proc. R. Dublin Soc. Ser. A,* 1, 163, 1960.

353. Thompson, S. O. and Chesters, G., Acidic properties of plant lignins and humic materials of soils, *J. Soil Sci.,* 20, 346, 1969.

354. Ragland, J. L., The use of thermometric titrations in soil chemistry studies, *Soil Sci. Soc. Am. Proc.,* 26, 133, 1962.

355. Khalaf, K. Y., MacCharty, P., and Gilbert, T. W., Application of thermometric titrations to the study of soil organic matter. II. Humic acids, *Geoderma,* 14, 331, 1975.

356. Choppin, G. R. and Kullberg, L., Protonation thermodynamics of humic acid, *J. Inorg. Nucl. Chem.,* 40, 651, 1978.

357. Perdue, E. M., Solution thermochemistry of humic subtances. I. Acid-base equilibria of humic acid, *Geochim. Cosmochim. Acta,* 42, 1351, 1978.

358. Perdue, E. M., Solution thermochemistry of humic substances. II. Acid-base equilibria of river water humic substances, in *Chemical Modeling in Aqueous Systems,* Vol. 93, Jenne, E. A., Ed., American Chemical Society, Washington, D.C., 1979, 94.

359. Kamprath, E. J. and Welch, C. D., Retention and cation-exchange properties of organic matter in coastal plain soils, *Soil Sci. Soc. Am. Proc.,* 26, 263, 1962.

360. Yuan, T. L., Gammon, N., Jr., and Leighty, R. G., Relative contribution of organic and clay fractions to cation-exchange capacity of sandy soils from several groups, *Soil Sci.,* 104, 123, 1967.

361. Broadbent, F. E. and Bradford, G. R., Cation-exchange groupings in the soil organic fraction, *Soil Sci.,* 74, 447, 1952.

362. Helling, C. S., Chesters, G., and Corey, R. B., Contribution of organic matter and clay to soil cation-exchange capacity as affected by the pH of the saturating solution, *Soil Sci. Soc. Am. Proc.,* 28, 517, 1964.

363. Schnitzer, M., Contribution of organic matter to the cation exchange capacity of soil, *Nature,* 207, 667, 1965.

364. Aiken, G. R. and Gillam, A. H., Determination of molecular weights of humic substances by colligative property measurements, in *Humic Substances, Vol. 2, In Search of Structure,* Hayes, M. H. B., MacCarthy, P., Malcolm, R. L., and Swift, R. S., Eds., John Wiley & Sons, New York, 1989, chap. 18.

365. Chen, Y. and Schnitzer, M., Sizes and shapes of humic substances by electron microscopy, in *Humic Substances Vol. 2, In Search of Structure,* Hayes, M. H. B., MacCarthy, P., Malcolm, R. L., and Swift, R. S., Eds., John Wiley & Sons, New York, 1989, chap. 22.

366. Clapp, C. E., Emerson, W. W., and Olness, A. E., Sizes and shapes of humic substances by viscosity measurements, in *Humic Substances, Vol. 2, In Search of Structure,* Hayes, M. H. B., MacCarthy, P., Malcolm, R. L., and Swift, R. S., Eds., John Wiley & Sons, New York, 1989, chap. 17.

367. Swift, R. S., Molecular weight, size, shape, and charge characteristics of humic substances: some basic considerations, in *Humic Substances, Vol. 2, In Search of Structure,* Hayes, M. H. B., MacCarthy, P., Malcolm, R. L., and Swift, R. S., Eds., John Wiley & Sons, New York, 1989, chap. 15.

368. Swift, R. S., Molecular weight, shape, and size of humic substances by ultracentrifugation, in *Humic Substances, Vol. 2, In Search of Structure,* Hayes, M. H. B., MacCarthy, P., Malcolm, R. L., and Swift, R. S., Eds., John Wiley & Sons, New York, 1989, chap. 16.

369. Wershaw, R. L., Sizes and shapes of humic substances by scattering techniques, in *Humic Substances, Vol. 2, In Search of Structure,* Hayes, M. H. B., MacCarthy, P., Malcolm, R. L., and Swift, R. S., Eds., John Wiley & Sons, New York, 1989, chap. 19.

370. Tanford, C., *Physical Chemistry of Macromolecules,* John Wiley & Sons, New York, 1961.

371. Eisenberg, H., *Biological Macromolecules and Polyelectrolytes in Solution,* Oxford University Press, London, 1976.

372. Williams, J. W., *Ultracentrifugal Analysis in Theory and Experiment,* Academic Press, New York, 1963.

373. Stevenson, F. J., van Winkle, Q., and Martin, W. P., Physicochemical investigations of clay-adsorbed organic colloids. II, *Soil Sci. Soc. Am. Proc.,* 17, 31, 1953.

374. Piret, E. L., White, R. G., Walther, H., and Madden, A. J., Some physicochemical properties of peat humic acids, *Sci. Proc. R. Dublin Soc. Ser. A,* 1, 69, 1960.

375. Flaig, W. and Beutelspacher, H., Investigations of humic acids with the analytical ultracentrifuge, in *Isotopes and Radiation in Soil Organic Matter Studies,* International Atomic Energy Agency, Vienna, 1968, 23.

376. Orlov, D. S., Ammosova, Ya. M., and Glebova, G. I., Molecular parameters of humic acids, *Geoderma*, 13, 211, 1975.

377. Cameron, R. S., Thornton, B. K., Swift, R. S., and Posner, A. M., Molecular weight and shape of humic acid from sedimentation and diffusion measurements on fractionated extracts, *J. Soil Sci.*, 23, 394, 1972.

378. Posner, A. M. and Creeth, J. M., A study of humic acid by equilibrium ultracentrifugation, *J. Soil Sci.*, 23, 333, 1972.

379. Cameron, R. S. and Posner, A. M., Molecular weight distribution of humic acid from density gradient, ultracentrifugation profiles corrected for diffusion, *Trans. 10th Int. Congr. Soil Sci. Moscow*, 2, 325, 1974.

380. Ritchie, G. S. P. and Posner, A. M., The effect of pH and metal binding on the transport properties of humic acids, *J. Soil Sci.*, 33, 233, 1982.

381. Rafikov, S. R., Pavlova, S. A., and Iverdokhlebova, I. I., *Determination of Molecular Weights and Polydispersity of High Polymers*, Israel Program for Scientific Translations, Davey, New York, 1964, 345.

382. Flory, P. J. and Fox, T. G., Treatment of intrinsic viscosities, *J. Am. Chem. Soc.*, 73, 1904, 1951.

383. Kumada, K. and Kawamura, Y., Viscosimetric characteristics of humic acid, *Soil Sci. Plant Nutr.*, 14, 190, 1968.

384. Chen, Y. and Schnitzer, M., Viscosity measurements on soil humic substances, *Soil Sci. Soc. Am. J.*, 40, 866, 1976.

385. Visser, S. A., Viscosimetric studies on molecular weight fractions of fulvic and humic acids of aquatic, terrestrial and microbial origin, *Plant Soil*, 87, 209, 1985.

386. Datta, C. and Mukherjee, S. K., Viscosity behavior of natural humic acids isolated from diverse soil types, *J. Indian Chem. Soc.*, 47, 1105, 1970.

387. Ghosh, K. and Schnitzer, M., Macromolecular structures of humic substances, *Soil Sci.*, 129, 266, 1980.

388. Glover, C. A., Absolute colligative property measurements, in *Polymer Molecular Weights, Pt. I*, Slade, P. E., Jr., Ed., Marcel Dekker, New York, 1975, 79.

389. Bonnar, R. V., Dimbat, M., and Stross, F. H., *Number Average Molecular Weights*, Wiley-Interscience, New York, 1958.

390. Hansen, E. H. and Schnitzer, M., Molecular weight measurements of polycarboxylic acids in water by vapor pressure osmometry, *Anal. Chim. Acta*, 46, 247, 1969.

391. Wilson, S. A. and Weber, J. H., A comparative study of number-average dissociation-corrected molecular weight of fulvic acids isolated from water and soil, *Chem. Geol.*, 19, 285, 1977.

392. Gillam, A. H. and Riley, J. P., Correction of osmometric number-average molecular weights of humic substances for dissociation, *Chem. Geol.*, 33, 355, 1981.

393. Reuter, J. H. and Perdue, E. M., Calculation of molecular weights of humic substances from colligative data: application to aquatic humus and its molecular size fractions, *Geochim. Cosmochim. Acta*, 45, 2017, 1981.

394. Shapiro, J., Effect of yellow organic acids on iron and other metals in water, *J. Am. Water Works Assoc.*, 56, 1062, 1964.

395. Wood, J. C., Moschopedis, S. E., and Elofson, R. M., Studies in humic acid chemistry. I. Molecular weights of humic acids in sulfolane, *Fuel*, 40, 193, 1961.

396. Schnitzer, M. and Desjardins, J. G., Molecular and equivalent weights of organic matter of a podzol, *Soil Sci. Soc. Am. Proc.*, 26, 362, 1962.

397. DeBorger, R. and DeBacker, H., Détermination du poids moleculaire moyen des acides fulviques per cryoscopie en milieu aqueux, *C.R. Acad. Sci. Ser. D*, 266, 2052, 1968.

398. Wright, J. R., Schnitzer, M., and Levick, R., Some characteristics of the organic matter extracted by dilute inorganic acids from a podzolic B horizon, *Can. J. Soil Sci.*, 38, 14, 1958.

399. Rodbard, D., Estimation of molecular weight by gel filtration and gel electrophoresis. I. Mathematical Principles, in *Methods of Protein Separation*, Vol. 2, Castimpoolas, N., Ed., Plenum Press, New York, 1976, 145.

400. Kasparov, S. Y., Tikhomirov, F. A., and Fless, A. D., Use of disk electrophoresis to fractionate humic acids, *Sov. Soil Sci.*, 36, 21, 1981.

401. Dover, S. D., The scattering of radiation by macromolecules, in *An Introduction to the Physical Properties of Large Molecules in Solution*, Richards, E. G., Ed., Cambridge University Press, Cambridge, 1980, chap. 7.

402. Guinier, A. and Fournet, G., *Small-Angle Scattering of X-Rays*, John Wiley & Sons, New York, 1955.

403. Wershaw, R. L., Burcar, P. J., Sutula, C. L., and Wiginton, B. J., Sodium humate solution studied with small-angle X-ray scattering, *Science,* 157, 1429, 1967.

404. Lindqvist, I., A small angle X-ray scattering study of sodium humate solutions, *Acta Chem. Scand.,* 24, 3068, 1970.

405. Wershaw, R. L. and Pinckney, D. J., Determination of the association and dissociation of humic acid fractions by small angle X-ray scattering, *J. Res. U.S. Geol. Surv.,* 1, 701, 1973.

406. Wershaw, R. L., Pinckney, D. J., and Booker, S. E., Chemical structure of humic acids. Part I, A generalized structural model, *J. Res. U.S. Geol. Surv.,* 5, 565, 1977.

407. Wershaw, R. L. and Pinckney, D. J., Chemical structure of humic acids. Part II, The molecular aggregation of some humic acid fractions in N,N-dimethylformamide, *J. Res. U.S. Geol. Surv.,* 5, 571, 1977.

408. Swift, R. S., Thornton, B. K., and Posner, A. M., Spectral characteristics of a humic acid fractionated with respect to molecular weight using an agar gel, *Soil Sci.,* 110, 93, 1970.

409. Dawson, H. J., Hrutfiord, B. F., Zasoski, R. J., and Ugolini, F. C., The molecular weight and origin of yellow organic acids, *Soil Sci.,* 132, 191, 1981.

410. Visser, S. A., Electron-microscopic and electron-diffraction patterns of humic acids, *Soil Sci.,* 96, 353, 1963.

411. Wiesemüller, W., Untersuchung uber die Fraktionierung der organischen Bodensubstanz, *Albrecht-Thaer-Arch.,* 9, 419, 1965.

412. van Dijk, H., Colloidal chemical properties of humic matter, in *Soil Biochemistry,* Vol. 2, McLaren, A. D. and Peterson, G. H., Eds., Marcel Dekker, New York, 1972, 16.

413. Schnitzer, M. and Kodama, H., An electron microscopic examination of fulvic acid, *Geoderma,* 13, 279, 1975.

414. Chen, Y. and Schnitzer, M., Scanning electron microscopy of a humic acid and a fulvic acid and its metal and clay complexes, *Soil Sci. Soc. Am. J.,* 40, 682, 1976.

415. Chen, Y., Banin, A., and Schnitzer, M., Use of the scanning electron microscope for structural studies on soils and soil components, in *Proc. Int. Symp. Scanning Electron Microscopy,* Johari, O., Ed., IIT Research Institute, Chicago, 1976, 425.

416. Chen, Y., Senesi, N., and Schnitzer, M., Chemical and physical characteristics of humic and fulvic acids extracted from soils of the Mediterranean region, *Geoderma,* 20, 87, 1978.

417. Ghosh, K. and Schnitzer, M., A scanning electron microscopic study of effects of adding neutral electrolytes to solutions of humic substances, *Geoderma,* 28, 53, 1982.

418. Stevenson, I. L. and Schnitzer, M., Trasmission electron microscopy of extracted fulvic and humic acids, *Soil Sci.,* 133, 179, 1982.

419. Stevenson, I. L. and Schnitzer, M., Energy-dispersive X-ray microanalysis of saturated fulvic acid-iron and -copper complexes, *Soil Sci.,* 138, 123, 1984.

420. Tan, K. H., Scanning electron microscopy of humic matter as influenced by method of preparation, *Soil Sci. Soc. Am. J.,* 49, 1185, 1985.

421. Mandelbrot, B. B., *The Fractal Geometry of Nature,* W.H. Freeman, San Francisco, 1982.

422. Pfeifer, P. and Obert, M., Fractals: basic concepts and terminology, in *The Fractal Approach to Heterogeneous Chemistry: Surfaces, Colloids, Polymers,* Avnir, D., Ed., John Wiley & Sons, Chichester, England, 1989, 11.

423. Schmidt, P. W., Use of scattering to determine the fractal dimension, in *The Fractal Approach to Heterogeneous Chemistry: Surfaces, Colloids, Polymers,* Avnir, D., Ed., John Wiley & Sons, Chichester, England, 1989, 67.

424. Horne, D. S., Determination of the fractal dimension using turbidimetric techniques. Application to aggregation protein systems, *Faraday Discuss. Chem. Soc.,* 83, 259, 1987.

425. Teixeira, J., Experimental methods for studying fractal aggregates, in *On Growth and Form,* Stanley, H. E. and Ostrowsky, N., Eds., Martinus Nijhoff, Dordrecht, The Netherlands, 1986, 145.

426. Senesi, N., The fractal approach to the study of humic substances, in *Humic Substances in the Global Environment and Implications on Human Health,* Senesi, N. and Miano, T. M., Eds., Elsevier, Amsterdam, 1994, 3.

427. Senesi, N., Fractals in general soil science and in soil biology and biochemistry, in *Soil Biochemistry,* Vol. 9, Stotzky, G. and Bollag, J.-M., Eds., Marcel Dekker, New York, 1996, 415.

428. Österberg, R. and Mortensen, K., Fractal dimension of humic acids, a small angle neutron scattering study, *Eur. Biophys. J.,* 21, 163, 1992.

429. Österberg, R. and Mortensen, K., Fractal geometry of humic acids. Temperature-dependent restructuring studied by small-angle neutron scattering, in *Humic Substances in the Global Environment and Implications on Human Health*, Senesi, N. and Miano, T. M., Eds., Elsevier, Amsterdam, 1994, 127.

430. Österberg, R., Szajdak, L., and Mortensen, K., Temperature-dependent restructuring of fractal humic acids: a proton-dependent process, *Environ. Int.*, 20, 77, 1994.

431. Rice, J. and Lin, J. S., Fractal nature of humic materials, in *Proc. ACS Natl. Meeting 203rd, Div. Environ. Chem.*, San Francisco, 5-10 April, 1992, American Chemical Society, Ann Arbor, MI, 1992, 11.

432. Rice, J. and Lin, J. S., Fractal nature of humic materials, *Environ. Sci. Technol.*, 27, 413, 1993.

433. Rice, J. and Lin, J. S., Fractal dimension of humic materials, in *Humic Substances in the Global Environment and Implications on Human Health*, Senesi, N. and Miano, T. M., Eds., Elsevier, Amsterdam, 1994, 115.

434. Senesi, N., Lorusso, G. F., Miano, T. M., Maggipinto, G., Rizzi, F. R., and Capozzi, V., The fractal dimension of humic substances as a function of pH by turbidity measurements, in *Humic Substances in the Global Environment and Implications on Human Health*, Senesi, N. and Miano, T. M., Eds., Elsevier, Amsterdam, 1994, 121.

435. Senesi, N., Rizzi, F. R., Dellino, P., Acquafredda, P., Maggipinto, G., and Lorusso, G. F., The fractal morphology of soil humic acids, *Trans. 15th World Congr. Soil Sci.*, 3b, 81, 1994.

436. Senesi, N., Rizzi, F. R., Dellino, P., and Acquafredda, P., Fractal dimension of humic acids in aqueous suspension as a function of pH and time, *Soil Sci. Am. J.*, 60, 1773, 1996.

437. Senesi, N., Rizzi, F. R., Dellino, P., and Acquafredda, P., Fractal humic acids in aqueous suspensions at various concentration, ionic strength, and pH, *Colloids Surfaces A Physicochem. Eng. Aspects*, 127, 57, 1997.

438. Meakin, P., Simulations of aggregation processes, in *The Fractal Approach to Heterogeneous Chemistry: Surfaces, Colloids, Polymers*, Avnir, D., Ed., John Wiley & Sons, Chichester, England, 1989, 131.

439. Schnitzer, M., Chemical, spectroscopic, and thermal methods for the classification and characterization of humic substances, in *Proc. Int. Meetings on Humic Substances*, Pudoc, Wageningen, 1972, 293.

440. Stevenson, F. J., Reductive cleavage of humic substances, in *Humic Substances, Vol. 2, In Search of Structure*, Hayes, M. H. B., MacCarthy, P., Malcolm, R. L., and Swift, R. S., Eds., John Wiley & Sons, New York, 1989, 121.

441. Parsons, J. W., Hydrolytic degradations of humic substances, in *Humic Substances, Vol. 2, In Search of Structure*, Hayes, M. H. B., MacCarthy, P., Malcolm, R. L., and Swift, R. S., Eds., John Wiley & Sons, New York, 1989, 99.

442. Griffith, S. M. and Schnitzer, M., Oxidative degradation of soil humic substances, in *Humic Substances, Vol. 2, In Search of Structure*, Hayes, M. H. B., MacCarthy, P., Malcolm, R. L., and Swift, R. S., Eds., John Wiley & Sons, New York, 1989, 69.

443. Hayes, M. H. B. and O'Callaghan, M. R., Degradations with sodium sulfide and with phenol, in *Humic Substances, Vol. 2, In Search of Structure*, Hayes, M. H. B., MacCarthy, P., Malcolm, R. L., and Swift, R. S., Eds., John Wiley & Sons, New York, 1989, 143.

444. Bracewell, J. M., Haider, K., Larter, S. R., and Schulten, H.-R., Thermal degradation relevant to structural studies of humic substances, in *Humic Substances, Vol. 2, In Search of Structure*, Hayes, M. H. B., MacCarthy, P., Malcolm, R. L., and Swift, R. S., Eds., John Wiley & Sons, New York, 1989, 181.

445. Duff, R. B., The occurrence of methylated carbohydrates and rhamnose as components of soil polysaccharides, *J. Sci. Food Agric.*, 3, 140, 1952.

446. Jakab, T., Dubach, P., Mehta, N. C., and Deuel, H., Abbau von Huminstoffen. I. Hydrolyse mit Wasser und Mineralsäuren, *Z. Pflanzenernähr. Düng. Bodenk.*, 96, 213, 1962.

447. Neyroud, J. A. and Schnitzer, M., The alkaline hydrolysis of humic substances, *Geoderma*, 13, 177, 1975.

448. Schnitzer, M. and Neyroud, J. A., Alkanes and fatty acids in humic substances, *Fuel*, 54, 17, 1975.

449. Riffaldi, R. and Schnitzer, M., Effects of 6 N HCl hydrolysis on the analytical characteristics and chemical structure of humic acids, *Soil Sci.*, 115, 349, 1973.

450. Cranwell, P. A. and Haworth, R. D., The chemical nature of humic acids, in *Proc. Int. Meet. Humic Substances, Nieuwersluis* Povoledo, D. and Gotterman K., Eds., Pudoc, Wageningen, 1975, 13.

451. Schnitzer, M. and Preston, C. M., Effects of acid hydrolysis on ^{13}C NMR spectra of humic substances, *Plant Soil*, 75, 201, 1983.

452. Preston, C. M. and Ripmeester, J. A., Application of solution and solid-state ^{13}C NMR to four organic soils, their humic acids, fulvic acids, humins and hydrolysis residues, *Can. J. Spectrosc.*, 27, 99, 1982.
453. Jakab, T., Dubach, P., Mehta, N. C., and Deuel, H., Abbau von Huminstoffen. II. Abbau mit Alkali, *Z. Pflanzenernähr. Düng. Bodenk.*, 102, 8, 1963.
454. Piper, T. J. and Posner, A. M., Sodium amalgam reduction of humic acid. I. Evaluation of the method, *Soil Biol. Biochem.*, 4, 513, 1972a.
455. Schnitzer, M. and Ortiz de Serra, M. I., The sodium-amalgam reduction of soil and fungal humic substances, *Geoderma*, 9, 119, 1973.
456. Schnitzer, M. and Ortiz de Serra, M. I., The chemical degradation of a humic acid, *Can. J. Chem.*, 51, 1554, 1973.
457. Martin, J. P., Haider, K., and Saiz-Jimenez, C., Sodium amalgam reductive degradation of fungal and model phenolic polymers, soil humic acids, and simple phenolic compounds, *Soil Sci. Soc. Am. Proc.*, 38, 760, 1974.
458. Burges, N. A., Hurst, H. M., and Walkden, B., The phenolic constituents of humic acid and their relation to the lignin of the plant cover, *Geochim. Cosmochim. Acta*, 28, 1547, 1964.
459. Matsui, Y. and Kumada, K., Aromatic constituents of humic acids released by Na-amalgam reductive cleavage, KOH fusion and zinc dust fusion, *Soil Sci. Plant Nutr.*, 23, 491, 1977.
460. Stevenson, F. J. and Mendez, J., Reductive cleavage products of soil humic acids, *Soil Sci.*, 103, 383, 1967.
461. Hurst, H. M. and Burges, N. A., Lignin and humic acids, in *Soil Biochemistry*, Vol. 1, McLaren, A. D. and Peterson, G. H., Eds., Marcel Dekker, New York, 1967, 276.
462. Cheshire, M. V., Cranwell, P. A., Falshaw, C. P., Floyd, A. J., and Haworth, R. D., Humic acid. II. Structure of humic acids, *Tetrahedron*, 23, 1669, 1967.
463. Cheshire, M. V., Cranwell, P. A., and Haworth, R. D., Humic acid. III, *Tetrahedron*, 24, 5155, 1968.
464. Hansen, E. H. and Schnitzer, M., Zinc dust distillation and fusion of a soil humic and fulvic acid, *Soil Sci. Soc. Am. Proc.*, 33, 29, 1969.
465. Hansen, E. H. and Schnitzer, M., Zinc dust distillation of soil humic compounds, *Fuel*, 48, 41, 1969.
466. Kumada, K. and Matsui, Y., Studies on the composition of the aromatic nuclei of humus. I. Detection of some condensed aromatic nuclei of humic acid, *Soil Sci. Plant Nutr.*, 16, 250, 1970.
467. Gottlieb, S. and Hendricks, S. B., Soil organic matter as related to newer concepts of lignin chemistry, *Soil Sci. Soc. Am. Proc.*, 10,117, 1945.
468. Kukharenko, T. A. and Savel'ev, A. S., Hydrogenation of humic acids on a nickel catalyst, *Dok. Akad. Nauk SSSR*, 76, 77; *Chem. Abstr.*, 45, 5451, 1951.
469. Kukharenko, T. A. and Savel'ev, A. S., Neutral products from the hydrogenations of humic acids of various origins, *Dok. Akad. Nauk SSSR*, 86, 729, 1952; *Chem. Abstr.*, 47, 8037c, 1953.
470. Maximov, O. B. and Krasovskaya, N. P., Action of metallic sodium on humic acids in liquid ammonia, *Geoderma*, 18, 227, 1977.
471. Barton, D. H. R. and Schnitzer, M., A new experimental approach to the humic acid problem, *Nature (London)*, 198, 217, 1963.
472. Schnitzer, M., Recent findings on the characterization of humic substances extracted from soils from widely differing climatic zones, in *Soil Organic Matter Studies, Proc. Symp. Braunschweig 1976*, International Atomic Energy Agency, Vienna, 1977, 117.
473. Chen, Y., Senesi, N., and Schnitzer, M., Chemical degradation of humic and fulvic acids extracted from Mediterranean soils, *J. Soil Sci.*, 29, 350, 1978.
474. Schnitzer, M., Alkaline cupric oxide oxidation of a methylated fulvic acid, *Soil Biol. Biochem.*, 6, 1, 1974.
475. Greene, G. and Steelink, C., Structure of soil humic acid. II. Some copper oxide oxidation products, *J. Org. Chem.*, 27, 170, 1962.
476. Schnitzer, M. and Skinner, S. I. M., The low temperature oxidation of humic substances, *Can. J. Chem.*, 52, 1072, 1974.
477. Schnitzer, M. and Skinner, S. I. M., The peracetic acid oxidation of humic substances, *Soil Sci.*, 118, 322, 1974.
478. Hansen, E. H. and Schnitzer, M., Nitric acid oxidation of Danish illuvial organic matter, *Soil Sci. Soc. Am. Proc.*, 31, 79, 1967.

479. Chakrabartty, S. K., Kretchmer, H. O., and Cherwonka, S., Hypohalite oxidation of humic acids, *Soil Sci.*, 117, 318, 1974.

480. Morrison, R. I., Products of the alkaline nitrobenzene oxidation of soil organic matter, *J. Soil Sci.*, 14, 201, 1963.

481. Savage, S. and Stevenson, F. J., Behaviour of soil humic acids towards oxidation with hydrogen peroxide, *Soil Sci. Soc. Am. Proc.*, 25, 35, 1961.

482. Swift, R. S., *Physico-Chemical Studies on Soil Organic Matter*, PhD thesis, University of Birmingham, 1968.

483. Hayes, M. H. B., Stacey, M., and Swift, R. S., Degradation of humic acid in a sodium sulfide solution, *Fuel*, 51, 211, 1972.

484. Craggs, J. D., *Sodium Sulfide Reactions with Humic Acid and Model Compounds*, PhD thesis, University of Birmingham, 1972.

485. Craggs, J. D., Hayes, M. H. B., and Stacey, M., Sodium sulfide reactions with humic acid and model compounds, *Trans. 10th Int. Congr. Soil Sci. (Moscow)*, 2, 318, 1974.

486. O'Callaghan, M. R., *Some Studies in Soil Chemistry*, PhD thesis, University of Birmingham, 1980.

487. Jackson, H. P., Swift, R. S., Posner, A. M., and Knox, J. R., Phenolic degradation of humic acid, *Soil Sci.*, 114, 75, 1972.

488. Schnitzer, M. and Hoffman, I., Pyrolysis of soil organic matter, *Soil Sci. Soc. Am. Proc.*, 28, 520, 1964.

489. Schnitzer, M. and Hoffman, I., Thermogravimetry of soil humic compounds, *Geochim. Cosmochim. Acta*, 29, 359, 1965.

490. Turner, R. C. and Schnitzer, M., Thermogravimetry of the organic matter of a podzol, *Soil Sci.*, 93, 225, 1962.

491. Kodama, H. and Schnitzer, M., Kinetics and mechanism of the thermal decomposition of fulvic acid, *Soil Sci.*, 109, 265, 1970.

492. Provenzano M. R. and Senesi, N., Differential scanning calorimetry of humic subtances of various origin, in *Journées Méditerranéennes de Calorimétrie et d'Analyse Thermique, Corte*, 1993, 377.

493. Schulten, H.-R., Relevance of analytical pyrolysis studies to biomass conversion, *J. Anal. Appl. Pyrol.*, 6, 251, 1984.

494. Irwin, W. J., Analytical pyrolysis, an overview, *J. Anal. Appl. Pyrol.*, 1, 3, 1979.

495. Irwin, W. J., *Analytical Pyrolysis, A Comprehensive Guide*, Marcel Dekker, New York, 1982.

496. Meuzelaar, H. L. C., Haverkamp, J., and Hileman, F. D., *Pyrolysis-Mass Spectrometry of Recent and Fossil Biomaterials — Compendium and Atlas*, Elsevier, Amsterdam, 1982.

497. Howarth, R. J. and Sinding-Larsen, R., Multivariate analysis, in *Statistics and Data Analysis in Geochemical Prospecting*, Howarth, R. J., Ed., Elsevier, Amsterdam, 1983, 207.

498. Saiz-Jimenez, C., Pyrolysis/methylation of soil fulvic acids: benzenecarboxylic acids revisited, *Environ. Sci. Technol.*, 28, 197, 1994.

499. Schulten, H.-R., Leinweber, P., and Schnitzer, M., Anaytical pyrolysis and computer modelling of humic and soil particles, in *Environmental Particles. Structure and Surface Reactions of Soil Particles*, Vol. 4, Huang, P. M., Senesi, N., and Buffle, J., Eds., IUPAC Book Series on Physical and Analytical Chemistry of Environmental Systems, John Wiley & Sons, New York, 1998, chap. 8, 281.

500. Saiz-Jimenez, C., Analytical pyrolysis of humic substances: pitfalls, limitations, and possible solutions, *Environ. Sci. Technol.*, 28, 1773, 1994.

501. Nagar, B. R., Examination of the structure of soil humic acids by pyrolysis-gas chromatography, *Nature (London)*, 199, 1213, 1963.

502. Wershaw, R. L. and Bohner, G. E., Jr., Pyrolysis of humic and fulvic acids, *Geochim. Cosmochim. Acta*, 33, 757, 1969.

503. Kimber, R. W. L., and Searle, P. L., Pyrolysis gas chromatography of soil organic matter. I. Introduction and methodology, *Geoderma*, 4, 47, 1970.

504. Kimber, R. W. L. and Searle, P. L., Pyrolysis gas chromatography of soil organic matter. II. The effect of extractant and soil history on the yields of products from pyrolysis of humic acids, *Geoderma*, 4, 57, 1970.

505. Martin, F., Pyrolysis-gas chromatography of humic substances from different origin, *Z. Pflanzenernähr. Bodenkd.*, 4/5, 407, 1975.

506. Martin, F., Effects of extractant on analytical characteristics and pyrolysis-gas chromatography of podzol fulvic acids, *Geoderma*, 15, 253, 1976.

507. Martin, F., The determination of polysaccharides in fulvic acids by pyrolysis-gas chromatography, in *Analytical Pyrolysis*, Jones, C. E. R. and Cramers, C. A., Eds., Elsevier, Amsterdam, 1977, 179.

508. Martin, F., Saiz-Jimenez, C., and Cert, A., Pyrolysis-gas chromatography-mass spectrometry of soil humic fractions. I. The low boiling point compounds, *Soil Sci. Soc. Am. J.*, 41, 1114, 1977.

509. Martin, F., Saiz-Jimenez, C., and Cert, A., Pyrolysis-gas chromatography-mass spectrometry of soil humic fractions. II. The high boiling point compounds, *Soil Sci. Soc. Am. J.*, 43, 309, 1979.

510. Bracewell, J. M., Robertson, G. W., and Welsh, D. I., Polycarboxylic acids as the origin of some pyrolysis products characteristic of soil organic matter, *J. Anal. Appl. Pyrol.*, 2, 239, 1980.

511. Wilson, M. A., Philip, R. P., Gilbert, T. G., Gillam, A. H., and Tate, K. R., Comparison of the structures of humic substances from aquatic and terrestrial sources by pyrolysis-gas chromatography-mass spectrometry, *Geochim. Cosmochim. Acta*, 47, 497, 1983.

512. Saiz-Jimenez, C. and de Leeuw, J. W., Pyrolysis-gas chromatography-mass spectrometry of soil polysaccharides, soil fulvic acids and polymaleic acid, *Org. Geochem.*, 6, 287, 1984.

513. Saiz-Jimenez, C. and de Leeuw, J. W., Chemical characterization of soil organic matter fractions by analytical pyrolysis-gas chromatography-mass spectrometry, *J. Anal. Appl. Pyrol.*, 9, 99, 1986.

514. Saiz-Jimenez, C. and de Leeuw, J. W., Lignin pyrolysis products: their structures and their significance as biomarkers, *Org. Geochem.*, 10, 869, 1986.

515. Saiz-Jimenez, C. and de Leeuw, J. W., Nature of plant components identified in soil humic acids, in *Advances in Humic Substances Research*, Becher, G., Ed., *Sci. Total Environ.*, 62, 115, 1987.

516. Saiz-Jimenez, C. and de Leeuw, J. W., Chemical structure of a soil humic acid as revealed by analytical pyrolysis, *J. Anal. Appl. Pyrol.*, 11, 367, 1987.

517. Tegelaar, E. W., de Leeuw, J. W., and Saiz-Jimenez, C., Possible origin of aliphatic moieties in humic substances, *Sci. Total Environ.*, 81/82, 1, 1989.

518. Saiz Jimenez, C., Applications of pyrolysis-gas chromatography/mass spectrometry to the study of soils, plant materials and humic substances. A critical appraisal, in *Humus, its Structure and Role in Agriculture and Environment*, Kubat, J., Ed., Elsevier, Amsterdam, 1992.

519. Saiz-Jimenez, C., Application of pyrolysis-gas chromatography/mass spectrometry to the study of humic substances: evidence of aliphatic biopolymers in sedimentary and terrestrial humic acids, *Sci. Total Environ.*, 117/118, 13, 1992.

520. Schulten, H.-R., Pyrolysis and soft ionization mass spectrometry of aquatic/terrestrial humic substances and soils, *J. Anal. Appl. Pyrolysis*, 12, 149, 1987.

521. Saiz-Jimenez, C., Martin, F., Haider, K., and Meuzelaar, H. L. C., Comparison of humic and fulvic acids from different soils by pyrolysis-mass spectrometry, *Agrochimica*, 22, 353, 1978.

522. Saiz-Jimenez, C., Haider, K., and Meuzelaar, H. L. C., Comparison of soil organic matter and its fractions by pyrolysis-mass spectrometry, *Geoderma*, 22, 25, 1979.

523. MacCarthy, P., DeLuca, S. J., Voorhees, K. J., Malcolm, R. L., and Thurman, E. M., Pyrolysis-mass spectrometry pattern recognition on a well-characterized suite of humic samples, *Geochim. Cosmochim. Acta*, 44, 2091, 1985.

524. Zech, W., Hempfling, R., Haumaier, L., Schulten, H.-R., and Haider, K., Humification in subalpine rendzinas: chemical degradation, IR and ^{13}C NMR spectroscopy and pyrolysis-field ionization mass spectrometry, *Geoderma*, 47, 123, 1990.

525. Hempfling, R. and Schulten, H.-R., Chemical characterization of the organic matter in forest soils by Curie-point pyrolysis-GC/MS and pyrolysis-field ionization mass spectrometry, *Org. Geochem.*, 15, 131, 1990.

526. Schnitzer, M. and Schulten, H.-R., Pyrolysis-soft ionization mass spectrometry of aliphatics extracted from a soil clay and humic substances, *Sci. Total Environ.*, 81/82, 19, 1989.

527. Hempfling, R. and Schulten, H.-R., Selective preservation of biomolecules during humification of forest litter studied by pyrolysis-field ionization mass spectrometry, *Sci. Total Environ.*, 81/82, 31, 1989.

528. Hempfling, R. and Schulten, H.-R., Pyrolysis-(gas chromatography)/mass spectrometry of agricultural soils and their humic fractions, *Z. Pflanzenernähr. Bodenk.*, 154, 425, 1991.

529. Schulten, H.-R. and Schnitzer, M., A contribution to solving the puzzle of the chemical structure of humic substances: pyrolysis-soft ionization mass spectrometry, *Sci. Total Environ.*, 117/118, 27, 1992.

530. Schulten, H.-R. and Schnitzer, M., Structural studies on soil humic acids by Curie-point pyrolysis-gas chromatography/mass spectrometry, *Soil Sci.*, 153, 205, 1992.

531. Schulten, H.-R., Plage, B., and Schnitzer, M., A chemical structure for humic substances, *Naturwissenschaften*, 78, 311, 1991.

532. McCoustra, M. R. S., Electronic absorption spectroscopy: theory and practice, in *Perspective in Modern Chemical Spectroscopy*, Andrews, D. L., Ed., Springer-Verlag, Berlin, 1990, 88.

533. Scott, A. I., *Interpretation of Ultraviolet Spectra of Natural Products*, Pergamon Press, New York, 1964.

534. Kumada, K., Absorption spectra of humic acids, *Soil Sci. Plant Nutr.*, 1, 29, 1955.

535. Schnitzer, M., Characterization of humic constituents by spectroscopy, in *Soil Biochemistry*, Vol. 2, McLaren, A. D. and Skujins J., Eds., Marcel Dekker, New York, 1971, 60.

536. Ghosh, K. and Schnitzer, M., UV and visible absorption spectroscopic investigations in relation to macromolecular characteristics in humic substances, *J. Soil Sci.*, 30, 735, 1979.

537. MacCarthy, P. and Rice, J. A., Spectroscopic methods (other than NMR) for determining functionality in humic substances, in *Humic Substances in Soil Sediment and Water: Geochemistry, Isolation, and Characterization*, Aiken, G. R., McKnight, D. M., Wershaw, R. L., and MacCarthy, P., Eds., Wiley-Interscience, New York, 1985, 527.

538. Bloom, P. R. and Leenheer, J. A., Vibrational, electronic, and high-energy spectroscopic methods for characterizing humic substances, in *Humic Substances, Vol. 2, In Search of Structure*, Hayes, M. H. B., MacCarthy, P., Malcolm, R. L., and Swift, R. S., Eds., John Wiley & Sons, New York, 1989, 409.

539. MacCarthy, P. and O'Cinneide, S., Fulvic acid I. Partial fractionation, *J. Soil Sci.*, 25, 420, 1974.

540. Tsutsuki, K. and Kuwatsuka, S., Chemical studies on soil humic acids. VII, pH dependent nature of the ultraviolet and visible absorption spectra of humic acids, *Soil Sci. Plant Nutr.*, 25, 373, 1979.

541. Chen, Y., Senesi, N., and Schnitzer, M., Information provided on humic substances by E_4/E_6 ratios, *Soil Sci. Soc. Am. J.*, 41, 352, 1977.

542. Baes, A. U. and Bloom, P. R., Fulvic acid ultraviolet-visible spectra: influence of solvent and pH, *Soil Sci. Soc. Am. J.*, 54, 1248, 1990.

543. Guibault, G. G., *Practical Fluorescence. Theory, Methods and Techniques*, Marcel Dekker, New York, 1973.

544. Schulman, S. G., *Molecular Luminescence Spectroscopy. Methods and Applications*, John Wiley & Sons, New York, 1985.

545. Wehry, E. L., *Modern Fluorescence Spectroscopy*, Vols. 3 and 4, Plenum Press, New York, 1981.

546. Senesi, N., Molecular and quantitative aspects of the chemistry of fulvic acid and its interactions with metal ions and organic chemicals. II. The fluorescence spectroscopy approach, *Anal. Chim. Acta*, 232, 77, 1990.

547. Choudri, G. G., *Humic Substances. Structural, Photophysical, Photochemical and Free Radical Aspects and Interactions with Environmental Chemicals*, Gordon and Breach, New York, 1984.

548. Senesi, N., Fluorescence spectroscopy applied to the study of humic substances from soil and soil related systems: a review, *Proc. 199th Am. Chem. Soc. Meeting, Boston, Div. Environ. Chem.*, Vol. 30, American Chemical Society, Boston, 1990, 79.

549. Bachelier, G., Étude spectrographique de la fluorescence des acides humiques et des acides fulviques de divers sols, *Cah. ORSTOM Ser. Pedol.*, 18, 129, 1980-81.

550. Senesi, N., Miano, T. M., Provenzano, M. R., and Brunetti, G., Characterization, differentiation, and classification of humic substances by fluorescence spectroscopy, *Soil Sci.*, 152, 259, 1991.

551. Senesi, N., Application of electron spin resonance and fluorescence spectroscopies to the study of soil humic substances, in *Humus, its Structure and Role in Agriculture and Environment*, Kubát, J., Ed., Elsevier, Amsterdam, 1992, 11.

552. Ghosh, K. and Schnitzer, M., Fluorescence excitation spectra of humic substances, *Can. J. Soil Sci.*, 60, 373, 1980.

553. Saar, R. A. and Weber, J. H., Comparison of spectrofluorometry and ion-selective electrode potentiometry for determination of complexes between fulvic acid and heavy-metal ions, *Anal. Chem.*, 52, 2095, 1980.

554. Zepp, R. G. and Scholtzhauer, P. F., Comparison of photochemical behavior of various humic substances in water. III. Spectroscopic properties of humic substances, *Chemosphere*, 10, 479, 1981.

555. Miano, T. M., Sposito, G., and Martin, J. P., Fluorescence spectroscopy of humic substances, *Soil Sci. Soc. Am. J.*, 52, 1016, 1988.

556. Senesi, N., Miano, T. M., Provenzano, M. R., and Brunetti, G., Spectroscopic and compositional comparative characterization of I.H.S.S. reference and standard fulvic and humic acids of various origin, *Sci. Total Environ.*, 81/82, 143, 1989.

557. Levesque, M., Fluorescence and gel filtration of humic compounds, *Soil Sci.*, 113, 346, 1972.

558. Provenzano, M. R., Miano, T. M., and Senesi, N., Concentration and pH effects of the fluorescence spectra of humic acid-like soil fungal polymers, *Sci. Total Environ.*, 81/82, 129, 1989.

559. Olsen, E. D., *Modern Optical Methods of Analysis*, McGraw-Hill, New York, 1975.

560. MacCarthy, P. and Mark, H. B. Jr., Infrared studies on humic acid in deuterium oxide. I. Evaluation and potentialities of the technique, *Soil Sci. Soc. Am. Proc.*, 39, 663, 1975.

561. MacCarthy, P., Mark, H. B., Jr., and Griffiths, P. R., Direct measurement of the infrared spectra of humic substances in water by Fourier transform infrared spectroscopy, *J. Agric. Food Chem.*, 23, 600, 1975.

562. Griffith, P. R., *Chemical Infrared Fourier Transform Spectroscopy*, John Wiley & Sons, New York, 1975.

563. Willis, H. A., Van Der Maas, J. H., and Miller, R. G. J., *Laboratory Methods in Vibrational Spectroscopy*, John Wiley & Sons, Chichester, 1988.

564. Baes, A. U. and Bloom, P. R., Diffuse reflectance and transmission Fourier transform infrared (DRIFT) spectroscopy of humic and fulvic acids, *Soil Sci. Soc. Am. J.*, 53, 695, 1989.

565. Niemeyer, J., Chen, Y., and Bollag, J.-M., Characterization of humic acids, composts, and peat by diffuse reflectance Fourier-transform infrared spectroscopy, *Soil Sci. Soc. Am. J.*, 56, 135, 1992.

566. Bellamy, L. J., *The Infrared Spectra of Complex Molecules*, Chapman and Hall, London, 1975.

567. Rao, C. N. R., *Chemical Applications of Infrared Spectroscopy*, Academic Press, New York, 1963.

568. Orlov, D. S., *Humus Acids of Soils*, Oxonian Press, New Delhi, 1985.

569. Wagner, G. H. and Stevenson, F. J., Structural arrangement of functional groups in soil humic acid as revealed by infrared analyses, *Soil Sci. Soc. Am. Proc.*, 29, 43, 1965.

570. Theng, B. K. G., Wake, J. R. H., and Posner, A. M., The infrared spectrum of humic acid, *Soil Sci.*, 102, 70, 1966.

571. Mathur, S. P., Infrared evidence of quinones in soil humus, *Soil Sci.*, 113, 136, 1972.

572. Stevenson, F. J. and Goh, K. M., Infrared spectra of humic acids: elimination of interference due to hygroscopic moisture and structural changes accompanying heating with KBr, *Soil Sci.*, 117, 34, 1974.

573. Theng, B. K. G. and Posner, A. M., Nature of the carbonyl groups in soil humic acid, *Soil Sci.*, 104, 191, 1967.

574. Stevenson, F. J. and Goh, K. M., Infrared spectra of humic acids and related substances, *Geochim. Cosmochim. Acta*, 35, 471, 1971.

575. Kumada, K., *Chemistry of Soil Organic Matter* (English translation), Japan Scientific Societies Press, Tokyo/Elsevier, 1987.

576. Preston, C., Applications of NMR to soil organic matter analysis: history and prospects, *Soil Sci.*, 161, 144, 1996.

577. Malcolm, R. L., Application of solid-state ^{13}C NMR spectroscopy to geochemical studies of humic substances, in *Humic Substances, Vol. 2, In Search of Structure*, Hayes, M. H. B., MacCarthy, P., Malcolm, R. L., and Swift, R. S., Eds., John Wiley & Sons, New York, 1989, chap. 12.

578. Wershaw, R. L., Application of nuclear magnetic resonance spectroscopy for determining functionality in humic substances, in *Humic Substances in Soil Sediment and Water: Geochemistry Isolation and Characterization*, Aiken, G. R., McKnight, D. M., Wershaw, R. L., and MacCarthy, P., Eds., Wiley-Interscience, New York, 1985, 561.

579. Steelink, C., Wershaw, R. L., Thorn, K. A., and Wilson, M. A., Application of liquid-state NMR spectroscopy to humic substances, in *Humic Substances, Vol. 2, In Search of Structure*, Hayes, M. H. B., MacCarthy, P., Malcolm, R. L., and Swift, R. S., Eds., John Wiley & Sons, New York, 1989, chap. 10.

580. Wilson, M. A., Solid-state nuclear magnetic resonance spectroscopy of humic substances: basic concepts and techniques, in *Humic Substances, Vol. 2, In Search of Structure*, Hayes, M. H. B., MacCarthy, P., Malcolm, R. L., and Swift, R. S., Eds., John Wiley & Sons, New York, 1989, chap. 11.

581. Fyfe, C. A., *Solid State NMR for Chemists*, CFC Press, Guelph, Ontario, 1983.

582. Shaw, D., *Fourier Transform N.M.R. Spectroscopy*, 2nd ed., Elsevier, Amsterdam, 1984.

583. Wilson, M. A., Jones, A. J., and Williamson, B., Nuclear magnetic resonance spectroscopy of humic materials, *Nature*, 276, 487, 1978.

584. Lentz, H., Lüdemann, H. D., and Ziechmann, W., Proton resonance spectra of humic acids from the solum of a podzol, *Geoderma,* 18, 325, 1977.

585. Sciacovelli, O., Senesi, N., Solinas, V., and Testini, C., Spectroscopic studies on soil organic fractions. I. IR and NMR spectra, *Soil Biol. Biochem.,* 9, 287, 1977.

586. Hatcher, P. G., Schnitzer, M., Dennis, L. W., and Maciel, G. E., Aromaticity of humic substances in soils, *Soil Sci. Soc. Am. J.,* 45, 1089, 1981.

587. Ruggiero, P., Sciacovelli, O., Testini, C., and Interesse, F. S., Spectroscopic studies on soil organic fractions. II. IR and ^{1}H NMR spectra of methylated and unmethylated fulvic acids, *Geochim. Cosmochim. Acta,* 42, 411, 1978.

588. Ruggiero, P., Interesse, F. S., Cassidei, L., and Sciacovelli, O., ^{1}H NMR and IR spectroscopic investigations on soil organic fractions obtained by gel chromatography, *Soil Biol. Biochem.,* 13, 361, 1981.

589. Wilson, M. A., Collin, P. J., and Tate, K. R., ^{1}H nuclear magnetic resonance study of a soil humic acid, *J. Soil Sci.,* 34, 297, 1983.

590. Wilson, M. A., Applications of nuclear magnetic resonance spectroscopy to the study of the structure of soil organic matter, *J. Soil Sci.,* 32, 167, 1981.

591. Preston, C. M., Review of solution NMR of humic substances, in *NMR of Humic Substances and Coal: Techniques, Problems and Solutions,* Wershaw, R. L. and Mikita, M. A., Eds., Lewis, Chelsea, MI, 1987, chap. 2.

592. Preston, C. M. and Blackwell, B. A., Carbon-13 nuclear magnetic resonance for a humic and a fulvic acid: signal-to-noise optimization, quantitation, and spin-echo techniques, *Soil Sci.,* 139, 88, 1985.

593. Schnitzer, M., Selected methods for the characterization of soil humic substances, in *Humic Substances in Soil and Crop Sciences: Selected Readings,* MacCarthy, P., Clapp, C. E., Malcolm, R. L., and Bloom, P. R., Eds., American Society of Agronomy, Madison, WI, 1990, 65.

594. Hatcher, P. G., Breger, I. A., Dennis, L. W., and Maciel, G. E., Solid state ^{13}C-NMR of sedimentary humic substances: new revelations on their composition, in *Aquatic and Terrestrial Humic Materials,* Christman, R. F. and Gjessing, E. T., Eds., Ann Arbor Science, Ann Arbor, MI, 1983, 37.

595. Saiz-Jimenez, C., Hawkins, B. L., and Maciel, G. E., Cross-polarization, magic-angle spinning ^{13}C nuclear magnetic resonance spectroscopy of soil humic fractions, *Org. Geochem.,* 9, 277, 1986.

596. Malcolm, R. L., Variations between humic substances isolated from soils, stream waters, and groundwaters as revealed by ^{13}C-NMR spectroscopy, in *Humic Substances in Soil and Crop Sciences: Selected Readings,* MacCarthy, P., Clapp, C. E., Malcolm, R. L., and Bloom, P. R., Eds., American Society of Agronomy, Madison, WI, 1990, 13.

597. Preston, C. M. and Schnitzer, M., Effects of chemical modification and extractants on the carbon-13 NMR spectra of humic materials, *Soil Sci. Soc. Am. J.,* 48, 305, 1984.

598. Gonzales-Vila, F. J., Lüdemann, H. D., and Martin, F., ^{13}C-NMR structural features of soil humic acids and their methylated, hydrolyzed and extracted derivatives, *Geoderma,* 31, 3, 1983.

599. Hatcher, P. G., Breger, I. A., Maciel, G. E., and Szeverenyi, N. M., Geochemistry of humin, in *Humic Substances in Soil, Sediment, and Water. Geochemistry, Isolation, and Characterization,* Aiken, G. R., McKnight, D. M., Wershaw, R. L., and MacCarthy, P., Eds., John Wiley & Sons, New York, 1985, 275.

600. Skjemstad, J. O., Frost, R. L., and Barron, P. F., Structural units in humic acids from south-eastern Queensland soils as determined by C-13 NMR spectroscopy, *Aust. J. Soil Res.,* 21, 539, 1983.

601. Lobartini, J. C. and Tan, K. H., Differences in humic acid characteristics as determined by carbon-13 nuclear magnetic resonance, scanning electron microscopy, and infrared analysis, *Soil Sci. Soc. Am. J.,* 52, 125, 1988.

602. Krosshavn, M., Bjørgum, J. O., Krane, J., and Steinnes, E., Chemical structure of terrestrial humus materials formed from different vegetation characterized by solid-state ^{13}C NMR with CP-MAS techniques, *J. Soil Sci.,* 41, 371, 1990.

603. Piccolo, A., Campanella, L., and Petronio, B. M., Carbon-13 nuclear magnetic resonance spectra of soil humic substances extracted by different mechanisms, *Soil Sci. Soc. Am. J.,* 54, 750, 1990.

604. Kögel-Knabner, I., Hatcher, P. G., and Zech, W., Chemical structural studies of forest soil humic acids: aromatic carbon fraction, *Soil Sci. Soc. Am. J.,* 55, 241, 1991.

605. Novak, J. M. and Smeck, M. E., Comparisons of humic substances extracted from contiguous alfisols and mollisols of Southwestern Ohio, *Soil Sci. Soc. Am. J.,* 55, 96, 1991.

606. Wershaw, R. L., Mikita, M. A., and Steelink, C., Direct ^{13}C NMR evidence for carbohydrate moieties in fulvic acids, *Environ. Sci. Technol.,* 15, 1461, 1981.

607. Mikita, M. A., Steelink, C., and Wershaw, R. L., Carbon-13 enriched nuclear magnetic resonance method for the determination of hydroxyl functionality in humic substances, *Anal. Chem.*, 53, 1715, 1981.

608. Kögel-Knabner, I., Biodegradation and humification processes in forest soils, in *Soil Biochemistry*, Vol. 8, Bollag, J.-M., and Stotzky, G., Eds., Marcel Dekker, New York, 1993, 101.

609. Guggenberger, G., Christensen, B. T., and Zech, W., Land-use effects on the composition of organic matter in soil particle-size separates. I. Lignin and carbohydrate signature, *Eur. J. Soil Sci.,* 45, 449, 1994.

610. Zech, W. and Kögel-Knabner, I., Patterns and regulation of organic matter transformation in soils: litter decomposition and humification, in *Flux Control in Biological Systems*, Schulze, E.-D., Ed., Academic Press, New York, 1994, 303.

611. Steelink, C., Application of N-15 NMR spectroscopy to the study of organic nitrogen and humic substances in the soil, in *Humic Substances in the Global Environment and Implications on Human Health*, Senesi, N. and Miano, T. M., Eds., Elsevier, Amsterdam, 1994, 405.

612. Thorn, K. A., Arterburn, J. B., and Mikita, M. A., N-15 and C-13 NMR investigation of hydroxylamine-derivatized humic substances, *Environ. Sci. Technol.*, 26, 107, 1992.

613. Benzing-Purdie, L., Ripmeester, J. A., and Preston, C. M., Elucidation of the nitrogen forms in melanoidins and humic acids by N-15 cross polarization-magic angle spinning NMR spectroscopy, *J. Agric. Food Chem.*, 31, 913, 1983.

614. Almendros, G., Freund, R., Gonzales-Villa, F. J., Haider, K. M., Knicker, H., and Luedemann, H., Analysis of C-13 and N-15 CPMAS NMR spectra of soil organic matter and composts, *FEBS Lett.*, 282, 119, 1991.

615. Thorn, K. A. and Mikita, M. A., Ammonia fixation by humic substances: a N-15 and C-13 NMR study, *Sci. Total Environ.*, 113, 67, 1992.

616. Weber, E. J., Thorn, K. A., and Spidel, D. L., Kinetic and N-15 NMR spectroscopic studies of the covalent binding of aniline to humic substances, Proc., 203rd National ACS Meeting, San Francisco, CA, April 5-10, 1992.

617. Thorn, K. A. and Mikita, M. A., N-15 NMR investigations of ammonia and nitrite fixation by humic substances, Proc., 203rd National ACS Meeting, San Francisco, CA, April 5-10, 1992.

618. Su-Neng, Z. and Qi-Xiao, W., Nitrogen forms in humic substances, *Pedosphere*, 4, 307, 1992.

619. Su-Neng, Z. Qi-Xiao, W., Li-Juan, D., and Shun-Ling, W., The nitrogen form of nonhydrolyzable residue of humic acid, *Chinese Sci. Bull.*, 37, 508, 1992.

620. Zhuo, S. and Wen, Q., Nitrogen forms in humic substances, *Pedosphere*, 2, 307, 1992.

621. Knicker, H., Quantitative ^{15}N- und ^{13}C-CPMAS-Festkörper und ^{15}N-Flüssigkeits-NMR-Spektroskopie an Pflanzenkomposten und natürlichen, *Böden. Dissertation, Ragensburg*, 1993.

622. Knicker, H., Fründ, R., and Lüdemann, H.-D., The chemical nature of nitrogen in native soil organic matter, *Naturwissenschaften*, 80, 219, 1993.

623. Ogner, G., ^{31}P-NMR spectra of humic acids: a comparison of four different raw humus types in Norway, *Geoderma*, 29, 215, 1983.

624. Bedrock, C. N., Cheshire, M. V., Chudek, J. A., Goodman, B. A., and Shand, C. A., Use of ^{31}P-NMR to study the forms of phosphorus in peat soils, *Sci. Total Environ.*, 152, 1, 1994.

625. Frye, J. S., Bronnimann, C. E., and Maciel, G. E., Solid-state NMR of humic materials, in *NMR of Humic Substances and Coal: Techniques, Problems and Solutions*, Wershaw, R. L. and Mikita, M. A., Eds., Lewis, Chelsea, MI, 1987, chap. 3.

626. Forster, J. C. and Zech, W., Phosphorus status of a soil catena under Liberian evergreen rain forest: results of ^{31}P NMR spectroscopy and phosphorus adsorption experiments, *Z. Pflanzenernähr. Bodenk.*, 156, 61, 1993.

627. Hoff, A. J., *Advanced EPR. Applications in Biology and Biochemistry*, Elsevier, Amsterdam, 1989.

628. Alger, R. S., *Electron Paramagnetic Resonance: Techniques and Applications*, Wiley-Interscience, New York, 1968.

629. Ingram, D. J. E., *Biological and Biochemical Applications of Electron Spin Resonance*, Hilger, New York, 1969.

630. Abragam, A. and Bleaney, B., *Electron Paramagnetic Resonance of Transition Ions*, Clarendon Press, Oxford, 1970.

631. Swartz, H. M., Bolton, J. R., and Borg, D. C, *Biological Applications of Electron Spin Resonance*, Interscience, New York, 1972.

632. Weil, J. A., Bolton, J. R., and Wertz, J. E., *Electron Paramagnetic Resonance. Elementary Theory and Practical Applications*, John Wiley & Sons, New York, 1994.

633. Gordy, W., *Theory and Applications of Electron Spin Resonance*, Wiley-Interscience, New York, 1980.

634. Steelink, C. and Tollin, G., Free radicals in soil, in *Soil Biochemistry*, Vol. 1, McLaren, A. D. and Peterson, G. M., Eds., Marcel Dekker, New York, 1967, 147.

635. Steelink, C., Review of ESR spectroscopy of humic substances, in *NMR of Humic Substances and Coal. Techniques, Problems and Solutions*, Wershaw, R. L. and Mikita, M. A., Eds., Lewis, Chelsea, MI, 1987, 47.

636. Senesi, N. and Steelink, C., Application of ESR spectroscopy to the study of humic substances, in *Humic Substances, Vol. 2, In Search of Structure*, Hayes, M. H. B., MacCarthy, P., Malcolm, R. L., and Swift, R. S., Eds., John Wiley & Sons, New York, 1989, chap. 13.

637. Senesi, N., Application of electron spin resonance (ESR) spectroscopy in soil chemistry, *Adv. Soil Sci.*, 14, 77, 1990.

638. Senesi, N., Molecular and quantitative aspects of the chemistry of fulvic acid and its interactions with metal ions and organic chemicals. I. The electron spin resonance approach, *Anal. Chim. Acta*, 232, 51, 1990.

639. Senesi, N., Electron spin (or paramagnetic) resonance spectroscopy, in *Methods of Soil Analysis: Chemical Methods*, Sparks, D. L., Ed., ASA-CSSA-SSSA Publ., Madison, 1996, 323.

640. Cheshire, M. V. and Senesi, N., Electron spin resonance spectroscopy of organic and mineral soil particles, in *Environmental Particles. Structure and Surface Reactions of Soil Particles,* Vol. 4, Huang, P. M., Senesi, N., and Buffle, J., Eds., IUPAC Book Series on Analytical and Physical Chemistry of Environmental Systems, John Wiley & Sons, New York, 1998, chap. 9, 331.

641. Blois, M. S., Jr., Brown, H. W., and Maling, J. E., Precision g-value measurements of free radicals of biological interest, in *Free Radicals in Biological Systems*, Blois, M. S., Jr., Brown, H. W., Lemmon, R. M., Lindlom, R. O., and Weissbluth, M., Eds., Academic Press, New York, 1961, 117.

642. Schnitzer, M. and Skinner, S. I. M., Free radicals in soil humic compounds, *Soil Sci.*, 108, 383, 1969.

643. Riffaldi, R. and Schnitzer, M., Electron spin resonance spectrometry of humic substances, *Soil Sci. Soc. Am. J.*, 36, 301, 1972.

644. Atherton, N. M., Cranwell, P. A., Floyd, A. J., and Haworth, R. D., Humic acid. I. ESR spectra of humic acids, *Tetrahedron*, 23, 1653, 1967.

645. Cheshire, M. V. and Cranwell, P. A., Electron spin resonance of humic acids from cultivated soils, *J. Soil Sci.*, 23, 424, 1972.

646. Senesi, N., Chen, Y., and Schnitzer, M., Hyperfine splitting in electron spin resonance spectra of fulvic acid, *Soil Biol. Biochem.*, 9, 371, 1977.

647. Schnitzer, M. and Levesque, M., Electron spin resonance as a guide to the degree of humification of peats, *Soil Sci.*, 127, 140, 1979.

648. Cheshire, M. V. and McPhail, D. B., Hyperfine splitting in the electron spin resonance solution spectra of humic substances, *Eur. J. Soil Sci.*, 47, 205, 1996.

649. Wilson, S. A. and Weber, J. H., Electron spin resonance analysis of semiquinone free radicals of aquatic and soil fulvic and humic acids, *Anal. Lett.*, 10, 75, 1977.

650. Lagerkrantz, C. and Yhland, M., Photo-induced free radical reactions in the solution of some tars and humic acids, *Acta Chem. Scand.*, 17, 1299, 1963.

CHAPTER 7

Characterizing Soil Redox Behavior

Richmond J. Bartlett

CONTENTS

I. INTRODUCTION

Redox is one of those catchy names invented by somebody who was not hampered by commitment to the use of scientifically correct terminology. The name is perfect, however, and it is hereby declared correct. The *red* stands for reduction and it signifies gain of electrons by a chemical species; the *ox* connotes oxidation, or electron loss by a chemical species.

II. ELECTRONS AND PROTONS

A. pe and pH

A pH is the negative log (p for power) of proton activity, and pe, its energy or work analog, is the negative log of the electron potential. An electron isn't a full-fledged analog of a proton.

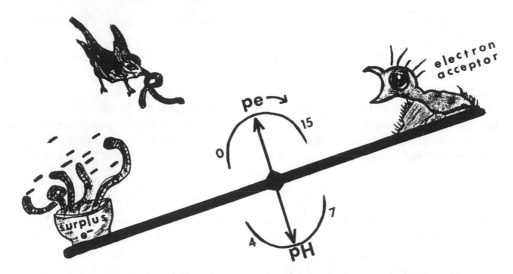

Figure 1   The soil electron seesaw.

Together, two equal but opposite charges make up a hydrogen atom, but that is about the extent of the equality between an electron and a proton. Without its proton, an electron is no longer an analog of H+, and it no longer has any claim to being part of a hydrogen atom. An electron doesn't bounce about by itself in the manner of an H⁺, and therefore it is probably not correct to try to characterize its "activity". Always it is either attached to an atom or radical or it is in the process of being transferred from one to another.

A proton is a cation. It can replace or be replaced by other cations and it is as good as any other cation when it comes to balancing a chemical equation. Electrons receive no recognition in balanced chemical equations because the donated and accepted electrons must always cancel one another on opposite sides of an equation. Electrons do not have anion status. They cannot trade places with other negatively charged species.

Usually we see release of H⁺ when metals are oxidized and H⁺ consumption with their reduction. Oxidation is furthered in a soil environment where protons and electrons are deficient; that is, where acidity and levels of easily decomposed electron donors are both low. But there must be a ready supply of available electron acceptors. Reduction is favored by surpluses of both protons and electrons. This means that low pH and high availability of organic substances both will promote reduction in soil. Reduction of Fe or Mn oxides, or of nitrate, uses up H⁺ and thereby increases pH of the soil and theoretically lowers the pe. Oxidation of Fe, Mn, or nitrite lowers the pH (measurable) and raises the pe (not measurable in most soils). Measuring changes in concentrations of redox species, or even using the seesaw (Figure 1), is more reliable for predicting these things in oxidized soils than is an attempted measurement of pe with a Pt electrode. For these reasons, much of the discussion in these early sections is theoretical rather than being grounded in empirical observations.

The farther apart the electrons, the more proportional work required to bring them together and the higher the respective pe. A low pe system has a surplus of electrons and, therefore, a big tendency to lose some of them and become more oxidized. A high pe system is hungry for electrons. As deficient electrons are replenished, the tendency for reduction to occur will increase. If we substitute pe and pH for their defined equivalents in a generic redox half-reaction in which activities of oxidized and reduced species are equal, we see, as follows, that the pe + pH sum is equivalent to the equilibrium constant of the half-reaction:

$$\text{Oxidized species} + e^- + H^+ = \text{reduced species} \tag{1}$$

$$\log K = \log red - \log ox - \log e^- - \log H^+ \tag{2}$$

$$\log K = pe + pH \tag{3}$$

If indeed their sum *is* constant, then, thermodynamically, pe and pH are on opposite ends of a seesaw. If behavior follows thermodynamic theory, then, if one goes up, the other will come down, like any sound seesaw (Figure 1). Lindsay[1] refers to this sum as the *redox parameter* because, if a soil is at internal equilibrium, the pe + pH represents the sums of all of the redox equilibrium constants in the soil.

B. Poise

In the real world, redox seesaws are not so simple. But the simple seesaw model can help guide us through the muck and the mire of the *Anoxic* soils and the corn fields and jungles of the *Manoxics* and *Suboxics* (see Section VII.B). The seesaw in Figure 1 shows the interaction between source/sink quantities and electron/proton intensities. If we add reduced substances such as plant residues or Mn(II) or Cr(III) to a *Manoxic* soil poised so that its easily reduced substances are in balance with its easily oxidized substances, some of the added reduced species will be quickly oxidized. On the other hand, adding Mn(IV) or Cr(VI) will result in immediate reduction of a portion of the added oxidants. There appears to be a tendency for a soil, if disturbed, to maintain a redox balance, that is, *poise*, by donating electrons to surplus electron acceptors or by accepting electrons from surplus electron donors.

A soil kept near field capacity moisture with occasional mixing, double bagged inside thin polyethylene for several months at 15 to 25°C will be close to internal equilibrium. If this metastable equilibrium is disturbed by adding an easily oxidized substance to it, say graham cracker crumbs, the *pe + pH* of the overall system will tend to remain fairly constant as the disturbed soil system moves back toward a new metastable equilibrium. In this instance, the pe will tend to go down, and to the extent that it does, the pH will tend to rise.

By adding increments of Cr^{3+} and $HCrO_4^-$, respectively, to separate subsamples of the same soil and then determining the amount of Cr reduced [loss of Cr(VI)] and the amount oxidized [gain of Cr(VI)], it is possible to find a *point of poise* or *buffered redox region*, where the electron donating and electron accepting tendencies cross. There the redox seesaw is balanced at dead-level.

C. Electron Dynamics

Oxidation of organic residues in the living soil depend largely upon the enormous profusion and incalculable enterprise of myriad microorganisms. These feast upon electron-rich organic substances, raising soil pe, and mineralizing metallic cations to elevate soil pH. Populations of these electron consumers also are aroused and stimulated by available forms of oxygen or Mn oxides, which serve as electron sinks or depositories for the surplus electrons. Sometimes the electron supply may exceed the electron demand. There is a shortage of electron parking places, and the soil remains reduced. Metabolism of microbes that feed on the surplus electron suppliers is more suppressed by waterlogging than is the growth of higher plants which have their tops in the atmosphere. This explains why production of reduced carbon exceeds its oxidation in wet soils and why organic matter accumulates in them (see Section VI.C). A soil at metastable equilbrium with itself will be brought closer to thermodynamic equilibrium by processes of acidification or drying, which cause oxidation of reduced carbon and concomitant reduction of oxidized species. Tillage particularly will foster an increase in internal thermodynamic equilibrium by drying and aerating a soil, increasing its surface temperature, and exposing soil surfaces to sunlight. These things bring about oxidation of soil organic matter accompanied by reduction of Mn oxides, oxygen species, nitrates, and nitrites.

A hydrogen atom doesn't lose a significant amount of weight when it gives up an electron, but the electron loses its whole world when it loses its proton. This doesn't sound like equality in any sense. Yet, in a redox reaction, the transfer of an electron is equivalent to the transfer of an H atom (which consists of an electron plus a proton). Oxidation of any chemical species means the removal of an electron with or without its proton; reduction is the addition of an electron with or without its proton. But only the proton half appears in a balanced redox chemical equation because chemical convention does not provide a way to show the whereabouts of electrons involved, except in a reduction or oxidation half-reaction. For example, in the following sum of two reversible half-reactions showing the reduction of MnO_2 and the oxidation of NO_2^-, respectively, the electrons have vanished, cancelled one another:

$$MnO_2 + 3H^+ + NO_2^- + OH^- \rightarrow Mn^{2+} + NO_3^- + 2H_2O \quad \log K = +27.3 \qquad (4)$$

By summing the ΔG values (free energy changes) on the right side of the equation and subtracting the sum of the left side and dividing the difference by $-2.3RT$, we obtain a positive log K for the reaction. This means that the reaction is possible in energy terms. The equation represents a thermodynamically spontaneous reaction in the direction written. It is theoretically possible for MnO_2 to oxidize nitrite to nitrate. There is no promise that it will (although it actually does[2]). It is essential to keep in mind that thermodynamics tells us what is possible or impossible in energy terms and says nothing about which reactions will occur or about possible rates of reactions. The word spontaneous does not refer to probability. Charcoal from campfires that burned thousands of years ago is still available for carbon dating in well-aerated soils, in spite of the fact that charcoal in the presence of O_2 is thermodynamically unstable.

D. pe/pH Equilibria

Thermodynamic diagrams using calculated equilibrium constants and assumed equilibrium concentrations can be especially helpful in comparing and thinking about possible or potential soil redox reactions. The following example shows the relation between pe and pH for a reduction half-reaction in which the free energies of formation of MnO_2, Mn^{2+}, and H_2O are -111.3, -55.1, and -56.7 cal mol^{-1}, respectively,[1] and the activities are assumed to be unity for the solid MnO_2 and 10^{-4} M for Mn^{2+}:

$$\tfrac{1}{2}MnO_2 + 2H^+ + e^- = \tfrac{1}{2}Mn^{2+} + H_2O \qquad (5)$$

$$G^\circ = -\tfrac{1}{2}\,55.1 - 56.7 + \tfrac{1}{2}\,111.3 = -28.6 \qquad (6)$$

$$\log K = \frac{-28.6}{2.3\,RT\,(\text{or } 1.364)} - \tfrac{1}{2}\log 10^{-4} - \log e^- - 2\log H^+ \qquad (7)$$

$$pe = 23 - 2pH \qquad (8)$$

Figure 2 shows pe vs. pH stability lines between oxidized and reduced species for several redox couples. Equilibrium diagrams can help us visualize energy relationships between redox species and boundaries and even patterns in the redox network. However, we must be very cautious about applying them to nonequilibrium soil systems. They will never enable us to predict what will

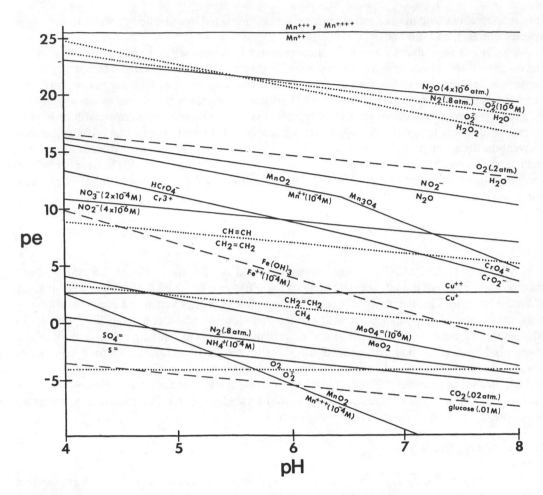

Figure 2 Stability lines between oxidized and reduced species for several redox couples. Solid phases of Mn and Fe are at unit activity and activities of other species are designated if not equal within a couple. Free energy data are taken from Lindsay[1] and from Garrels and Christ.[7]

happen, but they will help us to know sometimes, but not always, what can't happen or what can happen.

If thermodynamic equilibrium prevails, the pe and pH region above any given line in Figure 2 should favor the presence of the oxidized form of the couple — the region below the line, the reduced form. Frequently the iron line is considered the boundary between aerobic and anaerobic soil. In aerobic soils, oxidized species tend to remain oxidized, although the thermodynamic "tendency" is toward reduction. Sulfide is easily oxidized; nitrite is easily reduced. Uses of pe-pH diagrams and other thermodynamic relationships were discussed previously.[3]

A soil that has lost 99% of its oxygen shows only a slight change in the O_2/H_2O line. Thermodynamically, a soil with only one molecule of O_2 ha^{-1} is still aerobic, with the pe vs. pH line above the iron line. Insensitivity of soil behavior to O_2 changes is borne out by field observations. Greenwood[4] demonstrated that 1% of atmospheric O_2 was sufficient to prevent anaerobic metabolism in widely different soils. Of course, this small amount of O_2 will be used up very quickly if it is not continually replenished.

Thermodynamic predictions can never substitute for empirical evidence. Figure 2 shows as a possible reaction the reduction of MnO_2 coupled to the oxidation of Cr(III). Experience demonstrates that this is indeed a probable reaction.[5] From the diagram, oxidation of Cr by nitrate also is possible, but there is no experimental evidence that this ever happens in soils. The positions on the diagram tell us that both the superoxide free radical $O_2^-\cdot$ and the Mn^{3+} supermanganese free radical are powerful both as oxidizing agents and as reducing agents.

The tough reduction of N_2 to NH_3 down in the pe region of zero oxygen paradoxically requires expenditure of enormous amounts of O_2 somewhere else in a plant in order to mobilize the metabolic energy needed.

A pe is an intensity parameter. It indicates a potential, not a propensity for change. A high potential certainly does not mean a high likelihood of its becoming lower, any more than the state of being poor indicates that money will be soon available. A waterlogged, O_2-deficient soil will continue to have a low pe as long as O_2 is excluded. A well-aerated soil will continue to have a high potential for oxidation as long as it can draw at a sufficient rate on the inexhaustible atmosphere to poise it against reduction.

E. Empirical Measurements of Redox Potentials

Details of the EMpe measurement procedure are presented in Section IX. The EM for *empirical* reminds us that the potential measured by the electrode is not a thermodynamic pe and that it may not have any theoretical significance. A measured pe cannot have precise theoretical significance unless we know which species are in equilibrium, the equations of the reactions involved, and the values for the equilibrium constants. Most important of all, we must know that there indeed is truly internal equilibrium in the system and that all of the redox species are electroactive and of suficient solubility to affect the electrodes. These conditions are never met in aerated soil systems, not even approximately in *Redoxic, Suboxic, Manoxic,* and *Superoxic* redox categories (see Section VIII). However, in nearly closed *Sulfidic* or *Anoxic* soil systems, redox potential (pe) measurements can have some theoretical significance.[6,18]

Substantial changes in the partial pressure of O_2 are not reflected in measured pe, and there is a tendency for dissolved O_2 to act on a Pt electrode so that the observed potential only partially measures oxidation states of other dissolved components.[7] Generally, neither Mn or Fe oxides nor nitrate has the expected or even significant quantitative effect on electrode measurements — oxides because of insolubility and nitrate because it is not electroactive.[8] Since the major redox components of soils often have either ambiguous or minute effects on measured potentials, we might ask whether results of measurements in soils are a waste of time at best or blatantly misleading at worst. Is a soil potential measurement worth the effort? The answer is *yes*, in anaerobic systems, and a conditional *maybe* in all other soils. In these other soils, it must be recognized that results are empirical and qualitative, and are useless for theoretical interpretations. Another caveat: correcting measured EMpe values for pH in an aerobic soil may eliminate all of what first appeared to be interesting EMpe findings.[9] A potential measurement is reflecting pH rather than pe.

Bohn[10] pointed out the importance of clearly distinguishing between an equilibrium electrode potential (Eh), which is not amenable to determination in soils, and measured soil redox potential (E_{Pt}). The latter is a nonequilibrium mixed potential and it is not different from an EMpe. He stressed that the measurement is qualitative and its utility is in interpretation by a knowledgeable observer, a distinction often dealt with fuzzily in the literature. An E_{Pt} is converted to an EMpe or Ptpe by multiplying it by nF/RT.

Neither pH nor EMpe is measured directly but both are calculated by the Nernst equation from the electrical potential between a glass or Pt electrode and a reference electrode. The voltage between the electrode and the reference, corrected for the effect of the reference, is multiplied by Faraday's constant and divided by 2.3 RT to obtain the pH or the pe. In the case of pH, the meter does the calculating. Theoretically, a reduced system will tend to lose electrons to the inert Pt electrode, which will take on a negative charge.[11,19] The potential and the pe will be low. If the system becomes more oxidized, the charge on the Pt electrode will be more positive, and the pe measured will be higher.

The glass electrode is sensitive to protons but not electrons, while the Pt electrode is an all purpose electrode that will respond to pH as well as to pe or any other potential present.

In a system of constant pe, a one-unit decrease in pH will cause a one-unit increase in the apparent pe measured. Since the measured EMpe values are mainly to be used for making qualitative judgments only, correcting them for differences in pH can be done by subtracting from each measured value the difference between the measured pH and 7. Alternatively, the pH and EMpe values may be added together. The addition method will give larger numbers, but the relative relationship between values will be similar for both methods. Either correction requires the assumption that all half-reactions in the soils studied consume or liberate one proton for each electron. Of course this is not true, but untruth is a minor problem compared with the others inherent in measuring mixed potentials in aerobic systems.

1. Redox Changes in Experimentally Controlled Environments

Instead of characterizing the redox environment and the redox happenings in separate approaches, Patrick and Jugsujinda[12] devised a system for controlling the "redox potentials" in waterlogged soil systems by automatic additions of small increments of O_2 along with electronic monitoring so that the measured potentials could be controlled at predetermined set values. Each new system with a set potential was incubated for 15 d before sampling and introduction of the next increment of controlling oxygen. The results of a recent study showed that during the controlled incremental reduction of the soil, the reductions of nitrate and Mn(IV) were sequential, with no Mn reduced as long as nitrate was present. When reduced soil was oxidized stepwise, the oxidation of Mn(II) to Mn(IV) and the nitrification of NH_4^+ also were sequential, with no nitrate appearing until all of the Mn(II) had disappeared.

The critical measured potential at which all of the nitrate had disappeared from the soil solution and Mn(II) had begun to appear during reduction of the soil was about 200 mV. The potentials are comparable to EMpe or E_{Pt} discussed by Bohn,[10] and it is not likely that the values reflect the presence of either nitrate, which is not electroactive, or of insoluble Mn(IV).

III. FREE RADICALS IN SOILS AND WATER

A. Properties of Free Radicals

Free radicals are in the news these days as evil-acting chemical substances that must be scavenged from our bodies with vitamins C and E, beta-carotine, and selenium in order that we may be protected from cancer, failing hearts, and premature aging. Perhaps the news is misleading because it fails to tell us that free radicals not only are omnipresent in all living beings, including soils and plants, but also they are indispensable and vital to life. Free radicals make photosynthesis happen and probably also synthesis of humus. In other words — green plants and topsoil to grow them in, pretty good pluses for free radicals.

Free radicals abound wherever redox species are giving up or accepting electrons one at a time. Under soil conditions where O_2 is partially limited, available organic substances do not become completely oxidized. This means that O_2 often accepts electrons in stepwise fashion — one at a time — instead of acting as a four-electron acceptor with H_2O as the sole product of its reduction. The result of the single steps is the formation of odd electron intermediates, or free radicals[13] (Equations 9 to 12).

Single electrons, rather than electron pairs, characterize the outer shells of free radicals. A hydrogen atom is the simplest and most active of free radicals. Its single electron is its only electron. When it gives that up, the hydrogen atom is oxidized, and in the process it is acting as a powerful reducing agent. When a hydrogen nucleus, a proton, accepts a single electron, it is a powerful oxidizing agent. As H_2, hydrogen is stable because the electrons are paired.

Any species that can donate or accept a single electron is a free radical. A free radical is unstable and therefore extremely reactive, always ready to shuck off its odd electron or else accept a mate for it. And therein lies its potential for driving redox reactions and thereby either promoting or hindering life functions.

Because the spin of an odd electron is not cancelled by an electron of opposite spin, a free radical generates a magnetic moment that makes it paramagnetic, or attracted to an external magnetic field, amenable to detection and characterization by means of electron spin resonance spectrophotometry (ESR).

B. Formation and Behavior of Free Radicals

$$O_2 + e^- + H^+ = HO_2^{\cdot} \quad \{\text{protonated superoxide}\} \tag{9}$$

$$O_2 + 2e^- + 2H^+ = H_2O_2 \quad \{\text{hydrogen peroxide}\} \tag{10}$$

$$O_2 + 3e^- + 3H^+ = H_2O + OH^{\cdot} \quad \{\text{hydroxyl free rad.}\} \tag{11}$$

$$O_2 + 4e^- + 4H^+ = 2H_2O \quad \{\text{water}\} \tag{12}$$

Equations 9, 10, 11, and 12 are half-reactions showing reduction of O_2 by single electron additons. Thus, superoxide and hydroxyl, produced by 1 and 3 odd electron additions, are free radicals; whereas peroxide and water, with 2 and 4 electrons added, respectively, are not. Restricting conditions of interaction between the availabilities of soil O_2 and electron donors, for example at the interface between oxygenated water and anaerobic soil in a wetland, tends to favor transfers of electron in single steps, and thus such interfaces are likely to be sites for free radical formation. Free radical mechanisms appear to explain why kinetically slow and seemingly unlikely redox transformations often occur readily at interfaces (see Section VI). Oxygen free radicals are much more reactive than O_2 itself, and both superoxide and the hydroxyl free radical are especially reactive with H_2O_2, each one capable of being quickly transformed into the other (Equations 13 and 14).

C. Dissipation of Oxygen Free Radicals

The biodestructive hydroxyl free radical (OH·) is a powerful oxidant and is particularly dangerous to living systems, and H_2O_2 and $HO_2^{\cdot}$ together are just as dangerous because they react to form a hydroxyl free radical. Organisms that live in oxygen environments (aerobic and facultative anaerobic microorganisms, plant roots, and people) contain enzymes that destroy the peroxide and superoxide reactants and prevent the formation of OH·.[14] Fresh recently oxidized Mn oxides above pH 6 and other catalases will catalyze the oxidation of one of the oxygens in H_2O_2 by the other and thereby will destroy the peroxide by dismutation (Equation 16). Oxidized Mn will oxidize H_2O_2 in acid media (Equation 17).

Superoxide dismutase enzymes are usually activated by Cu, Fe, or Mn and are credited for making life possible in a world of oxygen. The dismutation of $HO_2^{\cdot}$ to H_2O_2 and O_2 by superoxide dismutase (SOD) enzymes[15] (Equation 15) followed by H_2O_2 dismutation to O_2 and H_2O or oxidation of H_2O_2 to O_2, make aerobic life possible[16,17] by preventing the formation of the biodestructive hydroxyl free radicals, by Equation 13. Reforming $HO_2^{\cdot}$ by Equation 14 is prevented.

$$HO_2^- + H_2O_2 \rightarrow H_2O + O_2 + OH^- \quad \log K = +22.0 \tag{13}$$

$$OH^- + H_2O_2 \rightarrow H_2O + HO_2^- \quad \log K = +15.0 \tag{14}$$

$$2HO_2^- \xrightarrow{\ SOD\ } H_2O_2 + O_2 \quad \log K = +42.1 \tag{15}$$

$$2H_2O_2 \xrightarrow{\ \text{catalase or MnO}_2 \text{ at pH} >6\ } 2H_2O + O_2 \quad \log K = +37.0 \tag{16}$$

$$H_2O_2 + MnO_2 + 2H^+ \xrightarrow{\ \text{low pH}\ } Mn^{2+} + 2H_2O + O_2 \quad \log K = +18.9 \tag{17}$$

$$MN^{2+} + 2HO_2^- \rightarrow MnO_2 + 2H^+ \quad \log K = +60.3 \tag{18}$$

The highly reactive supermanganese ion, Mn^{3+}, behaves as a free radical and is analogous to superoxide in many reactions. Both are swinging free radicals that serve as both powerful oxidizing and reducing agents. Unpaired electrons are part of the structures of NO, NO_2, MnO, Mn_3O_4, MnO_2, and no doubt they cause free radical behavior in these compounds.

IV. REDOX ENZYMES AND RELATED CATALYSTS

Some highly labile organic compounds are oxidized directly by atmospheric O_2, but the majority of apparently spontaneous oxidation reactions in soils occur with the aid of catalysts. These may be various catalytic agents, including organic enzymes, enzyme proteins with metal cofactors, metals alone or metal oxides, peroxides, or free radicals. Photochemical oxidation involving the reduction of catalytic Fe is important in soils exposed to light.[3] Proteins containing metals — usually Fe, Mn, or Cu — serve as enzymes, carrying electrons from reduced carbon to O_2. These enzymes are essential links in the respiration of all aerobic organisms. Flavoproteins without metals can transfer electrons to O_2, but the slow rate cannot account for aerobic respiration. To be considered a catalyst, a substance cannot be permanently altered by the reaction it catalyzes.

An electron carrier is a substance that will oxidize a reduced substance and will in turn be oxidized by a more highly oxidized species. The carrier first becomes reduced by receiving an electron from an electron donor and then is oxidized as it passes the electron on to the highly oxidized electron acceptor. For example, Fe(II) in a cytochrome is oxidized spontaneously by atmospheric O_2 and the Fe(III) formed then is reduced by an organic compound that becomes oxidized in the process.[20] Thus, the cytochrome Fe transfers an electron from the organic compound to O_2. Kinetic studies indicate that under aerobic conditions, the reduced enzyme oxidizes much faster than the organic compound so that most of the enzyme Fe is in the oxidized state at any given time.

Electron-transferring enzymes are given names according to their effects on the substrate end product, and the names are only slightly confusing. An enzyme that becomes reduced when it removes an electron from a substrate is termed an oxidase. One that donates an electron is a reductase. Enzymes that oxidize carbon compounds by transferring electrons from reduced organics to oxidized substances are termed reductases because they reduce the end products. An oxidase that removes a proton along with an electron from an electron-donating substance is called a dehydrogenase, since a proton plus an electron comprise a hydrogen atom free radical. A peroxidase that catalytically reduces a molecule of H_2O_2 by donating two H atom electrons is also a reductase. Each molecule of horseradish peroxidase contains one atom of Fe(III).[20] To pass along an electron to the peroxide, the Fe(III) must borrow an electron from a carbon atom to form for an instant an atom of Fe(II). The Fe(III) is restored as the H_2O_2 is reduced.

Molybdenum is an essential part of nitrogen reductase enzymes. Inhibition of nitrate reduction by Fe-binding compounds indicates that Fe is essential in nitrate reductase as well. Selenium is required for nitrate-reducing activity *E. coli*.[20] Copper-containing nitrite reductases have been identified. A superoxide dismutase containing Cu^{2+} and also Zn^{2+}, as a stabilizer was described by Fridovich.[15]

V. MANGANESE AND IRON: KEYS TO LIFE

Manganese and iron are by far the most pervasive and important of the metallic redox catalysts in soils that also include Cu, Mo, and Cr. The redox elements, oxygen and carbon, are basic to the building and behavior of soils and all living systems, and Mn and Fe are fundamental to the functioning of oxygen and carbon in life's overall redox stratagem.

A. On Our Knees to Manganese

1. Mn: Key, Originator, and Protector

Manganese provides preeminent parking for electrons. This transcendental transition metal even takes precedence over O_2 because it is the provider of O_2. As the key that unlocks oxygen from water by oxidizing water hydroxyl in the process of photosynthesis, Mn (probably as the super-manganese free radical, Mn^{3+}) is responsible for the presence of free O_2 in the atmosphere of our planet and in its soils and waters.[3,21-23] Furthermore, it is quite logical to hypothesize that before oxidized Mn introduced O_2 to the earth, Mn oxides were the first, the original, and the only electron sinks for the profusion of reduced gases, such as methane, ammonia, and hydrogen, that scientists agree probably constituted the earth's primeval atmosphere.

Colloidal Mn oxides have unique surface properties that suit them for adsorption of both inorganic and organic substances, cationic or anionic. Because they coat surfaces, they exert chemical influence far out of proportion to their total concentrations.[24] These oxides act as scavengers for heavy metals in soils and in waters by adsorption, complexation, and redox mechanisms.[25-28] Manganese, the protector of life, also mops up and demolishes, by oxidizing, reducing, or dismutating, biodestructive and death-dealing O free radicals formed at interfaces in soils. As a key to life and O_2 through photosynthesis, and the source of life, as the original electron parking place, and as protector of life, Mn stands out as an ultimate element. Still, life would empty, rather, nonexistent, without vital Fe, or O, or C, or N, etc.

2. Why Manganese?

We first became interested in soil Mn oxides in 1979, when we discovered[5] that they would readily oxidize Cr(III), then considered relatively harmless by regulatory officials, and still by some. Soil Mn(IV) converts soluble Cr(III) to toxic Cr(VI) chromate that can move through soils and into water supplies.[29,30] We continue to have a major interest in Cr, but now we see it in an additional role as an extremely valuable and useful research tool, the ideal electron tracer, with which to study behavior of soil Mn and all soil redox processes. And of course we are interested in Mn for its own sake — because of its astoundingly fundamental importance in soil and life.

3. Processes of Manganese Oxidation

Several researchers,[31-33] dismissed the concept that oxidation of Mn in soils could be nonmicrobial because their work showed that Mn oxides failed to form in the presence of heat sterilization and microbial inhibitors such as chloroform or sodium azide. However, Ross and Bartlett[34,38] found that either heat sterilization or these inhibitors reduced MnO_2 and destroyed the Mn oxidizing mechanism in the soil. Using Cr oxidation to evaluate Mn oxides in soil samples, Ross and Bartlett[34] showed that oxidation of added Mn was proportional to existing oxides. Arrhenius plots of rates of Mn oxidation at different temperatures indicated that the oxidation was nonbiological under the experimental conditions. They concluded that the oxidation was autocatalytic and that fresh Mn oxides tended to form on old oxide surfaces.

4. Reverse Dismutation of Manganese

Most Mn oxides in the humus-dominated surface horizons of soils are negatively charged amorphous Mn(IV) and Mn(III) colloids. Generally, the negative charges on the Mn oxide surfaces are occupied by adsorbed Mn^{2+} and Mn^{3+}, giving the overall surface a positive charge.[36] The positively charged surfaces tend to adsorb or be adsorbed by negatively charged colloidal humus.[3] This close approach of a negatively charged ligand to a Mn oxide surface can bring about and drive the reversal of the thermodynamically spontaneous dismutation reaction shown in the following equation:

$$2Mn^{3+} + 2H_2O \rightarrow Mn^{2+} + MnO_2 + 4H^+ \quad \log K = +9.1 \tag{19}$$

The reverse is driven by the energy of complex formation between Mn(III) and an organic ligand, for example, Mn(III)-oxalate.

5. Behavior of Mn(III) Complexes

To make a bright pink Mn(III)-oxalate complex by reverse dismutation, add 5 drops of a 0.1 M MnO_2 suspension (Section IX.H) to 10 ml of a solution that is 5 mM oxalic acid and 2.5 mM $MnSO_4$. As soon as the MnO_2 has nearly dissolved, add 15 ml of 50 mM ammonium oxalate to raise the pH. The pink of Mn(III) begins to fade soon after its formation, indicating that the reduction to Mn(II) is taking place. It is accompanied by the oxidation of the organic acid, at first to an oxalate free radical and then to CO_2 and H_2O. Formation and fading both are favored by low pH. Soluble Mn(III)-organic acids range in color from tawny to pink. Sometimes Mn(IV)-organic complexes are formed. As the catalytic decomposition of the organic acid takes place, the pH rises and the rate of the redox reaction is slowed. At pH 7.6, light brown Mn(III)-potassium citrate remained stable in the lab for a year. However, some of the citrate was decomposed by microbes and some dark-colored sediment, probably Mn(IV) oxide, formed, apparently the result of spontaneous dismutation of the Mn(III) at the high pH. This could be the mechanism for microbial "synthesis" of Mn(IV).

VI. INTERFACES: WHERE REDOX IS HAPPENING

A. Where Interfaces Are Found

A redox interface consists of the thin intersection or boundary between two regions of differing redox status. The contrast between the pe levels on opposite sides of an interface may range from moderate to extreme. As an example for study, the extreme case of a submerged wetland or paddy is most instructive. In searching for free radicals, we should avoid the anaerobic zone of surplus electrons, and we shouldn't expect to find them where aerobic microorganisms well endowed with dismutating enzymes are flourishing. Their place is at the redox interface where low pe and high pe conditions are cheek to cheek, with neither one overwhelming or destroying the other. On the interfacial plane roots are growing, microaerophiles multiplying and dying, humus synthesizing, pesticides degrading, Fe and Mn taking on electrons and giving them away, C, N, and S transforming from one redox state to another and back again — all in proximity to one another because thicknesses of interfaces are measured in millimeters or even fractions thereof.

The coming together of oxidative and reductive soil environments form interfaces in peds, paddies, manure storage pits, landfills, sewage lagoons,septic tank leach fields, sediments, clay or humus surfaces, and root-soil boundary regions of rhizospheres. Each thin interface is unique, consisting of many mini-interfaces, every one of which is likely to be optimum for a distinctive redox transformation. Who will believe that all of these can be simulated in a plastic beaker. Three separate redox entities are created when the soil in the beaker is flooded and left undisturbed for a period of time: saturated soil, the water above it, and the interface between them. A thin interface consists of soil immediately underneath partially oxygenated free water that is in contact with the atmosphere. This interface is immediately above saturated soil that is practically devoid of free O_2.

The interface is a standoff between extensive O_2 in the atmosphere above and the often huge reservoir of electron donors beneath. Shallow water in the wetland frequently contains accumulations of Mn and Fe oxides persisting at the interface in spite of the huge pool of spuming electrons beneath them in the anaerobic zone. When Fe^{2+} escapes into the water, it often oxidizes at the air/water interface, and shiny and fragile iron oxide films are commonly seen floating on still waters. Because these fragile films refract light and produce beautiful rainbow spectra, they are commonly mistaken for natural petroleum[39] or for polluting polyaromatic hydrocarbons. At a touch of a finger, however, the Fe film breaks and scatters into tiny fragments across the water surface. Oil or hydrocarbons will climb up your finger.

B. Demonstrating the Oxidative Power of an Interface

The soil of a submerged redox interface is commonly found to be more highly oxidized and more highly oxidizing than the solution above it, in equilibrium with O_2 of the atmosphere. This phenomenon is easily demonstrated. Place about 100 cm³ loose, moist A horizon with near neutral pH into a 1-l plastic beaker or pot (without holes), add 2 mmol $MnSO_4$ and water until its depth above the soil is about the same as the depth of the soil, and stir thoroughly. Keep flooded, adding distilled water occasionally, and allow to stand without further agitation, except for removal every few days of 0.1 to 0.2 ml samples of interface sediment to a spotplate using a pipet with the tip cut off to admit soil. Test the sample using tetramethylbenzidine (Section VII.C) After days or even weeks, depending on the redox poise of the system, you will find that the interface soil has become *Manoxic*, or better still, *Superoxic*. Next, add 0.1 mmol of $CrCl_3$ to the supernatant water, gently, so as not to disturb the interface, and then stir the supernatant very gently once only. After a few hours, sample the supernatant and test for Cr(VI), using about 5 drops of DPC (Section VII.D) 1 or 2 ml, in a small test tube. We have shown that Cr(VI) persists as long as 15 years in the undisturbed water above a *Manoxic* interface in a simulated flooded wetland.[29-38] However, mixing the phases across such an interface will quickly destroy the oxidized phase, and reoxidation of the reduced Cr usually will not occur.

C. Wetlands

1. Properties and Processes

No one is surprised to find a wetland that is wet. Wetlands are frequently thus, although often not continuously. A wetland owes its characteristics not to water directly, but to the effects of water-filled soil pores in restricting the supply of oxygen to surfaces in the soil body. Wetland soils are legally defined by agreement among the Soil Conservation Service (now officially Natural

Resources Conservation Service, NRCS), The Fish and Wildlife Service, the Environmental Protection Agency (EPA), and the Army Corps of Engineers.

A *Hydric* soil identifies a wetland. To be hydric, a soil must have a presence of free water and virtual absence of O_2 in it or on it for extended periods, but not necessarily continuously. During these periods, anaerobic conditions must develop in the upper profile. This means that decomposable organic residues are present, and temperatures are of the degree and duration required for microbial synthesis of reactive electron donors. Soil conditions must favor the production and regeneration of hydrophytic vegetation. Soils will tend to have high accumulations of organic matter, gley mottles near the surface, a predominance of gray or black colors below, and some Fe oxides reduced to Fe^{2+} during anaerobic periods. Recent NRCS nomenclature defines these features as *redoximorphic*.

Generally, all soils in the *Anoxic* and *Sulfidic* categories in the newly developed system for classifying redox status of field soils (discussed in Section VII) will be *Hydric* soils and many of the soils in the *Redoxic* category also will be *Hydric*. The chemical field test for easily reducible Fe(III) makes it possible to identify a soil in the *Redoxic* category after artificial drainage or during dry seasons. Redoximorphic features will become less obvious and may even disappear from a wetland after several years of permanent water table lowering. Rice paddies formed by flooding well-drained soils will gain redoximorphic features with time. Unpolluted sediments at the bottom of a cold lake with low biological activity usually will not constitute a wetland.

2. Wetland Conservation

We read and we hear that wetlands are crucial. They are endangered and must be preserved. Why? Many of the proponents may not know the one word answer. The one word is *interfaces*. An *Anoxic* horizon can be directly under a *Superoxic* horizon (Section VII) in the same flooded pedon, separated only by the interface between them. At a redox interface there are free radicals and wondrous and important redox chemical happenings resulting from the formation and dissipation of the free radicals. Redox interfaces occur in wetlands between water and colloidal soil surfaces in flooded soils and also between roots and moist soil in rhizospheres of the many plants in both the very wet and the less wet wetlands. All are major sites of humus synthesis. They are sites for preservation of life in the sense that life on earth depends on the thermodynamic non-equilibrium between organic and inorganic matter. Landfills are preservers of life. They are characteristic wetlands even if covered to prevent entry of rain water.

3. Creating Wetlands

Creating a natural wetland is an oxymoron, of course, but this doesn't mean that an "unnatural wetland" is by definition bad. It doesn't mean we can't mimic Mother Nature in giving natural birth to a desirable wetland. Constructed rice paddies have been responsible for feeding more people than any other enterprise on earth.

There are valid redox reasons for preserving any unique plant, animal, or soil ecosystem in its natural state, but this is not possible because every single one is unique. Soils, being used in productive agriculture, include huge areas of wetlands that already have been destroyed as parts of *natural ecosystems*. However, the continued preservation of these former wetlands as viable *cropland soil ecosystems* for producing food is now of extreme importance to the survival and health of humankind.

Manure and sewage lagoons are preferable to many alternative methods of storage or disposal. Preserving various chromium wastes anaerobically in wetlands prevents possible pollution of waters with toxic and carcinogenic Cr(VI). Septic leach fields can be tailored to assure that nitrates undergo denitrification rather than contaminate groundwater. Hydrocarbons not easily broken down by microorganisms when in high concentrations can be degraded to bite-sized portions on the oxygenated side of the interface. Bite sized may mean lower molecular weight soluble compounds or slightly soluble volatiles that escape across the redox interface in manageable quantities biologically. The interface can be of the paper-thin variety, typical of submerged soils, or it can be the matted rhizosphere of any vigorous wetland species such as *Fragmities*. The bulk of toxic material may remain intact and isolated below for a long, long time, but in the meantime, the micro and macro plants and animals are healthy above the interface.

Drainage of wetlands, personmade or natural, often results in accumulation of massive, semi-solid gel-like dark red, orange, or black blobs made up of the bodies and by-products of billions of bacteria along with partially reduced Fe and Mn oxides and organics. These blobs are beautiful or ugly, depending upon the eye of the beholder, but always they are inconvenient because they form almost overnight inside drains, drainage pipe, or tiles.

Large areas of soils in the U.S., developed on level lacustrine, marine, and ground moraine parent materials and also large areas of peat and muck soils have been artificially drained by ditching or tiling to create highly productive and valuable agricultural land. Drainage of these lands was encouraged by the Federal government for many years and has been accomplished to a considerable extent in the major humid agricultural regions of the country. Recently, tough new laws have been passed to prevent further drainage of wetlands. Consequently, and unfortunately, development emphasis is on diverting areas of precious well-drained farmland from food production to parking lots, highways, airports, and condos.

4. Cleanup

Cleanup seems to be a watchword in modern society. Destroy the toxic wastes, bury them deep, move them somewhere else, sweep them under a different rug, get them out of our site, and out of our sight. Spend big bucks. That's what the word *Superfund* means, and it certainly has accomplished that much. Cleaning up has more than one connotation. Certainly in some situations, environmental protection should mean protective custody of the toxic waste, preserving rather than cleanup. Protect us from it by means of a free radical-rich *redox interface*. It's an answer. An already polluted wetland that does not offer any means of escape to the groundwater could be the best place for testing this hypothesis. Wetlands commonly are well sealed against water movement through them, which is why they *stay* wet.

VII. SYSTEMATIC CATEGORIZATION OF SOIL REDOX STATUS BASED ON CHEMICAL FIELD TESTS[48]

A. Need for a New System

Because soils, and especially soils in the field, are never at thermodynamic equilibrium, redox potential, Eh or pe, is useless for the redox classification of a soil in the field or anywhere else. Because of the heterogeneity and instability of nonequilibrium aerated soil systems, an Eh or pe, measured either in the lab or in the field, is a confusingly mixed potential and has no quantitative utility. Furthermore, oxidized species of N, S, Mn, Fe, C, and H, and all major components of partially oxidized soils are not electroactive and therefore will not affect measured redox potentials. Excluding these species means that measured potentials are confusing at best and worthless at worst.

B. The Redox Status Categories

We propose a practical field oriented scheme for classifying soils qualitatively according to redox status.[48] Soils appear naturally to be separated into six categories of soil redox status, according to redox behavior in the field. Each category represents a range level of electron *lability* (*el*), meaning reactivity or availability. The commonly used kinetic term, *lability*, seems useful here for designating an electron reactivity level. The *el* could be considered the kinetic analog of *pe*, the thermodynamic electron potential, although the, *pe* (or *Eh*, from which it is calculated) is theoretical and cannot be measured. In contrast, the *el* is not a thermodynamic parameter. It is empirical and practical and amenable to qualitative estimation in all soils. Ranging from very low to very high, an *el* serves as a simple designation of a soil redox behavior level.

Proposed field chemical measurements provide a general framework that can be tied in with observations of redoximorphic features, botanical composition, and moisture and temperature regimes to contribute to and support the redox classification of field soils by category name. Following are the proposed redox category names and their *el* ratings.

Superoxic, *very low el*: soils contain very high levels of available oxidants, especially Mn oxides and Mn^{3+} or O free radicals. Many rhizosphere soils. Also, certain dark-colored A horizons developed from calcareous rock very high in Mn and Fe.

Manoxic, *low el*: highly oxidizing soil systems that are well drained, contain mature humus, are mineralizing and nitrifying when moist, and are free of gley mottles throughout the profile. Examples are tall corn or alfalfa soils, beech-maple climax forests, fertile A horizon material, and probably the garden of Eden.

Suboxic, *Medium el*: these soils are oxidized and have nitrates present, but they also have significant potentials for reduction. Reducing tendency is balanced against oxidizing propensity. *Manoxic* soils often are temporarily *Suboxic* when electron availability is increased by drying or by additions of easily oxidized organic residues, by heavy rain during a warm season, or by compaction. Most dried and stored soil samples ("lab dirt") are *Suboxic*, as are organic horizons in well- or moderately well-drained forest soils, well-cured composts and recently manured well- or moderately-well drained soils.

Redoxic, *medium high el*: reducing and oxidizing tendencies are balanced in these soils, slightly in favor of reduction. Fe(II) is absent, except in very acid materials, and Fe(III) is found in recently oxidized forms that are easily reduced by the dipyridyl test in the light. The pH is lower than in related *Anoxic* soils, and some nitrate may be present. *Hydric* soils during cold or dry seasons, those artificially drained, or those heavily fertilized with nitrates, generally will be found to be *Redoxic*. Dried samples of *Hydric* soils will usually test *Redoxic*.

Anoxic, *high el*: presence of ferrous Fe is diagnostic for these strongly reducing and highly reduced wetland soils. Nitrates absent; pungent odors of decomposition mild or absent. pH usually near neutral. Newly flooded rice paddies and beaver ponds are examples.

Sulfidic, *very high el*: soils are very strongly reducing and are recognized by distinct and pervasive odors of vegetative putrifaction, especially sulfides. Oxygen absent. CH_4 and H_2S are produced; pH is near neutral. Sulfidic horizons, garbage landfills, manure lagoons, septic leachfields, and sediments of eutrophic waters.

C. Chemical Field Test Procedures

1. Tetramethylbenzidine Oxidation

Slightly more than saturate a soil sample on a spotplate with TMB III reagent [1.0 M acetic acid solution, that is, 10 mM K_3-citrate, and 1.0 mM 3,3′,5,5′-tetramethylbenzidine, an apparently safe substitute for carcinogenic benzidine.[40] If out-of-doors, in bright daylight, cover the spotplate loosely with a piece of black plastic to shut out excessive light.

A blue color around the edges of the spotplate spot after 20 to 30 min indicates a one electron oxidation of TMB, usually by mixed Mn oxides. Development of a yellow color is the result of a two electron oxidation and is the probable indicator of the supermanganese Mn^{3+} free radical,[3,35,40] as the predominating Mn oxide. On addition of a few additional drops of TMB, a pale yellow color will give an intense blue color if the initial yellow was indicative of the two-step oxidation. The

TMB concentration in the testing solution is such that a blue color, without initial yellow, indicates a one electron oxidation of TMB by Mn oxides. If the TMB-to-soil ratio is too high, the electron demand of the soil will be met by single-step electron transfers, and the two-electron yellow may not develop. Thus, the yellow color indicates a very highly oxidized or *Superoxic* soil; and blue, a highly oxidized or *Manoxic* soil. The blue color develops slowly if soil Mn oxides are not freshly formed.

The small amount of citrate present is high enough to prevent TMB blue color formation with slightly soluble Fe hydroxides and low enough to allow a controlled low level of reverse dismutation of highly reactive Mn(IV) oxides to form superoxidizing Mn(III).[9] Extremely acid leached soils often are depleted in Mn and if so, will not exhibit positive TMB tests even when they are well aerated and highly aerobic.

2. Chromium Oxidation

The following is a simple and quick alternative Cr oxidation spotplate field test for Cr oxidation that is proving to be a successful and simpler substitute for the earlier published field method. To a small pinch (about 0.3 cm^3) of moist soil on a spotplate, add enough distilled water to saturate the sample and flood it so that free water is visible around the outside of the spot. Add two 0.05-ml drops of 0.1 M CrCl$_3$ and mix gently without dispersing the soil. After 5 min, add 1.5 ml H$_2$O, with very gentle mixing, and then eight drops of DPC (Section VII.D) reagent with gentle swirling. Unmistakably perceptible magenta color around the edges of the spot after 10 min is a *positive* test result. Significantly more color is a *positive-plus*.

A positive Cr(VI) net test is diagnostic for both *Superoxic* and *Manoxic* materials. In any field or laboratory test, added Cr(III) is oxidized by soil Mn-oxides to Cr(VI). Tests are termed *net* tests because only a part of the Cr(VI) formed is measured. Some Cr(VI) is reduced during a test by easily oxidizable organics.[5,42] Their effects are enhanced by acidification of the sample. Such easily oxidized organics are prominent in *Suboxic* soils, and their presence identifies the *Suboxic* category.

3. Ferrous Iron Test

Add five drops of 2,2′-dipyridyl (DIPY) reagent (10 mM reagent in pH 4.8 NH$_4$OAc, 1.25 M acetate) to a soil sample on a spotplate and note the color in the solution around the edges of the spot. Pink or red indicates ferrous Fe and is diagnostic for soil material that is highly reduced (*Anoxic*) or very highly reduced (*Sulfidic*).

4. Easily Reducible Iron

Add enough 0.1 M oxalic acid to saturate a pinch of field moist soil on a spotplate and place in full sunlight at noon on a cloudless day for 1 min or in very bright shade or artificial light for 2 to 15 min, or deeper shade for 15 min to as long as 1 h. After the light exposure, add five drops of DIPY reagent and note the absence or presence of pink and depth of pink, or presence of red color and its depth. A positive test for Fe(II) after oxalate and light identifies easily reducible Fe(III), such as might be found in a *Hydric soil* during a dry season. It is diagnostic for the *Redoxic* category. The test combines extraction of reactive Fe(III) by the organic acid and its reduction to Fe(II) by the organic/light interaction.[3,5,49] The test could be made more quantitative in terms of consistency in measuring ease of reducibility of Fe(III) if an artificial light, perhaps one plugged into the cigarette lighter of an automobile, were used as a standard light. Light intervals used might be calibrated against soils with known drainage/aeration characteristics. Artificial light and standardized light intervals would seem to be necessary in order to keep the test from being too subjective.

5. Common Scents

Detection entails sniffing, without addition of acid, of odoriferous emanations from wet soil remindful of anaerobic decomposition. In addition to these common scents, the characteristic and distinctive *rotten egg odor* of H_2S evolution upon addition of 1 M HCl must also be present if a soil is to be classified *Sulfidic*. The familiar anaerobic scents that disappear when sulfidic material is dried and oxidized are more important in diagnosing the *Sulfidic* redox category than is the H_2S release with acid, since sulfidic material recently dried and partially oxidized, so that it no longer belongs in the *Sulfidic* category, may still release H_2S on acidification.

D. Using Test Results to Place Soils into Redox Categories[48]

Test Results	Redox Status Classification
A. +TMB yellow or green, by 30 min; pos+Cr(VI)	Superoxic
B. +TMB blue; pos or pos+Cr(VI)	Manoxic
C. +TMB; posCr(VI); pH <5.3 (greenish brown ind)	Manoxic
D. +TMB or +TMB blue; negCr(VI); pH <5.3	Suboxic
E. +TMB or +TMB blue; pos+Cr(VI); pH >5.3 (brownish green ind)	Suboxic
F. +easily reducible Fe(III); +TMB; pos+Cr(VI)	Redoxic/Manoxic
G. +very easily reducible Fe(III); negTMB; negCr(VI)	Redoxic
H. pos Fe^{2+}; +TMB and/or posCr(VI)	Manoxic/Anoxic
I. pos Fe^{2+}; neg or pos H_2S with HCl; negTMB; negCr(VI)	Anoxic
J. pos Fe^{2+}; pos H_2S with HCl; pos odoriferousness without HCl	Sulfidic

Cr(VI) ratings: pos = unmistakably visually detectable; pos+ = estimated >0.1 mg l^{-1}. +TMB = faint or strong blue; +TMB blue = intense blue, not yellow before 30 min.

1. Key Manganese

Soil Mn, or its lack, appears to be the key to the entire spectrum of soil redox behavior and status. Except for the sniff tests, decisions can be made about all six categories of soil redox status using only TMB and the Cr oxidation tests, both of which identify oxidized forms of Mn.

2. Double Categories

Any soil sample, as a nonequilibrium redox system, is made up of a mixture of redox states. Generally, contrasting states tend to cancel the effects of one another because they exist in some sort of metastable equilibrium with an easily detectable predominating redox status. However, an occasional single soil sample can be shown by chemical tests to contain two or more discreet and contrasting redox status portions. In the table above, item (E), ferrous Fe and reactive Mn oxides are present together. Such a sample requires a double redox status name. In a double-named category, the dominant redox category name is second, with the first becoming the modifier. Whenever the Cr(VI) test is positive, *Manoxic* (or *Superoxic*) will be part of the status name, and whenever Fe^{2+} is positive, *Anoxic* will be part of the name. A soil testing positive for easily reduced Fe(III) would be *Suboxic* if it had a positive TMB test and *Redoxic* if TMB were negative.

3. Contrasting Redox Layers and Interfaces

When double-named categories include contrasting redox domains, the slash between two names may be indicative of a redox interface. In a classic flooded soil, it is common to find the TMB test indicating *Superoxic* soil material on the aerated side of an interface that has an *Anoxic* soil containing ferrous Fe below it. A Pt electrode responding only to the solution phase would

give an average value, weighted in favor of species that are both soluble and *electroactive*, generally those found only under anaerobic conditions.

4. Borderlines

The pH is useful in deciding the redox classifications of soils that are truly borderline between adjoining redox categories. An acid soil, compared with a neutral soil, tends to be more oxidized than field tests might indicate. For example, a soil with a pH <4 will oxidize Cr(III) only if it contains highly reactive Mn oxides or perhaps available free radicals; whereas a near-neutral pH soil is more likely to contain reactive Mn oxides even if the total Mn supply is low. If a soil is borderline and the pH is <5.3 (greenish brown Vermont indicator test), it should be placed in the category on the better oxidized side of the border, and, if the pH is >5.3 (brownish green), on the lesser oxidized side.

5. Variability of Redox Status with Time and Place

The distinctive features of a specific redox status category should be considered to refer only to the particular sample being tested at the point in time of testing. A redox species is transitory; a redox category designation is an intensity parameter. It will vary with time in minutes, hours, or seasons. It will vary with location, from pedon to pedon, from horizon to horizon, ped to ped, inside of a ped to the outside of the ped. The sediments at the bottom of a pristine lake usually will be *Manoxic*, whereas a well-drained "sanitary" landfill (not pristine) could be *Anoxic* or even *Sulfidic*. A well-aerated soil may produce a negative, low, medium, or high Cr net test. The variations are real and not sampling errors. Manganese oxides, reducing organics, and even Fe^{2+} can exist side by side with less than a millimeter of interfacial boundary separating them. A Mn oxide coating can exist for long periods on the outside of a soil ped without reducing adsorbed Fe(II) inside. However, because the sample is thoroughly mixed during the test equilibration, the Cr net test will integrate the redox across such an interface and supply a compromise answer.[81]

6. Field Chemical Identification of Hydric Soils

The low *el* categories of *Sulfidic, Anoxic,* and *Redoxic* soil materials, all are identifiable by simple chemical field tests as categories with characteristics of poorly or very poorly drained soils, that is, Hydric or wetland soils that develop wherever a water table is near or above the soil surface for a sufficient length of time during a season warm enough for reduction to take place.

By inducing changes in the availability of O_2, the ultimate electron acceptor, altering water table depths by natural or artificial drainage or flooding of a soil, can be expected to cause temporary or long lasting shifts in reducing tendency and reducing power and therefore in *Redox status* category. Sometimes, the simple presence of Fe^{2+} is used to indicate hydric character, but reliance on such a test is misleading because of the ephemeral nature of ferrous Fe. A soil testing positive will be hydric only at the moment of testing. The Fe^{2+} is quickly oxidized as the soil is partially dried or drained. However, even though Fe^{2+} is absent, the soil may still be Hydric, according to other Hydric soil criteria. Redoximorphic features such as accumulation of organic matter and the presence of gley mottles, along with presence of hydrophytic vegetation provide some positive identification of hydric character, even when the soils are dry. The light-induced Fe(II) positive test can provide strong reinforcement of this identification.

Drying an *Anoxic* soil, which most surely is also a Hydric soil, oxidizes the Fe(II) in it to Fe(III), converting it to a *Redoxic* soil, which is still *Hydric*, even though its Fe(II) has vanished. Thus, the field test for easily reduced Fe(III), specific for *Redoxic* status, also serves to identify a *Hydric* soil that has been drained or dried. Any dry soil containing easily reduced Fe(III) probably contained Fe(II) when it was last wet. However, in its present dry state, it may contain reactive Mn

oxide by the TMB test, meaning that its present classification would be *Suboxic*. On wetting, the Mn oxides will disappear by reduction, and the soil will become *Redoxic*, as it would have been classified when dry, if Mn oxides had been absent. Probably a soil should be considered *Hydric* if it had ferrous Fe in its recent past, regardless of its present redox status. The DIPY test in the light makes this easily discernible.

VIII. EMPIRICAL METHODS FOR CHARACTERIZING SOIL REDOX IN THE LAB

A. Soil Handling

Sieving, mixing, and storing moist soil samples is inconvenient and can be responsible for gross errors in measurements and poor repeatability. But if you are interested in the behavior of moist soils, to support growing plants, for example, then using moist soils is absolutely essential in the soil chemical laboratory. Evaluated high experimental error is preferable to misleading results. Of course, it is entirely appropriate to dry a soil if the purpose is to mimic a natural process in the field. Chemistry of frozen soils changes while they are frozen, but freezing is a natural process in many soils, and should not be ignored. The important thing is to be aware of the changes you are inducing by the treatments you are imposing. This means studying soils in the variety of states that occur in the field.

Moist soil samples are most suitable for handling and are most stable chemically if near field capacity moisture. It is difficult to get moist samples to pass through a 2-mm sieve, but a 4-mm polyethylene sieve is generally suitable and has the advantage of preserving some of the field soil crumb structure and aeration status during storage before analysis. Time is saved by presieving through an old tennis racquet and mixing in the field. Samples should be mixed individually before and again after sieving. Samples will remain most stable if they are stored in double, 25 µm thick polyethylene bags with moist paper towels between the layers in a refrigerator at 4°C. Freeze-drying causes soils to change during experimentation almost as much as air-drying; ordinary freezing changes them less. Drying in the sunlight is an unconditional *never*, unless you are mimicking a field condition and have considered its implications. Microbial activity appears to become stabilized after 2 or 3 months at 4°C, and most soil samples seem to reach an internal metastable equilibrium. It is safe at this time to transfer samples to tight heavy-walled plastic bags or garbage containers with lids to prevent all moisture loss. Samples should be kept in semidarkness, but temperatures as high as 10 to 12°C can be tolerated in metastable moist soils.

Well-mixed subsamples may be weighed for analyses after determining moisture on separate samples. For many determinations, a volume measurement, to be corrected later for dry weight, is more convenient, although somewhat less accurate. A packed and leveled teaspoon of soil is 5 cm^3, and is approximately 5 g dry weight for many topsoils.

B. Testing for Free Radicals in the Laboratory with TMB

The TMB solution for identifying powerful oxidants, including free radicals, for use in the lab is the same as the one used for field testing, 1 mM TMB in 1 M HOAc, without the Mn(II) and citrate added. Dissolve 240 mg of 3,3′,5,5′-tetramethylbenzidine in 28.6 ml glacial acetic acid and quickly dilute it to 500 ml with distilled water. The solution is somewhat irritating and is best used out-of-doors or in the hood, even though it is supposedly noncarcinogenic.[40] All Mn oxides react readily with this TMB reagent, and slightly soluble amorphous Fe, e.g., $Fe(OH)_2^+$, will give an intense blue color in some acid and wetland soils. Blue color indicates a one-electron oxidation of TMB; yellow, a two-electron oxidation.[39] The TMB itself will reduce many easily reducible species, causing a slow development of blue color. The TMB indicator produces a positive blue color rather

slowly with NO_2^- and still more slowly with H_2O_2. Probably the TMB is forming nitrogen and oxygen free radical species by reduction. If the supermanganese Mn^{3+} free radical is present to begin with, it will give an immediate yellow color reaction with the TMB, indicating a two-electron oxidation of the TMB. However, until time has been allowed for free radical formation by partial reduction of MnO_2 by TMB, the full blue color will not develop. Nitrite, Fe(III), and H_2O_2 each will produce full color immediately with TMB addition if first partially reduced to free radicals by hydroquinone, dipyridyl, orthophenanthroline, or peroxidase, respectively. Cr(VI) gives an immediate positive TMB test.

Thus, although TMB will readily form blue color in the presence of oxidizing free radicals, such as supermanganese or superoxide, TMB also is useful as an indicator of oxidized species that have a predilection for being very easily reduced by the TMB itself to oxidizing free radicals. The first increment of TMB added sensitizes the substrate by reducing it to a free radical so that when more TMB is added, it will change color quickly and intensely.

An intense blue color forming instantly upon addition of TMB to a spotplate sample of field soil usually indicates the presence of available Mn(III) supermanganese free radical, along with Mn(IV). If an intense blue color intensity builds up slowly, this means that reactive Mn(IV) and/or Fe(III) are present. Addition of three or four drops of 0.1 M citric acid before the TMB will prevent reaction of Fe with the TMB, and the citrate will drive the reverse dismutation of Mn(II) and Mn(IV) toward Mn(III) (Equation 20). Adding citrate or Mn(II), and especially both together, will greatly enhance the color development, causing first a green color (yellow and blue mix) and then yellow, indicating Mn(III). In other words, the Mn(III) is being formed during the test. Instant yellow color is the only indication that Mn^{3+} is actually present in the sample. Even TMB and acetic acid alone, if enough time is allowed, can cause reverse dismutation in some samples that appear to contain only Mn(IV) to start with, and the yellow color of Mn(III) will become apparent. It is best to set a time limit of about 30 min in order not to be misled by Mn(III) that is more apparent than real. There is an art to knowing when to repeat an experiment or when or whether to accept the apparent results.

C. Laboratory Incubations

Simple short-term laboratory incubations can be a boon to the experimental approach. Small amounts of soil at approximately field capacity moisture are treated and incubated in polyethylene bags. Light weight and thin polyethylene has the advantage of being quite permeable to O_2 and especially CO_2 but not to ions and solutions.[80] Most gases, including water vapor, will pass through slowly, however, and if you wish to prevent drying for long periods, place one bag inside another with a moist paper towel between the layers.

For monitoring CO_2 evolution, the soil can be left in a closed bag and placed inside a 4-l pickle jar with a tight lid, along with a 100-ml polyethylene beaker or test tube containing 1 M NaOH. The base can be titrated with HCl in the beaker after adding 2 M $BaCl_2$ and phenolphthalein. Some "Saran"-type plastic sandwich wrap material is permeable to light but not to CO_2 and is effective for light vs. dark studies. Whatever is used will need to be tested by running blanks.

Incubating soil in flooded beakers is useful for studying partially restricted O_2 availability with colloidal sediment solids containing oriented water and separated by gravity from freer oxygenated water above, oxygenated by air. A plastic beaker serves as a useful redox interface model of a pond, paddy, ped, rhizosphere, sewage lagoon, or leachfield. As the ultimate electron tracer, Cr is a most valuable tool for studying losses, gains, and transfers of electrons at interfaces. It is especially useful for demonstrating the oxidative power of the interface. Chromium(III), long bound in tannery waste, shredded moccasin leather, high chrome serpentine soil, and soil treated with high Cr Chicago sewage sludge was mixed with a high-Mn soil, flooded with 0.1 M K-citrate, and incubated in plastic beaker simulated wetlands. After 2 months, 10 to 20 μM concentrations of Cr(VI) were measured on the oxygenated side of the interfaces.

Flooded beakers should be kept in the dark, unless you plan to monitor and control the light and algae as part of the experiment. Samples may be removed from the supernatant solution, from interface microlayers, or from deep in the sediment. EMpe measurements made on opposite sides of an interface using the Pt electrode are quite helpful in confirming and demonstrating the existence of an interface in a wetland soil and for helping to characterize the redox extremes that occur only a few millimeters apart.

D. Standard Chromium Net Oxidizing Test

1. Shake 2.5 g of soil (dry weight basis or 2.5 cm^3 packed volume) for 15 min with 25 ml of 1 mM $CrCl_3$.
2. Add 0.25 ml of 1 M, pH 7.2, phosphate buffer ($K_{1.5}H_{1.5}PO_4$), shake 15 s longer only, and then filter or centrifuge.
3. Determine Cr(VI) by adding 0.5 ml diphenylcarbazide (DPC) reagent to 4 ml of extract or water, mix, and let stand 20 min before comparing color at 540 nm with that in standards (0.5 to 50 μM) at 540 nm. (Prepare the DPC reagent by adding 120 ml of 85% phosphoric acid, diluted with 280 ml distilled water, to 0.38 g of s-diphenylcarbazide dissolved in 100 ml of 95% ethanol.)
4. If cloudiness from precipitated organic or mineral matter is present, it is easily removed by filtering following color development using a 0.2 μm filter and syringe. Most of the colored complex remains in solution, and filtration of standards and unknowns improves sensitivity.

A portion of the Cr oxidized during the soil shaking is not measured as Cr(VI) because it is reduced almost as fast as it is oxidized. Depending on the availability of easily oxidizable organic matter, some or even most of the Cr(VI) formed is reduced during the 15-min period. Leaching a sample, especially a dry one, will remove some of the low-molecular-weight reducing organics and thereby will increase the Cr(VI) quantity measured.[42] With dried and stored "lab dirt" samples, the reduction frequently equals the oxidation and no net Cr(VI) is measured. This net test characterizes oxidation only to the extent that it exceeds reduction.

E. Soil Reducing Intensity

1. Shake intermittently 2.5 cm^3 of moist soil 18 h with 20 ml of pH 4.0 NH_4OAc, 0.6 M with respect to ammonium, containing 0.1, 0.5, or 2.5 mM $K_2Cr_2O_7$.
2. Filter or centrifuge, and determine concentration of Cr(VI) remaining (as in Section VIII.D above). If all of the Cr(VI) is reduced, repeat with increased concentrations until the Cr(VI) remaining is measurable. One mole of Cr is equivalent to 6 mol of Mn plus charge.
3. To measure the tendency of a particular oxidizable organic substance to reduce Cr(VI), repeat this test after adding varying measured amounts of the substance to the soil sample.

F. Available Reducing Capacity

1. Shake intermittently 2.5 g, dry weight basis, or 2.5 cm^3 of moist soil 18 h with 25 ml of 0.1, 0.5, 2.5, or 10 mM as $K_2Cr_2O_7$ in 10 mM H_3PO_4, filter or centrifuge, and determine Cr(VI) not reduced in the extract.
2. Begin with the lowest concentration of Cr(VI). If all of the Cr(VI) is reduced, repeat with increased concentrations until the Cr(VI) remaining is measurable but below 0.1 mM.

G. Measuring Ready Electron Supply and Demand

1. Soluble Soil Organics

1. Levels of highly reduced soluble organic matter can be determined by tests designed to measure phenols or tannins.[43] Any reducing agent strong enough to reduce phosphotungstate or phosphomolybdate to

the blue-colored complex will give a positive test. Reduced Fe, for example, will give the color, as will many organics that are not phenols. Still, the test appears to be a practical means for characterizing soluble humic and fulvic acid fragments and microbial by-products. Water extracts of dried soils give highly positive tests relative to extracts of well-aerated soils kept continuously moist for periods of time.

2. Probably most of the soluble, readily oxidizable organic substances in soils are phenolic compounds. Schnitzer and Levesque[44] showed that the ESR peaks indicating free radical contents of NaOH extracts of humified peat were directly proportional to the concentrations of the "phenolics" in the extracts.

3. The amber color of a soil extract is a redox parameter related to the phenol determination. Soluble humic substances can be estimated by amounts of absorbance of short wave lengths of light or UV by soil extracts obtained by shaking soils with 0.1 M $NaHCO_3$, pH 7, 10 mM pyrophosphate solutions. High values of E_4/E_6 ratios (absorbance at 465 nm, 665 nm) indicate high contents of carboxyl groups and stable free radicals.[45]

Other common measurements relating to electron supply quantities are soil organic matter determination[46] (a chemical oxygen demand method) and empirical incubation methods such as BOD (biochemical oxygen demand).[47] A quick method for measuring oxidizable organics in soil solutions utilizes the disappearance of color when Mn(III)-pyrophosphate is reduced.[37]

2. Manganese Electron Demand (MED)

1. As a direct titration of the reducible Mn oxides in a soil sample, the manganese electron demand (MED) determination is a direct way of estimating the oxidizing capability by the Mn oxides in a soil without consideration of the reduction that might take place under the conditions of a less direct oxidation measurement such as the net Cr(VI) test (Section VIII.D).

2. Shake intermittently for 18 h, 2 g, dry weight basis, or 2 cm³ moist soil with 12 ml of pH 4.0 NH_4OAc, 0.6 M with respect to ammonium, and 4 ml of 0.2 M KI. Add four drops of starch solution (0.3 g potato starch boiled with 50 ml water), and titrate to a colorless end point with 2 mM $Na_2S_2O_3$. Millimoles of thiosulfate per unit of soil are equivalent to mmol of e^- or plus charge of Mn.

3. Total Electron Demand (TED)

The TED test includes an empirical estimation of easily reducible Fe(III) along with Mn oxides. It is modified from MED as follows: 12 ml of 0.1 M HCl is added instead of pH 4.0 acetate, and the equilibration time, instead of an approximate 18 h, is a rigid 15 min with centrifuging and titration required *immediately* afterwards. The time is critical in soils that contain high amounts of recently oxidized Fe, because the amount of easily reducible Fe that will react may lie in an ambiguous valley rather than occur on an easily identifiable peak.

A more serious problem is that iodide is slowly oxidized by O_2 at low pH, and inflated TED values can result if time is not strictly limited. Thus, TED is very much an empirical measurement, although with careful control of time, the test is amenable to calibration for measuring a particular species or fraction of reducible Fe.

H. Preparing Manganese Oxides

1. Synthetic Amorphous Mn(IV)

Dissolve 40 mmol of $KMnO_4$ in about 40 ml of distilled water heated to approximately 60°C and transfer with mixing into 30 ml of 2 M $MnSO_4$. Add 80 mmol KOH dissolved in 10 to 20 ml of water. Mix and adjust the pH to 7.5 with KOH or sulfuric acid and let stand overnight with occasional stirring, and then adjust the pH to 6.0 with additional sulfuric acid. There should be no permanganate color remaining. Transfer into dialysis tubing and dialyze against repeated changes of distilled water until the outside solution is close to salt-free, as checked by barium precipitation or conductivity. Dilute the suspension to 500 ml, or any desired volume. The procedure may be carried out quantitatively so that there are exactly 100 mmol of MnO_2, or a diluted suspension can be standardized by iodine titration.

2. Stable Mn(III) Solutions

1. *Mn(III)-pyrophosphate*: to about 100 ml of distilled water in a 250-ml volumetric, add in order: 75 ml of 0.1 M $Na_4P_2O_7$, 11.5 ml of 0.5 M H_2SO_4, and 5.0 ml of 0.10 M $KMnO_4$. Mix, and immediately add with mixing 20 ml of 0.10 M $MnSO_4$, and bring to volume. The colored solution is 10 mM Mn(III)-pyrophosphate and should have a pH of 4.3.[45] *Keep away from light.*
2. *Mn(III)-citrate*: to 200 ml of distilled water, add 5.0 ml of 1.0 M $K_{2.5}$citrate, made from KOH and citric acid, 3.33 ml 1.0 M $MnSO_4$, and 0.10 M $KMnO_4$. Bring to a 250 ml volume. The tawny-colored solution is 17 mM Mn(III). The citrate will eventually oxidize until it is equivalent to K_3, and the pH will end up at about 7.5. If the solution is made without citrate, the Mn(III) will gradually dismutate, forming precipitated Mn(IV) with adsorbed Mn(II). If oxalate had been used instead of citrate, all of the organic would be oxidized and only precipitated Mn(II) and Mn(IV) would remain.

I. Dismutation of H_2O_2

Manganese oxides, at pH >6, and catalase enzymes both destroy H_2O_2 by catalyzing its dismutation. Rate of dismutation is a better indication of quantities and activities of catalytic substances present than is amount of H_2O_2 dismutated since a small amount of catalyst will act upon a large amount of substrate. Rate of dismutation is evaluated by adding 5 ml of 0.5 M H_2O_2 solution to 2.5 g of moist soil, dry weight basis, and clocking the time required for the soil to evolve enough bubbles to displace 24 ml of H_2O (1 mmol of O_2).

J. Measuring Soil and Water pH with Vermont Indicator

1. Preparing the Indicator

To 370 ml 95% ethanol, add 0.22 g of brom cresol green, 0.15 g brom cresol purple, 0.05 g methyl red, 0.10 g methyl orange, and 0.045 g phenolphthalein, stir, bring to about 980 ml with distilled water, and stir again until all residues are completely dissolved. Bring pH to 5.8 by adding 0.1 M NaOH a drop at a time (7.5 to 8.0 ml; the solution will reflect green light and transmit red). Adjust final volume to 1 l. Use directly for measuring the pH of solutions. Dilute 1:1 with 20 mM $CaCl_2$ for most pH measurements of field soils. For "water pH" comparisons of soils or for identifying variable charge soils, use salt-free indicator diluted 1:1 with distilled water.

2. Color Changes

The following colors will be observed when the indicator is added to pH buffers:

pH 3   light reddish orange
pH 4   darkish orange (slightly brownish)
pH 5   olive brown
pH 6   green
pH 7   greenish blue
pH 8   bluish purple

A portable water standard is easily made by adding 0.2 ml of undiluted indicator to 2 ml of pH buffer in a disposable plastic cuvette (1 cm^2 cross section) with a tight-fitting lid. Six of these are held securely side by side by a wide rubber band. (The colors will deteriorate toward the greenish with time and must be made fresh every couple of weeks.) With experience, you may find that pH 5 and 6 are the only standards you need.

3. Measuring Soil pH

Add enough diluted indicator to a pinch of moist field soil in a spotplate so that the solution volume is in excess by about 15%. More indicator will be needed if the soil is high in organic matter. Mix thoroughly using a glass or plastic rod or a flat toothpick. Read color around the edges as the soil settles after 1 or 2 min. Do not wait more than 5 min. Natural light from a northern exposure is best.

Because soils may adsorb individual dyes or may contain organic substances that interfere with color perception, the colors seen in soil supernatants frequently vary from those in buffer solutions of comparable pH. Manganese oxides or bright sunlight can decolorize dyes, especially in acid soils. Compare quality of color (hue) observed, rather than intensity or brightness. Measuring soil pH with indicators is exasperating at times, but the amount of frustration and inaccuracy probably is not significantly greater than those with a glass electrode.

K. Interferences

Since most colorimetric tests used for studying soil redox processes themselves also involve redox reactions, these tests are fraught with interfering reactions among themselves and with the soil. Interfering interactions are the name of the game, and we sometimes can learn a great deal about the overall soil redox by working with the interferences instead of fighting them. Manganese oxides produce positive interference in measuring nitrate by the modified hydrazine reduction method and also with brucine or diphenylamine, positive interference in nitrite by the diazonium salt method, but no interference with the diphenyl carbazide color for Cr(VI). Mn(III), nitrite, citrate, and hydroxylamine negatively interfere with the Cr(VI) test, and nitrite and Cr(VI) positively interfere with determination of nitrate by brucine.

Mn(III)-organic plus DPC produces an immediate positive Cr(VI) color which fades quickly, then after a few hours, the color returns with a brownish hue added. $Fe(OH)_2^+$ seriously and positively interferes with the diazonium nitrite test. Mn^{3+}, but not Mn(III)-citrate, forms a complex with the NH_2 group that mimics the diazonium compound color.

Highly oxidized hypochlorite and peroxides and highly reduced substances, such as thiosulfate, sulfides, amines, and ascorbic acid, interfere with everything imaginable.

Nitrate is too inert to interfere with most tests, which tells us something about the low lability of nitrate. Flexibility, ingenuity, and wariness, and an appreciation of nice colors, are required in the laboratory of redox chemistry.

L. Determination of Empirical pe

For many years, the platinum electrode has been the tool of fashion for studying redox in soils. As a result of this, study of redox characteristics of aerated soils has been neglected because platinum electrodes are not seen as very useful in such soils. The field of redox has become associated with wet soils because electrode potentials are seen to be useful only for anaerobic soils. Interpretation of redox potential measurements are thoroughly discussed in Section II of this chapter. Here, for what it's worth, we simply present a practical method for making an empirical pe, or EMpe,[42] measurement in any soil, regardless of Redox status category.

1. Attach a bright platinum electrode (in place of the glass electrode) to the plus terminal of a pH meter with a millivolt scale and a saturated calomel electrode to the negative terminal.
2. Before each reading, rinse the platinum electrode, but not the reference electrode, in a 1/1 6 M HCl/liquid detergent solution followed by 10% H_2O_2 and then thoroughly rinse with distilled water. Clean in aqua regia after a few hours of use.
3. Adjust the potentiometer to read +219 mV when the electrodes are in a pH 4 suspension of quinhydrone in 0.1 M potassium acid phthalate.
4. Add 30 ml of 2 mM NaNO$_3$ to 10 g of field moist soil (dry weight basis) in a polyethylene beaker, stir until soil and solution are well mixed and let stand for 20 to 30 min with occasional swirling. (The ionic strength for both EMpe and redox pH should be kept below that in the natural soil in order to prevent, to the extent possible, flocculation of soluble and electroactive organics. This is a change from our earlier methods.)
5. Insert electrodes so that the reference electrode is in the upper half of the supernatant solution and the platinum electrode is near the bottom of the suspension. Swirl for a few seconds, let stand for at least 5 min, and without jiggling or touching the cup, read Eobs in mV. Measure pH of the same suspension.
6. EMpe = (Eobs + 244 for sce)/59

REFERENCES

1. Lindsay, W.L., *Chemical Equilibria in Soils,* Wiley Interscience, New York, 1979.
2. Bartlett, R.J., Nonmicrobial nitrite-to-nitrate transformation in soils, *Soil Sci. Soc. Am. J.,* 45, 1058, 1981.
3. Bartlett, R.J. and James, B.R., Redox chemistry of soils, *Adv. Agron.,* 50, 151, 1993.
4. Greenwood, D.J., The effect of oxygen concentration on the decomposition of organic materials in soils, *Plant Soil.,* 14, 360, 1961.
5. Bartlett, R.J. and James, B.R., Behavior of chromium in soils. III. Oxidation, *J. Environ. Qual.,* 8, 31, 1979.
6. Ponnamperuma, F.N., The chemistry of submerged soils, *Adv. Agron.,* 24, 29, 1972.
7. Garrels, R.M. and Christ, C.L., *Solutions, Minerals, and Equilibria,* Freeman, Cooper, San Francisco, 1965.
8. Stumm, W. and Morgan, J.J., *Aquatic Chemistry,* 1st ed., Wiley-Interscience, New York, 1970.
9. Bartlett, R.J., Manganese redox reactions and organic interactions in soils, in *Manganese in Soils and Plants,* Graham, R.D., Hannam, R.J., and Uren, N.C., Eds. Kluwer Academic, Dordrecht, 1988, chap. 4.
10. Bohn, H.L., Redox potentials, *Soil Sci.,* 112, 39, 1971.
11. Hess, P.R., *A Textbook of Soil Chemical Analysis,* Chemical Publishing, New York, 1972, 437.
12. Patrick, W.H., Jr. and Jugsujinda, A., Sequential reduction and oxidation of inorganic nitrogen, manganese, and iron in flooded soil, *Soil Sci. Soc. Am. J.* 56, 1071, 1992.
13. Pryor, W.A., The role of free radical reactions in biological systems, in *Free Radicals in Biology,* Pryor, W.A., Ed., Academic Press, New York, 1976, chap. 1.
14. Morris, J.G., The physiology of obligate anaerobiosis, *Adv. Microb. Physiol.,* 12, 169, 1975.
15. Fridovich, I., Superoxide dismutases, *Annu. Rev. Biochem.,* 44, 147, 1975.
16. Fridovich, I., The biology of oxygen radicals, *Science,* 201, 875, 1978.
17. Halliwell, B., Manganese ions, oxidation reactions and the superoxide radical, *Neurotoxicology,* 5(1), 113, 1974.
18. Bartlett, R.J., Soil redox behavior, in *Soil Physical Chemistry,* Sparks, D.L., Ed. CRC Press, Boca Raton, FL, 1986, 179.
19. Zumdahl, S.S., *Chemistry,* D.C. Heath, Lexington, MA, 1986, 931.
20. Fruton, J.S. and Simmonds, S., *General Biochemistry,* John Wiley & Sons, New York, 1961.
21. Dismukes, G.C., The metal centers of the photosynthetic oxygen-evolving complex, *Photochem. Photobiol.,* 43, 99, 1986.
22. Brudvig, G.W. and Crabtree, R.H., Bioinorganic chemistry of manganese related to photosynthetic oxygen evolution, *Prog. Inorg. Chem.,* 37, 99, 1989.

23. Thorp, H.H. and Brudvig, G.W., The physical inorganic chemistry of manganese relevant to photosynthetic oxygen evolution, *New J. Chem.,* 15, 479, 1991.

24. Jenne, E.A., Controls on Mn, Fe, Co, Ni, Cu, and Zn concentrations in soils and water: the significant role of hydrous Mn and Fe oxides, in *Trace Inorganics in Waters,* Advances in Chemistry Series, Gould, R.F., Ed., 1968, 337.

25. Davis, J.A., III and Leckie, J.O., The effect of complexing ligands on trace metal adsorption at the sediment/water interface, in *Environmental Biogeochemistry and Geomicrobiology,* Vol. 2, Krumbein, W. E., Ed., Ann Arbor Science Publishers, Ann Arbor, MI, 1978, chap. 82.

26. Hem, J.D., Redox processes at surfaces of manganese oxide and their effects on aqueous metal ions, *Chem. Geol.,* 21, 199, 1978.

27. Jenne, E.A., Trace element sorption by sediments and soils, in *Molybdenum in the Environment,* Vol. 2, Chappell, W.R. and Peterson, K.K., Eds., Marcel Dekker, New York, 1977, chap. 5.

28. Murray, J.W., The interaction of metal ions at the manganese dioxide-solution interface, *Geochim. Cosmochim. Acta,* 39, 505, 1975.

29. Bartlett, R.J. and James, B.R., Mobility and bioavailability of chromium in soils, in *Advances in Environmental Science and Technology,* Nriagu, J., Ed., John Wiley & Sons, New York, 1988, 267.

30. Bartlett, R.J., Chromium cycling in soils and water: links, gaps, and methods. *Environ. Health Perspectives,* 92, 17, 1991.

31. Ehrlich, H.L., Manganese as an energy source for bacteria, in *Environmental Biogeochemistry and Geomicrobiology,* Vol. 2, Nriagu, J.O., Ed., Ann Arbor Science Publishers, Ann Arbor, 1975, chap. 40.

32. Silver, M., Ehrlich, H.L., and Ivarson, K.C., Soil mineral transformations by soil microbes, in *Interactions of Soil Minerals with Natural Organic and Microbes,* Huang, P.M. and Schnitzer, M., Eds., Soil Science Society of America, Madison, WI, 1986.

33. Sparrow, L.A. and Uren, N.C., Oxidation and reduction of Mn in acidic soils, effect of temperature and soil pH, *Soil Biol. Biochem.,* 19(2), 143, 1987.

34. Ross, D.S. and Bartlett, R.J., Evidence for nonmicrobial oxidation of manganese in soil, *Soil Sci.,* 132, 153, 1981.

35. McBride, M.B., Electron spin resonance investigation of Mn^{2+} complexation in natural and synthetic organics, *Soil Sci. Soc. Am. J.,* 46, 1137, 1982.

36. Loganathan, P., Burau, R.G., and Fuerstenau, F.W., Influence of pH on the sorption of Co^{2+}, Zn^{2+} and Ca^{2+} by a hydrous manganese oxide, *Soil Sci. Soc. Am. J.,* 41, 57, 1977.

37. Bartlett, R.J. and D.S. Ross, Colorimetric determination of oxidizable carbon in acid soil solutions, *Soil Sci. Soc. Am. J.,* 52, 1191, 1988.

38. Bartlett, R.J., Unpublished data, 1989–1994.

39. Goodman, B. and Petry, F., Oil spills: nature's own, *Audubon,* Nov.–Dec., 82, 1991.

40. Liem, H.H., Cardenas, F., Tavassoli, M., Poh-Fitzpatrick, M.B., and Muller-Eberhard, U., Quantitative determination of hemoglobin and cytochemical staining for peroxide using 3,3′,5,5′-tetramethylbenzidine dihydrochloride, a safe substitute for benzidine, *Annu. Biochem.,* 98, 388, 1979.

41. Bartlett, R.J., A biological method for studying aeration status, *Soil Sci.,* 100, 403, 1965.

42. Bartlett, R.J., Oxidation-reducton status of aerobic soils, in *Chemistry in Soil Environments,* Baker, D., Ed., Soil Science Society of America, Madison, WI, 1981.

43. Folin, O. and Ciocalteu, V., On tyrosine and tryptophane determinations in protein, *J. Biol. Chem.,* 72, 627, 1927.

44. Schnitzer, M. and Levesque, M., Electron spin resonance as a guide to the degree of humification of peats, *Soil Sci.,* 127, 140, 1979.

45. Chen, Y., Senesi, N., and Schnitzer, M., Information provided on humic substances by E_4/E_6 ratios, *Soil Sci. Soc. Am. J.,* 41, 352, 1977.

46. Walkley, A. and Black, I.A., An examination of the Degtjareff method for determining soil organic matter and a proposed modification of the chromic acid titration method, *Soil Sci.,* 37, 29, 1934.

47. *Standard Methods for the Examination of Water and Wastewater,* American Public Health Association, Washington, D.C., 1975, 448.

48. Bartlett, R.J. and James, B.R., System for categorizing soil redox status by chemical field testing, *Geoderma,* 68, 211, 1995.

49. Childs, C.W., Field tests for ferrous iron and ferric-organic complexes (on exchange sites or in water-soluble forms in soils, *Aust. J. Soil Res.,* 19, 175, 1981.

Index